ONTOGENY
AND
PHYLOGENY

STEINBERG
A
C
D
B

ONTOGENY
AND
PHYLOGENY

Stephen Jay Gould

THE BELKNAP PRESS OF
HARVARD UNIVERSITY PRESS
Cambridge, Massachusetts
London, England

Printed in the United States of America

20 19 18 17 16 15 14 13 12 11

Library of Congress Cataloging in Publication Data

Gould, Stephen Jay.
Ontogeny and phylogeny.

Bibliography: p.
Includes index.
1. Phylogeny. 2. Ontogeny. I. Title.
QH371.G68 575.01 76-45765
ISBN 0-674-63940-5 (cloth)
ISBN 0-674-63941-3 (paper)

To the Philomorphs of Cambridge,
the world, and beyond,
where D'Arcy Thompson must lie
in the bosom of Abraham

Acknowledgments

Ontogeny and Phylogeny

In the beginning is the end;
But ends unfold, becoming strange.
Lives—and generations—suffer change.
The tested metabolic paths will tend
To last and shape the range
Of future evolution from the past.

J. M. Burns, from *Biograffiti.*
Written for my seminar
on recapitulation.

Although the result is, I trust, tolerably ordered, this book arose in a haphazard way. Its genesis and execution were probably typical of most general treatises. We rarely separate the logical and psychological aspects of research and we tend to impute the order of a finished product to the process of its creation. After all, the abandoned outlines and unused note cards are in the wastebasket and the false starts are permanently erased from memory. It is for this reason that P. B. Medawar once termed the scientific paper a "fraud"; for it reflects so falsely the process of its generation and fosters the myth of rational procedure according to initial outlines rigidly (and brilliantly) conceived. I view this book as an organism. I have lived with it for six years. Perceptive comments from colleagues in casual conversation have provided almost all the crucial steps in its ontogeny. Those whom I acknowledge will probably not remember their contribution, but I want to record their inspiration. Likewise, I apologize for forgetting the sources of other insights; they did not arise *sui generis.* I am a very effective sponge (and a fair arranger of disparate information); I am not much of a creator.

Ernst Mayr, in a passing comment, suggested that I write this book. I only began it as a practice run to learn the style of lengthy exposition before embarking on my magnum opus about macroevolution. And I'm mighty glad I did because, in the meantime, my views on macroevolution have changed drastically and my original plan, had it been executed, would now be an embarrassment to me.

In addition, I wish to thank Tony Hallam for providing me office space in the very area of the Oxford Museum where Huxley debated Wilberforce; Gary Freeman for sharpening my perception of the essential differences between von Baer and Haeckel during a protracted argument; J. B. S. Haldane (posthumously) for being so brilliant and inserting major insights into the most conventional research reports; Jim Mosimann for taking time (as busy scientists so rarely do) to write a long letter explaining his views on the independent measurement of size and shape; Gary Sprules for his extra effort in responding to an annoyingly time-consuming inquiry about amphibian neoteny; Mary-Claire King and Allan C. Wilson for publishing their important paper on chimp-human differences just when I was floundering for want of an epilogue; Roy Britten and Eric Davidson for a rather inebriated argument about regulation (which I remember though they may not) at Jim Valentine's house; Jane Oppenheimer for serving as a preeminent model of an excellent scientist who can be an equally excellent historian and for reading with so much insight and at such short notice the historical chapters of this book; G. Ledyard Stebbins for his irrepressible enthusiasm about all things and for convincing me about "increasing precocity of gene action" at a crucial time; John Bonner for his quiet and eloquent campaign to unite the two biologies on their common field of development; Frank Sulloway and Robert McCormick for helping me through a terra incognita of Freudian studies; E. O. Wilson for the richness, order, and clarity of his thoughts and for access to his magnificent reprint collection on locusts; Michel Delsol for making this a transoceanic venture by sharing the concerns of French scientists and for his generosity in sending me the manuscript of his unpublished book on the same subject; Agnes Pilot (may everyone have such an intelligent and conscientious German-speaking secretary when 80 percent of the research is *auf Deutsch*); Saul Steinberg for permitting me to exploit his work in a frontispiece; Bill Coleman and Camille Limoges for their historical insights and ready references; Gordon Cantor for his extremely kind and unsolicited effort to guide me through literature on child development; Robert Fagen, Richard Estes, and Valerius Geist for their thoughts on neoteny in social mammals; Doug Gill for the same in amphibians; Tim Smock for his goodwill in wading through a dull literature on primary education. All these were acts of kindness for no personal reward; this is the true spirit of collegiality. Also, my thanks go to at least fifty historians and evolutionary biologists who responded in various ways to such open-ended inquiries as ". . . and what do you think about neoteny?"; to those who read parts of the manuscript and added their insights to this collective effort: Polly Winsor, David Kohn, Fred Churchill, Dov Ospovat, Joan Cadden, David Jablonski, my class in the history of embryology, and many others; to those who brought me references to recapitulation from their newspapers and novels; and, above all, to an institution that has its own humanity and seems to me more an organism than a place—the Library of the Marine Biological Laboratory at Woods Hole. Where else could an idiosyncratic worker like me find a library open all the time, free from the rules and bureaucracy that stifle scholarship and "protect" books only by guarding them from use. It is an anomaly in a suspicious and anonymous age. May it survive as it is, despite all the improbabilities. I finished this book in Gar Allen's house on Hyatt Road in Woods Hole, appropriately named after the man who first identified clearly the role of developmental timing in heterochrony—the major theme of this book (a meaningless coincidence, but I thought I'd mention it). I thank the American Philosophical Society for supporting my work in English libraries.

I don't know why authors feel constrained to say so—for it should be obvious—but I too hold all these dear friends and colleagues free, blameless, unencumbered, and innocent of responsibility for the errors in this book. How could it be otherwise; it is, after all, my work. I am responsible for all translations from non-English languages (except those few quoted from a secondary, English source). Finally, my thanks to G. G. Simpson for his intellectual breadth and for his ability to inspire a ten-year-old boy with his general writing, thereby eclipsing a previous worship of Joe DiMaggio. To my wife, Deborah, for being the kind of person about whom one could never write the conventional: "Thanks to my wife, whose patient understanding . . . and who typed the manuscript and kept the kids out of my hair." And to my parents for unflagging enthusiasm and encouragement, in the absence of any tradition for advanced education in our family, and against the bemusement of some older relatives who didn't know what paleontology meant and who, upon finding out, could only mumble (with an inflection that I cannot transcribe on paper): "That's a profession for a Jewish boy?"

Contents

PART TWO
HETEROCHRONY AND PAEDOMORPHOSIS

ONTOGENY
AND
PHYLOGENY

—1—

Prospectus

> A plausible argument could be made that evolution is the control of development by ecology. Oddly, neither area has figured importantly in evolutionary theory since Darwin, who contributed much to each. This is being slowly repaired for ecology . . . but development is still severely neglected.
>
> Van Valen, 1973

I am aware that I treat a subject currently unpopular. I do so, first of all, simply because it has fascinated me ever since the New York City public schools taught me Haeckel's doctrine, that ontogeny recapitulates phylogeny, fifty years after it had been abandoned by science. Yet I am not so detached a scholar that I would pursue it for the vanity of personal interest alone. I would not have spent some of the best years of a scientific career upon it, were I not convinced that it should be as important today as it has ever been.

I am also not so courageous a scientist that I would have risked so much effort against a wall of truly universal opprobrium. But the chinks in the wall surfaced as soon as I probed. I have had the same, most curious experience more than twenty times: I tell a colleague that I am writing a book about parallels between ontogeny and phylogeny. He takes me aside, makes sure that no one is looking, checks for bugging devices, and admits in markedly lowered voice: "You know, just between you, me, and that wall, I think that there really is something to it after all." The clothing of disrepute is diaphanous before any good naturalist's experience. I feel like the honest little boy before the naked emperor.

I began this book as an indulgent, antiquarian exercise in personal interest. I hoped, at best, to retrieve from its current limbo the ancient subject of parallels between ontogeny and phylogeny. And a rescue it certainly deserves, for no discarded theme more clearly merits the old metaphor about throwing the baby out with the bath water. Haeckel's biogenetic law was so extreme, and its collapse so spectacular, that the entire subject became taboo; otherwise no modern reviewer would begin with these words his account of a work that dared to mention it: "There are still those who would Haeckel biology" (Du Brul, 1971, p. 739).

But I soon decided that the subject needs no apology. Properly restructured, it stands as a central theme in evolutionary biology because it illuminates two issues of great contemporary importance: the evolution of ecological strategies and the biology of regulation. The starting point for a restructuring must be the recognition that Haeckel's theory requires a *change in the timing of developmental events* as the mechanism of recapitulation. For Haeckel, the change was all in one direction—a universal acceleration of development, pushing ancestral adult forms into the juvenile stages of descendants. Our current, enlarged concept does not favor speeding up over slowing down; all directions of change in timing are equally admissible. Paedomorphosis—the appearance of ancestral juvenile traits in adult descendants—should be as common as recapitulation.

Despite its baroque excrescences and digressions, this book is primarily a long argument for the evolutionary importance of *heterochrony*—changes in the relative time of appearance and rate of development for characters already present in ancestors. It is not a general discussion of the relationship between ontogeny and phylogeny. That some relationship exists cannot be denied. Evolutionary changes must be expressed in ontogeny, and phyletic information must therefore reside in the development of individuals. This, in itself, is obvious and unenlightening. This book emphasizes the importance of one kind of relationship—the *changes in developmental timing* that produce *parallels* between the stages of ontogeny and phylogeny. The greatest obstacle to understanding my theme is the lamentable confusion that exists in the literature between the ideas of von Baer and the strikingly different theory that generalizes Haeckel's recapitulation to encompass all directions of heterochronic change.

Haeckel interpreted the gill slits of human embryos as features of ancestral *adult* fishes, pushed back into the early stages of human ontogeny by a universal acceleration of developmental rates in evolving lineages. Von Baer argued that human gill slits do not reflect a change in developmental timing. They are not adult stages of ancestors pushed back into the embryos of descendants; they merely represent a stage common to the early ontogeny of all vertebrates (embryonic fish also have gill slits, after all).

The confusion between von Baer and Haeckel arises from an unfortunate tradition in natural history, the emphasis of results and their classification rather than processes and their explanation. It is true that both theories permit inferences about ancestors from embryonic stages of descendants—their utility in reconstructing phylogenetic trees does not differ very much. Does it matter whether we are actually repeating the *adult* stage of a fish-like ancestor (as the recapitulationists claimed), or only developing a common embryonic feature that fish, as primitive vertebrates, retain throughout life (as von Baer claimed)? The phyletic information is the same—we learn the same thing about our evolutionary relationship with fish in either case. If we are interested only in reconstructing family trees, the difference between these two theories of development is trifling.

If, however, we are interested in the mechanisms by which phyletic information appears in ontogeny, then the differences could scarcely be more important. For von Baer's theory of increasing differentiation calls only upon a conservative principle of heredity to preserve stubbornly the early stages of ontogeny in all members of a group, while evolution proceeds by altering later stages. Recapitulation, on the other hand, requires an *active mechanism* that pushes previously adult features into progressively earlier stages of descendant ontogenies—that is, it requires a change of developmental timing.

To decide between Haeckel or von Baer, one key question had to be answered: are *adult* stages of ancestors repeated by descendants? All the original participants in the debate knew perfectly well that this was the primary point; they argued incessantly about whether the undeniable phyletic content of juvenile stages had anything to do with *adult* ancestral forms. Thus, Thomas Hunt Morgan wrote: "To my mind there is a wide difference between the old statement that the animals living today have the original *adult* stage telescoped into their embryos, and the statement that the resemblance between certain characters in the *embryos* of higher animals and corresponding stages in the *embryos* of lower animals is most plausibly explained by the assumption that they have descended from the same ancestors" (1916, p. 23, my italics; see also Buckman, 1899, p. 116; Gegenbaur, 1874, in Russell, 1916, p. 262; MacBride, 1914, p. 649; 1917, p. 425; Garstang, 1922, p. 89; Temkin, 1950; Hadzi, 1952, p. 1019; Donovan, 1973, p. 2).

In this case those ignorant of history are not condemned to repeat it; they are merely destined to be confused. In recent years, a kindly and "liberal" tradition has tended to amalgamate von Baer and Haeckel. After all, we know that Haeckel was a bit extreme and we have had to drop his insistence on the telescoping of adult stages. But, since embryos do repeat the embryonic stages of their ancestors, why not call this recapitulation as well, thus effecting a sweeping synthesis of the two most contradictory views of developmental mechanics. Thus, Wald defines the biogenetic law only as the notion "that the evolution of organisms has left traces in their embryological development" (1963, p. 14; see also Johnson et al., 1972, p. 760; Lovejoy, 1959, p. 443). And de Beer advised: "If only the recapitulationists would abandon the assertion that that which is repeated is the *adult* condition of the ancestor, there would be no reason to disagree with them" (1930, p. 102). Indeed, but then they would not be recapitulationists.

I am, by the way, only making a theoretical point. I do not deny the cardinal importance of von Baer's laws and I will discuss them at length. In fact, I am convinced that the vast majority of supposed recapitulations represent nothing but the conservative nature of heredity, as expressed in von Baer's laws. I insist on the rigid separation of von Baer and Haeckel because it is only Haeckel's view—when properly expanded to include retardation as well as acceleration—that invokes the mechanism of changes in developmental timing. Since the importance of changes in timing is the primary theme of this book, I must make this distinction clear.

I wish to emphasize one other distinction. Evolution occurs when ontogeny is altered in one of two ways: when new characters are introduced at any stage of development with varying effects upon subsequent stages, or when characters already present undergo changes in developmental timing. Together, these two processes exhaust the formal content of phyletic change; the second process is heterochrony. If change in developmental timing is important in evolution, then this second process must be very common (if it is predominant in frequency, I will be in even better shape). I wish that I knew some way to make an estimate of relative frequencies for the two processes; for such data would be crucial in current debates on the evolutionary role of changes in gene regulation. Heterochronic changes are regulatory effects—they represent a change in rate for features already present. ("New" features may arise either from changes in structural genes or from shifts in regulation yielding complex morphological effects that we choose to designate as novel.) In any case, the relative frequency of heterochrony can be used to form a minimum estimate for the importance of changes in regulation. I believe that a study of heterochrony represents the most fruitful way to extract information about regulation from classical data of macroevolution and morphology.

Finally, I should comment on the unusual organization of this book. I have tried to present both a serious history of ideas and a reasonably complete analysis of current theory. In attempting this wide scope, I may have treated one of the halves inadequately, or made the book so long and expensive that no one will read it. I have further lengthened the book by quoting verbatim every important statement on the nature of relationships between ontogeny and phylogeny. The subject is so vexatious and confused that I could not adequately render a myriad of subtle distinctions in paraphrase. I hope that I will be forgiven for sacrificing literate prose to proper documentation.

Still, I do not regard the book as a hybrid of history and science, but as a coherent whole. I did not, as so often happens, intend to write a short historical introduction that subsequently grew. The final design was my original intention. I have several motives in writing a history of ideas at such a scale in a book intended largely for biologists.

History fascinates me for itself, but scientific utility set my plan. The argument for current significance of my subject required a historical treatment. The essential distinction between von Baer and Haeckel can hardly be appreciated except in historical context. The extraordinary persistence of a belief in recapitulation of adult stages cannot be a mass delusion. It has been with us since Aristotle, and it has insinuated itself into all theories, even those that would seem to abhor it (preformationism, for example). I refuse to believe that so many of the most brilliant scientists in the history of biology consistently placed at center stage a topic of merely peripheral importance.

Moreover, historical discussion is necessary if we are to sort out the various ways by which parallels between ontogeny and phylogeny arise. We often encounter only vague and fruitless analogies without causal implications. But more precise parallels can arise during evolution in two basic ways.

First, they may develop because a similar external constraint regulates both processes even though phylogeny has no direct effect upon ontogeny. The early nineteenth-century *Naturphilosophen,* for example, talked about a single developmental tendency that all dynamic processes must follow. In a later, mechanistic context, Hertwig (1906) argued that ontogeny seems to parallel phylogeny because only a few structural paths lead to the development of complexity from single cells. Still later, L. S. Berg (1926, p. 155) attributed recapitulation to the operation of identical developmental laws in the independent realms of ontogeny and phylogeny.

Second, parallels may arise by a direct effect of phylogeny upon ontogeny. In Haeckel's view, the phyletic law of acceleration pushes ancestral adult stages into the early ontogeny of descendants. Any heterochronic change is a phyletic event that alters the course of ontogeny directly.

With these distinctions, we can understand some current claims that invite confusion. Jean Piaget, for example, believes that the development of thought in children closely parallels the evolution of consciousness in our species. Yet he explicitly denies any theory of recapitulation in Haeckelian terms. In fact, he adheres to the principle of external constraints—in this case the boundary condition of the mind's basic structure.

Another motive for writing this book is my belief that the history of recapitulation illustrates some generalities about science that will surprise no historian but prove interesting to many scientists. The inadequacy, for example, of a "hard sciences" model for crucial experiments in proof and disproof has never been more evident. The data of natural history are so multifarious, complex, and indecisive that simple accumulation can almost never resolve an issue. Counter-cases can always be documented in large numbers, and no one can find and count enough unbiased cases to establish a decisive relative frequency. Theory must play a role in guiding observation, and theory will not fall on the basis of data accumulated in its own light. Recapitulation was largely impervious to empirical disproof by accumulated exceptions. It fell when it became unfashionable in practice, following the rise of experimental embryology, and untenable in theory, following scientific change in a related field (Mendelian genetics). As another example, my documentation in Chapter 5 of the decisive influence recapitulation has had in nonbiological areas as disparate as racism and Freudian analysis may convince some scientists that the arcane researches of cloistered academics may be full of social significance, and that scientific detachment and absolute objectivity are myths.

I have by now undoubtedly infuriated my historical colleagues with this rapist's approach to history—extraction from the past for thinly disguised ulterior purposes in explicating current issues. But my disingenuousness may run primarily in the other direction. I am probably using the threads of current relevance to justify to scientific colleagues the unsullied history that I wanted to write for the simplest of all intellectual reasons—personal excitement, fascination, and even awe (how can anyone read von Baer without a sense of exaltation for the sheer intellectual power of it all). This, at least, is irreducible and needs no defense.

Epitome

In cryptically simplified form, my primary argument runs:

1. The idea of a relationship between ontogeny and the history of life was not the invention of a nineteenth-century German evolutionary zealot. Aristotle defended an analogical relationship between human development and organic history. The notion of a parallel between stages of ontogeny and sequences of adults (either created or evolved) has been ubiquitous in biological theory. It even played a central role in the preformationistic theories of Charles Bonnet—a system that would, at first glance, seem to deny both ontogeny and phylogeny.

2. Two radically different concepts of relationships between the ontogeny of higher forms and sequences of adults arose in the early nineteenth century as concepts of motion, change, and progress replaced the static outlook in biology. Oken, Meckel, Agassiz, and others argued that stages of ontogeny repeat the *adult* forms of animals lower down the scale of organization. Von Baer retorted that no higher animal repeats any adult stage; development proceeds from undifferentiated homogeneity to differentiated heterogeneity—from the general to the special.

3. Agassiz's theory of recapitulation and von Baer's theory of embryonic similarity were both reread in the light of evolution. Darwin supported von Baer, while Haeckel, Müller, Cope, and Hyatt independently established the biogenetic law—ontogeny recapitulates the *adult* stages of phylogeny. The growing prestige of recapitulation soon eclipsed von Baer's alternative.

4. Evolutionary theory imposed a radical restructuring of mechanisms for recapitulation. Oken had invoked a single developmental tendency: gill slits in a human embryo and in an adult fish reflect the same stage in universal development. But in Haeckel's evolutionary reading, the human gill slits *are* (literally) the adult features of an ancestor. How then did they get from a large adult ancestor to a tiny, transient fetus? All evolutionary recapitulationists accepted a mechanism based on two laws: first, "terminal addition"—evolutionary change proceeds by adding stages to the end of ancestral ontogeny; second, "condensation"—development is accelerated as ancestral features are pushed back to earlier stages of descendant embryos. The

principle of condensation identified change in developmental timing as the mechanism that produces parallels between ontogeny and phylogeny.

5. The rise of a mechanistic experimental embryology presaged the death of recapitulation. The biogenetic law finally collapsed as Mendelian genetics repudiated the generality of its two necessary principles—terminal addition and condensation. All varieties of change in developmental timing became orthodox. The development of individual parts could be either accelerated or retarded relative to other parts. These accelerations and retardations engender the full set of parallels between ontogeny and phylogeny.

6. The collapse of Haeckel's law prompted a confusing variety of complex classifications for relations between ontogeny and phylogeny. These classifications treat the *results* of changes in developmental timing (recapitulation and paedomorphosis), not the *mechanisms* (acceleration and retardation). The classifications are static and complex. Very little fruitful use has been made of them and the general subject has lost almost all its previous popularity.

7. I replace these classifications of results with a simple classification of processes. There is no one-to-one correspondence between process and result. Retardation can lead to paedomorphosis (neoteny by slowing down of somatic development) or to recapitulation (hypermorphosis by retardation of maturation). Acceleration may also generate paedomorphosis (progenesis by speeding up of maturation) or recapitulation (by acceleration of somatic organs). The processes are more fundamental than the results; they determine the evolutionary significance of heterochrony. (Classifications of results are further confused by inconsistencies in the criteria used to compare ancestor and descendant—size, age, or developmental stage. I develop a "clock model" to depict all criteria simultaneously and to display the complexity as a result of an underlying simplicity involving only acceleration and retardation.)

8. I develop a new context for considering the evolutionary importance of recapitulation and paedomorphosis. Classical arguments are based upon the macroevolutionary significance of morphology—paedomorphosis as an escape from specialization, recapitulation as a motor of evolutionary progress by addition of organs. I focus upon the immediate significance of acceleration and retardation in the evolution of life-history strategies for ecological adaptation. In this context, the timing of maturation assumes special importance.

9. I illustrate the primacy of process by showing that paedomorphosis (a result) mixes two very different phenomena that superficially share the common property of juvenilized morphology. One is

the result of acceleration, the other of retardation. *Progenesis* reflects the truncation of ontogeny by precocious sexual maturation (acceleration). It represents a life-history strategy for *r*-selective regimes, where early reproduction is highly favored. Selection is for precocious maturation or small size; juvenilized morphology is often a secondary consequence. *Neoteny*, on the other hand, represents the retardation of somatic development for selected organs and parts. It occurs in *K*-selective regimes, where morphology is fine tuned to immediate ecological conditions.

10. Progenesis and neoteny have different roles in macroevolution. New higher taxa may occasionally arise by progenesis, because selection upon morphology relaxes in a developmental context mixing juvenile and adult characters (precocious maturation usually leaves some characters in their juvenile state, while accelerating others that are more strongly correlated with maturation itself). Evolution of a new higher taxon may be very rapid and largely "fortuitous" (in the sense that selection does not operate directly to produce the new morphology). Neoteny provides a flexible alternative to the hypermorphic overspecialization that usually accompanies a delay in maturation. It has been important in the evolution of complex social behavior in higher vertebrates. Delayed growth and development can establish the ranks in a dominance hierarchy or lead to an increase in cerebralization by prolonging into later life the rapid brain growth characteristic of fetuses.

11. Neoteny has been a (probably *the*) major determinant of human evolution. When we recognize the undeniable role of retardation in human evolution, the data of neoteny can be rescued from previous theories that made them so unpopular. Human development has slowed down. Within this "matrix of retardation," adaptive features of ancestral juveniles are easily retained. Retardation as a life-history strategy for longer learning and socialization may be far more important in human evolution than any of its morphological consequences.

12. Humans and chimps are almost identical in structural genes, yet differ markedly in form and behavior. This paradox can be resolved by invoking a small genetic difference with profound effects—alterations in the regulatory system that slow down the general rate of development in humans. Heterochronic changes are regulatory changes; they require only an alteration in the timing of features already present. If the frequency of heterochronic change were known, it would provide a good estimate for the importance of regulation as an evolutionary agent.

This epitome is a pitiful abbreviation of a much longer and, I hope, more subtle development. Please read the book!

PART ONE

RECAPITULATION

—2—

The Analogistic Tradition from Anaximander to Bonnet

The Seeds of Recapitulation in Greek Science?

The microcosm: ontogeny. The macrocosm: cosmic history, human history, organic development. This comparison may be the most durable analogy in the history of biology (Kleinsorge, 1900). It seems, to use another ontogenetic metaphor, as inevitable as aging.

I have chosen, for this book, just one of these pervasive comparisons —that between stages of ontogeny and a sequence of *adult* organisms, either created or evolved. As Haeckel's biogenetic law, this comparison provided an argument second to none in the arsenal of evolutionists during the second half of the nineteenth century. Moreover, the relation envisaged by Haeckel and, in a very different manner, by the earlier Naturphilosophen transcended mere analogy to become an intimate and necessary causal connection. Yet the basic argument is as old as recorded biology; in pre-evolutionary thought, its formal treatment as an analogy played an important role in the systems of men as different in time and belief as Aristotle and Charles Bonnet. Its appearance in Aristotle as an argument for epigenesis offers no surprise and testifies only to its antiquity; on the other hand, its role in the preformationism of Bonnet—a system that would seem to deny not only phylogeny but ontogeny as well—attests to its remarkable ubiquity.

The analogy of individual to cosmic history was favored by many pre-Socratic thinkers. The nascent cosmos of Anaximander, Anaximenes, and Democritus was surrounded by an envelope resembling the amniotic membrane (Wilford, 1968, p. 109). Empedocles' cos-

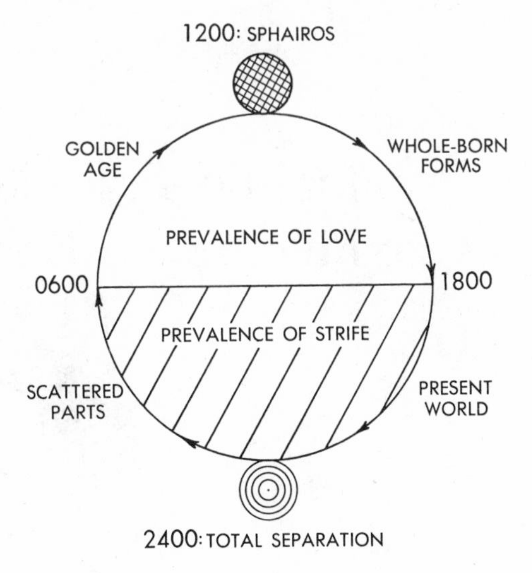

Fig. 1. The cosmogony of Empedocles as interpreted by de Santillana (1961).

mogony (circa 450 B.C.) rested largely on ontogenetic analogies. In de Santillana's account (1961, pp. 108–128), the cyclical history of the cosmos depends upon the alternating domination of love and strife (Fig. 1). Love drives like to unlike; strife cleaves and separates the cosmos, ultimately into the four primal elements in unmixed, concentric layers (Wilford disputes the common interpretation of layering but affirms the separation into basic elements). We are at present in the second quadrant of strife's increasing domination. Human embryology is comparable to that portion of the cycle leading from love's unity to our present condition.[1] The human embryo begins in a formless state (comparable to the unity of love); the differentiation, elaboration, and separation of its parts reflects the dominance of strife in our period and forms the microcosm of universal history. "Empedocles' leading idea here is that the growth of the embryo rehearses in a foreshortened way the cosmogonic process" (de Santillana, 1961, p. 114). Moreover, the "whole-born forms" of the first quadrant are not unlike human embryos in their lack of differentiation. They have no sex. They have, Empedocles writes, "neither the lovely forms of limbs nor voice." "They are imagined as slow-growing foetuses incubated by the earth's heat" (de Santillana, 1961, p. 114).

Anaximander, in Osborn's interpretation, compares stages of human ontogeny with "ancestors" of the historical development he describes. In a debate on historical primacy among the four elements,

Anaximander asserted the preeminence of water and envisaged an originally fluid earth. "His hypothetical ancestors of man were supposed to be first encased in horny capsules, floating and feeding in water; as soon as these 'fish-men' were in a condition to emerge, they came on land, the capsule burst, and they took their human form" (Osborn, 1929, pp. 47–48). To support both the primacy of water and this "phyletic" sequence, Anaximander pointed to the early fluidity and transparency of the embryo, its residence in the amniotic fluid, and its long period of helplessness after birth.

Empedocles and Anaximander had compared the stages of human embryology with hypothetical forms in previous stages of the cosmic cycle. Aristotle—in a comparison more congenial with later formulations—drew an analogy between actual, living adults and human ontogeny. Since Aristotle recognized embryology as epigenetic and classified organisms by increasing degrees of "perfection," such a comparison could scarcely be avoided.

In *De generatione animalium,* Aristotle classified animals into five groups: (1) mammals; (2) ovoviviparous sharks; (3) birds and reptiles; (4) fish, cephalopods, and crustaceans; (5) insects. This sequence itself is based upon ontogeny, for its criterion is increasing perfection in generation: "We must observe how rightly Nature orders generation in regular gradation. The more perfect and hotter animals produce their young perfect . . . The third class do not produce a perfect animal, but an egg and this egg is perfect [that is, it does not increase in size]. Those whose nature is still colder than these produce an egg, but an imperfect one, which is perfected outside the body" (*De generatione,* 733b, lines 1–10). Ontogeny also progresses towards greater complexity: "For nobody would put down the unfertilized embryo as soulless or in every sense bereft of life . . . That then they possess the nutritive soul is plain. As they develop they also acquire the sensitive soul in virtue of which an animal is an animal" (736a, lines 33–38).

Aristotle portrayed his epigenetic view of ontogeny as a series of increasingly higher "souls"—nutritive, sensitive, and rational—entering the human embryo during its development. (This claim formed the basis of later debates on such topics as the theological status of aborted fetuses; see Oppenheimer, 1975, for a discussion of Aristotle's view on times of permissible abortion based on development of sequential souls by the fetus.) In several short statements, he compared this threefold sequence of souls with adult organisms: nutritive to plants, sensitive to animals, and rational to humans. For example: "At first all such embryos seem to live the life of a plant" (736b, lines 13–14). "Since the embryo is already potentially an animal but an imperfect one, it must obtain its nourishment from elsewhere; accordingly it makes use of the uterus and the mother, as a plant does of the earth, to get nourishment, until it is perfected to the point of being now an animal potentially locomotive" (740a, lines 24–28).

These several statements have won for Aristotle the role of great-great-grandfather to the theory of recapitulation. Needham, for example, writes: "He foreshadowed the theory of recapitulation in his speculations on the order in which the souls came to inhabit the embryo during its growth" (1959, p. 55). In one sense, this is absurd. Aristotle's comparisons are analogies, drawn to reinforce his belief in the epigenetic nature of development. They are separated by two millenia of time and belief from the causal theories of the Naturphilosophen and Haeckelian evolutionists. Moreover, I reject an approach to the history of science that rapes the past for seeds and harbingers of later views; such a perspective only makes sense within the abandoned faith that science progresses by accumulation towards absolute truth.

In another sense, Aristotle's special role in the history of western thought guaranteed a larger role for these incidental analogies. His words were studied assiduously in later centuries; and they were recalled when the epigenetic nature of embryology was affirmed anew. Many later comments, often cited as seeds of recapitulation, are simple restatements of the Aristotelian position. In his *Religio Medici* of 1642, Sir Thomas Browne used Aristotle's argument to support the analogy of microcosm (man) and macrocosm (universe):

> To call ourselves a Microcosm, or little World, I thought it only a pleasant trope of Rhetorick, till my neer judgment and second thoughts told me there was a real truth therein. For first we are a rude mass, in the rank of creatures which onely are, and have a dull kind of being, not yet priviledged with life . . . ; next we live the life of Plants, the life of Animals, the life of Men, and at last the life of Spirits, running on in one mysterious nature those five kinds of existences, which comprehend the creatures, not only of the World, but of the Universe.

Diderot and Harvey have also been assigned paternity for recapitulation, simply because they affirmed Aristotelian epigenesis (Diderot, by Crocker, 1959; Harvey, by many authors, ranging from Serres, 1827, to Meyer, 1935). Harvey, for example, writes in his *Anatomical exercitations* of 1653: "About the fourth day the egg beginneth to step from the life of a plant to that of an animal."

In the meaningless quest for supposed harbingers of the theory of recapitulation, only one item deserves special attention—a passage in John Hunter's "Progress and peculiarities of the chick" (the goose chick), posthumously published by Richard Owen:

> If we were capable of following the progress of increase of the number of the parts of the most perfect animal, as they first formed in succession, from the very first to its state of full perfection, we should probably be able to compare it with some one of the incomplete animals themselves, of every order of animals in the Creation, being at no stage different from some of the inferior orders; or, in other words, if we were to take a series of animals, from the more imperfect to the perfect, we should probably find an imperfect animal, corresponding with some stage of the most perfect. (in Owen, 1841, p. 14)

But two observations vitiate Hunter's claim to paternity for the idea of recapitulation as a sequential repetition of "lower" adults: first, although secondary sources (Meyer, 1935, p. 381)[2] usually give the date of this passage as about 1755, when Hunter began his work on development of the goose, Owen himself dates the manuscript between 1775 and Hunter's death in 1793—well within the period when basic tenets of *Naturphilosophie* were being disseminated, though before the traditional date, 1793, for Kielmeyer's inception of the specific argument. Second, Hunter's passage is short, disconnected, and speculative, and is not based on observation. Hunter merely conjectured about the course of development *before* any embryonic structure could be resolved by the best microscopes then available. The next sentence in Hunter's manuscript, although never cited in the literature on recapitulation, makes this clear: "But all our observations can only begin at a visible stage of formation, prior to which we are left to conjecture" (Owen, 1841, p. 14).

Ontogeny and Phylogeny in the Conflict of "Evolution" and Epigenesis: The Idyll of Charles Bonnet

Needham spoke of epigenesis and preformationism as "an antithesis which Aristotle was the first to perceive, and the subsequent history of which is almost synonymous with the history of embryology" (1959, p. 40).

What greater mystery can there be than the growth of something so complex as a human baby from humble beginnings in an essentially formless egg or, as Aristotle would have it, the menstrual blood? The two extreme solutions have their strengths and problems. One can believe what one sees and argue that parts are formed sequentially by external forces acting upon matter only potentially capable of normal development (epigenesis). But what natural force could then regulate ontogeny? Indeed, the eighteenth-century epigeneticists often took refuge in vitalism or outright mysticism. Or one can label what one observes as mere appearance and contend that the complexity of the final product is present from the first, though the germ and young embryo may be too tiny or too transparent to show it (preformationism). Ontogeny, then, is the evolution*—literally the unrolling—of this preformed complexity. This position avoids the dilemma of mystical forces, but it compels us to postulate what we do not perceive.

We know that the epigeneticists had "won" by von Baer's time. One might think that the theme of this chapter should be the record of this victory—for what can be a stronger prerequisite for the triumph of recapitulation than the defeat of a theory that denied both development in embryology and any true organic change through time. As Etienne Geoffroy Saint-Hilaire wrote: "According to the system of unlimited encapsulation [*emboîtement indéfini*], organisms are and remain through the centuries what they have always been: from this, men have concluded that the forms of animals are unalterable" (1833, p. 89). Yet it is not so. The first extensive parallel between ontogeny and the history of life—albeit a version in the analogical tradition—came from the pen of Charles Bonnet, the leading spokesman for preformation. To understand this paradox, we must dispel some common myths about "evolution."

The solution to great arguments is usually close to the golden mean, and this debate is no exception. Modern genetics is about as midway as it could be between the extreme formulations of the eighteenth century. The preformationists were right in asserting that some preexistence is the only refuge from mysticism. But they were mistaken in postulating preformed structure, for we have discovered coded instructions. (It is scarcely surprising that a world knowing nothing of the player piano—not to mention the computer program—should have neglected the storage of coded instructions.) The epigeneticists, on the other hand, were correct in insisting that the visual appearance of development is no mere illusion.

With this rapprochement, one might think that the two original views would be equally respected in the historical paragraphs of modern textbooks in embryology. Yet these paragraphs almost invariably

* It is one of the ironies of history that a word coined by preformationists to describe their static view of life should now have the opposite meaning of organic *change* (see the appendix to this chapter). Etymology favors the original meaning since the growth of a preformed homunculus is a true unrolling of a scroll already written, while Darwinian evolution is both unpredictable and under the control of external selective forces. Spencer's original definition of "evolution," the seed of modern usage, was not restricted to organic change and did prescribe a tendency toward greater heterogeneity and complexity.

read like the script of a western movie, with epigeneticists as "good guys," preformationists as "bad guys," and the supposed triumph of epigenesis as the foundation of modern embryology. In so doing, preformationism is rendered as an absurd caricature of itself. Were this caricature correct, no preformationist could ever have constructed an analogy between ontogeny and the history of life, and this chapter could have no meaning.

The caricature involves three statements:

1. The preformationists believed that ontogeny involved only the increase in size of a perfectly proportioned miniature enclosed in the egg or sperm.

2. This belief leads to absurdity because the homunculus-in-the-egg must have a homunculus in its egg, and so on ad infinitum. The entire history of mankind resided in the ovaries of Eve (or the sperm of Adam).

3. Since anyone can see that a chick embryo is no miniature hen, the preformationists must have been fanatically dogmatic anti-empiricists.

These three charges are by no means new; in expounding his system of evolution, Charles Bonnet dealt with all of them.*

1. The miniature homunculus. The standard text reproduces Hartsoeker's little man with a big head neatly curled up within a sperm, or Dalenpatius' tiny animals, or D'Agoty's homunculus suspended in a glass of water and visible "even without the help of lenses" (1752). But the vivid imagination of a few peripheral figures should not be allowed to distort the common opinion of all leading evolutionists. Malpighi,[3] Haller, and Bonnet had all studied the development of the chick embryo and knew perfectly well that it seemed to progress from transparent homogeneity to differentiated complexity. Yet they claimed that this appearance was illusory.

Visual appearance, they argued, cannot be equated with true exis-

* Bonnet (1720–1793) received his doctorate in law in 1744. At the age of 26, he discovered parthenogenesis in aphids. As a young man, he also did important work on the regeneration of worms and on respiration in insects. Burdened with deafness and increasing blindness, he assumed in his later work a philosophical tone and became a leading spokesman for the theory of preformation. He explored its implications—biological, ethical, and theological—in three important works: *Considerations sur les corps organisés* (1762), *Contemplation de la nature* (1764), and *La palingénésie philosophique* (1769). By "palingenesis" (repeated generation), Bonnet means the evolution of organisms already preformed in the germ—particularly the resurrection of each individual at the end of time by evolution of his germ of restitution. Haeckel later used the term in a completely different sense, also consistent with etymology. To Haeckel, "repeated generation" referred to the recapitulation of previous phyletic stages in ontogeny.

tence. The young embryo is so tiny and so transparent that we cannot see features which must be there. When we consider that the best microscopes of the day were no match for student models in modern elementary biology courses, this argument gains some force. Bonnet wrote:

In its first beginnings, the animal is almost fluid. By degrees it assumes a gelatinous consistency. When the lung of the chick begins to become visible, its size is already 1/10 of an inch. We can prove that it would have been visible at 4/100 of an inch had it not been for its perfect transparency. (1764, p. 156)

Do not, therefore, mark the time when organized beings begin to exist by the time when they begin to become visible; and do not constrain nature by the strict limits of our senses and instruments. (1762, p. 169)

Moreover, to argue that parts are preformed is not to maintain that the early embryo is already a miniature adult. The parts, to be sure, are there from the first, but in proportions and positions different from those displayed in adults. Even if we could see all the parts from the very first, ontogeny would present a panoply of change by rearrangement, differential increase in size, and alteration in shape:

While the chick is still a germ, all its parts have forms, proportions and positions which differ greatly from those that they will attain during evolution . . . If we were able to see the germ enlarged, as it is when small, it would be impossible for us to recognize it as a chick . . . All the parts of the germ do not develop at the same time and uniformly. (1762)

2. The reductio ad absurdum of encapsulation. From our perspective, no idea could be more absurd than the encapsulation of all human history in the ovaries of Eve—the homunculus in the egg, the next homunculus in the egg of the homunculus, box within box within box, to prospective future generations of incredible tininess. Epigeneticists of the eighteenth century did raise this argument, but preformationists did not find it insuperable.

Indefinite encapsulation of lineages that persist through millions of generations may strain the imagination; yet a world slated to survive but a few thousand years might permit sufficiently few generations to inspire belief. Furthermore, the preformationists had no cell theory setting a lower limit of size and rendering absurd a human homunculus smaller than the smallest single cell. It was not clear to Bonnet that size had any upper or lower limit, that it was any more than a relative concept. Infusorians had been discovered only recently, with the help of microscopes known to be poor compared to what might be developed. Again, in the absence of cell theory, it was widely believed that these protozoans had organs and perhaps parasites with their own organs, and so forth, ad infinitum. "Nature works as small as it wishes," Bonnet wrote (1764, p. xxxii).

We know not at all the lower boundary of the division of matter, but we see that it has been prodigiously divided. From the elephant to the mite, from the whale to the animalcule 27 million times smaller than the mite, from the globe of the sun to the globule of light, what an inconceivable multitude of intermediate degrees! This animalcule can perceive light; it penetrates into its eye; it traces there the image of objects; what could be more incredibly small than the size of that image; what could be still more incredibly small than the size of a globule of light, of which several thousand, perhaps several million, enter all at once into the eye. (p. 162)

3. The most famous statement of Bonnet, and the one chosen most often for ridicule, proclaims: "This hypothesis of encapsulation is one of the greatest victories that pure understanding has won over the senses" ("Cette hypothèse de l'Emboîtement est une des plus belles victoires que l'Entendement pur ait remporté sur les Sens"; 1764, pp. 162–163). Wrenched from its context by a modern experimentalist, Bonnet's statement invites disdain. Yet in its time, it marked a necessary claim in support of a position that modern experimentalists would regard as "scientific."

Charles Lyell argued strenuously that the key to a true geology lay in a commitment *not* to read literally the evidence of the strata. Cuvier, the great literalist, had postulated a series of catastrophic faunal extinctions (based on missing strata and local shifts in environment within the Paris Basin). Lyell emphasized the extreme imperfection of the geologic record. He argued that we must interpolate between the bits of surviving evidence the events demanded by a true theory of geology—the uniformity of process. Bonnet's claim was within the same tradition: we know that the tininess and transparency of the embryo and the crudeness of our instruments preclude any direct observation of what is really there in early stages of ontogeny; we must therefore either accept the epigenetic evidence of our raw senses or postulate what we cannot yet perceive. If we choose the former course, we avoid the Scylla of unseen entities but confront instead the Charybdis of a general theory that substitutes mystical forces for the mechanics of Newtonian science. Bonnet's theory of evolution was designed to protect the general attitude that a modern experimentalist would recognize as "scientific" from the vitalism implied by the evidence of raw sensation. The preformationists were the mechanists of their time. (And their basic argument is none the worse today. If the egg were truly unorganized, how could it yield such consistent complexity without a directing entelechy. It does so, and can only do so, because the information—not merely the raw material—needed to build this complexity already resides in the egg.) As Bonnet wrote: "We must either undertake to explain mechanically the sequential formation of organs . . . or we must admit that the germ contains in miniature all the parts essential to the plant or animal that it represents" (1762, p. 20).

The link of preformationism to the Newtonian world view has been emphasized by many writers (Adelmann, 1966, p. 875; Gasking, 1967; Meyer, 1939, p. 310; Whitman, 1894a, p. 219; 1894c, p. 260; Huxley and de Beer, 1934, p. 2). Wilkie has expressed the dilemma particularly well:

> By the canons of Cartesianism they [preformationists] were not allowed to appeal to souls or substantial forms as agents in embryology . . . The laws of motion seemed far too general to account for the particularities of the embryology of different species, and too simple to explain the extreme complexity of any one animal form. They were not allowed to attribute the formation of an animal *hic et nunc* to God, whose creative activity was held to have terminated with formation of the universe at an unique moment in the remote past. Hence they were left with the notion that each individual organism must somehow have existed from the first, having been created along with all other things. (1967, pp. 140–141)

In 1824, Prévost and Dumas confessed: "At first sight this hypothesis is startling, but gradually the mind gets used to it and soon comes to prefer it to any other. It seems easier to imagine a time when nature, as it were, labored and gave birth all at once to the whole of creation, present and future, than to imagine a continual activity" (in Gasking, 1967, p. 144).[4]

Bonnet's philosophical writings encompass nearly everything from the properties of infusorians to the nature of the Godhead. As a dominant theme in all these works, Bonnet extended to the entire universe the basic philosophical tenet of his preformationistic beliefs about ontogeny—development is only apparent; it represents the unfolding of structures preformed at the creation itself. God, the clockwinder, had not only ordained the laws of the universe; he had created all its structures as well: one creation followed by the complete evolution of all preordained structure to the appointed end of time.

This grand conception of universal design inspired Bonnet to rapture:

> I delight in considering this magnificent succession of organized beings, encased as so many little worlds, one inside the other. I see them moving away

from me by degrees, diminishing according to certain proportions, and finally losing themselves in an impenetrable night. I taste a secret satisfaction in discovering . . . the germ of the hero who will found, in several thousand years, a great empire; or perhaps that of a philosopher who will then discover the cause of gravity, the mystery of generation and the mechanics of our being. (1762, p. 104)

With his preformationist tenet, Bonnet could reconcile two notions of God that might otherwise clash: a conviction that the benevolent God of Christianity had ordained a universal history marked by increasing perfection, progress, and happiness, and a belief that God had acted but once, and that the world's order implied the construction of its entire history by this single act.

Both notions are implicit in Bonnet's most famous conception: his chain of being. The chain extended without interruption from its "first term, the atom" to "the highest of the CHERUBIM" (1764, p. 29).

Between the lowest and highest degree of spiritual and corporal perfection, there is an almost infinite number of intermediate degrees. The succession of degrees comprises the *Universal Chain.* It unites all beings, ties together all worlds, embraces all the spheres. One SINGLE BEING is outside this chain, and this is HE who made it. (p. 27)

The identification of intermediate stages between fairly discrete groups (minerals, plants, animals, for example) posed some difficulty, but Bonnet strove to close the gaps. Pointing to the fibrous structure of higher plants, he picked asbestos and sedimentary strata as transitional forms between minerals and plants—though, he confessed, "this transition is not as happy as those which we observe in other classes of terrestrial beings; nature seems to make a jump here; but this jump will, without a doubt, disappear when our knowledge attains greater extent and precision" (p. 37). Other transitional forms included the hydra between plants and animals (p. 47), the eel between reptiles and fish (p. 64), flying fish and aquatic birds between fish and birds (p. 66), and the bat and ostrich between birds and mammals. The glaring gap between monkey and man he filled with the customary gradation—"savages" on the bottom, one's own group on the pinnacle: "We may oppose the impropriety of the Hottentot to the propriety of the Hollander. From the cruel cannibal, we pass rapidly to the humane Frenchman . . . We mount from the Scottish peasant to the great Newton" (p. 82).

Though the chain marked a complete sequence of increasing perfection, it was entirely static. It had been created all at once, and its constituent links could not be transformed one into the other. Thus did Bonnet reconcile his notions of perfectibility and of created order.

He used a similar argument to explain how the appearance of increasing perfection in history could be reconciled with a creation of all at the outset. And he based this reconciliation upon a complex and extended analogy between ontogeny and the history of life.

Bonnet began by using the general principle of embryological evolution to argue that the world (like the germ) had been created with its entire history encapsulated within it.

> How can we suppose that this WILL, which was able to preordain everything by a single act, would intervene immediately and unceasingly in space and time? Does IT first create the caterpillar, then the chrysalis, and then the butterfly? Does IT create new germs at each moment? (1769, 2:133)

Bonnet then repeats for the cosmos the same argument that explained the epigenetic *appearance* of ontogeny and demolished the red herring of a perfectly miniaturized homunculus.

> I do not suppose that, at the first moment of the creation, all the celestial bodies were precisely disposed, one with regard to the other, as they are today . . . But, DIVINE WISDOM had foreseen and approved these changes, as IT had foreseen and approved the almost infinite number of diverse modifications that would arise from the structure or primitive organization of the beings inhabiting each world. All the pieces of the universe are, therefore, contemporaneous. By a single act, the CREATING WILL realized everything that could be. (1769, 1: 246–247)

Bonnet then made a more specific analogy between the actual sequence of stages in ontogeny and the history of life. Both seem to progress from simplicity to complexity. The appearance is illusory in each case.

Moses, Bonnet argued, could not have described the original creation in the Book of Genesis (p. 237). Genesis places the formation of the sun on the fourth day; but it cannot be younger than an earth subservient to it. Moses is describing a mere appearance: as a former world evolved to the next, it passed through a period of chaos recorded in the first statements of Genesis. The Bible discusses only our current world, and the apparent immutability of the life it describes applies only to the latest of a long series of worlds.

The physical and biological are intimately entwined. If the earth has undergone a series of revolutions in its apparent form, we may infer a parallel sequence of illusory changes for life: "Who can deny that ABSOLUTE POWER could enclose in the first germ of each organized being the succession of germs corresponding to the diverse revolutions that our planet would be called to undergo? Do not the microscope and scalpel show us generations encapsulated one inside the other?" (p. 261).

But what might life have looked like on these former worlds? The principle of perfectibility decrees that the successive appearances of any species be arrayed in a progressive series: "Animals are called to a perfection, for which the organic principle existed at the creation, and whose complete development is reserved for the future state of our globe" (p. 199). Embryology affirms such a series, since the preformed homunculus evolves from illusory simplicity to its full and final complexity. Bonnet continually employs an ontogenetic metaphor to justify his views on the history of life's apparent progress: "Some species undergo a large number of metamorphoses which reclothe individuals in forms so varied that they appear to represent different species. Our world was previously in the form of a worm or caterpillar; it is now in that of a chrysalis; the last revolution shall reclothe it in the form of a butterfly" (p. 262). The embryology of the chick provides more specific clues to the form of animals inhabiting previous worlds:

> We cannot view without astonishment . . . the strange revolutions that the chick undergoes from the moment it begins to become visible to the moment at which it displays its final form. When the chick begins to become visible, it appears in a form very similar to that of a very small worm. Its head is large, and to this head is attached a sort of tapering appendage. It is, however, in this appendage, so similar to the tail of a small worm, that the trunk and limbs of the animal are contained . . . If the imperfection of our vision and of our instruments permitted us to see the earlier stages of the chick's development [*remonter plus haut dans l'origine du poulet*], we would, without doubt, discover much more still disguised. The different phases in which the chick is successively revealed to us allows us to judge the diverse revolutions that organized bodies have undergone to reach this last form by which they are known to us. All this helps us to conceive the new forms that animals will attain in that future state to which, I conjecture, they are called. (pp. 178–179)

In fact, Bonnet does describe as "embryonic" the appearance of organisms in past worlds:

> How astonished we would be if we could penetrate into these depths, and cast our glances into this abyss! There we would discover a world very different from our own, a world whose bizarre decorations would throw us into a confusion that would grow without cease. There, a Réaumur, a Jussieu, a Linnaeus, would be lost. There we would look for our quadrupeds, our birds, our reptiles, our insects, and we would see in their place only bizarre figures, whose irregular and incomplete traits would leave us uncertain that we were looking at what would become quadrupeds and birds . . . We conceive that this first state of all organized beings is the state of the germ, and we have said

that the germ contains in miniature all the parts of the future animal or vegetable. It does not acquire organs that it never had; but organs which have not yet appeared begin to become visible. (1764, pp. 161–162)

Thus, organisms have reached their present state through a series of changes in form comparable to the illusory progression of the embryo from simplicity to complexity. But what of the future? What happens to the essence of a creature after the decay of its earthly body? The principle of perfectibility cannot permit an irrevocable disappearance: "What philosophical reasons impose upon us the obligation to believe that death is the end of an animal's course? Why should a being, so perfectible, be extinguished forever while it possesses a principle of perfectibility to which we cannot assign the limits" (1769, 1: 182). Resurrection of the body at the end of time is proclaimed by scripture. By the principle of perfectibility, the body housing the soul at the last judgment must be superior to the body it inhabited during a previous earthly existence. Yet the principle of preformation dictates that all structure, including this superior body, be present at the creation. Where, then, does the soul reside with its better body while it awaits the sound of the trumpet?

Bonnet proposes that man's soul resides within a miniature homunculus located in the corpus callosum of the brain. Again, with an ontogenetic metaphor, he writes:

The resurrection will only be the prodigiously accelerated development of this germ, presently hidden in the corpus callosum. Could not THE AUTHOR of nature, who preordained all beings at the beginning, enclose the spiritual body in the animal body—as he enclosed originally the plant in the seed, the butterfly in the caterpillar, future generations in present generations? . . . O Christians who support this doctrine of life, do you fear death? Your immortal soul is tied to immortality by physical bonds and these bonds are indissoluble. United to an imperishable germ, your soul sees only a happy transformation in death: a transformation which, in freeing the seed from its envelope, will give to the plant a new being. O death where is thy sting! O grave where is thy victory! (1764, pp. 88–90)

But if man shall be so bounteously rewarded with a better body for his immortal soul, what of God's lower creatures? Does not the principle of perfectibility also demand that animals, perhaps plants as well, have souls? Do not these souls also share the promise of an improved body at the resurrection?

If the ADORABLE WISDOM which presided at the formation of the universe wanted the greatest perfection for all sentient beings (and how can one doubt this desire in SUPREME GOODNESS), IT would have preformed in this little indestructible body, true seat of the soul of beasts, new senses, more exquisite senses, and parts appropriate to these senses. (1769, 1: 181–182)

Even plants may expect a resurrection as improved creatures capable of motion (p. 225).

This general concept carries some intriguing implications, and Bonnet does not shrink from them. What, for example, happens to the "germ of restitution" (p. 180) when the body enclosing it dies:

I do not think that it is very difficult to answer this question. The indestructible germs can be dispersed, without inconvenience, into all the individual bodies that surround us. They can sojourn in this or that body until the moment of its decomposition; they can then pass without the least alteration into another body; from there to a third, etc. I conceive with greatest ease that the germ of an elephant can lodge itself first in a molecule of earth, pass from there to the seed of a fruit, from there to the leg of a mite, etc. (p. 207)

What of the germs of restitution preformed in bodies that never lived—in stillborn fetuses and the numerous homunculi encapsulated in the ovaries of these fetuses:

But if all these organized beings have been preformed from the beginning, what will become of the so many billions of germs which never developed in the present state of our world . . . My reader will have already guessed my response: each of these germs encloses another imperishable germ of restitution which will develop only in the future state of our planet. Nothing is lost in the immense storehouse of nature; all has its use there, its goal, and its best possible end. (pp. 206–207)

All creatures shall be rewarded with improved bodies to enclose their soul at the resurrection. And that increased perfection will be so adjusted to present status that all organisms will retain their relative position in the chain of being while the entire chain is transported upwards:

In the future state of our globe, we will observe, no doubt, the same progression that we discover today among the different orders of organized beings . . . Man, transported then to another plane more suited to the eminence of his faculties, will leave to the monkey or to the elephant this first place that he occupied among the animals of our planet. [Since the current chain of being extends beyond man to the celestial angels, man may no longer reside on earth in the future world.] In this universal restitution of animals, we will be able to find the Newtons and Leibnizes among the monkeys or elephants, the Perraults and Vaubans among the beavers, etc. The most inferior species the oyster, the polyps, etc., will be the most elevated species of the new hierarchy, as birds and quadrupeds are to man in our present hierarchy. (pp. 203–205)

We have, in summary, an illusory appearance of transmutation in the history of life. If we could watch the complete history of a species from the creation, we would see a series of improvements in design correlated with physical revolutions of the globe,[5] and a final perfection at the last judgment. And yet, this entire history is nothing more than the successive display of preformed structures hidden by encapsulation at the creation. Only in this way could a preformationist postulate the appearance of development that a principle of perfectability demanded. Why, after all, should the succession of encapsulated generations in the ovaries of a primal Eve all bear the same form. We might open the Russian doll in this primal ovary and find only ten dolls of identical form; inside the tenth we might discover a vastly superior creature, and after ten similar boxes another being of still more perfect design. In Bonnet's system, the illusory transmutation of each lineage occurs in two stages: first, a succession of forms encapsulated in programmed sequence within the ovaries of its first representative; finally, the emergence of perfected germs of restitution at the end of time.[6] Bonnet's constant analog for this "phylogeny" is ontogeny. There is an illusory appearance of development in ontogeny; the stages advance from simplicity to complexity. Yet all is preformed from the start. The limbs of the chick lie hidden in the embryo's worm-like sheath; the perfect body of our immortal soul waits patiently for the second coming.

I have developed Bonnet's "enchanting picture" (as Cuvier called it) at some length because it illustrates so well the extraordinary influence of the parallel between ontogeny and the history of life. If any system were to be immune to this influence, preformationism would surely be the most likely candidate; for it would seem that Haller, Bonnet, and their followers denied both ontogeny and phylogeny. Yet there was one possible way to construct a parallel—based upon a dual illusion to be sure—and Bonnet not only found it, he based a theory of universal history and divine resurrection upon it.

Appendix: The Revolution in "Evolution"

Bonnet is often credited with the first use of "evolution" as a biological term (Osborn, 1929; Carneiro, 1972). Yet Haller coined it in 1744 as a name for preformationism:

> But the theory of evolution proposed by Swammerdam and Malpighi prevails almost everywhere [*Sed evolutionem theoria fere ubique obtinet a Swammerdamio et Malphighio proposita*] . . . Most of these men teach that there is in fact included in the egg a germ or perfect little human machine . . . And not a few of them say that all human bodies were created fully formed and folded up in

the ovary of Eve and that these bodies are gradually distended by alimentary humor until they grow to the form and size of animals. (Cole, 1930, p. 86; Adelmann, 1966, pp. 893–894)

Haller had made a sound etymological decision, for the Latin "evolutio" denotes an unrolling of parts already existing in compact form, as in a scroll or the fiddlehead of a fern (Bowler, 1975).

The transformation of this word to its opposite meaning of organic change is an interesting tale. The seeds of ambiguity were present from the start, for "evolution" had also been a widely understood, albeit uncommon, word in the English vernacular for some time. *The Oxford English Dictionary* traces its first use to mid-seventeenth-century poetry. In this general and figurative sense, "evolution" could refer to "almost any kind of connected series of events" (Bowler, 1975, p. 99). Moreover, some English epigeneticists occasionally used the vernacular meaning to describe ontogeny (Bowler cites passages from J. T. Needham, 1745, and from Erasmus Darwin, who spoke, in his *Botanic Garden* of 1791, of "the gradual evolution of the young animal or plant from the seed"). But by the 1820s and 1830s, as the theory of preformationism moved towards extinction, confusion began to surround the technical meaning of evolution in embryology as well. Serres (1827a, p. 57) noted that Haller had not imagined the homunculi in human ova as perfectly proportioned miniatures of adults (he seems unaware that no serious preformationist held this extreme view). He therefore supposed, quite incorrectly, that Haller had coined "evolution" to characterize a view midway between preformationism and epigenesis: "This word was a formal protest against preexistences. For, by these 'evolutions,' the embryo was no longer the exact miniature of the completed animal; it passed through diverse stages which were no longer its original state; in a word, it changed." A. J. L. Jourdan (1835), in translating C. G. Carus into French, used "evolution" to signify the epigenetic aspects of development.

De Beer (1969) and Bourdier (1969) have credited Etienne Geoffroy Saint-Hilaire (1833) with the first usage of "evolution" in the modern sense of transmutation, in his fourth monograph on the teleosaurians of Caen. But Geoffroy's words are both ambiguous and contradictory. In one sentence, he allies "evolution" to epigenesis: "two theories on the development of organs; the one supposes the preexistence of germs and their infinite encapsulation; the other admits their successive formation and their evolution in the course of ages" (p. 89). But a few lines later, in the same footnote, he refers to preformationism as the "système de l'évolution." The only hint of a reference to transmutation comes in a passage affirming epigenesis in ontogeny: "In the evolution of a being which has passed through all the phases of its life, you have in miniature, in some respects, the spectacle of the evolution of the terrestrial globe" (p. 81). I am by no means convinced that this passage has anything to do with transmutation; for it is part of a section describing the direct effect of the environment in transforming organisms through time. The "evolution of the terrestrial globe" may refer only to the historical sequence of these environmental agencies.

We find occasional uses of "evolution" for transmutation during these decades, but they all carry implications of a progressive directionality and may only reflect the vernacular meaning. In the second volume of the *Principles of Geology,* Lyell writes, paraphrasing Lamarck: "The testacea of the ocean existed first, until some of them by gradual evolution were improved into those inhabiting the land" (1832, p. 11). But this is the only use of "evolution" in the *Principles;* in other passages, Lyell talks of "transmutation."

In any event, well into the 1860s, "evolution" still was not in vogue as a term for organic change. Lamarck did not use it at all. Darwin spoke of "descent with modification" and used "evolved" only once in the first edition of the *Origin*—as the last word of the book, and clearly in the vernacular sense. No major review of the *Origin* used the word evolution (Bowler, 1975). Haeckel (1866, 2:148) gives a complete list of synonyms of transmutation, and does not include "evolution": *Descendenz-Theorie, Abstammungs-Lehre, Transmutations-Theorie, Transformations-Theorie, Umwandlungs-Lehre,* and *Umbildungs-Lehre.* Haeckel, in fact, uses "evolution" only in the original sense of preformation. The following statement contrasts "evolution" with phylogenesis: "Die continuirliche Phylogenesis ist ebenso eine wirkliche Epigenesis (und nicht eine Evolution), wie die continuirliche Ontogenesis" (1866, 2:418).

Herbert Spencer was clearly the primary instigator for a transformation of "evolution" into a term for organic change (Taylor, 1963; Oppenheimer, 1967; Carneiro, 1972; Bowler, 1975). Spencer wrote in his autobiography:

> I came across von Baer's formula expressing the course of development through which every plant and animal passes—the change from homogeneity to heterogeneity . . . This phrase of von Baer expressing the law of individual development, awakened my attention to the fact that the law which holds of the ascending stages of each individual organism is also the law which holds of the ascending grades of organisms of all kinds. (1904, pp. 445–446)

Since von Baer led Spencer to a concept that he chose to call "evolution," the modern usage is based on an ontogenetic metaphor (from the epigenetic camp). Spencer first used evolution for organic change in his 1852 essay on *The Development Hypothesis,* though he did not offer his famous general definition until the *First Principles* of 1862: evolution is "a change from an indefinite, incoherent homogeneity, to a definite, coherent heterogeneity; through continuous differentiations and integrations" (in Carneiro, 1972, p. 249). He did not confine the word to organic change and he did insist, as the ontogenetic metaphor demanded, that the word be restricted to progressive change towards increasing complexity. But Spencer's real break with previous usage—and the aspect of his definition that permitted a later extension to *all* organic change—came with his contention that evolution was not controlled by a preset, internal program, but depended upon interaction with external forces. He wrote in the *Principles of Sociology:* "Evolution is commonly conceived to imply in everything an intrinsic tendency to become something higher. This is an erroneous conception of it. In all cases it is determined by the cooperation of inner and outer forces" (in Carneiro, 1972).

We may say, in conclusion, that the transformation of "evolution" to an opposite meaning proceeded through three stages:

1. As the theory of preformation collapsed, some authors began to use "evolution" for epigenetic aspects of development. This was consistent with a vernacular meaning that long predated Haller's technical definition. One crucial usage in the epigenetic sense occurred in the 1851 edition of W. B. Carpenter's *Principles of Physiology,* the book that Herbert Spencer was reviewing when he discovered von Baer's principle (Bowler, 1975). In fact, Carpenter uses "evolution" to describe both embryology (with praise for von Baer) and the fossil record (in a progressivistic and creationist interpretation). Carpenter clearly intended the vernacular meaning as a synonym for progress, in noting, for example, the "evolution of structure and the complication of function . . . both in the ascending scale of creation, and in the growth of embryos" (1839, p. 170).

2. In analogy with epigenetic views, Spencer defined evolution as progressive change. We must assume that he chose the word because Carpenter had used it in praising von Baer's principle of progressive differentiation—the catalyst for Spencer's general definition of evolution (Spencer never explained his choice—Bowler, 1975). Thus, however ironically and indirectly, von Baer is the father of our modern usage.

3. Biologists appropriated his word and applied it to all organic change. Spencer had discussed organic evolution extensively in his widely read *Principles of Biology* (1864–1867). Thus, "evolution" was available when many scientists felt a need for a term more succinct than Darwin's "descent with modification." Moreover, since most evo-

lutionists (though not Darwin) saw organic change as leading to greater complexity (that is, to us), the appropriation of Spencer's term did no violence to his definition.

Bowler (1975) has traced the spread of evolution as a synonym for transmutation (Lyell in the tenth edition of the *Principles of Geology* [1867–1868]; Wallace in his 1869 review of Lyell; Darwin in the Introduction to the *Descent of Man* [1871—though Darwin used it very rarely thereafter, as did Wallace until its general acceptance late in his life]; Huxley in 1868 and frequently thereafter). By 1878, it had gained sufficient orthodoxy for an entry in the *Encyclopedia Britannica* (written by Huxley). However, throughout the late nineteenth century, evolution referred to a general development of life by transmutation, not to specific cases of adaptation. Only in this century have we extended it to any genetic change in populations, thus completing the severance of usage from Spencer's original notion of general progress.

— 3 —

Transcendental Origins, 1793–1860

It took Bonnet's ingenuity to insinuate a notion of recapitulation into preformationism, a system of thought fundamentally opposed to such dynamic ideas. In the last chapter, I defended preformationism as a reasonable theory for its time; yet I do not deny the traditional view that both modern embryology and Darwinian evolution required its downfall. For Bonnet had written amidst his musings on illusory perfectability: "No change; no alteration; perfect identity. Victorious over the elements, time, and the grave, species preserve themselves, and the term of their duration is unknown to us" (1762, p. 123). And Whitman has described Bonnet's system in vivid terms:

> "Progress" that discloses nothing but a succession of preformed hierarchies; a "law of continuity" . . . without any bond of connection whatever; . . . a "genealogy" of contemporaneous beings; "heredity" that transmits nothing; "births," "evolutions," and "revolutions" that bring nothing new, and so on through all the negations that a fertile genius could invent against the intrusion of epigenesis. (1894, p. 257)

It is a cliché of intellectual history that progressivist, historical thinking replaced cyclic or static views of nature during the late eighteenth century. I have neither the space nor competence to assess the reciprocal roles of science and society in fashioning this change.[1] I wish merely to identify it as a precondition for the theories of epigenesis and evolution, and for the common acceptance of recapitulation.

Collingwood distinguishes three sequential views of nature, each based upon a compelling analogy: the Greek comparison between nature the macrocosm and man the microcosm; the Renaissance analogy between nature as God's handiwork and machines as the creation of man; and the modern view, finding its first expression towards the end of the eighteenth century, and "based on the analogy between the processes of the natural world as studied by natural scientists and the vicissitudes of human affairs as studied by historians" (1945, p. 9). Collingwood identifies three major components of the modern view: (1) change is now viewed as progressive, not cyclical; (2) nature is no longer conceived in mechanical terms; and (3) teleology is reintroduced. This philosophy had inspired Wolff's famous dictum of 1759: "Qui igitur systemata praedelineationis tradunt, generationem non explicant, sed, eam non dari affirmant" ("therefore, he who defends the system of predelineation does not explain generation, but affirms that it does not exist"). By the end of the century, the triumph of this philosophy had guaranteed the victory of epigenesis.

> Again a philosophical need had created a demand which again an observational embryologist—this time Caspar Friedrich Wolff—was to fulfill . . . Without this background, it is unlikely that Wolff would have found a homogeneous blastoderm under his microscope as it was inevitable that Malpighi should have denied one a century before. (Oppenheimer, 1967, pp. 132–133)

Many authors have affirmed the importance of this dynamic view. Bury, in his classic work of 1920, argued that a belief in the progressive nature of human history did not flower until the eighteenth century (though the obstacles to its acceptance began to disappear in the sixteenth century); he links this flowering to the growth of science, rationalism, and the struggle for religious and political liberty. Lovejoy (1936) notes that the late eighteenth century retained its allegiance to the ancient idea of a chain of being with its principles of plenitude, continuity, and gradation. But the new, progressivist thinking inspired scientists to "temporalize" the chain and view it as a ladder that organisms might climb rather than a rigid ranking of immutable entities.

This new view of nature penetrated everywhere. Hegel constructed a new logic to grasp change and motion through the recognition of contradiction (Jordan, 1967). And Condorcet, in hiding from the government that had decreed his death, wrote in his *Esquisse,* in 1793, "that the perfectibility of man is really boundless, that the progress of this perfectibility, henceforth independent of any power that would arrest it, has no other limit than the duration of the globe where nature has set us."

The influence of this view was surely felt in embryology. By 1810, Oken felt he needed no justification for epigenesis beyond this epigram: "Die Präformations-theorie widerspricht den Gesetzen der Naturentwicklung" ("The theory of preformation contradicts the laws of nature's development"—1810, p. 28).

Naturphilosophie: An Expression of Developmentalism

In Germany, a group of late-eighteenth- and early-nineteenth-century biologists combined a progressivist view of nature with the romantic thought then current in philosophy and literature to produce the controversial school of *Naturphilosophie.* It is among the Naturphilosophen that recapitulation first became a central theory.

Although the origin of recapitulation among the Naturphilosophen has long been acknowledged, there has been much debate about its initiator. Many cite Goethe, others the historian Herder; most prefer Kielmeyer[2] (1793), as did Meckel (1821) in the first attempt I know to establish a chronological list of recapitulation's supporters. Others, noting that Kielmeyer speaks only of physiology, identify Autenrieth (1797) as the first to apply recapitulation to morphology (Temkin, 1950). Kohlbrugge (1911) industriously catalogued 71 pre-Haeckelian supporters of recapitulation. Yet the entire inquiry is at worst futile, at best of antiquarian interest only.

Debates about the priority of ideas are usually among the most misdirected in the history of science. This is surely true here, for a fundamental reason: recapitulation was an inescapable consequence of a particular biological philosophy. Its spread among the Naturphilosophen bears no analogy to procreation (with extinction as a threatened consequence of early parental death), but rather to the invention of a simple machine whose parts are ubiquitous and whose use is obvious.

Naturphilosophie was the scientific incarnation of German romanticism. Gode von Aesch prescribed the following "comprehensive program of all romantic thought":

1. The establishment of a universal order of metaphysical, not just pragmatic, validity.
2. The determination of a place for man compatible with the faith in a human superiority of more than relative importance.
3. A substantiation of the belief in man's brotherhood and even identity with all of life and thus with all existence. (1941, p. 207)

The Naturphilosophen transcribed this program for biology. Most of their conclusions, including recapitulation, sprang from a small set of common assumptions. Most important among these were an uncompromising developmentalism and a belief in the unity of nature and its laws.

1. An uncompromising developmentalism. All nature is in flux; this motion, not the momentary configuration of matter, is nature's irreducible property. Furthermore, the flux is unidirectional, moving ever from lower to higher, from initial chaos to man. To Schelling, the history of the universe is the striving of spirit (*Geist*), originally unconscious, gradually and progressively to reach self-consciousness in man. (Note, however, that this conviction need not entail a belief in the physical continuity of organisms through organic evolution.[3] Similarly, recapitulation does not require a belief in evolution; for embryonic stages may parallel a static but ascending sequence along the chain of being as well as the steps of an actual lineage.)

2. A belief in the unity of nature and its laws. Man is the highest configuration of matter on earth, but we are indissolubly linked to all objects as the goal toward which they strive. Nature and spirit, the inorganic and organic, are one; the universe itself is a single organism (see Walzel, 1932, p. 52, on Schelling). The laws of nature operate in the same way upon all processes and all objects. All previous dualisms are dissolved into a "biocentric universalism" (Gode von Aesch, 1941, p. 185).

Goethe's "insistent perception of unity"[4] led the Naturphilosophen to link all objects (Ritterbush, 1964, p. 208). Since they thought in developmental terms and saw but a single, progressive direction of motion, this linking took the form of a single, ascending chain. As Herder wrote in his *Ideen zur Philosophie der Geschichte* (1784–1785): "From stones to crystals, from crystals to metals, from these to plants, from plants to animals, and from animals to man, we see the form of organization ascend; and with it the powers and propensities of the creature become more various, until finally they all, so far as possible, unite in the form of man" (in Lovejoy, 1959, pp. 208–209).

The development of complexity during ontogeny (so evident that preformationists affirmed it, if only as an illusion), and the recognition that there are "higher" and "lower" species are two inescapable phenomena of biology. If there is but a single direction to organic development, and if all processes are governed by the same laws, then the stages of ontogeny must parallel the uniserial arrangement of adult forms. If there is but one path of ascent to man, and if a human embryo must begin in Oken's "initial chaos," then the stages of human ontogeny must represent the completed forms of lower organisms. As Oken stated in his colorful metaphor, what are the lower animals but a series of human abortions? Or, as Robinet described them, "the apprenticeship of nature in learning to make man" (in Lovejoy, 1936). One can scarcely hold the basic premises of Naturphilosophie without accepting recapitulation as a consequence. As Gode von Aesch wrote:

> If it is metaphysically true that man is an epitome of the universe, then it must also be physically true, for one law rules throughout all the realms of existence. If it is metaphysically necessary to conceive of man as the last and most perfect link in the chain of the animal kingdom, then it must also be physically observable that he repeats its various stages. Thus we know *a priori* that the basic law of biogenetics was part of the intellectual equipment of every good romantic thinker. (1941, pp. 120–121)

As the following samples show, the recapitulationists (and their opponents) gave explicit recognition to the major a priori beliefs of romantic biology:

1. Nature displays a single developmental tendency and a single sequence of forms. Milne-Edwards identified this necessary assumption in order to ridicule it: "If these latter [lower animals] were in some way permanent embryos of the former [higher animals], it would be necessary to admit, at least for the types, a progressive and linear series extending from the monad to man" (1844, p. 70).

2. The same laws regulate ontogeny and the historical progression of species. Russell (1916, p. 236) designated this early nineteenth century version of recapitulation as the "Meckel-Serres Law" to distinguish it from Haeckel's evolutionary formulation. Both Meckel and Serres cited the unity of nature's laws to explain recapitulation: "The development of the individual organism obeys the same laws as the development of the whole animal series; that is to say, the higher animal, in its gradual development, essentially passes through the permanent organic stages that lie below it" (Meckel, 1821, p. 514). "The animal series and man seem to perfect themselves by the same laws" (Serres, 1860, p. 352). In what most authors have taken as the first scientific formulation of recapitulation, Kielmeyer invoked the identity of law: "Since the distribution of powers [*Kräfte*] in the series of organisms follows the same order as their distribution in the developmental stages of given individuals, it follows that the power by which the production of the latter occurs, namely the reproductive power, corresponds in its laws with the power by which the series of different organisms of the earth were called into existence" (1793, p. 262)

3. The animal kingdom is an organism. This metaphor led many Naturphilosophen to view lower animals as the intermediate stages of a developmental process leading to man; the comparison of "lower" animals with ontogenetic stages of the human fetus becomes unavoid-

able: "Even as each individual organism transforms itself, so the whole animal kingdom is to be thought of as an organism in the course of metamorphosis" (Tiedemann, 1808, in Russell, 1916, p. 215). "The entire animal kingdom can, in some measure, be considered ideally as a single animal which, in the course of formation and metamorphosis of its diverse organisms, stops in its development, here earlier and there later" (Serres, 1860, p. 834).

In order to consider seriously the contribution of Naturphilosophie to the study of ontogeny, it is necessary to sweep away an old prejudice based on an erroneous conception of how science works. In 1947, Cohen stated the prejudice in order to refute it:

> It has become a tradition among those who talk glibly about science that the romantic *Naturphilosophie* of Schelling and his followers represents the lowest degradation of science and that only by completely freeing themselves from that nightmare were modern biology and medical science able to resume their scientific progress. The incident has been used by empiricists as a moral to warn us against speculative philosophy in the natural sciences. (1947, p. 208)

If scientific progress were motivated only by the accumulation of information under the single, fruitful aegis of "the scientific method," then speculation based on a different metaphysics would be vain and harmful, for it would chain facts to false theory and direct inquiry along incorrect lines. But facts never exist outside theory, and imaginative theory may be even more essential than new information in yielding scientific "progress," We must treat Naturphilosophie seriously as a creative and comprehensive attempt to understand nature through the common beliefs of a prevailing culture; like any good theory, it generated a host of fruitful hypotheses about specific phenomena.[5]

Recapitulation was only one of the influential concepts derived from Naturphilosophie. Oersted ascribed his discovery of electromagnetism to the stimulus he received from Schelling, particularly Schelling's concept of an underlying unity among nature's forces. In the *Edinburgh Encyclopedia* of 1830, Oersted wrote about his own discovery: "He was not so much led to this by the reasons commonly alleged for his opinion, as by the philosophical principle, that all phenomena are produced by the same original power" (in Stauffer, 1957, p. 48). As one of the first popular scientific movements that incorporated the new spirit of a pervasive developmentalism, Naturphilosophie helped to spread dynamic views throughout science (Raikov, 1968, pp. 382–404). Since so much of nineteenth-century science, and especially nineteenth-century biology, hinged upon this spread, Naturphilosophie must be counted as an influential movement in the history of science. As Louis Agassiz testifies:

> The young naturalist of that day who did not share, in some degree, the intellectual stimulus given to scientific pursuits by physio-philosophy [Naturphilosophie[6]] would have missed a part of his training . . . The great merit of the physio-philosophers consisted in their suggestiveness. They did much in freeing our age from the low estimation of natural history as a science which prevailed in the last century. They stimulated a spirit of independence among observers; but they also instilled a spirit of daring, which, from its extravagance, has been fatal to the whole school. (in E. Agassiz, 1885, pp. 152–153)

Two Leading Recapitulationists among the Naturphilosophen: Oken and Meckel

Lorenz Oken's *Lehrbuch der Naturphilosophie* appeared in three parts from 1809–1811. It is a listing of 3,562 statements, taking all knowledge for its province, and filled with bald, oracular pronouncements of the engaging sort that feign profundity but dissolve into emptiness upon close inspection. It is also responsible for Oken's bad reputation as the most idle (if cosmic) speculator of a school rife with unreason.[7] In fact, Oken was one of the best comparative anatomists and embryologists of his day; his works on the embryology of the pig and dog (1806) are classics (he was also an influential, if naive, political thinker of liberal to radical bent—see Raikov, 1969). Russell called him "a careful student of embryology" (1916, p. 90). Von Baer, an implacable foe of recapitulation and much else dear to Oken's system, wrote that his observations "are often among the most accurate that we possess about mammals, and the general statements, although a majority of them must now appear erroneous, have, nonetheless, infinitely furthered [*unendlich gefördert*] our knowledge of development" (1828, p. xvii). Louis Agassiz attended Oken's lectures and wrote:

> Among the most fascinating of our professors was Oken. A master in the art of teaching, he exercised an almost irresistible influence over his students. Constructing the universe out of his own brain, deducing from *a priori* conceptions all the relations of the three kingdoms into which he divided all living beings, classifying the animals as if by magic, in accordance with an analogy based on the dismembered body of man, it seemed to us who listened that the slow laborious process of accumulating precise detailed knowledge could only be the work of drones, while a generous, commanding spirit might build the world out of its own powerful imagination. (in E. Agassiz, 1885, pp. 151–152)

Oken's Classification of Animals by Linear Addition of Organs

Oken renders all of nature with the aid of his philosophical principles. The unity of law and structure dictates that the entire mineral kingdom assume the same form: the crystal. The earth itself is a giant crystal, a rhomboidal dodecahedron. Its strata are cleavages, its mountains are edges: "The land cannot therefore have an equal elevation everywhere above the water, because the crystal consists of edges, angles, and surfaces or sides. The mountain tops are probably the angles, the mountain ridges or chains the edges, and plains the lateral surfaces of the crystal" (1847, p. 123).

Yet Oken's most pervasive principle is his own version of the single developmental tendency: all development begins with a primal zero and progresses to complexity by the successive *addition* of organs in a determined sequence. This law holds for all developmental processes: human ontogeny, the historical sequence of species, the evolution of the earth itself: "If we take a retrospective glance at the development of the planet, we find that it commenced with the simplest actions, and then assumed a more elevated character by gradually drawing together several actions and letting them work in common" (p. 178).

The sequence of additions follows Oken's ordering of the four Greek elements. Translated into the organs of animals, this sequence includes:

1. Earth processes—nutrition.
2. Water processes—digestion.
3. Air processes—respiration.
4. Aether (fire) processes—motion.

Man contains all organs within himself; thus he represents the entire world; "in the profoundest, truest sense . . . a microcosm" (p. 202). "Man is the summit, the crown of nature's development, and must comprehend everything that has preceded him . . . In a word, Man must represent the whole world in miniature" (p. 12). All lower animals, as imperfect or incomplete humans, contain fewer than the total set of organs. "The animal kingdom," wrote Oken in his most famous pronouncement, "is only a dismemberment of the highest animal, i.e. of Man" (p. 494). The position of any animal upon the single chain of classification depends upon the *number* of organs it possesses: "Animals are gradually perfected, entirely like the single animal body, by adding organ unto organ . . . An animal, which e.g. lived only as an intestine, would be, doubtless inferior to one which with the intestine were to combine a skin" (p. 494).

From this simple (if fanciful) premise, Oken's system of classification becomes extraordinarily complex. Since this complexity obscures its basic constitution as a single, linear chain (and since recapitulation depends upon this linearity), we shall examine Oken's system in some detail. Table 1 lists Oken's series of animal classes and their correspondence with three criteria for classification: elements, organs, and senses. Oken's taxonomic hierarchy contains the following levels:

1. *Province.* Animals are divided into two provinces: invertebrates and vertebrates.

2. *Circles.* There are four circles corresponding to the elements and their representative organs. (Note that the circles correspond exactly to the four *embranchements* of Cuvier's orthodox classification: Radiata, Mollusca, Articulata, and Vertebrata.)

3. *Classes.* There are thirteen classes corresponding to organs, three classes in each of the first three circles, four in the last. There is an inconsistency in the criterion for classes, one of many in the system and

Table 1. Oken's linear classification of animals.

Criteria				Taxa	
Element	Organ		Sense	Class	Circle
Earth	Intestinal	(gastric)	Feeling	1. Infusorians	I
		(intestinal)		2. Polyps	
		(absorbent)		3. Acalephs	
Water	Vascular	(venous)		4. Clams	II
		(arterial)		5. Snails	
		(cardiac)		6. Squids	
Air	Respiratory	(reticular [skin])		7. Worms	III
		(branchial)		8. Crustaceans	
		(trachial)		9. Flies	
Fire	Osseous		Taste	10. Fish	IVa
(aether)	Muscular		Smell	11. Reptiles	
	Nervous		Hearing	12. Birds	
	(Sensory)	1. Feeling	Sight	13. Mammals	IVb
		2. Taste		* { 13. Feeling	
		3. Smell		14. Taste	
		4. Hearing		15. Smell	
		5. Sight		16. Hearing	
				17. Sight }	

* Alternative system with five classes rather than one in Circle IVb.

the only one that Oken recognizes himself (1847, p. 511). Nine classes of invertebrates (the first three circles) correspond to but three organs. To encompass the diversity of invertebrates, the first three organs must be subdivided and shared. Oken also presents an alternate system of seventeen classes. Here, he divides the thirteenth class (the Mammalia) into five separate classes, each representing one of the five senses (since the key organ for Mammalia is sensory).

4. *Orders.* Each class has as many orders as there are circles in and below it; the class Infusoria has but one order, Mammalia has five (one for each of the first three circles, two for the fourth circle, which Oken subdivides).

5. *Families.* Each class has as many families as there are classes in its circle and below.

6. *Genera.* Each family has five genera, corresponding to the organs of sense.

The later levels of the hierarchy are presented in Table 2, a classification of the Mammalia. The five orders correspond to the circles of the entire system (the fourth circle being divided in two); the mammalian families correspond to the classes of the whole. Each family has

Table 2. Oken's classification of mammals. (From Oken, 1847.)

Order	Family	Genera
1. Intestinal	1. Infusorians	Rats, beavers
	2. Polyps	Squirrels
	3. Acalephs	Rabbits
2. Vascular	4. Clams	Sloths
	5. Snails	Herbivorous marsupials
	6. Squids	Carnivorous marsupials
3. Respiratory	7. Worms	Moles
	8. Crustaceans	Shrews
	9. Insects	Bats
4. Animal systems	10. Fish	Whales
	11. Reptiles	Pachyderms, pigs, horses
	12. Birds	Ruminants
5. Sensory systems	13. Feeling	Carnivores
	14. Taste	Seals
	15. Smell	Bears
	16. Hearing	Apes
	17. Sight	Man

five genera corresponding to the senses. Thus, the genera of the twelfth family are: camels (skin), deer (tongue), goats (nose), giraffes (ear), and oxen (eye). Oken's arguments for why animals correspond to particular organs are usually fanciful and specious; they need not concern us here.

It is the infolding of systems within systems that engenders the greatest complexity and obscures the linear character of this classification. Thus, the five organs of sense determine a primary division of all classes (Table 1), but they also govern a secondary division of classes within the Mammalia (Table 1), a tertiary division of families within the fifth order of mammals (Table 2), and a quaternary division of genera within each family. Each subdivision is itself separated according to the same criterion used to divide the entire system. At this point, I can only invoke a figure (Fig. 2) and an analogy to resolve confusion. The outstanding feature of Christian history is its linearity: events are sequential in time and occur but once (Haber, 1959). Yet, as the iconography of any medieval cathedral displays, there are detailed correspondences between Old and New Testament events. Mary stands for the burning bush because she held the fire of God within her yet was not consumed; the resurrection of Christ represents the deliverance of Jonah because each liberation followed three days of captivity. The New Testament replays the Old while adding constantly to it in a linear sequence. Similarly, the five classes of mammals replay the ascent of all animal classes according to the addition of senses, yet each new mammalian class is a progressive, terminal addition to a single sequence.[8]

Oken's sequence of additions to perfection—earth, water, air, fire—not only regulate the order of animals (where they stand for organs); they govern all developmental sequences and dictate the arrangement of their parts. In the ordering of cultural progress, for example, they stand for human achievements. In the closing paragraphs of his work, the radical Oken casts an apotheosis in terms that place him among the intellectual antecedents of German fascism:

> The first science is the science of language, the architecture of science, the earth.
> The second science is the art of rhetoric, the sculpture of science, the river [water].
> The third science is philosophy, the painting of science, the breath [air].
> The fourth science is the art of war, the art of motion, dance, music, the poetry of science, the light [fire].
> As all arts are united in poetry, so are all arts and all sciences united in the art of war.
> The art of war is the highest, most exalted, godly [*göttliche*] art.

The hero [*Held*] is the highest man
The hero is the God of mankind
Through the hero is mankind free
The hero is the prince
The hero is God.[9] (1811, pt. 3, pp. 373–374)

It seems almost superfluous to add that recapitulation is an automatic consequence of these beliefs. All development proceeds along the same path by adding elements to an original nothingness. Higher

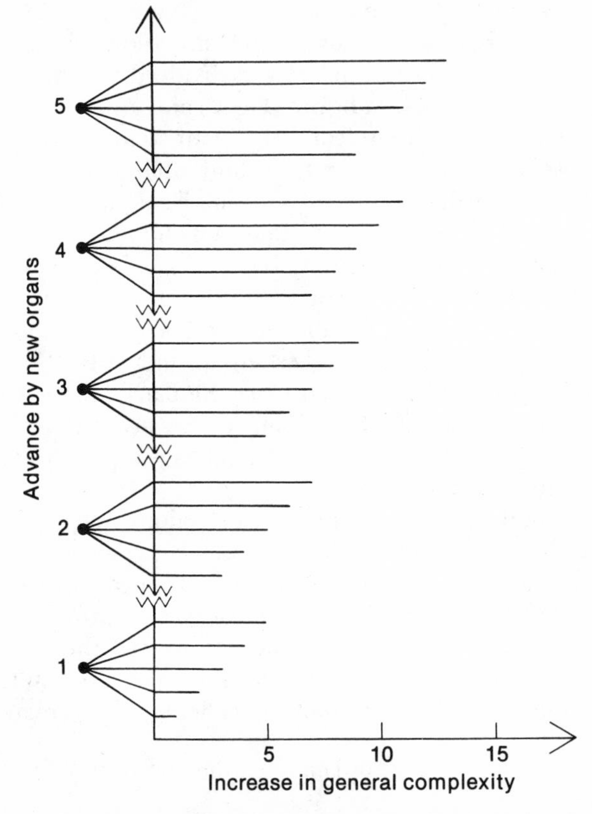

Fig. 2. A précis of Oken's scheme of classification. Ascent up the scale of perfection (vertical) depends upon the addition of organs; increase in general complexity occurs within each level but permits no ascent. Higher forms within a level are more complex (higher values on the horizontal axis) than lower forms of the next level. The breaks between levels indicate that no physical tie of evolution connects them. The animals of each new level must be reformed anew in the primal zero.

animals must, as they add organs in their own development, pass through the permanent stages of those lower on the scale: "The whole animal kingdom is none other than the representation of the several activities or organs of Man; naught else than Man disintegrated" (1847, p. 19).

During its development the animal passes through all stages of the animal kingdom. The foetus is a representation of all animal classes in time.
At first it is a simple vesicle, stomach, or vitellus, as in the Infusoria.
Then the vesicle is doubled through the albumen and shell, and obtains an intestine, as in the Corals.
It obtains a vascular system in the vitelline vessels, or absorbents, as in the Acalephae.
With the blood-system, liver, and ovarium, the embryo enters the class of bivalved Mollusca.
With the muscular heart, the testicle, and the penis, into the class of Snails.
With the venous and arteriose hearts, and the urinary apparatus, into the class of Cephalopods or Cuttle-fish.
With the absorption of the integument, into the class of Worms.
With the formation of branchial fissures, into the class Crustacea.
With the germination or budding forth of limbs, into the class of Insects.
With the appearance of the osseous system, into the class of Fishes.
With the evolution of muscles, into the class of Reptiles.
With the ingress of respiration through the lungs, into the class of Birds. The foetus, when born, is actually like them, edentulous.* (1847, pp. 491–492)

J. F. Meckel's Sober Statement of the Same Principles

It is often assumed that to be a Naturphilosoph one had to engage in the kind of mystical and cosmic pronouncement that Oken favored (Mayr, 1965). But in fact, many scientists who shared Oken's philosophy wrote in a very dry and controlled fashion; allegiance to a school

* It is important to understand what Oken means by these statements, lest he be dismissed as a madman. The human embryo at the time it forms branchial fissures is surely not a crustacean. Oken is not concerned with the external appearance of crabs, their size, their shape, the arrangement of their parts. The crab merely represents or symbolizes an ideal step in universal progression by addition of organs: the respiratory organ in its branchial form (all else about a crab is irrelevant). When the human fetus develops gill slits, it has reached the stage of ontogeny that crabs symbolize in the historical sequence of adults (all else, again, being irrelevant). C. G. Carus, an influential Naturphilosoph and supporter of Oken, wrote: "Each degree of development of a superior animal constantly recalls a determined form of an inferior organism; but between the two there is no complete identity, but only a resemblance of fundamental nature or essence" (1835, 2: 438).

must be measured more by shared ideas than by styles and rhetoric.[10] J. F. Meckel was probably more influential among scientists than Oken, but he wrote no popular or all-encompassing work, and his technical studies are largely unknown today. Yet Russell (1916) argued that he did more than any other Naturphilosoph to advance the theory of recapitulation.

Meckel accepts the premise that nature is unified by a single set of laws.[11] After comparing the arrangements of organs with the strengths of electric currents he writes: "The agreement between the laws that direct the formation of organisms and those according to which more general powers operate seems, therefore, to be proved by these facts" (1811b, pp. 67–68). He agrees with Oken that a single developmental tendency must govern all processes, but he does not relate it, as Oken did, to an addition of organs or powers. To Meckel, the law of development is coordination and specialization: simple animals have many similar but poorly coordinated parts;[12] advanced creatures have highly distinct and specialized organs that function together in an integrated body: "The operation and proof of a more perfect organization is the union of many individual parts into a whole" (1811b, p. 69). (The reduction of many identical parts to fewer, more specialized, and more coordinated organs is an ancient truth of biology that has often been rediscovered. It is generally known to evolutionary biologists today under a later incarnation as "Williston's Law.")

Meckel's essay of 1811, "Entwurf einer Darstellung der zwischen dem Embryozustande der höheren Tiere und dem permanenten der niederen stattfindenen Parallele" ("Sketch of a portrayal of the parallels that exist between the embryonic stages of higher animals and adults of lower animals"), is Naturphilosophie's major statement of recapitulation. The styles of Meckel and Oken were as different as their ideas were similar. Meckel begins with a two-page lament about the difficulty of obtaining specimens for dissection, writes a short introductory paragraph stating that all good physiologists have noted the resemblances between lower animals and the embryos of higher forms, and then fills 57 pages with systematic examples of recapitulation, treated organ by organ. There are no statements about the universe, none even about the nature of biology, just a technical listing of examples. To be sure, many of Meckel's cases are forced or fanciful by modern standards. He compares, for example, the mammalian placenta to the gills of clams, since these, like the placenta, envelop the body (the gill slits of the human embryo had not yet been discovered—Rathke and von Baer did so in the late 1820s—though Meckel [p. 25] predicts that they will be found). But his list also includes some examples still cited today. He writes, for example, that the mammalian heart is first simple and tubular, as in insects; it then acquires a single chamber like that of crustaceans; later, when it possesses an auricle, ventricle, and aortic bulb, it represents the heart of fishes; when the auricle becomes divided, it adopts the reptilian form. At the end of his list, Meckel then appends this simple, concluding paragraph: "These few pages will suffice to prove that the analogy between the human embryo and the lower animals is unmistakable [*unverkennbar*], and that the completion of this parallel by exact and careful investigations of the human embryo and that of other animals . . . is one of the most desirable objectives of a rational anatomy, physiology, and zoology."

Serres and the French Transcendentalists

Naturphilosophie was not the only biological translation of the new, developmental view of nature. The French transcendental morphologists, under the leadership of Etienne Geoffroy Saint-Hilaire, shared a set of assumptions with their German colleagues, including the belief that all animals are built upon a single, structural plan; the idea of a chain of being; and a belief in recapitulation (Russell, 1916).

Geoffroy himself, though he gave the matter no particular attention, supported recapitulation. The development of this doctrine was left to his "chief follower" (Russell, 1916, p. 79), the medical anatomist Etienne Serres. Serres championed recapitulation in his monographs on the comparative anatomy of the vertebrate brain (1824–1826) and in a series of articles collectively titled "Recherches d'anatomie transcendante," published in the *Annales des sciences naturelles* during the 1820s and 1830s. As late as 1860, long after Geoffroy's time had passed, Serres wrote a thousand-page paean to his mentor, upholding Geoffroy's doctrines in scarcely modified form. In this work of his old age, Serres recalls the delight of his first demonstration of recapitulation forty years before.

> I did not know how to express the feeling of admiration that I felt for the grandeur of the creation in general, and for that of man in particular, when I saw that, at a first stage [of ontogeny], the human brain resembled that of a fish; that at a second stage, it resembled that of reptiles; at a third, that of birds; and at a fourth, that of mammals, in order finally to elevate itself to that sublime organization that dominates all nature. (1860, pp. 398–399)

Geoffroy and his school took as their guiding belief the notion that all animals share a single plan of construction. The greatest challenge to this idea, so effectively exploited by Cuvier in his famous debate with Geoffroy (Amlinskii, 1955), is the apparent dissimilarity of adult vertebrates and invertebrates. Geoffroy tried to compare the exoskeleton of arthropods with the internal skeleton of vertebrates (relegating insects to a life within their own vertebrae); he sought identity in the location of parts by likening the basic design of vertebrates to a worm turned over (yielding both the happy circumstance of dorsal nerve cords and such problems as a mouth above the brain). Serres agreed, attributing the inversion to a reversed position of the embryo relative to the yolk (1860, pp. 825–826).

Yet Serres acknowledged the difficulty of comparing adults and set out to prove the unity of plan on another basis: by the fact of recapitulation. The nervous systems of vertebrates and invertebrates have a common design (though this may shock some physiologists since it implies that invertebrates have a will). This identity is not apparent in adult vertebrates, but transient stages of the vertebrate fetus repeat the permanent configurations of invertebrate systems and display thereby a unity of plan. Serres claims that his work "has proven that lower animals are, for certain of their parts, permanent embryos of higher classes" (1824, p. 378). Later, he states his own resolution of Geoffroy's dilemma even more forcefully:

> The discordance that we observe between vertebrates and invertebrates is only relative; it is incontestable if we compare invertebrates with adult vertebrates. But if we consider them for what they appear to be, *permanent embryos,* and if we compare their organization to the embryogeny of vertebrates, the differences disappear and we see, from their analogies, a host of unsuspected resemblances. (1834, p. 247)

Serres allied his belief in the unity of plan to the same uncompromising developmentalism that characterized German Naturphilosophie. If animals must be arranged in a single sequence from lower to higher, and if that sequence is inherent in all organic development, then recapitulation must occur. "A natural classification is nothing but a table of organogeny, indicating step by step the march to perfection. Now, this gradual perfection of organogeny in the animal kingdom is only a copy of the successive perfection of the organogeny of man. The one repeats the other" (1860, p. 352). Or, as his oft-repeated epigram proclaimed: "Human organogeny is a transitory comparative anatomy as, in its turn, comparative anatomy is the fixed and permanent state of human organogeny" (1842, p. 90; see also 1827b, pp. 126–127; 1860, pp. 370–371).

But what is the mechanism of recapitulation? This question, which was to obsess evolutionists in fifty years, agitated the transcendentalists and Naturphilosophen scarcely at all.[13] Serres touched lightly upon the issue that most of his colleagues had ignored completely: Why do the lower animals stop their development at an intermediate station on the single track leading to man? Serres argued that they must simply contain less of whatever it is that propels development: "Since the *formative force,* whatever it is, has less energetic impulse [in lower animals] than in higher animals, the organs run through only a part of the transformations that they undergo in superior creatures. From this it follows that they offer to us, in a permanent manner, the organic configurations that are only transitory in the embryo of man and the higher vertebrates" (1830, p. 48).

Recapitulation and the Theory of Developmental Arrests

We have, thus far, spoken of recapitulation as an almost passive consequence of early nineteenth-century biological philosophies. But was it only a deduction, albeit a colorful one, from prior principles? Were the examples that illustrated it useful only as reflections of these principles, or did recapitulation serve the function of any fruitful scientific hypothesis: did it suggest new ideas and help to generate new data?

Teratology, the study of abnormal development, has always exerted a strange fascination over scientists. Many French and German anatomists, Serres included, had been trained in medicine and had opportunities to receive and dissect seriously deformed fetuses. Oken had spoken of the lower animals, metaphorically to be sure, as so many human abortions. Since the human fetus passes through stages representing lower animals, many abnormalities might be explained as arrests of development. If different parts of the fetus can develop at different rates, then "monstrosities" will arise when certain parts lag behind and retain, at birth, the character of some lower animal. And if, as Serres believed, development is regulated by a formative force of some kind, then a local arrest indicates a local deficiency of force; it might, in principle, be curable. "If the formative force of man or the higher vertebrates is arrested in its impulse, it reproduces the organic arrangements of lower animals . . . These cases of pathologic anatomy are only a prolonged embryogeny" (Serres, 1830, pp. 48–49).

Serres (1860, pp. 534–549) dissected a seriously deformed fetus that lacked a head (Fig. 3). Since clams are the highest acephalous invertebrates, Serres sought other points of resemblance with mollusks in attempting to identify the stage of arrest for this monstrosity. He

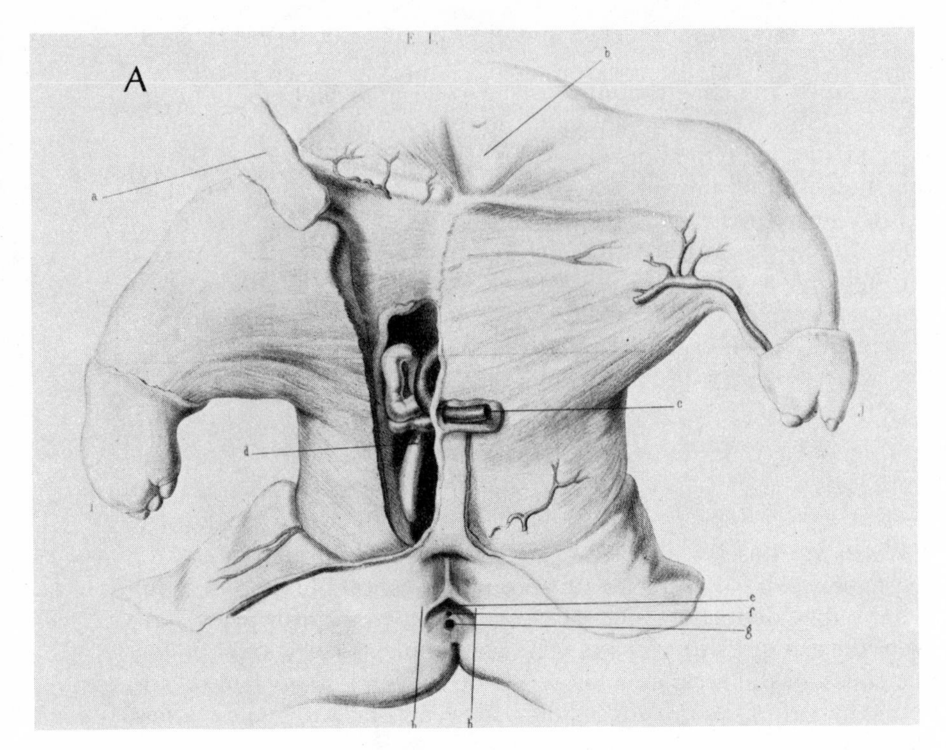

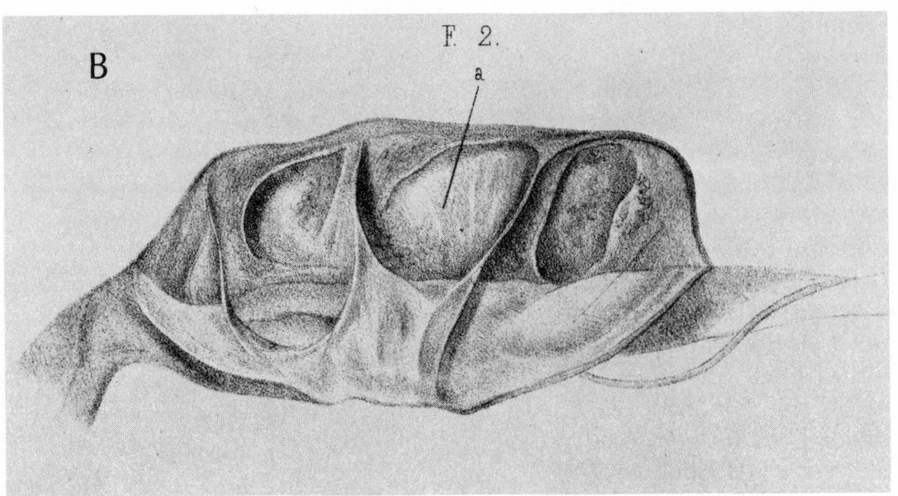

Fig. 3. (*A*) Seriously deformed, headless, spontaneously aborted fetus dissected by Serres. (*B*) Pockets and sinuses on back of fetus identified by Serres as organs of accessory cutaneous respiration and used by him to identify the stage of abortion as molluscan. (From Serres, 1860.)

noted that its placenta was too small for adequate respiration and identified some pockets and sinuses on the back, shoulder, and arms as organs of accessory cutaneous respiration; these were filled with liquid and surrounded by arteries and veins. Thus, recapitulation served as a useful hypothesis to direct inquiry: the lack of a head suggested a position within the Mollusca; this dictated a search for cutaneous respiration.[14] (The Mollusca, as presently defined, rely very little upon their mantle for respiration; Serres had in mind the brachiopod *Lingula* and the tunicates. Both are placed in different phyla today.)

Serres also applies his theory of developmental arrests to less bizarre and more frequently encountered anomalies. He compares the undescended testicles of many men to the permanent state of fishes (1860, p. 493). He arranges malformations of the heart in a long sequence representing the entire animal kingdom: "In the numerous anomalies caused by the arrest of this organ in man and the higher vertebrates, do we not recognize at first the simple heart of crustaceans and insects, then finally that of birds. Are not these cases, so frequently encountered in teratology, also engendered by a suspension in the course of development" (p. 496). Serres waxes so enthusiastic about his theory of teratology that at several points he reverses his perspective and identifies living invertebrates as human malformations: "In effect, invertebrates are often only living monstrosities, if we compare them to perfect vertebrates" (p. 368).

But what of malformations that cannot be compared with lower animals; what, in particular, of the duplication of parts and organs, from polydactyly to siamese twinning. Serres classifies malformations in two categories: *monstres par défaut* lack parts or powers and are comparable with lower animals; *monstres par excès* have extra parts. But extra parts are always duplications; they are never higher stages. A malformation may represent a lower animal because this stage must be traversed in normal development; but when an animal has too much of the formative force, it can only duplicate the final products of normal development. It cannot transcend its type (p. 751).

J. F. Meckel also supported the theory of developmental arrests (*Hemmungsbildungen*), but he tried to give recapitulation a wider scope in teratology by classifying some duplications as retentions of a lower state. In Meckel's view, the single tendency of development leads from a multiplicity of similar parts to a smaller number of specialized and better integrated organs. From this standpoint, he discusses polydactyly (1808, p. 95). The presence of extra fingers or toes is often correlated with other malformations that obviously represent arrested development. But polydactyly is an arrest (*Stehenbleiben*) in itself: the bud from which fingers differentiate is large when it first develops in the fetus. In normal ontogeny part of it disappears before the fingers differentiate. If the bud is arrested at its original size, too much material remains when the fingers develop and too many fingers appear.

The theory of developmental arrests was both successful and influential (see work of Etienne's son Isidore Geoffroy Saint-Hilaire, 1833); it added much prestige to the concept of recapitulation. Even von Baer, recapitulation's nemesis, had to praise this application: "It [recapitulation] won more influence, as it proved itself fruitful; a series of malformations could be understood when they were considered as the consequence of a partial arrest of development at earlier structural stages" (1828, p. 200). Two things were incontestable: (1) many malformations are arrested embryonic states; (2) this conclusion had been reached with the aid of recapitulation. But, von Baer noted, a third statement, crucial to biological theory, does not necessarily follow—that arrested embryonic stages are comparable to permanent conditions of lower animals (p. 232). Von Baer set out to demolish this third proposition.

Von Baer's Critique of Recapitulation

The Direction of Development and Classification of Animals

Ernst Haeckel's writings are sparing of praise and generous in skilled rhetoric of withering intensity against opponents. Yet he called von Baer's *Entwickelungsgeschichte* "the most significant work in the entire ontogenetic literature" (1866, 2:14). And while the aged von Baer was attacking Darwin from his outposts in St. Petersburg and Dorpat, Huxley was referring to him as Darwin's equal (Oppenheimer, 1959).

Karl Ernst von Baer (1792–1876) was a paragon of nineteenth-century science (Raikov, 1968). After studying with Burdach in Dorpat and with Döllinger in Würzburg, he received a professorship at Königsberg in 1819. There he published the first part of his *Entwickelungsgeschichte der Thiere* in 1828 and reported his discovery of the mammalian ovum in 1827. In 1834 he gave up embryology and moved to St. Petersburg. This sudden decision recalls Rossini's abandonment of opera at the height of his fame and may have had a similar cause: nervous breakdown and the threat of ill health. In Russia, von Baer led expeditions to Novaya Zemlya and the Caspian Sea, founded Russian anthropology, made notable advances in ecology, established the law relating erosion of river banks to the earth's rotation, and, at the end of his long life, wrote some essays attacking the new Darwinian theory.

Part 1 of the *Entwickelungsgeschichte* consists of two sections. The first, a masterpiece of descriptive science, meticulously done and beautifully presented, contains his observations on the embryology of the chick. The second is a set of six "scholia and corollaries to the development of the chick in the egg." These contain his philosophy of biology and life, presented as a set of commentaries on the earlier descriptive material. This preeminent observer and foe of unsupported speculation tried to maintain that his general comments flowed from his observations; his great work is subtitled "Beobachtung und Reflexion." Recalling the fate of Oken's bare speculations, he expressly requested that subsequent editors not separate the commentary from the descriptions (1828, p. xviii). Yet, as with any scientist of worth, von Baer's brilliant thoughts often preceded and directed his inquiry. His intransigent opposition to recapitulation arose more from a general philosophy than from his observations on the embryology of the chick.[15]

Von Baer devotes his fifth scholium to an attack on recapitulation: *Ueber das Verhältniss der Formen, die das Individuum in den verschiedenen Stufen seiner Entwickelung annimmt* (pp. 199–262). He begins by acknowledging the influence of recapitulation, referring to "the dominant idea [*herrschende Vorstellung*] that the embryo of higher animals runs through the permanent forms of lower animals." He then presents a series of six short "objections" (*Einwürfe*) that must stand as mere debating points compared to the two powerful refutations of later pages (Haeckel later skirted them all with reasonable success by acknowledging that recapitulation had exceptions, but held in most cases):

1. Many features of embryos are not present in adult animals. The placenta is a special adaptation to uterine life. Although the incisors of mammals erupt first in ontogeny, they are never the only teeth in an adult.[16]

2. The mode of life of an embryo often precludes any complete repetition of lower forms: the mammalian embryo, lying in its placental fluid, can never be a flying bird or an air-breathing insect.

3. There is never a complete morphological correspondence between an embryo and any lower adult. Indeed, the chick embryo, at one stage, has a heart and circulation very much like that of a fish, but at the same time it lacks "a thousand other things" that all adult fishes possess (p. 205).[17]

4. In a reversal of what recapitulation predicts, transitory features in the ontogeny of *lower* animals often appear in adult stages of higher

creatures. Von Baer presents a list of features that are fixed in adult mammals, but transitory in bird embryos. In a brilliant bit of fun, von Baer gives them a recapitulationist interpretation in the only possible way: by letting birds write the textbooks.

Let us only imagine that birds had studied their own development and that it was they in turn who investigated the structure of the adult mammal and of man. Wouldn't their physiological textbooks teach the following? "Those four and two-legged animals bear many resemblances to embryos, for their cranial bones are separated, and they have no beak, just as we do in the first five or six days of incubation; their extremities are all very much alike, as ours are for about the same period; there is not a single true feather on their body, rather only thin feather-shafts, so that we, as fledgelings in the nest, are more advanced than they shall ever be . . . And these mammals that cannot find their own food for such a long time after their birth, that can never rise freely from the earth, want to consider themselves more highly organized than we?" (pp. 203–204)

Von Baer also cites the relatively large brain of vertebrate embryos and adult humans, and the presence of but three pairs of legs in young myriapods (a state resembling adults of a "higher" group: the insects—Fig. 4).[18]

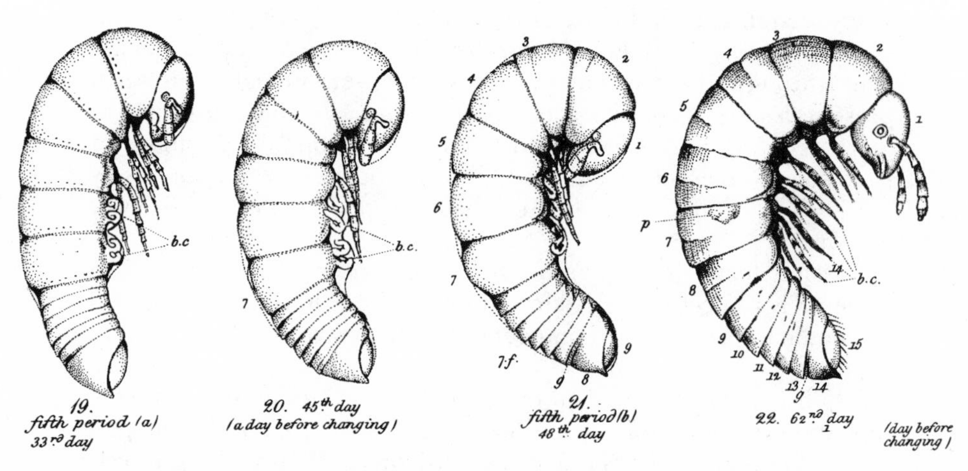

Fig. 4. Stages in myriapod development. Note the six-legged stage of no. 19. Some modern biologists use this evidence to postulate a paedomorphic origin of insects from myriapod larvae. It has also been frequently cited as one of the many accumulating cases of paedomorphosis that dethroned the theory of recapitulation. In fact, it and many other classic cases were well known and appreciated before Haeckel's birth. (From Newport, 1841.)

5. Structures possessed in common by higher embryos and lower adults do not always develop in embryos in the same order as their appearance in a sequence of lower adults would imply.

6. Parts that characterize higher groups should appear late in embryology, but often do not. The vertebral column of the chick, for example, develops very early.

After playing the picador, von Baer unleashes his major attack in two parts:

The first argument. Von Baer had previously cited many specific examples that did not conform to the predictions of recapitulation. These are not, he now argues, exceptions to a possibility: in theory, recapitulation cannot occur. Development is individualization; it proceeds from the general to the special; it is a true differentiation of something unique from an initial state common to all. Von Baer recalls his observations on the chick. The essential components of vertebrate design appear at the outset; all subsequent development is the increasing differentiation of a particular species of vertebrate, indeed of a particular individual. "The type [*Typus*] of each animal seems to fix itself in the embryo right at the beginning and then to govern all of development" (p. 220). At first the chick embryo possesses only a few of the most basic features that identify it as a vertebrate; at this stage, one cannot tell what kind of a vertebrate it will become. Similar limb buds produce bird wings, human hands, and horse hooves. Later, the embryo is recognizable as a bird, then as a gallinaceous bird, then a member of the genus *Gallus,* then of the species *Gallus domesticus,* finally as Joe or Henry the rooster.

> The further we go back in the development of vertebrates, the more similar we find the embryos both in general and in their individual parts . . . Therefore, the special features build themselves up from a general type. (p. 221)

The extent of individualization is the criterion of progressive development:

> The grade of development [*Grad der Ausbildung*] of an animal body consists of the greater or lesser extent of heterogeneity in the parts that compose it . . . The more homogeneous [*gleichmässiger*] the entire mass of the body, the lower the stage of development. We have reached a higher stage if nerve and muscle, blood, and cell-material [*Zellstoff*] are sharply differentiated. The more different they are, the more developed the animal. (p. 207)

From this view of development, recapitulation cannot possibly occur. The embryonic vertebrate, at every stage, is an undeveloped and imperfect vertebrate; it can represent no adult animal whatever.

Embryology is differentiation, not a climb up the ladder of perfection.

The vertebrate embryo is, at the beginning, already a vertebrate; at no time is it identical with an invertebrate animal. An adult [*bleibende*] animal possessing the vertebrate type and exhibiting as little histological and morphological differentiation as the embryos of vertebrates is not known. Thus, in their development, the embryos of vertebrates pass through no (known) adult stage of another animal.

On this basis, von Baer enunciates his famous laws of development, the epitome of his contribution (and probably the most important words in the history of embryology);

1. The general features of a large group of animals appear earlier in the embryo than the special features.
2. Less general characters are developed from the most general, and so forth, until finally the most specialized appear.
3. Each embryo of a given species [literally *Thierform*], instead of passing through the stages of other animals, departs more and more from them.
4. Fundamentally therefore, the embryo of a higher animal is never like [the adult of] a lower animal, but only like its embryo. (p. 224)

The second argument. Von Baer asserts that the occurrence of recapitulation is "necessarily bound" to "the view of a unilinear scale of animals" (p. 231). Recapitulation permits "only *one* direction of metamorphosis that reaches its higher stages of development either in an individual (individual metamorphosis) or through the different forms of [adult] animals (metamorphosis of the animal kingdom); abnormalities [of birth] had to be designated as retrogressive metamorphosis because unilinear metamorphosis is like a railway that moves only forwards or backwards, never to the side" (p. 201).[19]

The animal kingdom, von Baer argues, is not a graded series built upon a single theme, but a collection of four independent groups. In analyzing morphology, we must distinguish type of organization (*Typus der Organisation*) from grade of differentiation (*Grad der Ausbildung*) within each type. We may encounter graded series for certain organs within a type; members of a type may even be linked by physical evolution.* But there can be no transformation of any kind

* Raikov (1968) devotes most of his book to von Baer's views on evolution. Despite shifting emphases, von Baer's general opinion changed very little during his long life. He was a teleologist; he disliked the mechanistic aspects of Darwinian theory. He allowed for limited physical evolution within types, but no transformation among them. His early words on general advance in the universe refer not to physical descent, but to the same ideal progress that Schelling and other anti-evolutionists took as the universal law of nature.

between types, either in ontogeny or phylogeny. The four great groups cannot be arrayed in a progressive sequence; they are simply different from each other, not higher or lower.[20] The type is established in the very first stages of ontogeny and governs all subsequent development. A higher animal cannot pass through the adult stages of lower forms during its own development.

Von Baer's types correspond to the *embranchements* of Cuvier: peripheral type to the Radiata, longitudinal to the Articulata, massive to the Mollusca, and vertebrate to the Vertebrata. Yet, von Baer claimed that he had developed his classification independently, "in so far as a man can call anything that is a fruit of its time his own" (p. vii). Where Cuvier had based his system upon the morphology of adults, von Baer adopted the dynamic perspective of development: "Type is to be understood through its mode of development . . . Different conditions or building forces must work upon the germ [which is originally similar in all animals] in order to create this diversity" (p. 258; see also Milne-Edwards, 1844).

Even if von Baer's classification precluded Oken's cherished vision of a human embryo mounting "through *all* the spires of form," perhaps recapitulation could still work on the more modest scale of sequences within types.[21] But even this von Baer would not allow. The animals within a type form no ascending series for two reasons based on a common premise. The premise is Cuvier's greatest insight: the shapes of organs are adapted to their function, not arrayed in ideal series. First, a series established by the differing states of one organ will not hold for other organs. Organs are patterned to their function; a sequence in locomotion will not parallel one in feeding. Second, it is doubtful that animals within a type can be arrayed meaningfully by stages of development in a single organ. Functions come in clusters (swimming, running, flying), not sequences. Differing forms of an organ are variants about a central theme, not rungs of ladders.

Milne-Edwards combined both of von Baer's major arguments into a picture that renders inconceivable any thought of recapitulation in Oken's version of unilinear advance (Fig. 5):

> The metamorphoses of embryonic organization, considered in the entire animal kingdom, do not constitute a single, linear series of zoological phenomena. There are a multitude of these series . . . They are united in a bundle at their base and separate from each other in secondary, tertiary, and quaternary bundles, since in rising to approach the end of embryonic life, they depart from each other and assume distinctive characteristics. (1844, p. 72)

Darwin was not the first to use a tree as a biological metaphor; its earlier, nonevolutionary popularity in England can be traced directly to von Baer's influence (Ospovat, 1974).

Fig. 5. Milne-Edwards' Cuvierian (and von Baerian) classification of animals is basically incompatible with the unilinear, progressive schemes required in the version of recapitulation supported by Naturphilosophie. (From Milne-Edwards, 1884.)

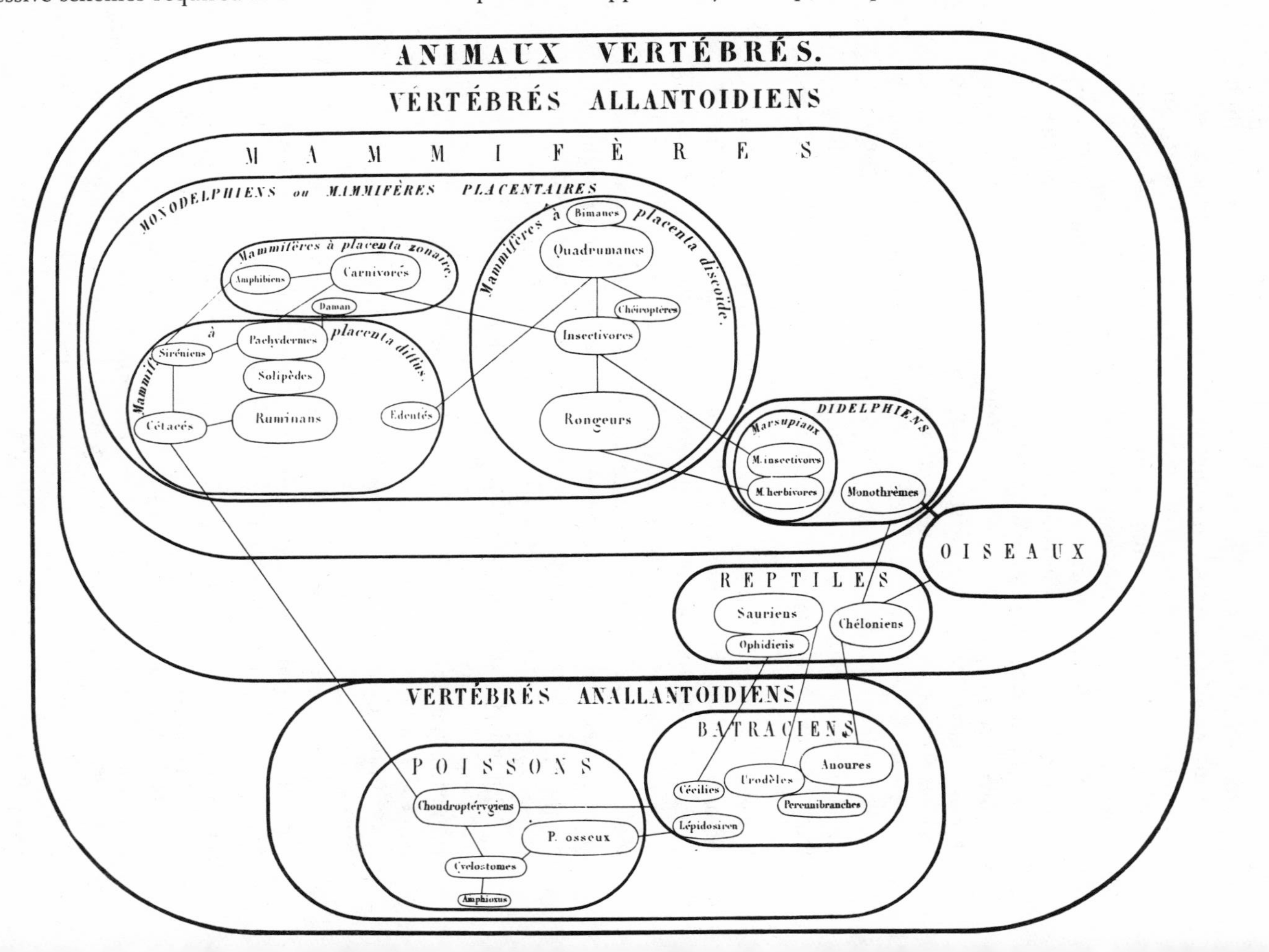

Von Baer and Naturphilosophie: What Is the Universal Direction of Development?

Was von Baer a Naturphilosoph? How shall we view his debate with Meckel and Oken over recapitulation? Did it represent a clash of two philosophies or a disagreement over the interpretation of a common framework? The subject has been discussed often and has given rise to a curious difference of opinion. Humanists (Cohen and Lovejoy) have detected an a prioristic Naturphilosoph where scientists (Oppenheimer, Meyer, and Severtsov) note a circumspect analyst of careful observations.[22] The disagreement has an easy explanation (as facile as its simplistic categories but, perhaps, basically sound nonetheless). I doubt that many historians or philosophers have paid much attention to von Baer's descriptive embryology of the chick; they have concentrated on the scholia (especially the sixth, with all its cosmic pronouncements) and some popular essays on development. On the other hand, it is hard for scientists to ignore (though they should) the anachronistic influence of von Baer's triumph; for his laws, in refurbished evolutionary dress, are now more widely accepted than ever before, and his descriptions mark the beginning of modern embryology. It is then tempting to reason: if von Baer led us away from the fantasies of Oken's school, he must have opposed its philosophy, and, in the inductivist bias that most scientists impose upon their own history, he must have substituted the careful objective study of facts for the priority of speculation.

I doubt that such a controversy could have arisen unless both positions were valid (though incomplete). Raikov has recently tried to resolve this dilemma by arguing that von Baer's thinking moved "in two different planes . . . each with its own inner logic: the plane of the Naturphilosoph's conception of reality that rests on intuitions . . . and does not ask for proof, and the plane of scientific thought . . . that bases itself strongly upon facts and demands tangible proof" (1968, pp. 397–402). This complexity, indeed this inconsistency, surely exists in von Baer's thought, but I would prefer to render it as an internal clash of two biological philosophies.

Von Baer's empirical work unmistakably bears the stamp of Cuvier's thought. No one argued more incessantly than Cuvier that science should move from observation to theory, and shun the excesses of unsupported speculation. No one insisted more strongly than Cuvier that organs should be studied functionally as shapes designed for performance, not as ideal series distributed to meet the requirements of philosophical visions. (A laudatory biography of Cuvier, written by von Baer, was published posthumously in 1897.)

Von Baer's second major argument against recapitulation, the classification of animals into four types, is in Cuvier's style. Moreover, the subsidiary parts of this argument are written from a functional standpoint as incompatible with Naturphilosophie as anything von Baer ever wrote. Opposing the idea that organs might be arrayed in series even within types, von Baer states:

Tiedemann unites the seal with the dugong; Pallas places them far from each other. The former has used the extremities [for his classification], the latter the teeth. What does such an example teach us, but that different organ systems vary in different ways. Moles and bats seek the same prey, the former in the earth, the latter in the air. Therefore, their organs of locomotion are different according to the environments in which they reside. The dugong and the seal are both in water; they have fin-like extremities. But what they seek in the water is entirely different, hence their dentition and stomach. (p. 241)

The other side of von Baer's thought is attuned to the dominant biology of his time and place—to Naturphilosophie. We know that von Baer was strongly influenced by Naturphilosophie at the beginning of his career. Both his eminent teachers, Burdach and Döllinger, though more sober than Oken or Carus, adhered to Schelling's philosophy (Raikov, 1968, pp. 23–41, 389). Raikov has examined two unpublished manuscripts displaying all the speculative charm of Oken's approach. Von Baer may have dismissed them later as a folly of youth. In one, part 2 of his popular essays "Anthropologie für den Selbstunterricht," von Baer depicts the human body according to a favorite scheme of the most speculative Naturphilosophen: the comparison of upper and lower halves of the body. The body is built by similar powers working as polar opposites; the brain corresponds to the organs of reproduction, the lung is "the organ of excretion for the blood" (Raikov, 1968, p. 57). Another manuscript, written in 1819, is an exposition of Schelling's view of nature as a product of the *Weltgeist* striving to reach self-consciousness (Raikov, 1968, pp. 390–391). The same thoughts, albeit in muted form, pervade several essays written in Königsberg and presented or published in the late 1820s and 1830s. The best known, his 1834 address on "the most general law of nature in all development," concludes that "the history of nature is only the history of the ever-advancing victory of spirit over matter" (1864, pp. 71–72)—Schelling's *Weltgeist* again striving for self-consciousness as man.

But we also know that von Baer turned from the excesses of Naturphilosophie. In his autobiography, he records his thorough study of Oken's *Lehrbuch,* his fascination with its pronouncements and his final rejection of its methods (Raikov, 1968, p. 390). In his last essay, *Über Darwins Lehre* of 1876, von Baer recalls his disagreement: "Naturphilosophie accustomed thinkers to treat certain similarities as unities, without emphasizing where the identities and where the differences lay. Who does not remember such sentences as: 'Architecture is frozen music'" (p. 242).

"We are of the opinion," writes Raikov (p. 59), "that from the entire arsenal of concepts in Naturphilosophie that Burdach, Döllinger and Oken taught, the concept of development exercised the most lasting influence upon von Baer." This is surely true. The search for a general law of development motivated the *Entwickelungsgeschichte;* the elucidation of it unites all parts of the book. As a whole, von Baer's treatise must be ranked among the works that Naturphilosophie inspired; for it sought, in a different way but with as much zeal as Oken's *Lehrbuch,* the universal law of development. In the closing words of his preface, von Baer belittles his own contribution by citing the goal of all biology: "The palm [of victory] shall be gained by the lucky man who traces back the developmental powers [*bildenden Kräfte*] of animal bodies to the general powers or directions of life [*Lebensrichtungen*] of the entire world."

Yet von Baer had achieved more of the victory than modesty allowed him to state, for he had posited a general law of all biological development and, through it, thought he had glimpsed the essence of all development (*Entwicklung*): the homogeneous, coarsely structured, general, and potential develops into the heterogeneous, finely built, special and determined. This law of differentiation is much more than a postulate brought forth in the fifth scholium to counter recapitulation; it is *the* law of biological development, the single tendency of all change.

This law of differentiation is the unifying theme of von Baer's entire work. Of his 140 pages of detailed description, he writes: "Our account of the development of the chick is only a long commentary upon this assertion" (p. 220). Each of the six scholia treats one of its consequences.

In the first scholium, von Baer recounts his opposition to preformation. Of all the arguments available to him, he chooses one based firmly on his law of differentiation: no one can argue that organs are truly preformed in young embryos, but merely too fine for resolution by our best microscopes. At its first appearance to an observer, an organ is simpler, grosser, and relatively larger than it will be later as differentiation proceeds.

The second scholium treats the control of differentiation. Development is not completed in mechanical fashion with each stage acting as cause of the next. Young embryos are more variable than older ones; if each step controlled the next, variability would increase with age. There must be some higher, teleologic control that regulates development by directing variation back to the normal path of differentiation. Von Baer then characterizes that normal path: "The most important result of development . . . is the increasing individuality [*Selbstständigkeit*] of the growing animal" (p. 148).

The third scholium discusses the general path of differentiation. It proceeds in three sequential stages: primary, or the formation of layers; histological; and morphological. It leads to increasing individualization according to the general principle "that the special and heterogeneous are built up from the general and homogeneous" (p. 153). Increasing complexity is always a true differentiation from something simpler. New organs develop not from former empty spaces, but from undifferentiated masses of matter. All differentiation is *Umbildung,* not *Neubildung.*

The fourth scholium applies the law of differentiation to the development of the vertebrate type in general.

The fifth scholium, as we have seen, uses the law of differentiation to refute recapitulation. It also presents a classification of animals into four types. Here the Cuvierian side of von Baer's thought challenges his desire, as a Naturphilosoph, to render differentiation as the completely general law of biological development. If the fixation of the type in the embryo is the very first step of development, then the initial state is not as generalized as it might be in theory. To give his law its full grandeur and sweep, all development must begin with a completely undifferentiated, homogeneous state retaining all its potential to follow any path of development, not with a form already constrained to follow one of four major routes. Von Baer argues that there is a "short moment" of initial agreement before establishment of type; for articulates (arthropods and annelids) as well as vertebrates begin development with a primitive streak: "In this short moment, there is agreement between them [the articulates] and the vertebrates. In the actual condition of the germ there is probably identity among all embryos that develop from a true egg."

Finally, in the sixth scholium, von Baer considers the relation between his law of all biological development and the desideratum of Naturphilosophie—a general law of all development. This scholium bears the title: "the most general result" (*Allgemeinstes Resultat*), and displays it in bold italics:

> The development of the individual is the history of growing individuality in every respect [*Die Entwickelungsgeschichte des Individuums ist die Geschichte der wachsenden Individualität in jeglicher Beziehung*]. (p. 263)

In the last paragraph of his work, von Baer reaches out from his law to the cosmos:

> If our most general result is true, then there is one fundamental thought [*Grundgedanke*] that permeates all the forms and stages of animal development and governs all their relationships. It is the same thought that, in the cosmos [*Welträume*], collects the separated masses into spheres and binds these together into a solar system; the same that allows the scattered dust on the surface of the metallic planet to develop into living forms. This thought, however, is nothing but life itself, and the words and syllables in which it expresses itself, are the different forms of the living. (pp. 263–264)

Von Baer attacked recapitulation from both sides of his thought. On the one hand, the functional perspective of Cuvier would not permit the unilinear classification that recapitulation required. Ironically though, von Baer's most effective argument lay in his particular version of a general principle that he shared with the Naturphilosophen—a principle that validated recapitulation for Oken and Meckel. Oken, Meckel, and von Baer all agreed that a single developmental tendency pervaded nature. For Oken, it was the progressive addition of organs or powers; for Meckel, the coordination and specialization of parts. Both yield recapitulation. For Oken, the human embryo begins in the primal chaos of zero and adds organs in a sequence reflecting the order of lower adults. For Meckel, the human embryo begins with uncoordinated parts and develops an integrated set of specialized organs in a sequence running parallel with the ascending series of lower adults. But for von Baer, the single tendency is differentiation, the development of the special from the general. This precludes recapitulation. The human embryo begins as a generalized vertebrate retaining the potential to become any species of its type; it cannot represent the completed adult of any lower animal.

Louis Agassiz and the Threefold Parallelism

Von Baer was not the only great biologist caught in a dilemma of allegiance to the contrasting schools of Naturphilosophie and Cuvierian functionalism. Von Baer's dilemma was particularly acute, for he held both viewpoints concurrently. Louis Agassiz espoused them sequentially. As a young man, he adopted many concepts of romantic biology, notably recapitulation. Later, when he had abandoned the easy explanations that Naturphilosophie provided for recapitulation, he had to supply new justifications consistent with the spirit of Cuvier's thought.

During the 1820s, Agassiz studied with several of Germany's leading Naturphilosophen. Lurie (1960) supposes that Tiedemann first acquainted Agassiz with recapitulation at Heidelberg in 1826,[23] but his teachers at Munich (1827–1830), Döllinger[24] and Oken, may have exerted a stronger influence. Agassiz spent a good part of the summer of 1827 reading Oken's *Lehrbuch der Naturphilosophie*; he stated that it gave him "the greatest pleasure" (Lurie, 1960, p. 27). He attended Oken's lectures on Naturphilosophie, though his fellow student Alexander Braun recorded an ulterior motive of no mean appeal: "We go once a week to hear Oken on Naturphilosophie, but by that means we secure a good seat for Schelling's lecture immediately after" (Lurie, 1960, p. 51).

Yet Agassiz had also fallen under the spell of Cuvier. In December 1831, he arrived in Paris to study fishes and to seek Cuvier's favor. In both endeavors, his success was unbounded. Cuvier admired the young naturalist greatly, accepting him both as a personal friend and a scientific equal. Cuvier was so impressed with Agassiz's work on fossil fishes that he abandoned his own study of these animals and entrusted all his notes and drawings to Agassiz's care (Lurie, 1960, p. 56). Though their personal friendship was short (Cuvier died in 1832), Cuvier's influence was decisive and permanent. Agassiz abandoned his short flirtation with the principles of Naturphilosophie and dedicated himself to Cuvier's vision of a permanent order that might be apprehended through patient observation.

All Agassiz's work on recapitulation appeared after Cuvier's death. To retain this principle of Naturphilosophie, Agassiz had to find an explanation for it within Cuvier's system of thought—a philosophy that had led von Baer to deny recapitulation completely. In this attempt he succeeded. Once more, as with Bonnet, we glimpse the extraordinary persistence of recapitulation, its ability to incorporate itself into philosophies that should, at first glance, have resisted it.

Agassiz first had to deal with von Baer's two fundamental objections to recapitulation. One of them he simply accepted, thus limiting severely the scope of recapitulation: there are but four fundamental plans of animal design, and there can be no transformation of any kind among them.[25] An embryo, therefore, can only repeat adult stages of lower animals *within its own type.*[26]

Von Baer's primary objection could not be so easily encompassed, for if development always proceeds from the general to the special, recapitulation is impossible. The adults of lower animals may be uncomplex, but they are not undifferentiated. Agassiz, therefore, simply denied to von Baer's law the absolute generality that von Baer's metaphysic had provided for it. Since Agassiz had not joined the Naturphilosophen in their search for the universal principle of development, he could treat von Baer's law as an unencumbered postulate and refute it with simple counter-cases. Against Martin Barry's presentation of von Baer's law, he writes: "This is very logical, but not in accordance with nature; we may frame such a system in our closets, but it does not answer our observations" (1849, p. 28). Agassiz then examines the development of the frog:

> Was it the character by which the frog is found to belong to the class of reptiles, which was first apparent? By no means. It appeared first, under the form and structure of a fish, and not under the form and with the characters of a reptile. The lowest form of vertebrated animals was first developed in the earlier changes of the egg, before the class to which that animal belonged could be recognized. (1849, p. 28)

Likewise, embryonic starfish do not first develop the plates and suckers that mark their group, but rather the "forms which would lead us to mistake them for Polypi or Medusae" (p. 28). (In Agassiz's system, the Coelenterata [including polyps and medusae] are a lower type of the *embranchement* Radiata, which contains Echinodermata [including starfish] as a higher group. This comparison does not violate the immutability of the four *Baupläne.*) Any competent naturalist can recognize whether a fetus will become a domestic cat (that is, he can determine the species of the embryo) before its generic characters appear (four molars in the upper jaw, three in the lower; retractile claws). Moreover, varietal characters (coat color) and even individual peculiarities (playfulness) precede the eruption of the molars. "In short, everything takes place in the reverse order from what it is supposed in this [von Baer's] system" (1849, p. 28).*

Agassiz then unveils his ambitious plans for recapitulation as a working doctrine:

> There is a gradation of types in the class of Echinoderms, and indeed in every class of the animal kingdom, which, in its general outlines can be satisfactorily ascertained by anatomical investigation; but it is possible to arrive at a more precise illustration of this gradation by embryological data . . . The most special comparisons of these metamorphoses [in ontogeny] with full grown animals of the same type, leads to the fullest agreement between both . . . These phases of the individual development are the new foundations upon which I intend to rebuild the system of zoology. (1849, p. 26)

Before Agassiz, recapitulation had been defined as a correspondence between two series: embryonic stages and adults of *living*

* The infelicities of phrase that often occur in these lectures on embryology do not only reflect Agassiz's unfamiliarity with English (he had arrived in 1846); they are a verbatim "phonographic report" of Agassiz's oral presentation, to which the stenographer has proudly appended a sample of his shorthand (p. 104).

species. Agassiz introduced a third series: the geologic record of fossils. An embryo repeats both a graded series of living, lower forms and the history of its type as recorded by fossils. There is a "threefold parallelism" of embryonic growth, structural gradation, and geologic succession.

It may therefore be considered as a general fact, very likely to be more fully illustrated as investigations cover a wider ground, that the phases of development of all living animals correspond to the order of succession of their extinct representatives in past geological times. As far as this goes, the oldest representatives of every class may then be considered as embryonic types of their respective orders or families among the living. (1857, 1962 ed., p. 114)

Eight years before, he had written: "To carry out these results in detail must now be, for years to come, the task of paleontological investigations" (1849, p. 27).

Agassiz's addition of the fossil record to form a threefold parallelism represented an inevitable implication of recapitulation. Tiedemann had stated the idea clearly in 1808 (Russell, 1916, p. 255) and many others had provided passing references (Agassiz, 1857, 1962 ed., p. 110). Yet, before Agassiz, it had gained little prominence for a simple reason: the ordering of fossils into a historical, geological sequence had not yet been achieved. Agassiz's name is rightly attached to this extension of recapitulation because he supplied its documentation.

Agassiz's evidence first appeared in his early work on fossil fishes (*Les poissons fossiles,* published in several installments from 1833 to 1843). In the fossil record, heterocercal tails appear before homocercal[27]; living lower fishes (sharks and their allies) have heterocercal tails, while advanced teleosts have homocercal tails: embryonic teleosts begin with a heterocercal tail, which becomes homocercal later in development (Fig. 6). Throughout his career, Agassiz catalogued other cases. He compared young polyps with fossil Rugosa, large-spined young echinoids with fossil *Cidaris,* embryonic bivalves with brachiopods, infant horseshoe crabs with trilobites, and embryonic elephants with mastodons. Crinoids supplied his favorite illustration: most living crinoids are unstalked and free-swimming as adults, but stalked and attached as embryos; many fossils of the Paleozoic are permanently stalked. Agassiz was confident that modern embryos would faithfully repeat the forms of ancient fossils: "If I am not mistaken, we shall obtain from sketches of those embryonic forms more correct figures of fossil animals than have been acquired by actual restoration" (1849, p. 104).

But why does recapitulation occur? Since he rejected the single developmental tendency of Naturphilosophie, Agassiz could not pro-

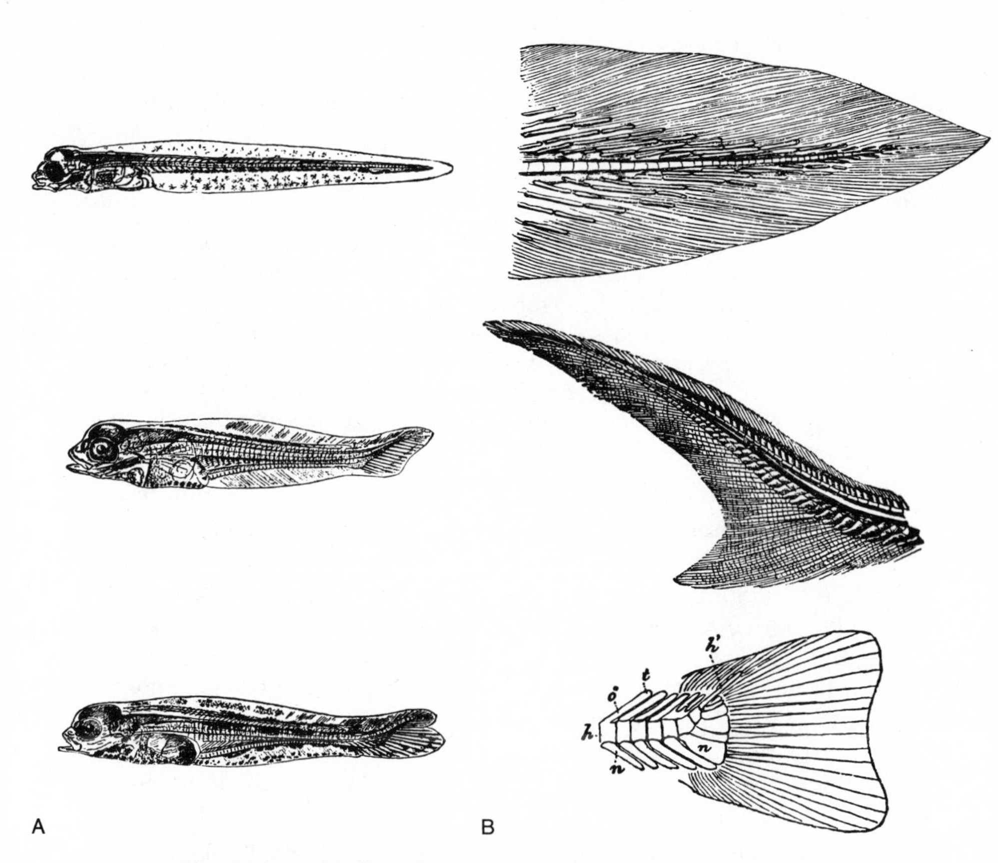

Fig. 6. Agassiz's favorite example of recapitulation (and an enduring classic for all later supporters). (*A*) The ontogeny of a "higher" teleost (the flatfish *Pleuronectes*) showing transition from diphycercal to heterocercal (upper lobe larger than lower) to homocercal (equal-lobed) tail. (*B*) Comparative anatomy of adult tails in sequence of primitive to advanced fish (also paralleled by their order of appearance in the geological record); from top to bottom: *Protopterus* with its diphycercal tail, a sturgeon with a heterocercal tail, and a salmon with a homocercal tail. (From Schmidt, 1909.)

pose the easy explanation of his teacher Oken. As Darwin's most implacable opponent, he could seek no aid from transmutationist doctrines. To Agassiz, the threefold parallelism reflected the unity of God's plan for His creation. It was also a fact of observation. What more need a Cuvierian empiricist say? "The leading thought which runs through the succession of all organized beings in past ages is manifested again in new combinations, in the phases of the development of the living representatives of these different types. It exhibits everywhere the working of the same creative Mind, through all times, and upon the whole surface of the globe" (1857, p. 115). Agassiz invoked his God specifically to forestall any evolutionary reading of recapitulation:

> There exists throughout the animal kingdom the closest correspondence between the gradation of their types and the embryonic changes their respective representatives exhibit throughout. And yet what genetic relation can there exist between the Pentacrinus of the West Indies and the Comatulae, found in every sea; what between the embryos of Spatangoids and those of Echinoids . . . what between the Tadpole of a Toad and our Menobranchus; what between a young Dog and our Seals, unless it be the plan designed by an intelligent creator. (1857, 1962 ed., p. 119)[28]

Yet, Agassiz's views contained an argument that no evolutionist could resist reinterpreting. If the fossil record is only a temporal display of the same divine plan that animals reflect in their own ontogeny, then the geologic component of Agassiz's threefold parallelism merely extends the scope of recapitulation and the generality of benevolent design. But if fossils record an actual history of physical descent, then the argument must be inverted. The geologic record is no mere addition to a twofold parallelism between embryonic stages and the structural gradation of living forms; it is the fundamental sequence that engenders the other two. The structural gradation of living forms is merely its artifact, because primitive animals have survived in each type. Embryonic stages are only its reflection, because an embryo must repeat the shapes of its ancestors before adding its own distinguishing features. Agassiz's parallelism, a divine union of three independent sequences, becomes the mechanical result of a single causal chain leading from the geologic record to the stages of embryology: ontogeny recapitulates phylogeny.[29]

—4—

Evolutionary Triumph, 1859–1900

Evolutionary Theory and Zoological Practice

In surrendering to Washington at Yorktown, Cornwallis' band played a ditty about the "world turned upside down." A reassembly of the chorus in 1859 would not have been inappropriate. Darwin's youthful essays of 1842 and 1844 record what must be the greatest of all intellectual delights: the systematic reconstruction of a body of knowledge according to novel principles of one's own invention. Intellectual historians have emphasized the profound impact of evolutionary theory upon social and political life; yet it is ironic that biologists often incorporated the new explanations without substantially altering their scientific practice. Systematists, for example, could easily explain homology by common evolutionary descent rather than similarity of divine thought; yet the procedures for recognizing homologies and constructing classifications from them were little disturbed by this explanatory reversal.[1] Alpheus Hyatt noted (with some surprise in hindsight) that he had been able to transfer bodily to evolutionary theory the taxonomic conclusions that Agassiz had based upon the creationist interpretation of recapitulation: "Although within a year after the beginning of my life as a student under Louis Agassiz I had become an evolutionist, this theoretical change of position altered in no essential way the conceptions I had at first received from him, nor the use we both made of them in classifying and arranging forms" (1897, p. 216).

The epigenetic character of embryology made it a field for phyletic speculation that no evolutionist could resist. There existed, in 1859, two major interpretations for the significance of embryonic stages. Each had been formulated under creationist tenets, but each could be easily restructured in evolutionary guise. These were, of course, von Baer's principle that development proceeds inexorably from the general to the special and the recapitulationist claim that embryonic stages represent adult forms of "lower" creatures. Both were quickly given their evolutionary meaning: Darwin accepted von Baer's principle but stood the original explanation on its head. F. Müller, Haeckel, Cope, and Hyatt independently recognized the irresistible promise of recapitulation as a key to the reconstruction of phylogeny.

Do old practices, like good bureaucrats, survive revolutions by enveloping their unaltered core of basic procedure in the appropriate window-dressing of a new theory? Did Darwin and Haeckel merely place old wine in new bottles by refurbishing some terminology? Or did evolutionary theory, in this case, prescribe a new way of proceeding? I shall argue that evolutionary theory transformed the workaday habits of comparative embryologists by posing problems and providing insights that earlier explanations for the same principles had not supplied.

Darwin and the Evolution of von Baer's Laws

On September 10, 1860, Darwin wrote to Asa Gray: "Embryology is to me by far the strongest single class of facts in favor of change of forms." It is often assumed that Darwin had recapitulation in mind,* but Fritz Müller's first evolutionary interpretation of recapitulation did not appear until 1864, and Darwin was not quoting the Naturphilosophen. In fact, Darwin had accepted the observations of von Baer—a flat denial of recapitulation and its obvious evolutionary meaning. Referring to von Baer, he wrote in his autobiography:

* This common assumption has two rather different bases. First, many authors simply don't read Darwin carefully and assume that his strong invocation of embryology must be based upon recapitulation; the phrases "evolutionary embryology of the nineteenth century" and "biogenetic law" are linked so closely that many authors simply do not recognize other evolutionary readings of embryology. This is an outright error and need detain us no longer. Second, many secondary sources (but no primary participants in the nineteenth century) have extended the word "recapitulation" beyond its original definition as the repetition of *adult* stages in ontogeny to encompass any belief that phyletic information resides in ontogeny—a proposition that can scarcely be denied by an evolutionist. This has produced the lamentable confusion that I document in the introduction. Lovejoy, for example, uses the phrase "Darwinian theory of recapitulation" in a title, though he recognizes that Darwin's views are von Baer's transformed. He contends that the laws of von Baer are a "denial, not of recapitulation itself, but simply of recapitulation of adult forms" (1959, p. 443).

"Hardly any point gave me so much satisfaction when I was at work on the Origin as the explanation of the wide difference in many classes between the embryo and the adult animal, and of the close resemblance of the embryos within the same class." Moreover, Darwin did not accept von Baer passively, as the only possible choice, for he clearly knew and considered the recapitulatory alternative as expressed by Naturphilosophie and the French transcendental morphologists. In the "B Transmutation Notebook," for example, he abstracts Etienne Geoffroy St. Hilaire's belief in "generation as a short process by which one animal passes from worm to man highest or typical of changes which can be traced in same organ in *different* animals in scale" (in de Beer, 1960, p. 54, italics original).

Von Baer had used his law of differentiation through ontogeny as proof that individuals could not repeat the adult stages of lower forms, as a guarantee that linear evolution could not occur, and as a defense of life's construction upon a set of immutable *Baupläne.* Darwin denied none of von Baer's observations, but he recognized an evolutionary interpretation of great potential. Evolutionary classification rests upon the identification of homologies linking diverse organisms to a common ancestral stock. Since most modifications appear at a "not very early" period of life, adult forms often hide their ancestry in a plethora of new adaptations (Darwin, 1859, p. 444). But early stages of ontogeny are generally resistant to change, especially in organisms leading a protected embryonic life in an egg or mother's body. Von Baer's law of progressive differentiation reflects no cosmic tendency of general development. It is a statement prescribing a course of action for the recognition of homology: look for similarity in embryos since evidence of common ancestry is so often obscured by highly particular adult modifications. Since evolutionary classification depends upon the identification of homologies linking diverse animals to common ancestors, von Baer's laws state a basic principle in phyletic reconstruction: "Community in embryonic structure reveals community of descent" (1859, p. 449). Darwin's favorite group, the barnacles, illustrates this principle particularly well. These animals had long been a zoological enigma because their curious adult form diverges so radically from the basic plan of the Arthropoda. Cuvier had classified them among the Mollusca. Their true status was affirmed only by the unmistakable similarity of their larvae to those of other arthropods. Moreover, the class contains some highly degenerate parasites with adult morphologies scarcely more elaborate than a simple bag of reproductive organs—but with larvae of obvious cirripede design.

The reinterpretation of von Baer's laws to yield criteria for the recognition of evolutionary homology must rank as Darwin's primary statement of the relationship between embryology and evolution. Yet, this interpretation did not revise the practice of classification. Von Baer could look to early ontogeny for the divine *Bauplan* in its common, albeit undeveloped, form. With Darwin, the *Bauplan* became the embryonic stage of a mortal common ancestor, but the procedures of taxonomy were not altered thereby: animals with similar embryos were classified together under either interpretation.

Darwin's principle also had a severe limitation: it could determine community of origin, but it offered no clues to actual evolutionary lineages. Fish, amphibians, reptiles, birds, and mammals shared a "community of descent" because all possessed gill slits as embryos. But what sequences of filiation existed among these groups? How could a phylogeny rather than just a grouping of common elements be generated from embryological data? The answer to this question provoked a change in embryological practice.

Von Baer had appended a statement to his fourth law: "It is only because the least developed animal forms are but little removed from the embryonic condition that they retain a certain similarity with the embryos of higher animal forms" (1828, p. 224). To von Baer, this was a mere corollary added only to dispel a recapitulatory interpretation of his beliefs: the embryo repeats no adult stage, but a "low" adult may resemble its own embryo simply because it fails to differentiate much further. Von Baer's incidental statement became, for Darwin, an embryological guide for the inference of evolutionary lineages. Darwin saw that ancestral groups in an established community of descent would differ least in their adult form from the embryonic state common to all members of the community. The gill slits of the human fetus represent no ancestral adult fish: we see no repetition of adult stages, no recapitulation. Yet adult fish, as primitive ancestors, have departed least from this embryological condition of all vertebrates. Thus, Darwin writes in his sketch of 1842: "It is not true that one passes through the form of a lower group, though no doubt fish more nearly related to foetal state" (1909, p. 42).[2] Later, he states explicitly that the idea of evolution had forced his reinterpretation of von Baer's results: "The less difference of foetus—that has obvious meaning on this view: otherwise how strange that a horse, a man, a bat should at one time of life have arteries running in a manner which is only intelligibly useful in a fish! The *natural system being on theory genealogical,* we can at once see why foetus, retaining traces of the ancestral form, is of highest value in classification" (1842, in 1909, p. 45, my italics). Expanding these views in the 1844 essay, Darwin first denies the fact of recapitulation explicitly (1909, p. 219); then, after stating von Baer's views, he presents his principle for the tracing of lineages by the similarity of ancestral adults to their own embryos.

It follows strictly from the above reasoning only that the embryos of (for instance) existing vertebrata resemble more closely the embryo of the parent-stock of this great class than do full-grown existing vertebrata resemble their full-grown parent stock. But it may be argued with much probability that in the earliest and simplest condition of things the parent and embryo must have resembled each other, and that the passage of any animal through embryonic states in its growth is entirely due to subsequent variations affecting *only* the more mature periods of life. If so, the embryos of the existing vertebrata will shadow forth the full-grown structure of some of these forms of this great class which existed at the earlier period of the earth's history. (1909, p. 230)

The same argument figures prominently in the embryological chapter of the *Origin:* "For the embryo is the animal in its less modified state; and in so far it reveals the structure of its progenitor" (1859, p. 449). Darwin applies it especially to Agassiz's claim that embryonic stages of modern forms resemble the adults of their fossilized ancestors: "As the embryonic state of each species and group of species partially shows us the structure of their less modified ancient progenitors, we can clearly see why ancient and extinct froms of life should resemble the embryos of their descendants—our existing species" (p. 381).

But is not Darwin perilously close to recapitulation at this point? Are we not splitting hairs in attempting to draw a distinction between the actual recapitulation of adult stages and the repetition of embryonic stages that resemble ancestral adults. What difference does it make? Both claims use embryonic stages to trace lineages in the same way.

If the goal of evolutionary theory is only to set up a series of pragmatic guidelines for the construction of evolutionary trees, then it makes no difference. But this would be an impoverished notion of evolutionary theory indeed. The two views imply radically different concepts of variation, heredity, and adaptation—the fundamental components of any evolutionary mechanism. If related animals merely repeat their ancestral embryonic stages without alteration, we have a simple case of evolutionary conservatism. If, on the other hand, the tiny human fetus with gill slits *is* (in essence) an adult fish, then we must seek an active mechanism to "push" the adult shapes of ancestors into early embryonic stages of descendants. The search for a mechanism of recapitulation dominated the theoretical side of late nineteenth-century comparative embryology and provoked a major debate within evolutionary theory. In a revised form, this mechanism forms the basis for modern views of the relationship between ontogeny and phylogeny.

Evolution and the Mechanics of Recapitulation

The fact of evolution recast Agassiz's threefold parallelism as a mirror image of its former self: the fossil sequence that he had added as a third illustration of divine wisdom became the primary cause for the other two series. Evolution also provoked a profound change in the mechanics of recapitulation. The single developmental tendency of Naturphilosophie no longer sufficed. Evolution implied a true, physical continuity of forms through time. This raised a host of questions that had never occurred to the Naturphilosophen. The need to solve them provided comparative embryology with an entirely new subject. Many evolutionists, Cope and Hyatt in particular, spent a major part of their careers trying to solve these problems and exploring the consequences of their solutions.

To Oken and Serres, embryonic stages of higher forms had "represented," "stood for," or been in some way "symbolic of" living, lower adults. No physical tie connected a fish to a human fetus with gill slits. But after 1859, recapitulationists had to view embryonic stages of descendants as the *actual, physical* remnants of previous ancestors. How had this remnant been transferred from an ancestral adult of large size (where it developed late in life as a permanent stage) to a tiny embryo (where it appeared early and endured but a short time)? What, in other words, is the mechanism of recapitulation? *There is only one way to make recapitulation work under a theory of evolution by physical continuity. Every recapitulationist, from the staunchest Darwinian (Weismann) to the most militant neo-Lamarckist (Cope and Hyatt), upheld this mechanism; there is no other (Fig. 7). It involves two assumptions.*

1. *Evolutionary change occurs by the successive addition of stages to the end of an unaltered, ancestral ontogeny.*[3] This assumption provokes two problems. First, since many lineages involve thousands of steps, ontogenies will become impossibly long if each step is a simple additon to a previous ontogeny. Second, embryonic stages usually occur much earlier in time and at much smaller sizes than the ancestral adult stage they represent. There must be some force continually operating to shorten ancestral ontogenies, thereby keeping the descendant's period of development within reasonable limits.

2. *The length of an ancestral ontogeny must be continuously shortened during the subsequent evolution of its lineage.*

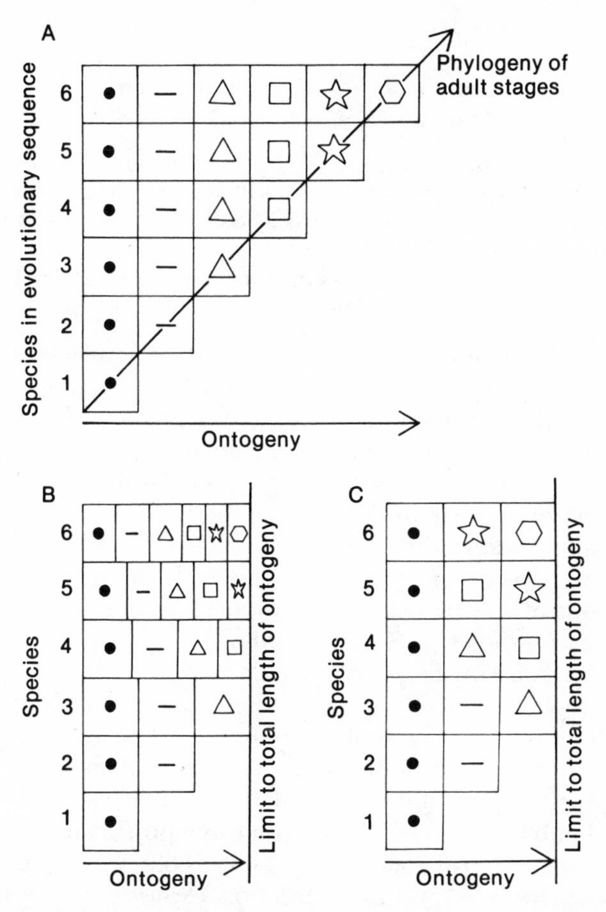

Fig. 7. The mechanism of recapitulation and its two principles. (*A*) The principle of terminal addition: new features are added in evolution to the end of ancestral ontogenies; the phylogeny of adult stages parallels the ontogeny of the most advanced descendant (species 6). (*B*) The principle of condensation (by acceleration): the length of ontogeny is limited and stages are shortened (accelerated) to make room for new features added terminally. (*C*) The principle of condensation (by deletion): stages of ontogeny are eliminated to make room for new features.

I shall call the first assumption "the principle of terminal addition," and the second "the principle of condensation." Of these, the principle of condensation inspired more debate and a greater variety of proposals for its implementation. I shall devote the rest of this chapter to a discussion of how evolutionists in opposing schools justified these principles.

Ernst Haeckel: Phylogeny as the Mechanical Cause of Ontogeny

Es ist ein ewiges Leben, Werden und Bewegen in ihr. Sie verwandelt sich ewig, und ist kein Moment Stillstehen in ihr. Für's Bleiben hat sie keinen Begriff, und ihren Fluch hat sie an's Stillstehen gehängt. Sie ist fest: ihr Tritt ist gemessen, ihre Gesetze unwandelbar.

Goethe on Nature, quoted by Haeckel
on title page of *Generelle Morphologie,* 1866

The law of recapitulation was "discovered" many times in the decade following 1859. Fritz Müller applied it in his masterful *Für Darwin* (1864), a short treatise on Darwinian explanations for crustacean morphology. He did not grant to recapitulation the universal status afforded it by his successors. Cope and Hyatt, the intellectual descendants of Agassiz in America, published their first works on recapitulation independently in 1866. In that same year, Haeckel's *Generelle Morphologie der Organismen* made its appearance; Huxley called it "one of the greatest scientific works ever published" (quoted in McCabe's footnotes to Haeckel, 1905).

Ernst Haeckel, son of a government lawyer, was born in Potsdam in 1834. He took a medical degree in 1858 and, after a short practice, moved to Jena to study zoology under the great anatomist Carl Gegenbaur. He became professor of zoology and comparative anatomy in Jena in 1862 and remained there until his death in 1919.

Haeckel published major treatises on three protist and invertebrate groups: Radiolaria (*Die Radiolarien,* 1862–1868), calcareous sponges (*Die Kalkschwämme,* 1872), and medusae (*Das System der Medusen,* 1879). But his dominating influence grew from two articles on the "gastraea theory" (1874 and 1875, though the idea was first promulgated in 1872, p. 467), and especially from three books: *Generelle Morphologie* (1866), *Natürliche Schöpfungsgeschichte* (1868), and *Anthropogenie* (1874). Haeckel conceived the books as popular works, but they contain a great amount of complex detail amidst speculation both bold and absurd. All deal heavily in phyletic reconstruction. His famous evolutionary trees first appear as plates in the second volume of *Generelle Morphologie*; Haeckel's trees have their roots (and most of their branches) in the principle of recapitulation—the "biogenetic law."* "Ontogeny is the short and rapid recapitulation of phylogeny . . .

* Haeckel was an inveterate coiner of terms; many words, common to scientists and laymen alike, were his invention: ecology, ontogeny, phylogeny. But most died with him, among them "biogeny"—the genesis of the history of organic evolution. Thus his phrase "biogenetic law" is often misunderstood, or at least not granted the force that Haeckel intended; for, under his definition, it is *the* law of the history of evolution.

During its own rapid development . . . an individual repeats the most important changes in form evolved by its ancestors during their long and slow paleontological development" (1866, 2: 300). Haeckel waxed ecstatic about its possibilities: "This is the thread of Ariadne; only with its aid can we find any intelligible course through this complicated labyrinth of forms" (1874, p. 9).

These works exerted immense influence. Haeckel was the chief apostle of evolution in Germany. Nordenskiöld (1929) argues that he was even more influential than Darwin in convincing the world of the truth of evolution. Yet influential as Haeckel was among scientists, his general impact was even greater. Nordenskiöld writes: "There are not many personalities who have so powerfully influenced the development of human culture—and that, too, in many different spheres—as Haeckel" (p. 505). From the 1880s onward, he focused increasing attention on the political, social, and religious implications of his biological views—a set of ideas that he amalgamated into his "monistic" philosophy. His major popular work, *Welträtsel* ("The Riddle of the Universe"; 1899), was among the most spectacular successes in the history of printing. It sold 100,000 copies in its first year, went through ten editions by 1919, was translated into twenty-five languages, and had sold almost half a million copies in Germany alone by 1933. One follower wrote that his name "will become a shining symbol that will glow for centuries. Generations will pass, new ones will arise, nations will fall, thrones will topple, but the wise old genius of Jena will outlast all" (quoted in Gasman, 1971, p. 16).

Haeckel's "monism" was viewed in many lights. His fulminations against religion and established privilege appealed to the left. His promise that science could release humanity from the shackles of ancient superstition endeared him to many "enlightened" liberals.

> On one side spiritual freedom and truth, reason and culture, evolution and progress stand under the bright banner of science; on the other side, under the black flag of hierarchy, stand spiritual slavery and falsehood, irrationality and barbarism, superstition and retrogression . . . Evolution is the heavy artillery in the struggle for truth. Whole ranks of dualistic sophistries fall together under the chain shot of this monistic artillery, and the proud and mighty structure of the Roman hierarchy, that powerful stronghold of infallible dogmatism, falls like a house of cards. (1874, pp. xiii–xiv)[4]

But, as Gasman argues, Haeckel's greatest influence was, ultimately, in another, tragic direction—national socialism. His evolutionary racism; his call to the German people for racial purity and unflinching devotion to a "just" state; his belief that harsh, inexorable laws of evolution ruled human civilization and nature alike, conferring upon favored races the right to dominate others; the irrational mysticism that had always stood in strange communion with his brave words about objective science—all contributed to the rise of Nazism. The Monist League that he had founded and led, though it included a wing of pacifists and leftists, made a comfortable transition to active support for Hitler.

Our narrow subject impinges upon these wider implications of Haeckel's beliefs, for Haeckel buttressed many of his political claims by references to recapitulation. He refutes the innate superiority of aristocrats, for example, by stating that all men are lowly creatures during their early development:

> Even in our day, in many civilized countries, the idea of hereditary grades of rank goes so far that, for example, the aristocracy imagine themselves to be of a nature totally different from that of ordinary citizens . . . What are these nobles to think . . . when they learn that all human embryos, those of nobles as well as commoners, are scarcely distinguishable from the tailed embryos of dogs and other mammals during the first two months of development. (1905, p. 337)

The Mechanism of Recapitulation

"Phylogenesis," Haeckel wrote in *Anthropogenie,* "is the mechanical cause of ontogenesis" ("Die Phylogenese ist die mechanische Ursache der Ontogenese"—1874, p. 5). "The connection between them is not of an external or superficial, but of a profound, intrinsic, and causal nature" (1874, p. 6); the two processes stand "in dem engsten mechanischen Causalnexus" (1866, 2: xix). These strong words, reflecting the aggressively mechanistic attitude of Haeckel's time, have often been ridiculed in our more cynical age.[5] Yet, although Haeckel was almost addicted to obfuscation by using fashionable words in meaningless contexts, it is important to recognize that when he said "phylogeny is the mechanical cause of ontogeny" he really meant it. The mechanism of recapitulation, as Haeckel envisaged it, provided just such a causal link.

The vitalistic forces of Naturphilosophie could be invoked no longer as the cause of recapitulation. Instead, Haeckel declared his allegiance with physiology in seeking the new path of mechanistic causation:

> Phylogenesis . . . is a physiological process, which, like all other physiological functions of organisms, is determined with absolute necessity by mechanical causes. These causes are motions of the atoms and molecules that comprise organic material . . . Phylogenesis is therefore neither the foreordained, purposeful result of an intelligent creator, nor the product of any sort of

unknown, mystical force of nature, but rather the simple and necessary operation of . . . physical-chemical processes. (1866, 2: 365)

This theme is invoked in all his popular works with an ardor and insistency that demands assent by sheer repetition, rather than by any increment of profundity. The "right" words—"mechanical," "physical-chemical laws," "absolutely necessary causal nexus"—abound in his commentary, and they are equated with all the common virtues of reason and rectitude. A sober and trusting scientist like Darwin was led to despair. After reading *Die Perigenesis der Plastidule* (1876), Haeckel's speculations on the mechanism of heredity, Darwin wrote to Romanes: "Perhaps I have misunderstood him, though I have skimmed the whole with some care . . . His views make nothing clearer to me, but this may be my fault. No one, I presume, would doubt about molecular movements of some kind." Romanes replied: "I do not see that biology gains anything by a theory which is really but little better than a restatement of the mystery of heredity in terms of the highest abstraction" (Romanes, 1896, pp. 51, 98). Haeckel's program for the reduction of biology to laws of physics and chemistry may have been crude and confused, but it did condition his search for laws that would display ontogeny as the necessary, mechanical result of phylogeny.[6]

Haeckel's reductionism not only included the familiar faith in basic laws of physics and chemistry, it also involved a sequence of strata within biology itself. Each new level is an aggregate of "individuals" in the next lower level. "Tectology," the science of organic composition, proclaims six ascending levels of aggregation: plastids (cells and other basic constituents), organs, antimeres ("homotypic" parts—halves and rays), metameres ("homodynamic" parts—segments), persons (individuals in our usual sense), and corms (colonies). The person has no special place within this hierarchy; it is composed of metameres, just as organs are compounded from plastids, or corms from persons (see Russell, 1916, p. 249, for a lucid analysis of these views). Just as corms are colonies of "bonded persons," so is every level but the first a true colony of its constituent parts; moreover, activities at any level can be explained by laws governing constituents, ultimately by the plastids themselves. Two of the "tectological theses" read:

22. Only the plastid (either cytodes or cells), as the morphological individual of the first and lowest order, is therefore a true, simple individual; all remaining morphological individuals (second to sixth order) are, rather, aggregated individuals or colonies.

29. All morphological and physiological unions of aggregated individuals (second to sixth order) are the necessary result [*Wirkung*] of the simpler individuals that compose it (plastids) and, to be sure, in the last instance of its active constituents (plasma and nucleus). (1866, 1: 361–368)

In analogy with this hierarchy of organic matter, Haeckel envisaged a hierarchy of developmental processes, including ontogeny and phylogeny. (Since we are conditioned to viewing ontogeny and phylogeny as distinct, we find it difficult to comprehend Haeckel's notion that they are but two steps in a continuum of developmental processes in nature.) As reducible levels in a hierarchy, ontogeny and phylogeny must be united under a single set of causes:

> Both ontogeny and phylogeny deal with the knowledge of a sequence of changes that the organism (in the first case, the individual, in the second case the stem or type) passes through during its developmental motions.[7] (p. 50)

> Phylogeny and ontogeny are, therefore, the two coordinated branches of morphology. Phylogeny is the developmental history [*Entwicklungsgeschichte*] of the abstract, genealogical individual; ontogeny, on the other hand, is the developmental history of the concrete, morphological individual.[8] (p. 60)

In short, Haeckel approached ontogeny and phylogeny with a predisposition towards their union and with a commitment to explain that union by mechanical, efficient causes. All he needed in addition was a defense for the two necessary premises of evolutionary recapitulation.

How did Haeckel defend the first premise, that evolutionary change occurs by the successive addition of stages to the end of an unaltered, ancestral ontogeny? Since Haeckel is so often cited as Darwin's apostle in Germany, it is generally assumed that he preached a Darwinian interpretation of evolution. In fact, he was only evolution's apostle. Though Haeckel acclaimed Darwin, he ranked Goethe and Lamarck as his equals in the origination of evolutionary theory (vol. 2 of *Generelle Morphologie* is dedicated to them jointly). Haeckel's own view of evolution is a curious and inseparable mixture of all three, each in about the same proportion.

To Lamarck, he owed his intense belief in the inheritance of acquired characters. He spoke of this principle as one "auf welcher die ganze Stammes-Entwicklung beruht" (1876, p. 47). Though Darwin accepted it as well, he preferred to explain the origin of most variations in other ways (Vorzimmer, 1970). To Haeckel, however, virtually every useful variation is actively acquired by parents during their life and passed on by heredity to their offspring (natural selection then accumulates and compounds these variations to produce new species). This Lamarckian principle "is an indispensable foundation of the theory of evolution" (1905, p. 863). "The origin of thousands of special arrangements remains perfectly unintelligible without this supposition" (1892, p. 221).

The heritability of acquired characters also explains why most evolutionary changes are additions to the end of an unaltered ancestral ontogeny. Haeckel, and most Lamarckians, did not base their belief on the voluminous folklore concerning inheritance of accidental mutilations; likewise, they rejected attempts to disprove the heritability of characters acquired by tail amputation, leg excision, and other dubious examples of vivisection. They insisted, rather, that an acquired character would tend to be inherited in proportion to the strength of the force imposing the character upon the organism, the persistence and continuity of that force, and the number of generations upon which the force acted.[9] Now, preadult stages of ontogeny are transient; they do not persist long enough to render transmissible whatever they acquire. But the adult stage, once reached, is permanent; it is therefore subject to the influence of strong and persistent forces that impose upon it (or call forth from it) acquired characters that can be inherited.[10] These acquired characters, the material of evolutionary change, appear as additions to the ancestral adult. "In the course of individual development, inherited characters appear, in general, earlier than adaptive ones, and the earlier a certain character appears in ontogeny, the further back must lie the time when it was acquired by its ancestor" (1866, 2: 298).*

Haeckel knew perfectly well that this principle of terminal addition had no absolute status; "laws" for the *results* of complex, evolutionary processes simply do not operate so inexorably.[11] In fact, his use of recapitulation was based (in theory at least) on a careful recognition and separation of exceptions.[12] Exceptions to recapitulation can arise in many ways, but the majority occur when larvae and juveniles acquire adaptations to their own environments. Haeckel acknowledged these exceptions in the *Generelle Morphologie* of 1866: "The true and complete repetition of phyletic development by biontic [ontogenetic] development is falsified and changed by secondary adaptation . . . thus, the more alike the conditions of existence under which the bion [individual] and its ancestors have developed, the more true will be the repetition" (2: 300). In later works, Haeckel expanded these views and finally (or, given his predilection for terminology, inevitably) bestowed a series of names upon them (Haeckel, 1875): characters

* Haeckel's definitions of heredity and adaptation do not follow modern usage. Heredity refers to characters that an animal receives from its parents, adaptation to those acquired during its lifetime. The adaptation of one generation may, of course, be the next generation's inheritance. Haeckel defines adaptation as "the fact that the organism . . . as a consequence of influences from the surrounding outer world, assumes certain new peculiarities in its vital activity, composition [*Mischung*], and form which it has not inherited from its parents" (1868, p. 173).

added by terminal modification and inherited in proper sequence are "palingenetic"; these alone reflect the true course of phylogeny. Characters added in juvenile stages or inherited out of proper sequence are "cenogenetic"; they falsify the history of lineages.

All of ontogeny falls into two main parts: first *palingenesis* or "epitomized history" [*Auszugsgeschichte*], and second, *cenogensis* or "falsified history" [*Fälschungsgeschichte*]. The first is the true ontogenetic epitome or short recapitulation of previous phyletic history; the second is exactly the opposite: a new, foreign ingredient, a falsification or concealment of the epitome of phylogeny. (1875, p. 409)

It is of the same importance to the student of evolution as the careful distinction between genuine and spurious texts in the works of an ancient writer, or the purging of the real text from interpolations and alterations, is for the student of philology . . . I regard it as the first condition for forming any just idea of the evolutionary process, and I believe that we must, in accordance with it, divide embryology into two sections—palingenesis, or the science of repetitive forms; and cenogenesis, or the science of supervening structures. (1905, p. 7)

Thus, Haeckel reformulated the biogenetic law in these terms:

The rapid and brief ontogeny is a condensed synopsis of the long and slow history of the stem (phylogeny): this synopsis is the more faithful and complete in proportion as palingenesis has been preserved by heredity and cenogenesis has not been introduced by adaptation. (1905, p. 415)

Haeckel also distinguished among the phenomena of cenogenesis. By far the most important were embryonic and juvenile adaptations. His favorite examples included the adaptations of free-swimming larvae to their own environments,[13] and the superficial differences in cleavage and gastrulation that arise from variations in yolk content and obscure the unity of early development. But Haeckel also established a second category of cenogenesis: temporal and spatial dislocations in the order of inherited events. These include: (1) "heterochrony"—displacement in time, or dislocation of the phylogenetic order of succession (in the ontogeny of vertebrates, for example, the notochord, brain, eyes, and heart arise earlier than their appearance in phylogeny would warrant); (2) "heterotopy"—displacement in place. Heterotopies arise when differentiating cells move from one germ layer to another in the course of phylogeny. Haeckel's favorite example involved the differentiation of reproductive organs from mesoderm in modern organisms, for these organs must have arisen historically in one of the two primary layers. On the evidence of ontogeny (the gastrula) and comparative anatomy (the coelenterates), Haeckel argued that the earliest Metazoa contained no mesoderm. Yet they must have possessed reproductive tissues and these could only have been generated from ectoderm or endoderm.

The second premise of evolutionary recapitulation is that the length of ancestral ontogenies must be continuously shortened during subsequent evolution of the lineage. Nature must make room for the new features added to the end of ontogeny. Recapitulationists, from Haeckel onwards, have offered a standard explanation for this condensation: it occurs as the result of a law of heredity; the law's causes are as unknown as its results are manifest.[14] But what law of heredity? Here the recapitulationists disagreed. Some spoke of a universal tendency towards acceleration of the developmental rate: descendants would pass through stages more quickly than their ancestors had (Fig. 7b). Others, Haeckel included, favored a law of "deletion"—certain stages would be excised, allowing the remaining ones to complete their appearance more rapidly (Fig. 7c).

Haeckel tied the condensation of ontogeny to three of his hereditary "laws." "The parallel between phyletic (paleontological) and biontic (individual) development is explained simply and mechanically by the laws of heredity, especially by the laws of homochronic, homotopic and shortened inheritance" (1866, 2: 372; see also p. 265, and 1868, pp. 166–167).

The laws of homotopic and homochronic inheritance[15] proclaim that an offspring will undergo the ancestral sequence of development in an unaltered spatial arrangement and temporal order. "With these laws, we explain the remarkable fact that the different successive stages of individual development always appear in the same order of succession [*Reihenfolge*], and that modifications [*Umbildungen*] of the body always develop in the same parts" (1868, p. 172). Once this is assured, condensation can occur simply by the deletion of certain steps. New adult features can now be added to the shortened ancestral ontogeny: "The chain of inherited characters, which follow each other in a determined sequence during individual development, . . . is shortened in the course of time, while certain links of the chain are deleted" (1866, 2: 186).

But there is a curious aspect to Haeckel's presentation of recapitulation: the argument has to be reconstructed from bits and pieces scattered throughout his work. The bits and pieces are explicit enough, and they are never contradicted in other passages. Yet, Haeckel never makes a complete and sustained argument for a mechanism of recapitulation. Although he inundates us with assurances that recapitulation has a "simple," "inevitable," and "mechanical" explanation, he seems singularly uninterested in it.

Haeckel's treatment of condensation creates a paradox that we can only resolve by recognizing that he was far more interested in tracing lineages than in establishing the mechanism of recapitulation. Haeckel often states that condensation, or shortened inheritance, is the most important cause of recapitulation. Nevertheless, he almost always unites condensation with cenogenesis, a process that confounds recapitulation by adding new stages in the midst of development. Cenogenesis and condensation are the two factors that make it most difficult to find the stages of phylogeny in ontogeny, that is, to demonstrate recapitulation. To be sure, condensation merely destroys good evidence, while cenogenesis actually falsifies phylogeny, but they both hinder the tracing of lineages. The last two "ontogenetic theses" of *Generelle Morphologie* read:

43. The true and complete repetition of phyletic development by biontic development is reduced and shortened by secondary condensation, since ontogeny strikes out on an ever straighter course. Thus, the longer the sequence of successive juvenile stages, the more true will be the repetition.
44. The true and complete repetition of phyletic development by biontic development is falsified and changed by secondary adaptation, since the bion [individual] adapts to new conditions during its individual development. Thus, the more alike the conditions of existence under which the bion and its ancestors have developed, the more true will be the repetition.[16] (1866, 2: 300)

Russell, with his usual insight, wrote: "From the point of view of the pure morphologist the recapitulation theory is an instrument of research enabling him to reconstruct probable lines of descent; from the standpoint of the student of development and heredity the fact of recapitulation is a difficult problem whose solution would perhaps give the key to an understanding of the real nature of heredity" (1916, pp. 312–313). Haeckel united condensation with cenogenesis because his primary interest lay in the tracing of lineages, and these were the two phenomena that impeded the recognition of phylogeny in ontogeny.[17] That they functioned so differently in the mechanism of recapitulation—condensation as the cause, cenogenesis as the rebuttal —interested him very little indeed.* For a man who spoke so rever-

* Russell's correlation applies throughout the history of ontogenetic studies: lineage tracers speak of condensation as a hindrance; searchers for the cause of evolution identify it as a mechanism of recapitulation and seek a wider relation between it and the operation of heredity. Thus, Cope and Hyatt, incessant students of evolution's cause, grant to condensation the primary interest that Haeckel denied. Perrier and Gravier, following Haeckel, explain condensation in a paragraph and use it to trace lineages for 200 pages: "Heredity, *because it is essentially tachygenetic* [their term for condensation], instead of preserving the long series of ancestral portraits in a state of purity, is an incessant cause for the alteration of these portraits" (1902, p. 348).

ently of mechanics and inviolable causes, Haeckel showed remarkably little concern for the way things worked. He was primarily a taxonomist of results—though not the mere arranger so often dismissed as a stamp collector, but a builder of vision who tried to render all the world's complexity in well-measured order.

Haeckel's main interests lay elsewhere, but his mechanism for recapitulation is clear nonetheless. New features are added to the end of ontogeny; condensation makes room for them by deleting earlier stages. Addition and deletion are phylogenetic processes; ontogeny is a sequence of stages under their direct control. Ontogeny has no independent status. Phylogeny, indeed, is the mechanical cause of ontogeny.

The American Neo-Lamarckians: The Law of Acceleration as Evolution's Motor

Progressive Evolution by Acceleration

We have seen, in Haeckel's case, how easily recapitulation fits with a belief in the heritability of acquired characters. Since this belief was the foundation of America's first major evolutionary school—that of the self-proclaimed "Neo-Lamarckists"—it is not surprising that the school's leaders, the paleontologists E. D. Cope and Alpheus Hyatt, exalted recapitulation to a higher status than it had enjoyed before or has achieved since.[18]

Edward Drinker Cope, though remembered more for the bombast of his feud with Marsh than for his substantial contributions to science, was America's first great evolutionary theoretician.[19] Cope published his evolutionary views in the *American Naturalist* and other journals during the 1870s and 1880s. He collected these essays in *The Origin of the Fittest* (1887) and reworked others to write *The Primary Factors of Organic Evolution* (1896).

Cope was interested more in the mechanics of evolution than in the tracing of lineages. He did not accept Darwin's emphasis on natural selection, for, although he saw how selection eliminated the unfit, he could grant it no role in the creation of the fit—hence the sardonic title of his 1887 work:

> The doctrines of "selection" and "survival" plainly do not reach the kernel of evolution, which is, as I have long since pointed out, the question of "the origin of the fittest." The omission of this problem from the discussion of evolution, is to leave Hamlet out of the play to which he has given the name. The law by which structures originate is one thing; those by which they are restricted, directed, or destroyed, is another thing. (1880, in 1887, p. 226)

Lamarck had made a primary distinction between two types of evolutionary events: progressive changes mediated by "the force that tends incessantly to complicate organization" and specific adaptations to definite environments (eyeless moles, long-necked giraffes—side branches on what would otherwise be a ladder to perfection, or at least to man). Cope makes an analogous separation.[20] New species represent the modification of existing structures; they produce the deflections or side-branches of evolution. New genera arise by addition to or subtraction from the sequence of ontogenetic changes; they alone are responsible for progressive evolution.[21]

Species and genera—horizontal branches and vertical steps on the tree of life—are not only distinguished by their physical position on a botanical metaphor; they are also produced by different causes. In early works, Cope argues that most new species (within a genus) may arise by the Darwinian process of fortuitous variation and natural selection (1870, in 1887, p. 144).[22] How, then, do new genera evolve? Cope argues that generic characters originate as additions to the end of ancestral ontogeny (although genera can also evolve retrogressively, by the loss of stages). They originate, moreover, as acquired characters in Lamarck's sense. Although the body's tissues (soma) are most easily modified during adolescence, the reproductive cells are most affected by constant repetition of an act during adulthood. Thus, new characters are impressed as additions to the adult stage: "Habits formed during adolescence are now practiced with special energy and frequency. The influence on the constantly renewed germ-plasma is correspondingly greater, and transmission is of course more certain" (1896, p. 447). The steps of progressive evolution—the generic changes—are stages added as acquired characters to the end of ontogeny: "Every change by complication of structure is by addition; every simplification is by subtraction" (1872, in 1887, p. 18).

But this principle of addition is not a complete evolutionary mechanism. If progressive evolution proceeds by addition, then descendant ontogenies will eventually become impossibly long (while, in retrogressive evolution by deletion, they will become disadvantageously short). Cope therefore provides a motor to reset the timing of ancestral ontogeny in order to permit the addition and subtraction of new stages. In progressive evolution, the speed of individual development is increased. The stages of ancestral ontogenies are repeated in successively shorter intervals, leaving time for the addition of newly acquired characters (Fig. 7b). This is the law of "acceleration"; it is responsible for all progressive evolution.

> The higher conditions have been produced by a crowding back of the earlier characters and an acceleration of growth, so that a given succession in order of advance has extended over a *longer range of growth* than its predecessor *in the same alotted time* . . . As all the more comprehensive groups present this relation to each other, we are compelled to believe that acceleration has been the principle of their successive evolution during the long ages of geologic time. (1870, in 1887, p. 142; my italics)

These principles of terminal addition and acceleration are the preconditions of recapitulation. In progressive evolution, the adult stages of ancestors are crowded back or "accelerated" into the juvenile stages of descendants. Recapitulation is the necessary result of progressive evolution.

In retrogressive evolution, on the other hand, individual development slows down. The later stages of ontogeny are not reached in the time alloted, and these are deleted. This is the law of "retardation."

> Retrogressive evolution may be accomplished by a retardation in the rate of growth of the taxonomic characters, so that instead of adding, and accumulating them, those already possessed are gradually dropped; the adults repeating in a reversed order the progressive series, and approaching more and more the primitive embryonic stages. This process I have termed "retardation." (1896, p. 201)

Retardation produces retrogressive evolution: "Acceleration implies constant addition to the parts of an animal, while retardation implies continual subtraction from its characters, or atrophy" (1876, in 1887, p. 126). "Retardation continued terminates in extinction" (1872, in 1887, p. 13).

Thus, in Cope's scheme, recapitulation is one result of the process propelling the more important of evolution's two modes: the production of new genera through movement up or down the main branch of a lineage. This movement proceeds by the law of acceleration and retardation, the speeding up or slowing down of development relative to age. As a consequence, the stages of ontogeny and phylogeny are related. This relationship—which Cope calls the "law of parallelism" —has two aspects: retrogressive evolution by subtraction of terminal stages and progressive evolution by recapitulation.

The law of acceleration and retardation plays a much more vital role in Cope's earlier beliefs than in his subsequent modifications. In later works, he attributes the acquisition of new characters to the activity of animals themselves—a favorite Lamarckian argument. "There are two alternative propositions expressive of the relations of the structures of animals to their uses. Either the use or attempt to use preceded the adaptive structure, or else the structure preceded and gave origin to the use . . . Many facts render the first of these propositions much the more probable of the two" (1878, in 1887, p. 352). Acceleration must still make room for a new character by pressing earlier ones back, but the primary impetus for its origin is the animal's own activity. In early works, however, Cope seems to have held that the sequence of characters added in progressive evolution is foreordained and out of the animal's control: "Genera have been produced by a system of retardation or acceleration in the development of individuals; the former on pre-established, the latter on preconceived lines of direction" (1869, in 1887, p. 123). In direct contradiction to his 1878 statement, quoted above, Cope had argued in 1870 that the introduction of a feature precedes its use:

> We look upon progress as the result of the expenditure of some force forearranged for that end. It may become, then, a question whether in characters of high grade the habit or use is not rather the result of the acquisition of the structure than the structure the result of the encouragement offered to its assumed beginnings by use, or by liberal nutrition derived from the increasingly superior advantages it offers. (1870, in 1887, pp. 145–146)

Acceleration, in this early reading, is the true motor of evolutionary progress. New features passively await their turn for expression; when acceleration has "made room" by crowding back the previous adult characters, these foreordained improvements make their automatic appearance.

The Extent of Parallelism

Cope granted his law of parallelism a much wider scope than Haeckel attributed to his own biogenetic law; for Cope took Haeckel's exception—cenogenesis—and tried to ignore it or render it as a variety of parallelism. Cope did consider the two most important aspects of cenogenesis: embryonic adaptation and heterochronism.[23] In later works, he simply admits the existence of embryonic and juvenile adaptation (1896, pp. 202–203), but in his early articles, he attempts to explain it away by an argument that seems sophistic even in its own context.* How can we say that a human embryo represents an ancestral fish; after all, it displays so many definite adaptations to the fetal state that it closely resembles, in toto, no fish living or extinct. Cope

* It is rendered more intelligible by a consideration of the long debate that taxonomists have endured and propagated about "key" characters in the definition of groups. Are orders defined, for example, by differing states of a designated "ordinal" character (rather than by some assessment of overall morphology). Cope, in his statement on

argues that mammals and fish are classes; therefore, we need to consider only the class characters separating vertebrate groups: Is the skeleton bone or cartilage? Is breathing by gills or lungs? In these characters, one stage of the human embryo is identical to the adult shark. Therefore, since key characters (and not total morphology) define a group, that human embryo *is* a fish:

When we reach species as far removed as man and a shark, which are separated by the extent of the series of vertebrated animals, we can only say that the infant man is identical in its numerous origins of the arteries from the heart, and in the cartilaginous skeletal tissue, with the class of sharks, and in but few other respects. But the importance of this consideration must be seen from the fact that it is on single characters of this kind that the divisions of the zoologist depend. Hence we can say truly that one order is *identical* with an incomplete stage of another order, though the species of the one may never at the present time bear the same relation in their entirety to the species of the other. (1872, in 1887, p. 8, my italics)

Cope then renders Haeckel's "heterochrony" as a variety of parallelism by redefining the argument. Haeckel thought in terms of the whole organism: the condensation of ontogeny proceeds equally for all characters and brings the total configuration of the ancestral adult into earlier and earlier stages. Cope applied his concepts to *individual organs* and recognized that they may be accelerated (or retarded) at different rates. The heart appears earlier in ontogeny than its origin in phylogeny would warrant—Haeckel's favorite example of "heterochrony." But this only indicates that the heart has been accelerated more intensely than other organs. All the organs are accelerated; all, considered individually, are examples of recapitulation: Haeckel's equal acceleration produces "exact parallelism"; Cope's unequal acceleration yields "inexact parallelism." By redefining the problem in terms of individual organs, Cope widely extended the range of recapitulation to include Haeckel's major exception to it.

The phenomena of exact parallelism or palingenesis are quite as necessarily accounted for on the principle of acceleration or retardation as are those of inexact parallelism or cenogenesis. Were all parts of the organism accelerated or retarded at a like rate, the relation of exact parallelism would never be disturbed; while the inexactitude of the parallelism will depend on the number of variations in the rate of growth of different organs of the individual. (1876, in 1887, p. 126)

---

generic characters (note 20), ranks himself among the defenders of the key-character concept. This idea is still reflected in the work of some brachiopod paleontologists who name new genera when they find differences in cardinalia, but only new species when they discover differences in surface ornament.

Why Does Recapitulation Dominate the History of Life?

For one committed, as Cope was, to the preeminent importance of recapitulation in evolution, this system leaves one point unanswered: If the law governing recapitulation is that of acceleration and retardation, why does recapitulation dominate? The law provides equally for recapitulation by acceleration and for its opposite, paedomorphosis by retardation. Moreover, it offers no reason for believing that cases of recapitulation should exceed those of paedomorphosis.

On one level, this dilemma has a simple resolution. We grant more emphasis to recapitulation because it is the necessary result of *progressive* evolution, and progressive evolution is more interesting and important, if only because it led to us. But this is not enough. Cope believed that cases of recapitulation far exceeded those of paedomorphosis in frequency as well as in importance. There must be some force impelling the general speed of development to accelerate through geologic time. In later works, Cope seems to favor an internal explanation, a type of energy that inheres in organic matter and accelerates its speed of development through time. In 1896 (p. 448), he speaks of "the phenomena of the building or growth of the added characters which constitute progressive evolution as evidence of the existence of a peculiar species of energy, which I termed bathmism."

In an earlier work (1870), however, he proposed an external explanation of great ingenuity. Acceleration may predominate because a directional change in atmospheric composition entails the speeding up of developmental rates. Development is tied to metabolism, metabolism to respiration and oxygen. The great coal deposits of Carboniferous times reflect the removal of vast amounts of carbon dioxide from the earth's atmosphere. This removal probably implies a rise in the level of oxygen. If oxygen has increased with time, so has respiration, metabolism, speed of development, and frequency of acceleration over retardation. Cope then cites the great thickness of fossil coals and mentions that the most luxurious vegetation today takes 50 tons of carbon from the atmosphere per century per acre, but produces from this a layer of coal only ⅓ inch thick:

> The atmosphere, thus deprived of a large proportion of carbonic acid, would in subsequent periods undoubtedly possess an improved capacity for the support of animal life. The successively higher degree of oxidation of the blood in the organs designed for that function, whether performing it in water or air, would certainly accelerate the performance of all the vital functions, and among others that of growth. Thus it may be that acceleration can be ac-

counted for, and the process of the development of the orders and sundry lesser groups of the Vertebrate kingdom indicated.[24] (1870, in 1887, p. 143)

Alpheus Hyatt and Universal Acceleration

Alpheus Hyatt learned the principle of recapitulation from his teacher Louis Agassiz; thereby, he continued an intellectual lineage extending back directly to Oken.

> I must have got directly from him, subsequently to 1858, the principles of this branch of research, and through this and the abundant materials furnished by the collections he had purchased and placed so freely at my disposal, I soon began to find that the correlations of the epembryonic stages and their use in studying the natural affinities of animals were practically an infinite field for work and discovery . . . The so-called Haeckelian "law of biogenesis" is really Agassiz's law of embryological recapitulation restated in the terms of evolution. (Hyatt, 1897, p. 216)

Alpheus Hyatt, Boston's celebrated invertebrate paleontologist, concentrated his work on cephalopods and wrote two major evolutionary treatises: *Genesis of the Arietidae* (1889) and "Phylogeny of an Acquired Characteristic" (1893). His publications ranged widely and included a monograph of the freshwater snails of Steinheim (1880); death interrupted his work on the famous Hawaiian tree snails.

His views on recapitulation run in remarkable parallel to those of Cope. They both developed the law of acceleration in 1866 (Cope, 1866, p. 398; Hyatt, 1866, p. 203). Both altered their concept of progressive evolution from a belief in foreordained stages to a conviction that animals acquire new characters by their own activity. Although they did not publish jointly, each lavished praise upon the other and happily shared credit for the major concepts of recapitulation.[25]

An epitome of the major argument in Hyatt's most famous treatise (1893) displays his manner of thinking and working. Nautiloids begin ontogeny with a straight shell. At a very small size, the shell begins to coil loosely; the whorls are not yet in contact. Finally, as the coil tightens, the whorls come into contact and remain in contact throughout growth (Fig. 8). A groove, running along the inside (dorsal) surface of each whorl, is called the impressed zone. In phylogeny, this zone arose mechanically from pressure exerted by contact of the inner surface with the outer keeled edge of the preceding whorl. This acquired character was then inherited and accelerated to earlier and earlier stages. Finally, the impressed zone appeared on the earliest, loosely coiled and uncoiled portions of the shell. It cannot have been imposed there by direct pressure since there is no contact

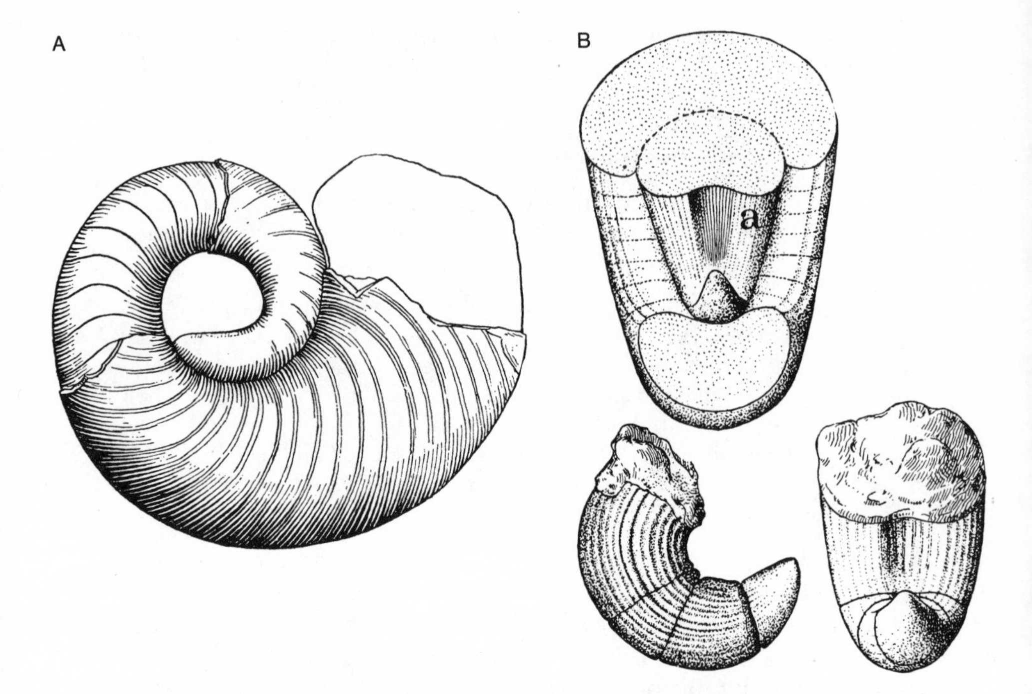

Fig. 8. (*A*) The earliest ontogeny of a nautiloid: the earliest portion is straight; coiling becomes progressively tighter until whorls come into contact with the previous whorl. (*B*) The impressed zone (*a*) as an accelerated character. It arose by pressure of contact with previous whorl later in ontogeny and was accelerated into this early stage where there is no contact with previous whorl (see lower left). (From Hyatt, 1893.)

with the previous whorl. It must have originated on the later whorls and been accelerated to the earliest stages of ontogeny.

In one respect, however, Hyatt's views differ markedly from those of Cope. Whereas Cope invoked acceleration and retardation as the agents of progressive and regressive evolution, Hyatt managed to render both progress and decline as the result of acceleration alone.[26] In his first definition of acceleration, Hyatt identifies its role in progressive evolution:

> There is an increasing concentration of the adult characteristics of lower species in the young of higher species, and a consequent displacement of other embryonic features which had themselves, also, previously belonged to the adult periods of still lower forms. This law . . . produces a steady upward advance of the complication. The adult differences of the individuals or

species being absorbed into the young of succeeding species, these last must necessarily add to them by growth greater differences, which in turn become embryonic and so on. (1866, p. 203)

But how can acceleration produce retrogressive changes? To explain this, Hyatt invoked the dominating idea of his career—his "old age theory," as he fondly called it. Hyatt believed that the sequence of new stages added to a lineage during the course of phylogeny runs parallel to the stages of an individual's ontogeny. Early in its history, a lineage advances by adding to its own ontogeny the youthful and then the mature stages of its total phylogeny (Fig. 9, stages 1–5). Later, as a lineage begins its phyletic decline, it adds old-age stages of the total phylogeny (Fig. 9, stages 6–13). These old-age stages, on the analogy of "second childhood," are similar to features of youth (though they signify decline and coming extinction, rather than youthful exuberance, since they now appear in an exhausted stock). The complete sequence of adult stages in phylogeny is a double staircase leading up to the platform of success and down the other side to senility and extinction. Phylogeny is programmed as ontogeny.[27]

When Cope detected juvenile features of ancestors in adult stages of descendants, he invoked retardation and admitted an effect opposite to recapitulation. To Hyatt, however, these seemingly "youthful" features of adult descendants are not youthful at all: they are the senile characters of second childhood, introduced at the end of previous ontogenies late in the phyletic life cycle (Fig. 9, stages 9–13). Moreover, they can be added only because ontogeny has been condensed by acceleration. The law of acceleration is universal; it regulates the sequence of stages in all lineages.

But one point seemed to argue for Cope's retardation and against Hyatt's acceleration of senile features that mimic youth: animals with "youthful" features in adult descendants have fewer total stages of ontogeny than their ancestors (Fig. 9, stages 10–13). Does this not support Cope's derivation by deletion of stages and contradict Hyatt's acceleration—for acceleration would seem to require an increasing number of stages? Hyatt replied: when condensation is so intense that senile features begin to appear in phylogeny, then acceleration, so to speak, has overrun its bounds. Stages are accelerated so rapidly that they begin to drop out entirely. But the earliest embryonic stages are the most stubbornly persistent of all. By acceleration, the newly introduced, senile features push back the older progressive traits until these encounter the persistent juvenile features. Pushed at one end, pressed, against an impenetrable wall at the other, the progressive features finally tumble off the treadmill. Old-age characters now

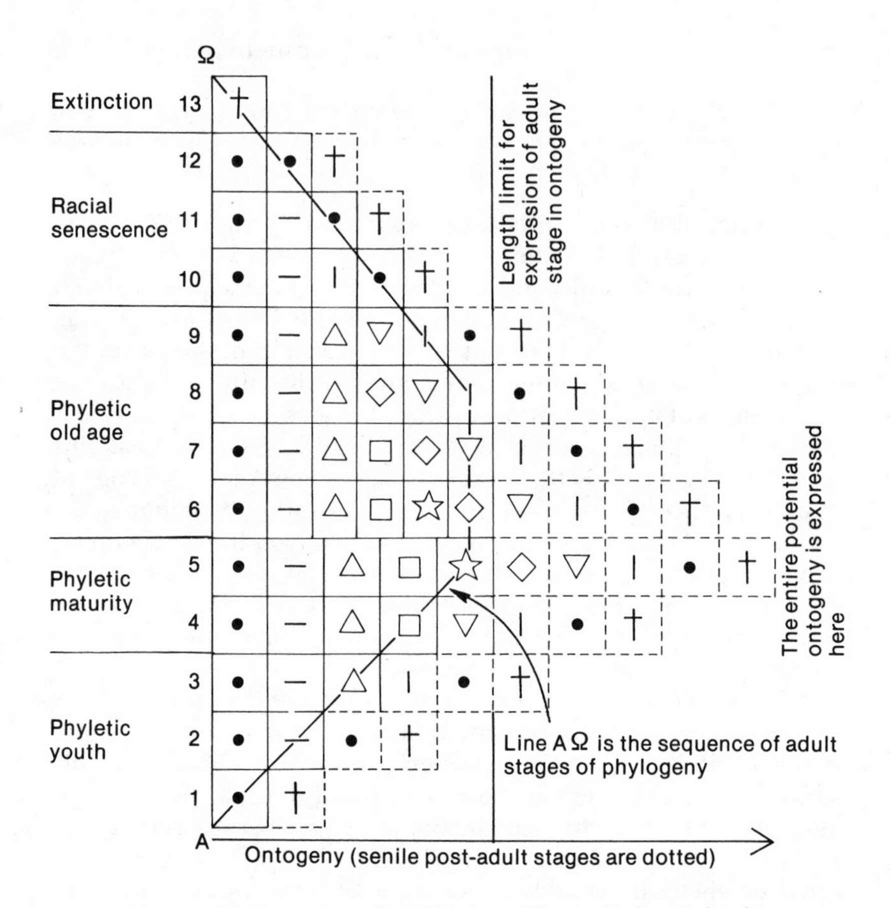

Fig. 9. Alpheus Hyatt's "Old-Age Theory." As extinction nears, the senile stages of phyletic youth and maturity become the adult stages of a waning stock. In racial senescence, ontogeny is so shortened by acceleration and deletion that senile stages merge with persistent juvenile stages to produce a greatly simplified ontogeny.

merge with juvenile features (Fig. 9). Ontogeny is both shortened (by the excision of intermediate stages) and simplified (because the remaining juvenile and old-age stages are so similar in external appearance).[28]

> Acceleration produces first, the earlier development of some of the progressive characteristics combined with geratologous characteristics; secondly, the earlier development of geratologous characteristics and their fusion with larval characteristics, which occasions the complete replacement of progressive characters, and occurs only in the extreme forms of retrogressive series, and in parasites. (1889, p. x)

In his first paper of 1866, Hyatt faced the problem of a Silurian nautiloid that maintained a smooth shell throughout life. Since the normal ontogenetic sequence moves from initial smoothness to final ornamentation, persistent smoothness could be viewed as "a retention of embryonic characters throughout life" (1866, p. 207). But this would contradict the universality of acceleration, and Hyatt seeks another interpretation. The normal ontogeny of a coiled nautiloid begins with a straight "orthoceras" stage, but the Silurian species was coiled from the outset. Now smoothness is not only a character of undeveloped youth; it is also a degenerative sign of old age. If this nautiloid lacked an orthoceras stage, its acceleration must have been remarkably intense—so intense that even the persistent juvenile stages were crowded out. Smoothness, in this animal, was an accelerated feature of old age.

To Cope, human evolution had been partly regressive because we retain certain embryonic features as adults (Chapter 5). Hyatt shared this unhappy view of our estate, but ascribed it instead to the acceleration of senile features:

> Perhaps the most remarkable instance of the loss of progressive characters correlating with a highly accelerated mode of development is man himself; and his example will serve a good purpose in making clear what we mean by a geratologous retrogression, which is often evidently due to a great change in habits, bringing about specialization in certain parts, enlarging and prematurely developing them at the expense of many of the normal progressive characters of the ancestral type. The Caucasian type, in losing the prognathism of the Anthropoids, which is certainly a highly specialized characteristic of the adult forms among the apes, has in a morphological sense made a step backwards instead of forwards.[29] (1889, pp. 45–46)

Hyatt was neither the first, nor the most strident of evolutionary pessimists, but he did add a new twist to the argument by branding humans as "the most remarkable of these phylogerontic types" (1897, p. 224). To the larger question of what produced this universal tendency towards acceleration, Hyatt had no answer, except to state that it must relate to the mystery of heredity: "The law of acceleration in development seems, therefore, to express an invariable mode of action of heredity" (1889, p. x).

If acceleration is universal, then its result, recapitulation, is also ubiquitous. "Cenogenesis" is not really an exception. A juvenile adaptation is merely a character introduced in the midst rather than at the end of ontogeny; it too will be accelerated backwards from its point of origin. Heterochronism, as Cope also argued, only indicates that organs are accelerated at different rates; but all are accelerated. Recapitulation is the mode of all evolution. In the Darwinian period, no one, not even in the brightest days of Haeckel's triumph, surpassed Hyatt in the exaltation of recapitulation.

Lamarckism and the Memory Analogy

Although Lamarckians[30] easily identified the processes that could produce recapitulation, their attempts to explain how these processes operated were surrounded by self-acknowledged frustration and defeat. They avoided the issue assiduously, and contented themselves with displaying the processes and making causal appeals to the mystery of heredity. This frustration merely reflected a more general problem of evolutionary biology before the early years of this century —ignorance of the mechanisms of inheritance. Among the numerous theories posed between Mendel's original work and its rediscovery, one was both attractive to Lamarckians and particularly conducive to an explanation of recapitulation: the analogy of memory and heredity. The general form of the argument was simple and acceptable to all adherents: the acquisition of a character is like learning; since characters so acquired are inherited in proportion to the intensity of their producing stimuli, inheritance is like memory (learning is retained through memory; memory is enhanced by constant repetition over long periods; actions invoked at first by conscious thought become automatic when repeated often enough). Instincts are the unconscious remembrance of things learned so strongly, impressed so indelibly into memory, that the germ cells themselves are affected and pass the trait to future generations. If behavior can be first learned and then inherited as instinct, then morphological features might be acquired and inherited in an analogous way. Thus, ontogeny is the sequential unfolding of characters in the order of their phyletic acquisition: it is the organism's memory of its past history. As Samuel Butler wrote:

> The small, structureless, impregnate ovum from which we have each one of us sprung, has a potential recollection of all that has happened to each one of its ancestors prior to the period at which any such ancestor has issued from the bodies of its progenitors—provided, that is to say, a sufficiently deep, or sufficiently often-repeated, impression has been made to admit of its being remembered at all.[31] (1877, p. 297)

But what is the physical ground of memory, indeed of all inheritance, if thoughts and things follow the same laws of transmission? What is impressed upon the germ cells to allow them to reproduce a sequence of acquired characters in the proper order? This question divided adherents to the general view: some spoke of vibrations and wave motions (Hering, 1870, in Butler, 1880; Haeckel, 1876), others of electrical potentials (Rignano, 1911), still others of chemical changes (Hartog, 1920; see Russell, 1916, pp. 335–344, and Hartog, 1920, for a review of the interesting, yet imprecise and often vacuous debate). The originator of the argument, the German physiologist Ewald Hering, advocated vibrations of some arcane sort.[32] The nervous system, as a united entity, pervades and interconnects the whole body. The vibrations of an external stimulus are transferred to the nervous system and hence to all other organs, especially to the developing gametes: "The organ of reproduction stands in closer and more important relation to the remaining parts, and especially to the nervous system, than do the other organs . . . both the perceived and unperceived events affecting the whole organism find a more marked response in the reproductive system than elsewhere" (Hering, 1870, in Butler, 1880, p. 77).

Haeckel, Cope, and Hyatt all accepted the general line of Hering's analogy between memory and inheritance. Haeckel's own theory of heredity—embodied in his curious work *The Perigenesis of Plastidules* (1876)[33]—is merely a restatement and formalization (replete with the usual array of new terms) of Hering's vibration theory. The special vibration of life, declares Haeckel in drawing analogies between cell division and speciation, is a branching wave movement (*eine verzweigte Wellenbewegung*) named "perigenesis." These motions govern all levels in his hierarchy of biological organization, but they reside ultimately in the basic building blocks, or "plastidules." "We name this true and ultimate efficient cause of the biogenetic process perigenesis, the periodic generation of waves [*Wellenzeugung*] by the atoms of life [*Lebenstheilchen*] or plastidules" (p. 65). Each plastidule inherits a sequence of these vibrations, carried as unconscious memory (*unbewusstes Gedächtniss*) from generation to generation. But plastidules are also affected by surrounding conditions that impose new wave motions upon them, usually as additions to the inherited sequence (Fig. 10). Ontogeny is the unfolding of these motions in the order of their acquisition in phylogeny.

Haeckel, Cope, and Hyatt all used the memory theory to attack Darwin's pangenesis as an explanation for Lamarckian inheritance. They did this by declaring a preference for the transmission of energy, rather than physical particles, from modified soma to the germ. Thus, Cope writes:

> It appears to me that we can more readily conceive of the transmission of a resultant form of energy of this kind to the germ-plasma than of material particles or gemmules . . . We may compare the building of the embryo to the unfolding of a record of memory, which is stored in the central nervous organism of the parent, and impressed in greater or less part on the germ-plasma

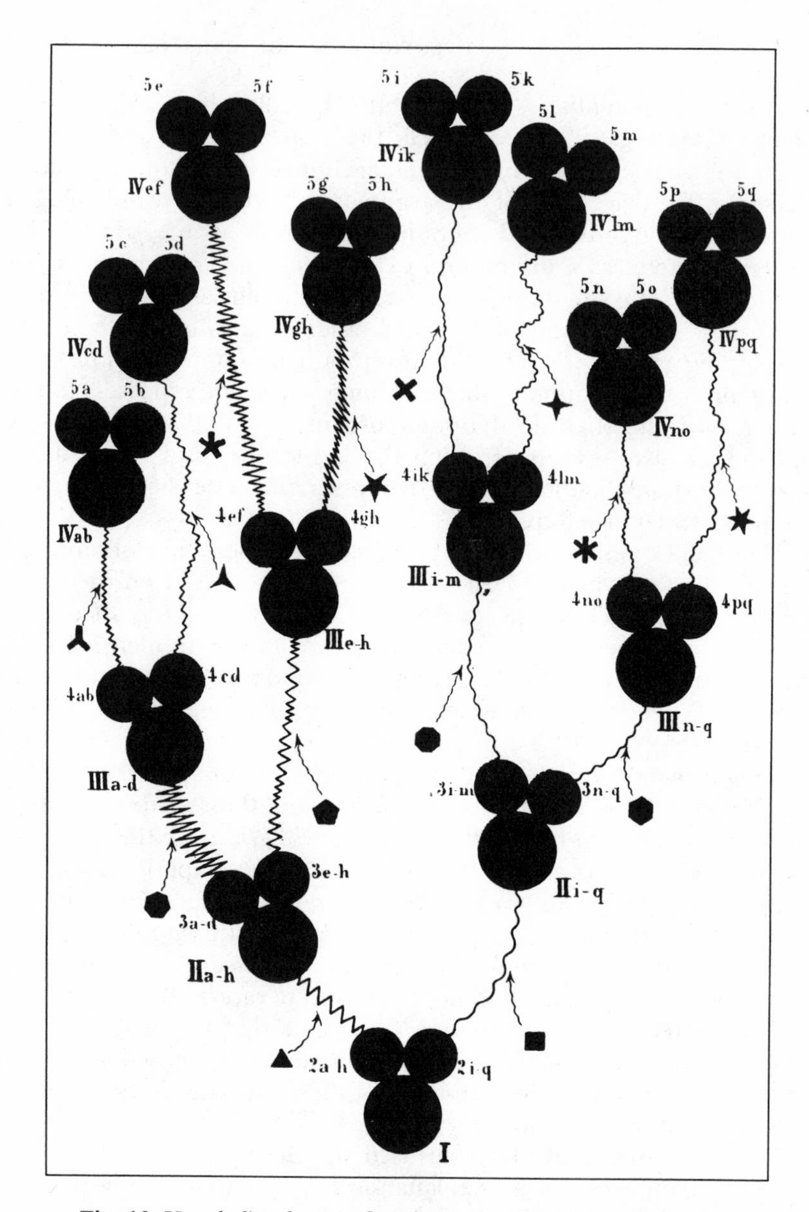

Fig. 10. Haeckel's scheme of perigenesis. Inherited differences in wave motions (due to previous acquisition in a Lamarckian mode) are portrayed by divergent patterns emanating from each of the pair of smaller spheres in each generation. New motions are imposed in each generation as different environments work their influence (the small, geometrical figures intersecting the wave patterns); these new motions are inherited in the next generation. (From Haeckel, 1876.)

during its construction in the order in which it was stored. (1896, p. 451, but an almost verbatim restatement of Cope, 1889)

While Hyatt declares:

The corpuscular theories, whether gemmules or biophors or pangenes are assumed, assert the need of minute bodies for the transmission of characters, while on the other hand the dynamic theories, more in accord with physical phenomena assume that there is a transmission of molecular energy through growth and some of these views support Hering's theory of what may be called mnemegenesis. Namely, that heredity is a form of unconscious organic memory and this from my point of view is the only satisfactory one yet brought forward. (1893, p.4)

The analogy to memory was particularly appealing to recapitulationists because it provided a ready explanation for their two required principles: terminal addition and condensation. If ontogeny is "a record of memory," then its course will not be deterred until that record is completely played. Anything new in evolution will have to be added to the end of ontogeny as an acquired character. "It [ontogeny controlled by memory] is incapable of a new design, except as an addition to its record" (Cope, 1896, p. 453). "Every organism," Hering wrote in his original formulation, "imparts to the germ that issues from it a small heritage of acquisitions which it has added during its own lifetime to the gross inheritance of its race" (1870, in Butler, 1880, p. 76).

As for condensation, the memory theory provided an easy explanation for either of its proposed mechanisms—Haeckel's deletion of stages or Cope's and Hyatt's acceleration. Hering, in fact, supported both. He linked deletion to forgetting and spoke of "all new germs transmitting the chief part of what had happened to their predecessors, while the remaining part lapsed out of their memory, circumstances not stimulating it to reproduce itself" (p. 80). Darwin's son Francis declared: "The blurred and imperfect character of the ontogenetic version of the phylogenetic series may at least remind us of the tendency to abbreviate by omission what we have learned by heart" (1908, p. 15).

The commoner explanation for condensation invoked the law of acceleration and relied on an analogy to learning and habit (usually portrayed by a simile to piano lessons)—the more you practice and repeat, the more quickly and automatically you perform. James Ward, professor of mental philosophy at Cambridge, said it succinctly in a running head: "the recapitulatory process briefer because it is all routine" (1913, pp. 18–19). And Hering argued: "How could all this be if every part of the central nerve system by means of which movement is effected, were not able to reproduce whole series of vibrations, which at an earlier date required the constant and continous participation of consciousness . . . if it were not able to reproduce them the more quickly and easily in proportion to the frequency of the repetitions" (1870, in Butler, 1880, p. 81). I doubt that a more charmingly simple explanation for the universality of acceleration has ever been offered: ontogeny is continually condensed because each generation practices it one more time. But it is a minute waltz without a lower limit; the correlation of evolutionary duration and speed of ontogenetic development is perfect, linear and inverse:

As a complicated perception may arise by means of a rapid and superficial reproduction of long and laboriously practiced brain processes, so a germ in the course of its development hurries through a series of phases, hinting at them only. Often and long foreshadowed in theories of varied characters, this conception has only now found correct expression from a naturalist of our own time.[34] For truth hides herself under many disguises from those who seek her, but in the end stands unveiled before the eyes of him whom she has chosen. (Hering, 1870, in Butler, 1880, p. 81)

Recapitulation and Darwinism

Although recapitulation quickly became the common property of all evolutionists, it achieved greatest popularity among Lamarckian thinkers. Its two necessary principles—terminal addition and condensation—received easy explanations within theories supporting the inheritance of acquired characters. Moreover, the two principles usually received a common explanation from Lamarckians—for both heritability and developmental rate of an acquired character increased as the producing stimulus became more intense and operated more frequently.

Most Darwinians, although they supported recapitulation as a guide to the tracing of lineages, had no such convenient explanation. They were constrained to identify natural selection as the efficient cause of structures added terminally, but why did terminal addition occur so much more frequently than interpolation within an ontogeny? The principle of condensation presented additional difficulties. One could reap the advantages of simplified causation by trying to link it with natural selection, but these arguments, advanced by F. Müller, Balfour, Neumayr, and Würtemberger, were weak and unpopular. Alternately, one could argue that structures are added by natural selection but shunted back by another process. But this version of condensation not only seemed to require an unknown principle of development—what Weismann (1881, p. 277) called "the innate law of growth which rules every organism"—but it also had to work outside the influence of natural selection (though Weismann argued that natural selection might check its action if it produced inadaptive forms). Yet the pervasive influence of natural selection was the primary ingredient of Darwinian beliefs.[35]

Fritz Müller, first to resurrect recapitulation in Darwin's light, offered a selectionist interpretation of condensation. He insisted that the shunting back of ancestral characters to earlier ontogenetic stages of descendants could not be "the result of an innate mystical drive" (*Folge eines inwohnenden mystischen Triebes*—1864, in 1915, p. 250). Intraspecific variation supplies a complete spectrum of developmental rates (he cites differences in times of tooth eruption among children of the same parents); natural selection can work upon this spectrum in any advantageous direction. Müller then argues that, in general, the advantageous direction will be a shortening of ancestral ontogeny "as development strikes out upon an ever straighter course from the egg to the mature animal" (p. 250). The general advantages of rapid development include earlier attainment of larger size for increased protection and sexual maturity for earlier reproduction.[36]

> In general it will be useful for an animal to express as early as possible those advantages by which it sustains itself in the struggle for existence. A precocious appearance of features first acquired at a later period will usually be advantageous, their retarded appearance disadvantageous. The former [the earlier appearance of features acquired late in ontogeny], when it occurs by chance, will be preserved by natural selection. (p. 250).

Ontogeny must be condensed by selection rather than by innate laws of inheritance because closely related forms living in very different habitats display widely varying rates of condensation—and each is related to the selective situation of its own habitat. (If innate inheritance were in control, degree of condensation would reflect genetic similarity.) If larvae live in the same environment and perform the same functions as adults, condensation will generally be advantageous and ancestral ontogenies will be strongly condensed. If juveniles inhabit different regions and play different roles than adults, condensation will be retarded and larval adaptations may be interpolated (as in planktonic, marine larvae of sedentary adults; in adapting for dispersion, these larvae may require a *longer* life and may evolve special adaptations for floating that falsify the record of ancestral ontogeny). Still, recapitulation predominates among all cases because condensation is usually advantageous. "The embryological record," argued F. M. Balfour, "is almost always abbreviated in accordance with the tendency of nature (to be explained on the principle of survival of the fittest) to attain her ends by the easiest means" (1880, pp. 3–4).

But the greatest of late nineteenth-century Darwinians, the strongest advocate of natural selection during the nadir of its general popularity, rejected natural selection as the cause of condensation. For August Weismann, in his brilliant studies of the color markings of caterpillars, found too many cases in which advantageous patterns of adult ancestors provided no benefit for descendants at the small, juvenile sizes to which they had been accelerated.

Weismann, a strong supporter of recapitulation, called it "the first important discovery which was made on the basis of the Darwinian Doctrine of Descent" (1904, p. 159). He applied his views in a series of detailed papers on the ontogeny and adaptive significance of color patterns in caterpillars, particularly of the Sphingidae (hawk moths). (These works were published in the 1870s and twice summarized, at some length in 1881 and much more briefly in 1904.)

Haeckel's gastraea was the hard salesman of recapitulation; it assaulted science and the public with the rhetoric of its implications and the sheer fascination of its potential existence. But the scientific debate on recapitulation did not center upon this imaginary animal, this inference based upon the precisely equal acceleration of all organs. The legions of sober, descriptive anatomists, embryologists, and paleontologists who supported recapitulation did not base their careers upon such a "mageres Thiergespenst."[37] Gastraea was a front; the real debate centered upon the hundreds of specific, documented cases involving the acceleration of individual characters, not the spectacular persistence of complete and remote ancestors in the early ontogeny of higher forms.[38]

Weismann's work epitomizes the discussion of recapitulation as it occurred among practicing scientists in the course of their normal, professional work, rather than the rhetoric used by popularizers as ammunition to advance the new science of evolutionary biology.*

Weismann's three conclusions on development reflect the importance of recapitulation in his thinking, for two of them correspond to the necessary principles of recapitulation—terminal addition and condensation:

* The history of recapitulation has been badly distorted because commentators do not read the detailed, primary literature. The arguments for Haeckel's hypothetical ancestors were very weak. If (as is customary) the entire subject is framed in their terms, it becomes very hard to understand why whole sciences fell under the sway of recapitulation. When we see how it helped Weismann unravel the phylogeny of caterpillars, its importance becomes clear. It was primarily a tool applied to individual organs for the tracing of specific lineages.

1. The development commences with a state of simplicity, and advances gradually to one of complexity.
2. New characters first make their appearance in the last stage of ontogeny.
3. Such characters then become gradually carried back to the earlier ontogenetic stages, thus displacing the older characters, until the latter disappear completely. (1881, p. 274)

Color patterns of sphingid caterpillars fall into three categories, each of adaptive significance to animals in particular habitats. Caterpillars with longitudinal stripes live on plants among grass or in grass itself (Fig. 11); their general body color corresponds to their background, but larger animals exhibit white, longitudinal lines "which, by mimicking the sharp light reflections of the grass stems, heighten the protective resemblance" (1904, p. 177). Caterpillars with transverse stripes live on trees and bushes; their color pattern imitates the lateral veining of leaves (Fig. 12). Finally, some caterpillars display spots of various sizes and in various positions. These spots have a variety of adaptive significances: they serve as warning colors in unpalatable species; in others, they produce a "terrifying effect" by imitating the eyes of larger animals; rarely, as when they mimic berries, they may enhance the caterpillar's resemblance to its surroundings.

All three patterns can protect the soft and easily-wounded animal in some way. Yet two of them, at least, would confer no value upon very young, very small caterpillars: "The transverse striping only makes the caterpillar look like a leaf when the stripes bear about the same relation to each other as those on the leaf [on tiny caterpillars, they would be too close together to imitate a leaf effectively], and eyespots can only scare away lizards and birds when they are of a certain size" (1904, p. 178).

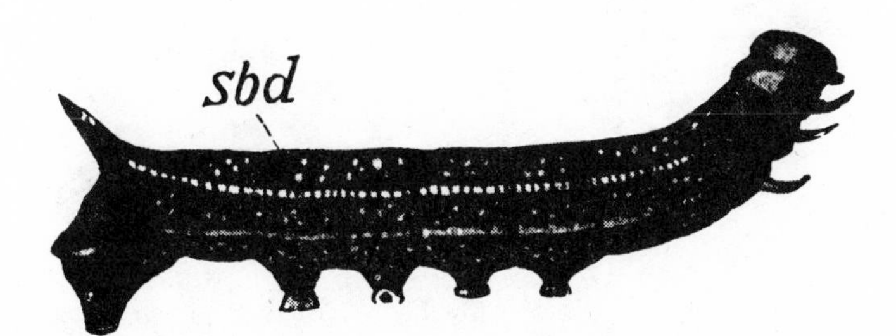

Fig. 11. A longitudinally striped caterpillar—the humming-bird hawk-moth *Macroglossa stellatarum*; *sbd* is the subdorsal line. (From Weismann, 1904.)

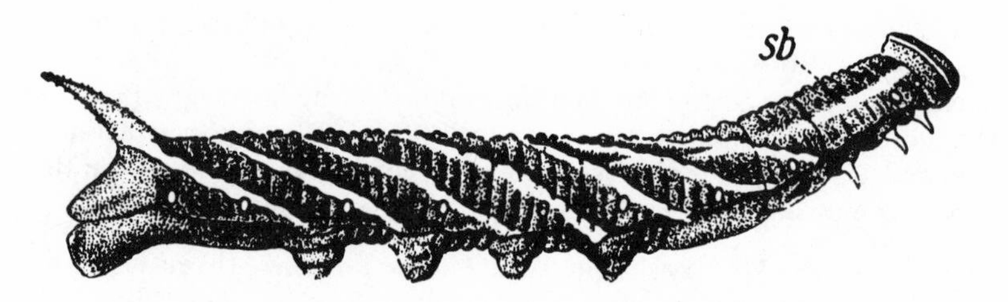

Fig. 12. A transverse-striped caterpillar—the eyed hawk-moth, *Smerinthus ocellatus*; *sb* is the subdorsal stripe. (From Weismann, 1904.)

The general ontogeny of coloration conforms to Weismann's first principle of increasing complexity through growth: tiny hatchling caterpillars generally lack any pattern (they are entirely green, perhaps in imitation of a single leaf vein); the complexity of patterning then increases during growth (Fig. 13).

Consider, for example, the "eye spots" of the elephant hawk-moth

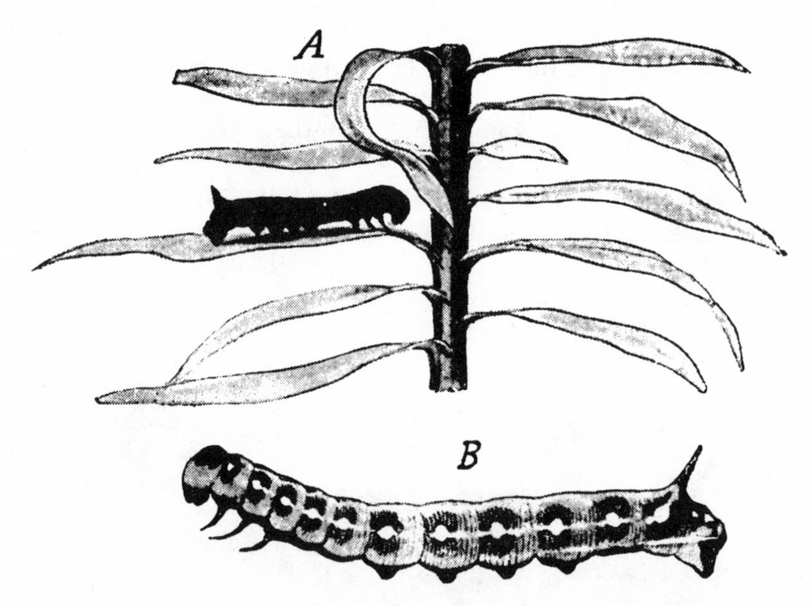

Fig. 13. Two stages in the life history of the spurge hawk-moth, *Deilephila euphorbiae.* (*A*) First stage—caterpillar is dark blackish-green and without marking. (*B*) Second stage—row of spots connected by vestige of subdorsal stripe. (From Weismann, 1904.)

Chaerocampa elpenor.[39] Patterning begins on the young caterpillar as a longitudinal marking, the subdorsal stripe (*sbd*—Fig. 14B). As growth proceeds, this stripe curves upward on the fourth and fifth segments (*Au*—Fig. 14C); a black line is then laid down at the lower edge of these curves (Fig. 14D). This line advances to the upper side. Finally, the black line completely encloses a segment of white stripe, producing a white-centered, black-framed "eye" (Fig. 14G). Eventually, the subdorsal stripe disappears from most of the rest of the body (Fig. 14F). Weismann argues that the subdorsal stripe, in itself, can have no selective value, for this caterpillar lives on large vine leaves or on the obliquely ribbed willow herb. The stripe must be the vestige of an ancestral adult that lived in grass (where longitudinal striping is advantageous). When eye spots developed as terminal additions in phylogeny, the longitudinal stripe was shunted back to the juvenile stage, in which it now appears. The caterpillar, now protected by its spots,

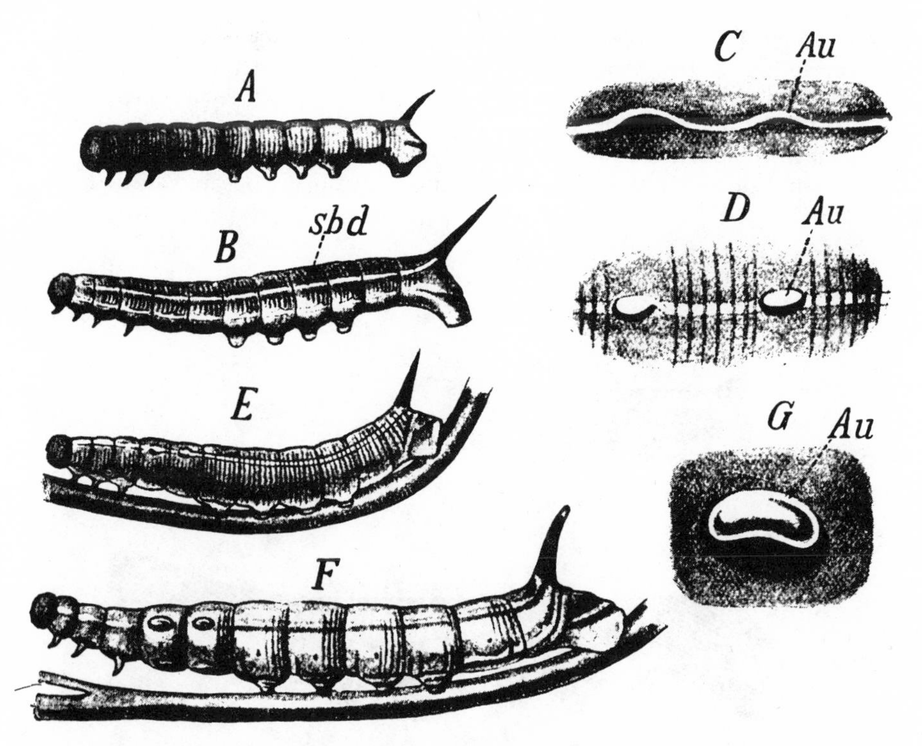

Fig. 14. The ontogenetic transformation of the subdorsal stripe into "eye spots" in the elephant hawk-moth, *Chaerocampa elpenor.* (From Weismann, 1904.)

changed its habitat to leaves. To buttress his argument that the subdorsal stripe is an accelerated stage of an ancestral adult, Weismann discusses its development in caterpillars bearing stronger or more numerous spots than *Chaerocampa elpenor.* Within the genus *Deilephila,* for example, the subdorsal stripe appears earlier and earlier as adult spotting becomes more and more intense (Figs. 15 and 16). In *Deilephila euphorbiae* (the most strongly spotted member of that genus), the subdorsal line never appears as a discrete element; signs of spot formation are present from the beginning of ontogeny (Fig. 13). Weismann draws three conclusions from his study of comparative spotting: (1) Spots always form from an initial longitudinal stripe. (2) In ontogeny, this stripe appears when the caterpillar is too small for it to be of selective significance; it is therefore the accelerated stage of an ancestral adult. (3) The greater the intensity of spotting, the more strongly the subdorsal line is accelerated (the greater the number of terminal additions, the further the shunting back of an ancestral pattern).

It can hardly be doubted that the biogenetic law is guiding us aright when we conclude . . . that the oldest ancestors . . . possessed only the longitudinal stripes, and that from these small pieces were cut off as ring-spots, and that these were gradually perfected and ultimately duplicated, while at the same time the original marking, the longitudinal stripe, was shunted back further and further in the young stages, until it finally disappeared altogether. (1904, p. 183)

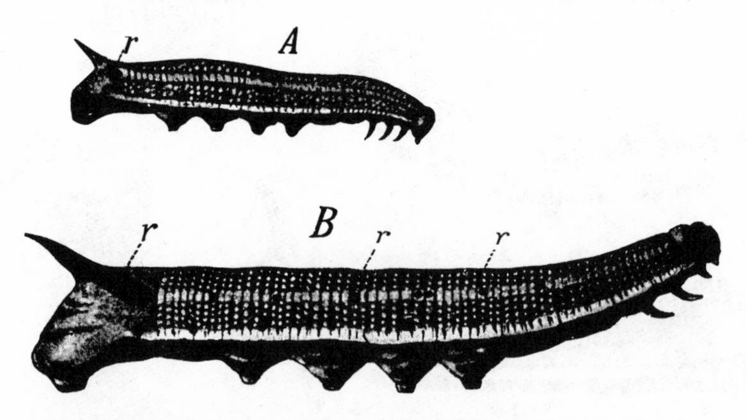

Fig. 15. Weakly spotted (*r*) caterpillar of the buck-thorn hawkmoth, *Deilephila hippophaes.* Subdorsal stripe remains strong. (From Weismann, 1904.)

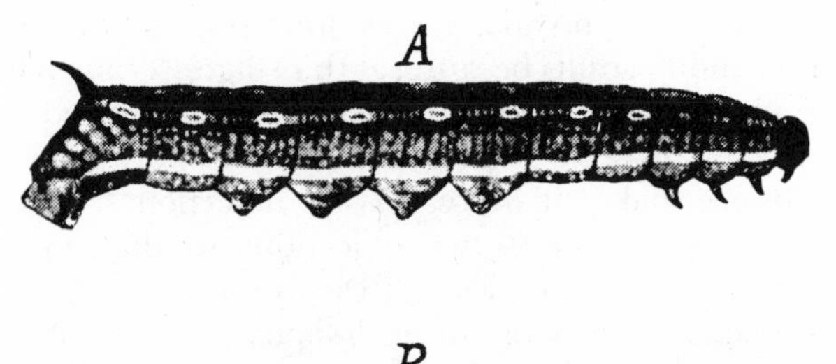

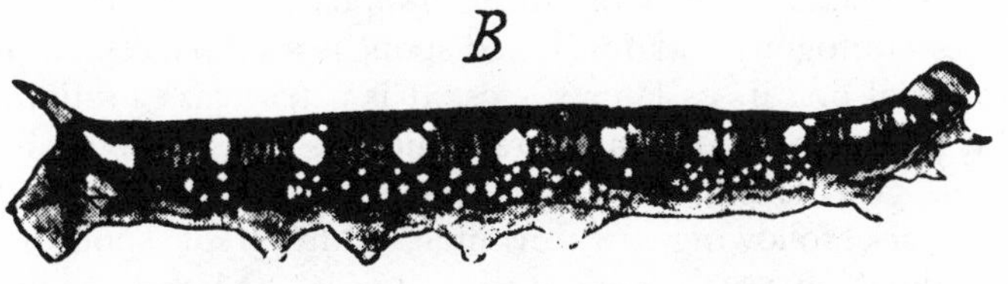

Fig. 16. More strongly spotted caterpillar of the bed-straw hawk-moth, *Deilephila galii*. (*A*) Subdorsal stripe still visible. (*B*) Fully formed caterpillar, subdorsal stripe gone, ten strong annular spots present. (From Weismann, 1904.)

Weismann then considers the ontogeny of transverse striping. In adults of *Smerinthus,* most of the body is covered with transverse stripes only, although the old subdorsal stripe persists on the anterior three segments (Fig. 12). The juvenile is born with no markings, but the subdorsal line soon appears; at the same time, all the transverse stripes arise simultaneously (cutting through the subdorsal line—Fig. 17). The subdorsal line then disappears from most of the body.

Weismann uses these examples to illustrate the principles of recapitulation: terminal addition and condensation. He is convinced that spots and transverse stripes arose in adults, because they are products

Fig. 17. First stage caterpillar of the poplar hawk-moth, *Smerinthus populi.* Subdorsal stripe and oblique stripes appear at the same time. (From Weismann, 1904.)

of selection and confer no advantages upon smaller sizes. Selection is more intense among adults because, at this stage, "a caterpillar is for a longer period exposed to the danger of being discovered by its foes; and since, at the same time, its enemies become more numerous, and its increased size makes it more easy of detection" (1881, p. 283).

Weismann then considers two objections to the application of condensation to these examples: (1) Do we observe a true shunting back of characters? Perhaps ancestors possessed a subdorsal line throughout ontogeny and developed spots by a simple replacement of the subdorsal line in its later stages. It is a true acceleration, Weismann argues, because the subdorsal line arose by selection, but only benefits large caterpillars. It must have arisen in adults and been shunted back following the terminal addition of spots. (2) But perhaps initial patterns appear at sizes too early to be advantageous in themselves because they are the necessary precursors to later adaptive developments. This explanation might apply to spotting. If spots must arise from the complex folding of an initial line, that line might have to appear at small sizes in order to provide enough time for its transformation. But it will not apply, Weismann argues, to transverse striping, for in *Smerinthus* the stripes are fully formed at their first appearance (Fig. 17), yet they can have no adaptive significance at this time, "for in the earliest stages of life the caterpillars are much too small to look like a leaf, and the oblique stripes stand much closer together than the lateral ribs of any leaf. Moreover, the little green caterpillars require no further protection when they sit on the underside of a leaf; they might then very easily be mistaken in toto for a leaf-rib" (1904, pp. 184–185). The first appearance of oblique stripes in *Smerinthus* represents the shunting back of a character that must have arisen as an advantageous trait in adults.

If the oblique stripes provide no advantage at the size of their first appearance, then Müller's argument linking condensation to natural selection must be wrong: "It is certainly not natural selection which effects the shunting back of the new characters" (1904, p. 185). Weismann is forced to a weak explanation involving an unknown and innate law of growth "acting independently of natural selection."

> An innate law of this kind, determining the backward transference of new characters, is deducible . . . from the fact that in many cases characters which are decidedly advantageous to the adult are transferred to younger stages, where they are at most of but indifferent value, and can certainly be of no direct advantage. This is the case with the oblique stripes of *Smerinthus*. (1881, pp. 277–278)

Weismann is stuck with a dilemma confronting almost all Darwinian recapitulationists: if condensation is nearly universal (as recapitulation requires if its occurrence be general), then condensation cannot be caused by natural selection because accelerated adult features often have no function in juvenile descendants. It must then be the result of some principle of inheritance. Not only does this principle work independently of natural selection (thereby compromising the general importance of Darwin's postulate), but it is also completely unknown:

> Newly acquired characters undergo, as a whole, backward transference, by which means they are to a certain extent displaced from the final ontogenetic stage by characters which appear later. This must be a purely mechanical process, depending on that innate law of growth, the action of which we may observe without being able to explain fully. (1881, p. 280)

Yet, through a glass darkly, Weismann glimpsed the direction of the coming solution. The laws of heredity must first be established: "If we could see the determinants, and recognize directly their arrangement in the germ-plasm and their importance in ontogeny, we should doubtless understand many of the phenomena of ontogeny and their relation to phylogeny which must otherwise remain a riddle" (1904, p. 189).

The determinants were soon elucidated as Mendelian genes; but Weismann's idea of universal recapitulation did not survive. He knew that no proper test could be made in the absence of a theory of inheritance; he could not know that the coming theory would invalidate his own conviction. We will return to the fate of recapitulation in Chapter 6, following a discussion in Chapter 5 of the remarkable influence that recapitulation exerted in fields as diverse as politics and primary education.

Appendix: The Evolutionary Translation of von Baer's Laws

In devoting most of this chapter to the mechanisms of recapitulation, I may have given the false impression that von Baer's alternative was nearly eclipsed or that Darwin was alone in his support of it. To be sure, von Baer's notion fared poorly among professional biologists in the late nineteenth century, only to be resurrected in our own. Yet it was adopted by two prominent evolutionists, neither in the main stream of professional science—Robert Chambers (author of the anomymous *Vestiges of the Natural History of Creation*) and Herbert Spencer. As the central theme of Spencer's cosmic defense of Victorian society, it became one of the most influential scientific ideas of the nineteenth century. Moreover, Ospovat (1974) has recently shown that von Baer's principle of increasing differentiation supplied the major ingredient for a widespread pre-Darwinian acceptance of the branching (rather than the linear) view of organic resemblances.

Robert Chambers developed his entire evolutionary theory as a metaphorical extension of von Baer's principle. Superficially, Chambers seems to be supporting the recapitulation of ancestral adults, for he writes of human ontogeny: "His first form is that which is permanent in the animalcule. His organization gradually passes through conditions generally resembling a fish, a reptile, a bird, and the lower mammalia, before it attains its specific maturity" (1844, p. 199). Yet Chambers clarifies his view with a specific rejection of recapitulation in favor of von Baer's embryonic repetition: "It has been seen that, in the reproduction of the higher animals, the new being passes through stages in which it is successively fish-like and reptile-like. But the resemblance is not to the adult fish or the adult reptile, but to the fish and reptile at a certain point in their foetal progress" (p. 212).

Chambers did not read von Baer in the original; he relied on the epitome in W. B. Carpenter's *Principles of General and Comparative Physiology*. From the 1841 edition of Carpenter's work, he copies (in evolutionary translation and without proper citation) the chart of von Baer's embryological theory (Fig. 18). Chambers used this chart to formulate a theory of evolutionary progress. He read the vertical line of fish-reptile-mammal-human as a foreordained path of evolutionary advance. But the fish, in its own ontogeny, moves vertically only to the point of its last identity with the (still prospective) higher reptile; thereafter, it diverges along its own lateral path. If, however, a fish were to extend its vertical ontogeny before striking out laterally, it might turn into a reptile all at once:

> It is apparent that the only thing required for an advance from one type to another in the generative process is that, for example, the fish embryo should not diverge at A, but go on to C before it diverges, in which case the progeny will be, not a fish, but a reptile. To protract the *straightforward part of the gestation over a small space*—and from species to species the space would be small indeed—is all that is necessary. (p. 213)

Ontogeny then becomes a metaphor for progressive evolution; the vertical path is a prospective cosmic ontogeny. Animals realizing only a part of it are arrested in their potential development. Organs may be arrested or advanced at their own rate. In what may well be the most striking bit of evolutionary nonsense ever propounded, Chambers imagines the derivation of mammals from birds in two easy steps through the intermediary of a duck-billed platypus:

> It is not great boldness to surmise that a super-adequacy [in whatever force propels organs along the vertical track] . . . would suffice in a goose to give

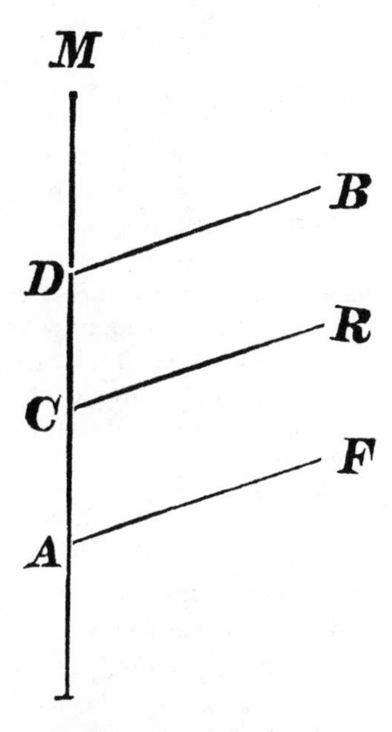

Fig. 18. The evolutionary translation of von Baer's laws. *F, R, B,* and *M* are, respectively, fish, reptile, bird, and mammal. *ACDM* is the ontogeny of a mammal, *AF* the ontogeny of a fish, *ACR* of a reptile and *ACDB* of a bird. (From Chambers, 1844.)

its progeny the body of a rat, and produce the ornithorynchus [*sic,* for the platypus, genus *Ornithorhynchus*], or might give the progeny of an ornithorynchus the mouth and feet of a true rodent, and thus complete at two stages the passage from the aves to the mammalia. (p. 219)

But why have animals been able, through time, to realize more and more of vertical ontogeny, thereby giving a progressive direction to evolution? Chambers invokes an early demonstration of neoteny—Edwards' experiment in which tadpoles, suspended in an opaque box in the Seine, reached a giant size without metamorphosis. If ontogenetic progress requires light (mammalian development in utero didn't seem to bother Chambers), then phylogeny must follow suit. External conditions determine how much of potential ontogeny can be realized at any time. Extensive forests of the Carboniferous period imply an atmosphere much richer in dense carbon dioxide. In addition, the residuum of matter that envelops the sun and causes zodiacal light must have been denser in the past. As oxygen increased and the air clarified, life advanced towards its current high estate (p. 229). Phylogeny progresses up the stages of a truly universal ontogeny. The cosmos is a mother and she regulates degrees of advance by setting physical constraints. All planetary bodies must follow the same course:

Thus, the production of new forms, as shewn in the pages of the geological record, has never been anything more than a new stage of progress in gestation, an event as simply natural . . . as the silent advance of an ordinary mother from one week to another of her pregnancy. Yet, be it remembered, the whole phenomena are, in another point of view, wonders of the highest kind, for in each of them we have to trace the effect of an Almighty Will which had arranged the whole in such harmony with external physical circumstances, that both were developed in parallel steps—and probably this development upon our planet is but a sample of what has taken place, through the same cause, in all the other countless theaters of being which are suspended in space. (pp. 222–223)

In his autobiography, Herbert Spencer chose "An Idle Year" as the title of his chapter for 1850–1851. Nonetheless, near the end, he records a minor item of special significance, "a piece, no less trivial . . . which I name not as in itself worth naming, but because it introduces an incident of moment" (1904, p. 445). The piece was a review of W. B. Carpenter's *Principles of Physiology* (1851 edition), the same work that had inspired Chambers. The incident of moment was Spencer's discovery of von Baer's principle and his recognition that it could serve as the central theme for a general theory of evolution, organic and inorganic.

In a section on the "law of progressive development," Carpenter had written:

It follows, therefore, that there is a greater variety of dissimilar parts in the higher organisms than in the lower; and hence the former may be said to be heterogeneous, whilst the latter are more homogeneous, approaching in some degree the characters of inorganic masses. This law is, therefore, thus concisely expressed by Von Bär, who first announced it in its present form. "A heterogeneous or special structure arises out of one more homgeneous or general; and this by a gradual change." (1839, p. 170)

The incident does not match Darwin's joy of discovery in 1838, but von Baer must still be counted as Spencer's Malthus. Spencer's formulation of general evolution—"a change from an indefinite, incoherent homogeneity, to a definite, coherent heterogeneity"—is no more than a paraphrase of von Baer's principle as Carpenter rendered it. In the *First Principles* of his synthetic philosophy (1881 edition, p. 337), Spencer records his debt to von Baer:

It was in 1852[40] that I became acquainted with von Baer's expression of this general principle. The universality of law had ever been with me a postulate, carrying with it a correlative belief, tacit if now avowed, in unity of method throughout Nature. This statement that every plant and animal, originally homogeneous, becomes gradually heterogeneous, set up a process of coordination among accumulated thoughts that were previously unorganized, or but partially organized. It is true that in *Social Statics,* written before meeting with von Baer's formula, the development of an individual organism and the development of the social organism, are described as alike consisting in advance from simplicity to complexity, and from independent like parts to mutually-dependent unlike parts—a parallelism implied by Milne-Edwards' doctrine of "the physiological division of labor."[41] But though admitting of extension to other super-organic phenomena, this statement was too special to admit of extension to inorganic phenomena. The great aid rendered by von Baer's formula arose from its higher generality; since, only when organic transformation had been expressed in the most general terms, was the way opened for seeing what they had in common with inorganic transformations.

Spencer did not accept von Baer's reading of embryology *faute de mieux.* He understood the debate between von Baer's law of progressive differentiation and the recapitulatory principle of terminal addition with acceleration of ancestral adult stages. In his *Principles of Biology* (1886, pp. 141–142, originally written in installments during 1863–1867), Spencer refers to von Baer's principle as "one of the most remarkable inductions of embryology." He adds: "The generalization here expressed and illustrated, must not be confounded with an erroneous semblance of it that has obtained considerable currency . . . that during its development, each higher organism passes through stages in which it resembles the adult forms of lower organisms" (p. 143).

Von Baer's principle did not convert Spencer to evolutionary thinking (he had been a transmutationist long before, and von Baer never was). Nor did it inspire his belief in a principle of universal progress, for Spencer had proclaimed this essential ingredient of his thinking before reading Carpenter. It did give him an epitome on which to base a systematization of all knowledge in his *Synthetic Philosophy*—one of the most influential intellectual achievements of the nineteenth century, despite Spencer's fall from grace in our own time. The differentiation of stars and planets from an incoherent gassy nebula; the complexity of function and division of labor in industrial society versus the uniform and repetitive practices of "primitive" agriculture—these represent the progressive differentiation of complexity from an initial poorly-bounded homogeneity, just as the chick develops from a uniform egg. It was von Baer who provided Spencer's warrant for a universal system embracing inorganic matter as well as life.

Young (1970) claims that Spencer generalized a law that von Baer had limited to the development of individual organisms. But the Naturphilosoph in von Baer had wanted nothing more than the establishment of his principle as the universal law of change. In fact, von Baer had used some of the same examples that Spencer would rediscover a half century later: "There is one fundamental thought that permeates all the forms and stages of animal development and governs all their relationships. It is the same thought that, in the cosmos, collects the separated masses into spheres and binds these together into a solar system" (von Baer, 1828, p. 263). Yet if Spencer completed von Baer's system, it was more a cruel joke than a fulfillment. Von Baer, the idealist, had envisioned a universal reign of thought; Spencer provided a cosmic apology for the materialistic spirit of Victorian capitalism—all in a thorough evolutionary framework, the first fruits of which von Baer lived to see and deplore (von Baer, 1876).

—5—

Pervasive Influence

Haeckel described the revolutionary power of Darwinian thought with a characteristic flourish:

> Dogma and authority, mutually dedicated to the suppression of all free thought and unfettered knowledge of nature, have erected a barrier of prejudice two or three times stronger than the Great Wall of China about the fortress of organic morphology—a citadel into which all kinds of distorted superstition have withdrawn as their last outpost. Nevertheless, we go into the battle without fear and sure of victory. (1866, 1:xv)

We may smile at the exaggeration or wince at a vigor rarely encountered in scientific treatises; still, we cannot deny that evolutionary theory was the most upsetting idea that later nineteenth-century science introduced to a world steadily retreating from traditional notions of static order. The influence of evolution in fields far removed from biology has been documented almost to exhaustion (at least to mine), but the impact of Haeckel's favorite weapon has not been widely noted by historians. Nonetheless, the theory of recapitulation played a fundamental role in a host of diverse disciplines; I suspect that its influence as an import from evolutionary theory into other fields was exceeded only by natural selection itself during the nineteenth century.

The historical chapters of this book deal almost exclusively with theories and debates about the mechanisms of relationships between ontogeny and phylogeny. I would need another volume even to begin an adequate treatment of how biologists used recapitulation in their daily work. This omission is unfortunate because we cannot judge a theory's impact merely by documenting lengthy debates about its mechanism. We grasp the importance of recapitulation only when we understand that it served as the organizing idea for generations of work in comparative embryology, physiology, and morphology. In pooh-poohing the biogenetic law, Conklin gave an excellent summary of its beguiling appeal: "Here was a method which promised to reveal more important secrets of the past than would the unearthing of all the buried monuments of antiquity—in fact nothing less than a complete genealogical tree of all the diversified forms of life which inhabit the earth. It promised to reveal not only the animal ancestry of man and the line of his descent but also the method of origin of his mental, social and ethical faculties" (1928, p. 70).

In my own field of paleontology, for example, it governed most studies in phyletic reconstruction from Haeckel's day right through the 1930s. At the turn of the century, the classification of almost every invertebrate phylum relied upon morphological criteria chosen for their ontogenetic value in constructing phylogeny from the biogenetic law (facial sutures of trilobites, suture patterns of ammonites, for example). As late as 1957, Jesse James Galloway wrote: "Ideally, a classification is built on the basis of comparative structure, and the application of the Law of Recapitulation, checked by the known geologic range of each taxonomic group" (p. 395). Faith in recapitulation was unbounded among paleontologists. In 1898, James Perrin Smith echoed a common, if extreme, claim:

> One can even prophesy concerning the occurrence of unknown genera in certain horizons when he finds their counterparts in youthful stages of later forms; in fact he could often furnish just as exact a description as if he had the adult genus before him. (p. 122)

With ample justification, a recent and popular textbook of invertebrate paleontology states: "It is no exaggeration to say that the theory of recapitulation has had more effect upon paleontologic thought than has any doctrine aside from that of organic evolution itself" (Easton, 1960, p. 33). My colleague Bernhard Kummel tells of an argument he had in the 1940s with R. C. Moore, the greatest "classical" paleontologist of our century. Kummel, as a bright young Turk, was expressing some gentle doubts about recapitulation over dinner one evening when Moore, his patience stretched to the limit, brought down his fist and exclaimed: "Bernie, do you deny the law of gravity!"

Though I have retreated shamefully before the task of documenting how biologists worked with the biogenetic law in their empirical studies, I cannot abandon the historical treatment of recapitulation without some discussion of its actual use. But I shall adopt a different tactic and explore the impact of recapitulation in fields far removed from biology. As a criterion for the importance of an idea, widespread and influential exportation to other disciplines must rank as highly as dominance within a field. With such an *embarras de richesses,* this chapter can be little more than a potpourri of citations. My files are bulging and my diligent scouts send me more material every week.

I shall present five essays on subjects strongly influenced by recapitulation: criminal anthropology, racism, child development, primary education, and psychoanalysis. Many other areas would have furnished equally impressive proof of influence. I have, for example, considered none of the arts, though recapitulation played a strong role in each branch. References abound in nineteenth-century literature. Tennyson, for example, wrote in the epithalamium of *In Memoriam* (1850) about a child who, in gestation, will be

> moved through life of lower phase
> Result in man, be born and think.

William Blake, in his *First Book of Urizen* (1794; cited in Oppenheimer, 1973), invoked a similar image while his scientific counterparts were advocating recapitulation as an essential ingredient of romantic biology:

> Many sorrows and dismal throes,
> Many forms of fish, bird and beast
> Brought forth an Infant form
> Where was a worm before.

Writers on the history and ethnology of art delighted in comparing the scribbles of "civilized" children with the finished products of ancient civilizations and modern "primitives" (Sully, 1895, 1896). John Wesley Powell, America's premier ethnologist, compared child development with human history in tracing the evolution of music "from dance to symphony" (1889).

Recapitulation intruded itself into every subject that offered even the remotest possibility of a connection between children of "higher" races and the persistent habits of adult "savages." One need only consult the 500-page compilation of A. F. Chamberlain's *The Child: A Study in the Evolution of Man* (1900)—an uncritical, but copious compendium of recapitulation. Consider the following:

The "counting out" rhymes used by children to begin games and make decisions: "a notable example of the survival in the usage of children of the serious practices of adults in primitive stages of culture . . . Children now select their leader or partner as once men selected victims for sacrifice" (p. 277).

A boy's practice of hunting and fishing with his hands, a reminiscence of prehistoric life before the evolution of tools: "The old proverb, 'a bird in the *hand* is worth two in the bush,' grew up quite naturally it would seem" (p. 279).

Name changing: "The child, the savage and the paranoiac love many names, like to change them, conceal them from strangers, etc." (p. 302).

Indiscriminate eating: "A prominent abdomen is a noticeable characteristic alike of children, women, and many primitive races; a 'pot-bellied' child and a 'pot-bellied' savage are common enough" (p. 315).

"Orophily," believe it or not: "the delight in being upon a mound, a height, a hill, and commanding the universe around or merging oneself into it" (p. 320).

Belief in the reality of dreams: "The child and the savage meet on this ground, some young boys and girls being as firmly impressed with the reality of their dreams as are the Brazilian Indians" (p. 336).

Love of adornment: "The child is one with the savage in picking up the pebble from the beach or the bright feather from the ground" (p. 453).

Finally, while still in the domain of the ridiculous, consider Havelock Ellis on posture: "The apes are but imperfect bipeds, with tendencies towards the quadrupedal attitude; the human infant is as imperfect a biped as the ape; savage races do not stand so erect as civilized races. Country people . . . tend to bend forward, and the aristocrat is more erect than the plebeian. In this respect women appear to be nearer to the infantile condition than men" (1894, p. 59).

Perhaps an even stronger testimony to the beguiling appeal of recapitulation can be found in its tenacious survival in casual references of modern humanists, more than half a century after scientists ditched it. "Ontogeny recapitulates phylogeny" is a literary epitome too appealing to resist, whatever its truth value.

Consider the following from a recent popular science-fiction novel (Edgar Rice Burroughs's *Out of Time's Abyss* should also be consulted):

> The brief span of an individual life is misleading. Each one of us is as old as the entire biological kingdom, and our bloodstreams are tributaries of the great sea of its total memory. The uterine odyssey of the growing foetus recapitulates the entire evolutionary past, and its central nervous system is a coded time-scale, each nexus of neurones and each spinal level marking a symbolic station, a unit of neuronic time. (Ballard, 1965, p. 43)

Or this from a work in literary criticism:

> The earliest age of mankind is associated with the verdure of springtime, with the spontaneity of childhood, and often with the awakening of love . . . Novelists, discovering for themselves the principle that ontogeny recapitulates phylogeny, have concentrated more and more intensively on the joys and pangs of adolescence. (Levin, 1969)

Or W. H. Auden writing on the death of Stravinsky:

> Stravinsky's life as a composer is as good a demonstration as any I know of the difference between a major and a minor artist . . . The minor artist, that is to say, once he has reached maturity and found himself, ceases to have a history. A major artist, on the other hand, is always re-finding himself, so that the history of his works recapitulates or mirrors the history of art. (1971, p. 9)

Or the Jungian poetry of Theodore Roethke (see pp. 161–163 on Jung's support of recapitulation):

> By snails, by leaps of frog, I came here, spirit.
> Tell me, body without skin, does a fish sweat?
> I can't crawl back through those veins,
> I ache for another choice.
>
> (from "Praise to the End!" 1951)

Or this, from a nuclear physicist:

> The Fermilab synchrotron employs a four-stage acceleration process. Protons, obtained by ionizing hydrogen in a discharge tube, receive their first push in the electric field produced by the 750,000 volts of a Cockcroft-Walton generator, a direct descendant of the first particle accelerator. Ontogeny recapitulates phylogeny for accelerators too! (Wilson, 1975, p. 20)

Or even Dr. Spock setting the child-rearing habits of a nation:

> Each child as he develops is retracing the whole history of mankind, physically and spiritually, step by step. A baby starts off in the womb as a single tiny cell, just the way the first living thing appeared in the ocean. Weeks later, as he lies in the amniotic fluid in the womb, he has gills like a fish. Towards the end of his first year of life, when he learns to clamber to his feet, he's celebrating that period millions of years ago when man's ancestors got up off all fours . . . The child in the years after six gives up part of his dependence on his parents. He makes it his business to find out how to fit into the world outside his family. He takes seriously the rules of the game. He is probably reliving that stage of human history when our wild ancestors found it was better not to roam the forest in independent family groups but to form larger communities. (1968, p. 229)

Criminal Anthropology

Evolutionary theory quickly became the primary weapon for many efforts in social change. Reformers argued that the social and legal systems of Western Europe had been founded on antiquated notions of natural reason or Christian morality; and did not face squarely the irrevocable biology of human nature. Proposals for change might shock traditional ethics, but if they brought social procedure into harmony with human biology, we might establish the beginning of a rational and scientific order freed from ancient superstition and therefore, in the long run, humane in the literal sense.

The late nineteenth-century school of "criminal anthropology" pursued this argument relentlessly. Previous systems of criminal law relied on social and ethical ideas of justice, fairness, protection, and retribution. They made no attempt to judge the biology of criminals—to learn if any recognizable peculiarity of their heredity might predispose them to lawlessness. Previous systems studied the crime; modern science would study the criminal. "Criminal anthropology," wrote Sergi, "studies the delinquent in his natural place—that is to say, in the field of biology and pathology" (in Zimmern, 1898, p. 744).

Criminal anthropology had its roots in Italy where Cesare Lombroso, its founding father, published the first edition of *L'uomo delinquente* ("Criminal Man") in 1876. It spread widely and became one of the most important scientific and social movements of the late nineteenth century.

Lombroso argued that many criminals were born with an almost irrevocable predisposition to lawlessness. These born criminals could be recognized by definite physical signs; they were indeed "a well characterized anthropological variety" (Ferri to Fourth International Conference on Criminal Anthropology, 1896, quoted in Parmelee, 1912, p. 80). At this point Lombroso's argument takes a phyletic turn. The stigmata of the born criminal are not anomalous marks of disease or hereditary disorder; they are the atavistic features of an evolutionary past. The born criminal pursues his destructive ways because he is, literally, a savage in our midst—and we can recognize him because he carries the morphological signs of an apish past. Lombroso records his moment of truth:

> In 1870 I was carrying on for several months researches in the prisons and asylums of Pavia upon cadavers and living persons, in order to determine upon substantial differences between the insane and criminals, without succeeding very well. Suddenly, the morning of a gloomy day in December, I found in the skull of a brigand a very long series of atavistic anomalies . . .

The problem of the nature and of the origin of the criminal seemed to me resolved; the characters of primitive men and of inferior animals must be reproduced in our times. (in Parmelee, 1912, p. 25)

Since ontogeny recapitulates phylogeny, a perfectly normal child must pass through a savage phase as well. The child, too, is a natural criminal at one stage of his development. The normal adult passes on to civilization as he mounts the phyletic scale in his own growth; the born criminal remains mired in his brutish past. This argument rings today with such patent absurdity (and viciousness) that I hasten to add a few disclaimers to explain why it attracted such a following during the late nineteenth century.

Lombroso and his school never attributed all criminal acts to innate depravity. In fact, their argument depended upon a strict separation between born criminals and those induced to commit their acts for social reasons such as poverty or extreme anger. Lombroso believed that no more than 40 percent of criminals bore the anthropometric stigmata of an inherent criminal disposition. His theories and recommendations applied to this group alone. Ferri (undated, p. 45) presented a graded classification, ranging from born criminals to criminals of passion via such intermediate categories as the criminal of contracted habit who transgresses by his own free choice but does it so assiduously that his children, through Lamarckian inheritance, become born criminals.

Although these distinctions might seem to mitigate the force of Lombroso's claims, they actually served to make the theory invincible to disproof. Lombroso did not tie specific criminal acts to criminal types—the types were defined by inferred biology, not by what they did. Murder might be the work of the most incorrigible born criminal or the most law-abiding cuckold. Hence, Lombroso's theory was invincible. Take the perpetrator of any criminal act: if he possesses the stigmata, he performed it by his biological nature; if not, by his social circumstance. All cases are covered. (We have seen this familiar strategy of "proof" in natural history before in this book—Haeckel's distinction between palingenesis and cenogenesis, Bolk's separation of primary and consecutive characters [see pp. 360–361]. You build a category to incorporate exceptions within your theory; all cases can then be allocated.)

Furthermore, the criminal anthropologists were, for the most part, neither petty sadists nor proto-fascists. The major figures in the Italian school were socialists who viewed their theory as the spearhead for a rational, scientific society based on human realities. They wished, for example, to reform criminal law by adapting punishment to the nature of criminals rather than to the severity of crimes. Social criminals should be jailed for the time needed to secure their amendment, but born criminals offered little hope for permanent cure. Lombroso and Ferri did not recommend the death penalty (though they did not oppose it—Ferri, p. 240; Lombroso, in letter to Zimmern quoted in Zimmern, 1893, pp. 600–601); human sensibility required some retreat from what might be biologically preferable. They tended to recommend irrevocable detention for life (in pleasant, but isolated surroundings) for any recidivist with the telltale stigmata (Lombroso, 1887, p. xvii). In addition, born criminals were not hopelessly doomed to a life of wrongdoing. Eyeglasses can cure inherited problems of vision, but they must be worn for life. If born criminals were identified early in childhood, they might be sent to bucolic retreats or shipped out to sea as cabin boys; they might be isolated and supervised in perpetuity, but they could not be cured: "Theoretical ethics passes over these diseased brains, as oil does over marble, without penetrating it" (Lombroso, 1895, p. 58).

The personal motivations of the Italian school may have been scientific, even humane, but their primary impact lay in another direction. And so it has always been with extreme versions of biological determinism—witness the conversion of some well-intentioned eugenics into a rationale for Hitler. Innate determination is a dangerous argument, for it is easily seized by men in power as a "scientific" excuse for preserving a status quo or eliminating any unfavored group as biologically perverse. Once Lombroso claimed a rationale for capital punishment in the operation of natural selection, he could not distance himself from Taine's implication: "You have shown us fierce and lubricious orang-utans with human faces. It is evident that as such they cannot act otherwise. If they ravish, steal, and kill, it is by virtue of their own nature and their past, but there is all the more reason for destroying them when it has been proved that they will always remain orang-utans." (Quoted, not unfavorably, in Lombroso, 1911, p. 428; in fact, Lombroso adds, p. 427: "The fact that there exist such beings as born criminals, organically fitted for evil, atavistic reproductions, not simply of savage men but even of the fiercest animals, far from making us more compassionate towards them, as has been maintained, steels us against all pity.") And once Ferri (p. 166) recommended that "tattooing, anthropometry, physiognomy . . . reflex activity, vaso-motor reactions and the range of sight" be used as criteria of judgment by magistrates, it was not a big step to Hitler's "final solution" for "undesirable" groups.

Lombroso, in the 1887 edition of *L'uomo delinquente,* presents his argument in a strict phyletic mode. Part 1, on the "embryology of crime," devotes its three chapters to demonstrating that what we call crime among civilized adults is normal behavior in animals (and even plants), adult savages, and children of civilized cultures—the threefold parallelism of classical recapitulation theory. Lombroso begins with a collection of animal anecdotes in the anthropomorphic tradition. Animals may kill from rage (an ant, made impatient by a recalcitrant aphid, killed and devoured it—p. 13); they murder to suppress revolts (some ant slaves, tired of being pushed around, seized the queen's leg and tried to pull her out of the nest; the queen seized the rebel by the head and killed her—p. 13); they do away with sexual rivals (an adulterous male stork and his lover found the rightful husband chasing frogs on a mud flat and killed him with their beaks—p. 16); they violate the young and undeveloped (male ants, deprived of sexual females, attempt to copulate with workers; but the workers, with their atrophied sexual organs, suffer greatly and often die); they form criminal associations (three communal beavers shared an area with one solitary individual; the three communals visited the solitary and were well-treated; the solitary returned the visit in a neighborly way and was killed for his solicitude—p. 18); even the voracious habits of insectivorous plants are considered as an "equivalent of crime," though I fail to see how this interspecific action differs from any other form of eating.

In part 2, on the "Pathological Anatomy and Anthropometry of Crime," Lombroso discusses the atavistic stigmata of born criminals—their physical links to a past represented today by living savages and our own children. The list of apish or primitive human features includes: relatively long arms, prehensile foot with mobile big toe, low and narrow forehead, large ears, thick skull, large and prognathous jaw, copious hair on the male chest, browner skin, and such physiological characters as diminished sensitivity to pain and absence of vascular reaction (criminals and savages do not blush). Atavisms do not stop at the primate level. Large canine teeth and a flat palate recall a distant mammalian past. The median occipital fossette of many criminals looks like that of rodents (and, by recapitulation, of three month old fetuses—1887, p. 181). Lombroso even compares the common facial asymmetry of criminals with the normal condition of flatfishes (1911, p. 373)! "The atavism of the criminal, when he lacks absolutely every trace of shame and pity, may go back far beyond the savage, even to the brutes themselves" (1911, p. 368).

But the stigmata are not only physical. The social behavior of the born criminal also allies him with savages and children. Lombroso and his school placed special emphasis on tattooing: "Tattooing is one of the essential characters of primitive man—one that still survives in the savage state" (1887, p. 284; see also Zimmern, 1898, p. 746). Lombroso analyzed the content of tattoos upon born criminals and found them lewd, lawless, or exculpating ("born under an unlucky star," "no luck," and "vengeance," for example, though one read, he had to admit, "long live France and french fried potatoes"—1887, p. 267). Criminal slang is a language of its own, markedly similar in such features as onomatopoeia and personification to the talk of children and savages: "Atavism contributes to it more than anything else. They speak differently because they feel differently; they speak like savages, because they are true savages in the midst of our brilliant European civilization" (1887, p. 467). Criminals are often as insensible to pain as savages (Ellis, 1910, p. 116, tells the tale of some Maoris who cut off their toes to wear European boots), and Lombroso notes that "Their physical insensibility well recalls that of savage peoples who can bear in rites of puberty, tortures that a white man could never endure. All travellers know the indifference of Negroes and American savages to pain: the former cut their hands and laugh in order to avoid work; the latter, tied to the torture post, gaily sing the praises of their tribe while they are slowly burnt" (1887, p. 319).[1]

In later years, Lombroso retreated steadily before criticism of his atavistic theory. While maintaining his strict adherence to the congenital nature of crime, he broadened both his criteria and his notion of cause. To the category of atavisms, he added the criterion of developmental arrest. Of course, under the biogenetic law, many arrests had the same phyletic significance as atavisms, since they brought the ancient phyletic states of embryology and childhood forward to the adult. But others could only be classed as anomalies and illnesses (Lombroso lays great stress on epilepsy). Stigmata are still to be found, but they need not have phyletic significance. We may therefore "see in the criminal the savage man and, at the same time, the sick man" (1887, p. 651).

The recapitulatory argument for natural criminality of children is one of the two or three central themes in Lombroso's fabric—not a mere collateral point. He devoted much of his work to cataloguing the criminal nature of child behavior in general and the statistics of legally criminal acts by children in particular. "One of the most important discoveries of my school is that in the child up to a certain age are manifested the saddest tendencies of the criminal man. The germs of delinquency and of criminality are found normally even in the first periods of human life" (1895, p. 53). Havelock Ellis proclaimed that "the child is naturally, by his organization, nearer to the animal, to the savage, to the criminal, than the adult" (1910, p. 211). An American follower added: "It is proved by voluminous evidence, easily accessible, that children are born criminals" (Morse, 1892, p. 438).

Lombroso cites the following traits as normal in children and criminal (or disposing towards it) in adults: anger, vengeance, jealousy, lying, lack of moral sense, lack of affection, cruelty, laziness, use of slang, vanity, alcoholism (if alcohol is made available),[2] predisposition to obscenity, imitation, and lack of foresight (1887, p. 99). Many members of his school made special studies to compare the expression of these traits in children, savages, and criminals. Crofton (1897), for example, considered the use of onomatopoeia in criminal slang, children's talk, and primitive language. The thief calls a train a "rattler," the child a "choo-choo."

But nature and education conspire to bring good from natural evil. The child mounts through his phyletic history and reaches the promise of his species as a young adult: "Now when the child becomes a youth, largely through the training of his parents and of the school, but more so by nature itself, when inclined to the good, all this criminality disappears, just as in the fully developed fetus the traces of the lower animals gradually disappear which are so conspicuous in the first months of the fetal life" (Lombroso, 1895, p. 56). In a nature not "inclined to the good," however, these traits persist and the arrested features of childhood produce a criminal adult. When Lombroso shifted his emphasis from atavisms to developmental arrests as criteria for the stigmata of criminality, he greatly increased the importance of a recapitulatory argument based upon children (as Havelock Ellis, 1910, emphasized so strongly). When the physical signs of an apish ancestry served as primary markers of the born criminal, children found no place in the argument (though they always played a major role in all discussions of behavioral signs from tattooing to slang)—humans are neotenous (Chapter 10), and our babies resemble adult ancestors in physical appearance even less than we do. In developmental arrest, however, children become the focus of inquiry. Crime is still a phyletic question, because children, by the biogenetic law, are close to ancestors.

The classical argument for recapitulation involves a threefold parallelism of paleontology, comparative anatomy, and ontogeny. Morphologists occasionally added a fourth source of evidence—teratology and the phyletic explanation of abnormalities as developmental arrests (Chapter 3). This fourth criterion—the abnormal individual as an arrested juvenile—forms an important part of the usage made by other disciplines of the biogenetic law. We have seen how Lombroso invoked it in his theory of criminality. We will encounter it again in Freud's theory of neurosis.

Racism

Vietnam reminds me of the development of a child.
Gen. William Westmoreland

Few connections are more intimate and pervasive than that between racism and statements by scientists about human diversity (I do not say scientific statements) made before the Second World War. Consequently, we should not be surprised that the very first sustained argument for recapitulation in morphology was cast in a racist mold. Autenrieth receives traditional credit for a work published in 1797 (Kielmeyer's article of 1793 spoke only of physiology, and earlier statements are either analogistic or incidental). After arguing that the completed forms of lower animals are merely stages in the ontogeny of higher forms, Autenrieth speaks of "certain traits which seem, in the adult African, to be less changed from the embryonic condition than in the adult European" ("quaedam, quae in adulto Afro minus, quam adulto Europaeo ex reliquiis embryonis mutata videntur"—1797, in Temkin, 1950).

For anyone who wishes to affirm the innate inequality of races, few biological arguments can have more appeal than recapitulation, with its insistence that children of higher races (invariably one's own) are passing through and beyond the permanent conditions of adults in lower races. If adults of lower races are like white children, then they may be treated as such—subdued, disciplined, and managed (or, in the paternalistic tradition, educated but equally subdued). The "primitive-as-child" argument stood second to none in the arsenal of racist arguments supplied by science to justify slavery and imperialism. (I do not think that most scientists who upheld the primitive-as-child argument consciously intended to promote racism. They merely expressed their allegiance to the prevailing views of white intellectuals and leaders of European society. Still, the arguments were used by politicians and I can find no evidence that any recapitulationist ever objected.)

Biological arguments based on innate inferiority spread rapidly after evolutionary theory permitted a literal equation of modern "lower" races with ancestral stages of higher forms. But similar arguments were far from unknown before 1859. Several of the leading pre-evolutionary recapitulationists ranked human races by the primitive-as-child argument. Schiller, a godfather of Naturphilosophie, wrote: "The discoveries which our European sailors have made in foreign seas . . . show us that different people are distributed around us . . . just as children of different ages may surround a grown-up man" (in Schmidt, 1909, p. 156).

Chambers, from an evolutionary perspective, wrote in his *Vestiges* of 1844 that "the varieties of his race are represented in the progressive development of an individual of the highest, before we see the adult Caucasian, the highest point yet attained in the animal scale." Agassiz endeared himself to the proponents of slavery when, as Europe's leading natural historian, he chose to settle in America and to maintain that blacks represent a separate and lower species. "The brain of the Negro," he claimed, "is that of the imperfect brain of a seven month's infant in the womb of the White" (in Stanton, 1960, p. 100).

Etienne Serres approached the subject from a more liberal perspective. Serres separated humans from other animals on the criterion of perfectibility. Animals are fixed in their created status; human races, however, can advance by evolution in both the physical and intellectual realms (1860, p. 771). The polygenist belief that human races are separate species relies on "dead anatomy which refutes the senses and disgusts the spirit by the aridity of its considerations . . . It replaces with multiple species the sublime idea of human unity—a tendency even more dangerous because it seems to lend scientific support to the enslavement of races less advanced in civilization than the Caucasian" (p. ii).[3] Nonetheless, however flexible in future movement, the scale of human races could still be ranked from lower to higher—and recapitulation provided the major criterion for ranking. Serres wrote: "May we not say in a general manner that, in its progress, the Ethiopian race has stopped at the beginning of Caucasian adolescence . . . and the Mongolian race at [Caucasian] adolescence itself—times of arrest which form degrees of development within the unity of the human species" (p. 765). Serres was hard pressed to find many morphological signs of a parallel between Caucasian ontogeny and the adult sequence Negro-Mongolian-Caucasian. Difficult though it be to take seriously today, Serres emphasized the relative position of the belly button (pp. 763–765). The relative distance between penis and navel increases during Caucasian ontogeny. Adult blacks have a navel as low slung as that of a Caucasian child; the adult Mongolian navel is a bit higher.

Biological arguments for racism may have been common before 1859, but they increased by orders of magnitude following the acceptance of evolutionary theory. The litany is familiar: cold, dispassionate, objective, modern science shows us that races can be ranked on a scale of superiority. If this offends Christian morality or a sentimental belief in human unity, so be it; science must be free to proclaim unpleasant truths. But the data were worthless. We never have had, and still do not have, any unambiguous data on the innate mental capacities of different human groups—a meaningless notion anyway since environments cannot be standardized. If the chorus of racist arguments did not follow a constraint of data, it must have reflected social prejudice pure and simple—anything from an a priori belief in universal progress among apolitical but chauvinistic scientists to an explicit desire to construct a rationale for imperialism.

Some recapitulationists ranked races by physical traits. Cope searched the human body to find a mere handful of characters that would affirm white superiority on the primitive-as-child argument:

Let it be particularly observed that two of the most prominent characters of the negro [*sic*] are those of immature stages of the Indo European race in its characteristic types. The deficient calf is the character of infants at a very early stage; but, what is more important, the flattened bridge of the nose and shortened nasal cartilages are universally immature conditions of the same parts in the Indo-European. (1883, in 1887, p. 146)

D. G. Brinton made a more comprehensive claim:

The adult who retains the more numerous fetal, infantile or simian traits, is unquestionably inferior to him whose development has progressed beyond them . . . Measured by these criteria, the European or white race stands at the head of the list, the African or negro at its foot . . . All parts of the body have been minutely scanned, measured and weighed, in order to erect a science of the comparative anatomy of the races. (1890, p. 48)

But arguments about the brain provided more direct ammunition for ranking by evolutionary status. Vogt wrote:

In the brain of the Negro the central gyri are like those in a foetus of seven months, the secondary are still less marked. By its rounded apex and less developed posterior lobe the Negro brain resembles that of our children, and by the protuberance of the parietal lobe, that of our females. (1864, p. 183)

If the primitive-as-child argument worked for the material ground of intelligence, it would perform even better for the mental traits themselves. Most racist recapitulationists relied more on the products of mind than on physical criteria for ranking. Cope argued that "some of these features have a purely physical significance, but the majority of them are . . . intimately connected with the development of the mind" (1883, in 1887, p. 293). "The intellectual traits of the uncivilized," claimed Herbert Spencer, "are traits recurring in the children of the civilized" (1895, p. 89). Lord Avebury compared "modern savage mentality to that of a child" (1870, p. 4), while the English leader of child study stated: "As we all know, the lowest races of mankind stand in close proximity to the animal world. The same is true for the infants of civilized races" (Sully, 1895, p. 5). And the

American leader of child study, G. Stanley Hall, maintained that "most savages in most respects are children, or, because of sexual maturity, more properly, adolescents of adult size" (1904, 2: 649).

My catalogue of specific examples is far too long to relate, but a sampling would include:

1. On the animism of children:

> Is it too fanciful to suppose that the belief of the savage in the occasional visits of the real spirit god to his idol has for its psychological motive the impulse which prompts the child ever and again to identify his toys and even his pictures with the realities which they represent. (Sully, 1895, p. 392)

> The child who says, "I am an engine" or "I am a tiger," making appropriate movements and sounds the while, is enjoining an imaginative identification of a genuinely primitive character. Is it not a justifiable distinction to say that the child is playing, while the savage is intensely serious at his rites. (Murphy, 1927, p. 109)

2. On aesthetic sensibility:

> In much of this first crude utterance of the aesthetic sense of the child we have points of contact with the first manifestations of taste in the race. Delight in bright, glistening things, in gay things, in strong contrasts of color, as well as in certain forms of movement, as that of feathers—the favorite personal adornment—this is known to be characteristic of the savage and gives to his taste in the eyes of civilized man the look of childishness. On the other hand, it is doubtful whether the savage attains to the sentiment of the child for the beauty of flowers. (Sully, 1895, p. 386)

3. On art:

> If we look at representation by drawing or sculpture, we find that the efforts of the earliest races of which we have any knowledge were quite similar to those which the untaught hand of infancy traces on its slate or the savage depicts on the rocky faces of hills.[4] (Cope, 1870, in 1887, p. 153)

The recapitulatory argument for ranking extended beyond races to any set of categories for which wealthy, Nordic males wished to assert their superiority. Lower classes within any society were a favorite target. Cope, for example, listed several simian characters among "the lower classes of the Irish" (1883, in 1887, p. 291).

Women fitted the argument especially well for two reasons—the social observation that men wrote all the textbooks and the morphological fact that skulls of adult women are more childlike than those of men. Since the child is a living primitive, the adult woman must be as well. In 1821, Meckel noted the lesser differentiation of women from a common (and primitive) embryonic type; he also suspected that women, with their smaller brains, were innately inferior in intelligence (1821, pp. 416–417). Again, recapitulationists quickly moved from morphology to mental traits. Cope, for example, discoursed on the "metaphysical characteristics" of women:

> The gentler sex is characterized by a great impressibility . . . , warmth of emotion, submission to its influence rather than that of logic, timidity and irregularity of action in the outer world. All these qualities belong to the male sex, as a general rule, at some period of life, though different individuals lose them at very various periods . . . perhaps all men can recall a period of youth when they were hero-worshippers—when they felt the need of a stronger arm, and loved to look up to the powerful friend who could sympathize with and aid them. This is the "woman stage" of character. (1870, in 1887, p. 159)

G. Stanley Hall argued that women's greater propensity for suicide expresses the primitive stage of submission to elemental forces.

> This is one expression of a profound psychic difference between the sexes. Woman's body and soul is phyletically older and more primitive, while man is more modern, variable, and less conservative. Women are always inclined to preserve old customs and ways of thinking. Women prefer passive methods; to give themselves up to the power of elemental forces, as gravity, when they throw themselves from heights or take poison, in which methods of suicide they surpass man. [Havelock] Ellis thinks drowning is becoming more frequent, and that therein women are becoming more womanly. (1904, 2: 194)

At this point, I hasten to add that I am not selecting the crackpot statements of a bygone age. I am quoting the major works of recognized leaders. The sway of biological determinism, the lack of sensitivity to environmental influence, and the blatant desire to crown one's own group as biologically superior are quite characteristic of the time—and scarcely extinct today.

But can we prove that these assertions emerged from the scientific literature to influence social and political life? They surely did, as two lines of evidence attest. First of all, many scientists drew explicit political conclusions in their widely read books. Consider Vogt's justification for colonialism:

> The grown-up Negro partakes, as regards his intellectual faculties, of the nature of the child, the female, and the senile white . . . Some tribes have founded states, possessing a peculiar organization; but, as to the rest, we may boldly assert that the whole race has, neither in the past nor in the present, performed anything tending to the progress of humanity or worthy of preservation. (1864, p. 192)

Tarde added that many criminals (equivalent to primitives in Lombroso's theory) "would have been the ornament and moral aristocracy of a tribe of Red Indians" (in Ellis, 1910, p. 254). The occasional anti-imperialists agreed completely with the primitive-as-child argument and only disputed the implied right of conquest:

> We are laboring to prevent the "big fist" of adults from breaking in upon childhood and its evolutional activities; we ought also to labor to prevent the bigger fist of "civilized races" from breaking in upon the like evolutional activities of primitive peoples with even more disastrous results . . . We ought to be as fair to the "naughty race" abroad as we are to the "naughty boy" at home. (Chamberlain, undated, p. 1)

Second, many politicians and statesmen borrowed the primitive-as-child argument with an explicit bow to science and recapitulation. The Rev. Josiah Strong, in his plea for an imperial America, noted that modern science had provided the rationale justifying colonialism as an ethical venture. In pre-evolutionary days, Henry Clay had voiced religious doubts about the concept of racial superiority. "I contend," Clay argued, "that it is to arraign the disposition of Providence Himself to suppose that He has created beings incapable of governing themselves" (in Strong, 1900, p. 289). Strong replied that recapitulation, with its primitive-as-child argument, had married imperialism with scientific respectability. We had not only the right, but also the duty, to annex the Philippines:

> Clay's conception was formed . . . before modern science had shown that races develop in the course of centuries as individuals do in years, and that an undeveloped race, which is incapable of self-government is no more of a reflection on the Almighty than is an undeveloped child who is incapable of self government. The opinions of men who in this enlightened day believe that the Filipinos are capable of self-government because everybody is, are not worth considering. (Strong, 1900, pp. 289–290)

In a similar vein, Kidd used recapitulation to justify the conquest of tropical Africa:

> The evolution in character which the race has undergone has been northwards from the tropics. The first step to the solution of the problem before us is simply to acquire the principle that [we are] dealing with peoples who represent the same stage in the history of the development of the race that the child does in the history of the development of the individual. The tropics will not, therefore, be developed by the natives themselves. (1898, p. 51)

The argument even turned up in the first verse of Kipling's most famous paean for colonialism:

> Take up the White Man's Burden
> Send forth the best ye breed
> Go, bind your sons to exile
> To serve the captives' need:
> To wait in heavy harness,
> On fluttered folk and wild—
> Your new-caught, sullen peoples,
> Half-devil and half-child.

Theodore Roosevelt, who had received an advance copy, wrote to Henry Cabot Lodge that it "was very poor poetry but made good sense from the expansion point of view" (in Weston, 1972, p. 35).

There should, in a view of history that motivates many scientists, be a happy ending to this sorry tale. By 1920, the theory of recapitulation was in disarray (Chapter 6). By the 1930s Haeckel's insistence on universal acceleration and the consequent pushing back of ancestral adult characters to the juvenile stages of descendants had yielded to an expanded version granting equal orthodoxy to the opposite process (Chapter 7): the juvenile traits of ancestors may, by retardation of development, become the adult stages of descendants. This appearance of ancestral juvenile traits in adult descendants is called paedomorphosis (child-shaped). It demands a conclusion exactly opposite to the primitive-as-child argument—for the child is now a harbinger of things to come, not a storehouse for the adult traits of ancestors. The child of a "lower" race should be like the adult of a "higher" group. In short, for racist arguments, two contradictory claims are necessary:

1. Under recapitulation, whites, as children, reach the level of black adults; whites then continue on to higher things during their ontogeny.
2. Under paedomorphosis, white adults retain the characteristics of black children, while blacks continue to develop (or rather devolve) during their ontogeny.

The irony of this conceptual change lies in the fact that our own species represents a most impressive case of paedomorphosis (Chapter 10). In feature after feature, we resemble the juvenile stages of other primates—and this includes such markers of intellectual status as our bulbous cranium and relatively large brain (see pp. 356–359). Even the nineteenth-century recapitulationists knew this in their heart of hearts, though they labored mightily to explain it away. Cope, for example, wrote at length about human features that display retarded development (see pp. 354–355). In fact, Cope tried to have it both ways by arguing that whites are superior in both recapitulated and paedomorphic traits:

> The Indo-European race is then the highest by virtue of the acceleration of growth in the development of the muscles by which the body is maintained in the erect position (extensors of the leg), and in those important elements of beauty, a well-developed nose and beard. It is also superior in those points in which it is more embryonic than the other races, viz., the want of prominence of the jaws and cheek-bones, since these are associated with a greater predominance of the cerebral part of the skull, increased size of cerebral hemispheres, and greater intellectual power. (1883, in 1887, pp. 288–290)

When this shift from recapitulation to paedomorphosis occurred, during the 1920s and 1930s, the available data on human evolution included 50 years of accumulated facts, virtually all supporting the claim that black (and other "primitive") adults were like white children. The men who collected these data—and it was always men—claimed that they had done so in the spirit of objective science, caring only for truth and untrammelled by political constraint. In fact, they often argued that their inegalitarian conclusions proved that hard science had triumphed over liberal or Christian sentimentalism. If their motivations were so simple and unsullied, then the replacement of recapitulation by paedomorphosis as an explanation for human evolution should have led to the following honest admission: hard facts prove that white children are like black adults; under paedomorphosis, children of primitives are like adult stages of advanced forms; therefore, blacks are superior to whites.

Needless to say, nothing of the sort happened. I presented the suggestion that it might have occurred not as a plausible hypothesis but merely as a bit of rhetoric illustrating the absurdity of any claim that scientists act "objectively" on matters so vital to the interests of their patrons (and their own privileged position). In fact, the founders of paedomorphosis quietly forgot all the old data on adult-blacks–as–white-children and set out to find some opposing information that would reaffirm racism on the opposite, paedomorphic model.

Louis Bolk, Dutch anatomist and primary proponent of human paedomorphosis, reversed the catechism of recapitulation and proclaimed: "In his fetal development the negro [*sic*] passes through a stage that has already become the final stage for the white man" (1926a, p. 473). Bolk made no attempt to hide his general interpretation: "It is obvious that I am, on the basis of my theory, a convinced believer in the inequality of races. All races have not moved the same distance forward on the path of human evolution [*Menschwerdung*]" (1926c, p. 38). Or, more explicitly:

> The question is of great importance from an anthropological as well as from a sociological point of view. For it need hardly be emphasized that a different degree of fetalization [Bolk's term for paedomorphosis] means a more or less advanced state of hominization. The farther a race has been somatically fetalized and physiologically retarded, the further it has grown away from the pithecoid ancestor of man. Quantitative differences in fetalization and retardation are the base of racial inequivalence. Looked at from this point of view, the division of mankind into higher and lower races is fully justified. (1929, pp. 26–27)

Bolk then scoured the human body for a selected list of traits to affirm the greater paedomorphosis of whites. He cites a more rounded skull, lesser prognathism of the jaw, slower somatic development, and longer life (without considering any environmental influence—p. 27). "The white race," he concludes, "appears to be the most progressive, as being the most retarded" (1929, p. 75).

The argument is by no means extinct today, despite the efforts of Ashley Montagu (1962) and other antiracists (scientific antiracism is largely a post-Hitler phenomenon). H. J. Eysenck (1971) notes that African and black American babies exhibit faster sensorimotor development than whites. By age three, however, whites surpass blacks in IQ. There is also a slightly negative correlation between first year sensorimotor development and later IQ. Eysenck invokes paedomorphosis to link these facts and imply an innate mental inferiority among blacks: "These findings are important because of a very general view in biology according to which the more prolonged the infancy the greater in general are the cognitive or intellectual abilities of the species. This law appears to work even within a given species" (1971, p. 79). (We have here a classic example of a potentially meaningless, noncausal correlation. Suppose that differences in IQ are completely determined by environment; then, rapid motor development does not cause low IQ—it is merely another measure of racial identification, and a poorer one than skin color at that.)

As a final proof of extrascientific motivation, I note the conspiracy of silence that has surrounded two aspects of the paedomorphic argument that are very uncomfortable for white males anxious to retain their exalted status. First, it is hard to deny that Mongoloids—not Caucasians—are the most paedomorphic of human groups. Bolk performed a song and dance about the facts he listed and ended up by arguing that Caucasian and Mongoloid differences were too close to call (1929, p. 28). But the iconoclastic Havelock Ellis, an early supporter of human paedomorphosis, [5] faced the issue squarely in 1894: "On the whole, it may be said that the yellow races are nearest to the infantile conditions; negroes and Australians are farthest removed from it, often although not always in the direction of the Ape; while the white races occupy an intermediate position" (p. 28). His generosity toward yellow skins did not extend to black, although it is easy to list an impressive set of features for which Africans are the most strongly paedomorphic of human groups (Montagu, 1962, p. 331). Ellis continues, presaging Bolk's argument: "The child of many African races is scarcely if at all less intelligent than the European child, but while the African as he grows up becomes stupid and obtuse, and his whole social life falls into a state of hide-bound routine, the European retains much of his childlike vivacity" (1894, p. 518).

Second, women are clearly more paedomorphic than men. Again, Bolk chose to ignore the issue and Ellis met it directly with an admission of inferiority: "The infant ape is very much nearer to Man than the adult ape. This means that the infant ape is higher in the line of evolution than the adult, and the female ape, by approximating to the infant type, is somewhat higher than the male" (p. 517). Women, Ellis affirms, are leading the direction of human evolution:

She bears the special characteristics of humanity in a higher degree than man . . . Her conservatism is thus compensated and justified by the fact that she represents more nearly than man the human type to which man is approximating. This is true of physical characters: the large-headed, delicate-faced, small-boned man of urban civilization is much nearer to the typical woman than is the savage. Not only by his large brain, but by his large pelvis, the modern man is following a path first marked out by woman. (p. 519)

Child Development

But if any biologist is willing to listen, he may care to recognize in the chorus of those who are singing the praise of the ruler of our time, the naturalist, and playing to him on instruments—the tibia of the archaic horse, the antennae of the hymenoptera, the many stops of the hydra's legs—the plaintive note of one who but tries to interpret the wail of the human babe.

J. M. Baldwin, 1906, p. x

Criminal anthropology and racist ideology used the primitive-as-child argument to reinforce their claims about adults—atavistic deviants or members of lower races, respectively. But the argument could be reversed, usually with more benevolence, to ask what comparative anatomy and evolutionary history had to say about the nature of children. Recapitulation supplied an obvious general answer: we understand children only when we recognize that their behavior replays a phyletic past.

Over and over again, we find an explicit appeal to biological recapitulation: since a human embryo repeats the physical stages of remote ancestors, the child must replay the mental history of more recent forebears. Consider one of Haeckel's more illustrious followers: Friedrich Engels:

> Just as the developmental history of the human embryo in his mother's womb is only an abbreviated repetition of the history extending over millions of years, of the bodily evolution of our animal ancestors, beginning from the worm, so the mental development of the human child is only a still more abbreviated repetition of the intellectual development of these same ancestors, at least of the later ones. (1876, in 1954, p. 241)

Or this, from an American popularist:

> It is a fundamental law in evolution that the individual in its development reproduces the life history of the race . . . what is true in the world of animals and physical structures is just as true in mental and culture history development. Every person who passes through a normal development represents the culture stages of man; the child at first is a savage, later he becomes a barbarian, still later it is possible that he may become a civilized being. The boy in the woods building his fire, baking potatoes in the ashes, roasting steak over the coals, is living over again the wild outdoor savage life of his ancestors. The same thought might be illustrated by a thousand other points in child life. (Starr, 1895, p. 32)

For such enthusiastic supporters as Bovet, the guidance of recapitulation had no bound:

> The theory assumes such a magnitude that no one has yet been found to expound it in all its wealth . . . [We find] relationships between the drawings of children and those of primitive peoples, between the grammar of baby-talk and that of certain well-worn idioms, between the dreams of the child's imagination and myths and folklore, and between so many other varied manifestations of mental activity in the beginnings of individual men and of mankind as a whole. (1923, p. 150)

Darwin's influence led to a surge of interest in children and their ways. In America, a semiformal organization, the Child Study movement, approached children with a strong recapitulationist bias (Ross, 1972). Two methods of inquiry often led to phyletic conclusions: the prolonged study of individual infants and the voluminous gathering of statistical information by questionnaire.

In Germany, Preyer had launched the evolutionary study of infancy with a detailed treatise on the first three years of his own son—*Die Seele des Kindes* (1884). Several scientific parents followed his lead. Many books of close observation were written by educated women, constrained by social convention to spend the early years of their children's lives at home. Millicent W. Shinn's *Biography of a Baby* (1900) is typical. She held an ambiguous attitude towards recapitulation (1907, p. 220), but found it unavoidable as a guide to the larger significance of her patient observations:

It has long been observed that there is a curious resemblance between babies and monkeys, between boys and barbaric tribes. Schoolboys administer law among themselves much as a tribal court does; babies sit like monkeys, with the soles of their little feet facing each other. Such semblances led, long before the age of Darwin, to the speculation that children in developing passed through stages similar to those the race has passed through; and the speculation has become an accepted doctrine since embryology has shown how each individual before birth passes in successive stages through the lower forms of life . . . If we can thoroughly decipher this ontogenic record, then what may we not hope to learn of the road by which we human beings came? (Shinn, 1900, pp. 7–8)

The British paleontologist S. S. Buckman studied his children's growth with the same techniques he applied to ammonites and brachiopods. Hyatt's law of acceleration should push the characters of adult monkeys into human childhood (Buckman, 1899, p. 92). Buckman regarded his own babies as miniature apes and sought to identify physical and behavioral characters of a simian past: greater prognathy,[6] flattened nose, puffy cheeks like the pouches of *Cercopithecus,* relatively long arms, a tendency to sleep on the stomach with limbs curled under (despite misguided maternal efforts to straighten them out), the soothing effect of rocking as an inducement to sleep by remembrance of a former life suspended in tree branches, and the child's urge to climb staircases as another vestige of an arboreal past (1894).

Buckman believed that babies pass first through a quadrupedal stage like that of pre-arboreal, and perhaps even pre-simian, ancestors. He pointed not to standard crawling (since quadrupeds do not move on their knees), but to the tendency shown by a minority of children for truly quadrupedal progression on all fours (Fig. 19; Hrdlicka, 1931, later devoted an entire book to "children who walk on all fours" as a proof of recapitulation[7]). An arboreal stage follows, replete with apish characters. Buckman argued, for example, that children could not straighten out their hands until they passed through the bough-grasping stage of their phyletic past, sometime after the age of five or six:

Two of my children . . . were told to hold out their hands as straight as they possibly could. I photographed their hands, and the bough-grasping curve is very apparent—both hands have a forward bending of the fingers: the children were unable to straighten them out. I stopped several village school chil-

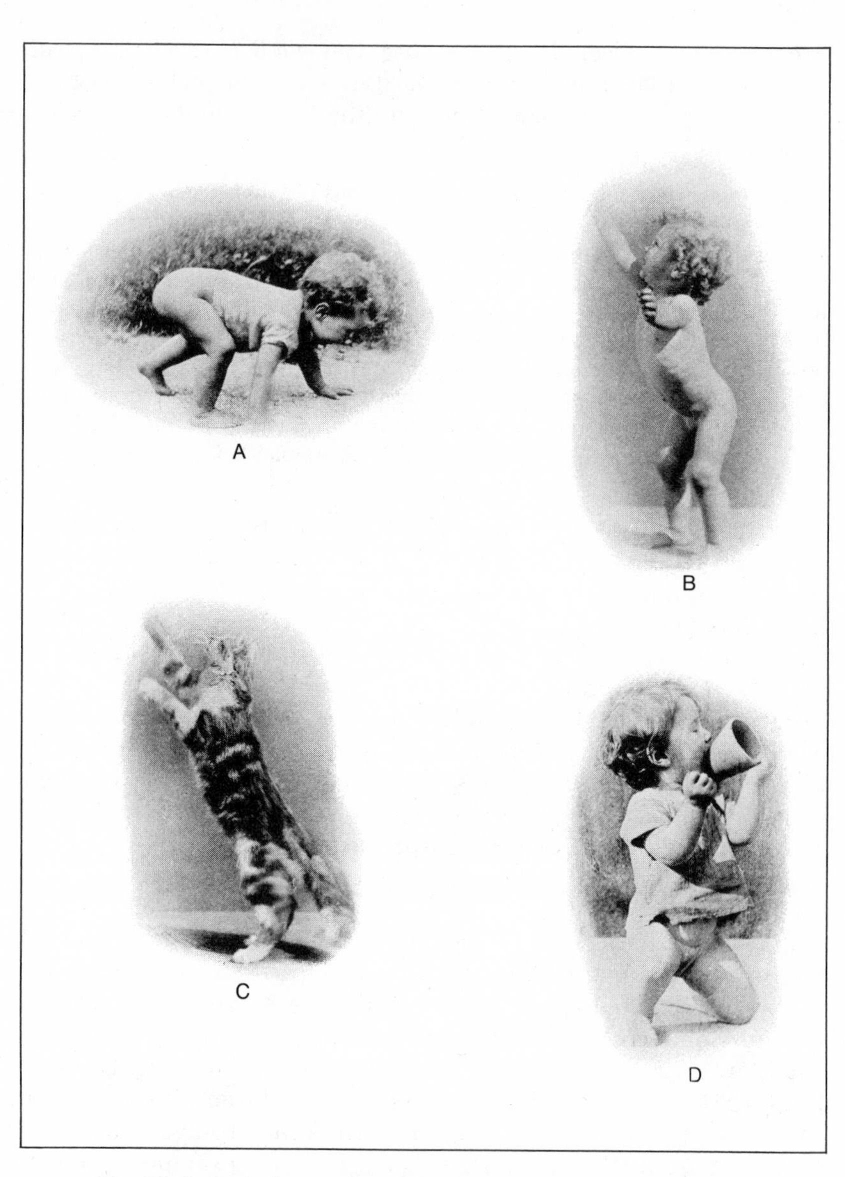

Fig. 19. Recapitulatory acts of Buckman's daughter (compared with his cat): (*A*) At ten months, quadrupedal progression (on all fours—not crawling). (*B*) Eleven months, not yet able to walk; note apish knee-flexure in precarious, first standing. (*C*) The family cat, for comparison with (*B*). (*D*) A different child at twelve months, grasping like an ape. (From Buckman, 1899.)

dren . . . on one occasion, and offered a prize to the one who could hold out the fingers the straightest . . . It was most interesting: the failure of some of them to straighten the fingers was ludicrous. Practically all but one showed a more or less definite curvature. (1899, p. 99)

Meanwhile, America's most diligent and famous student of child development, G. Stanley Hall, was conducting a massive, if uncritical, set of surveys by questionnaire. With their aid, he sought phyletic explanations for those distinctive behaviors of children that seemed to reflect nothing in their immediate environment and that disappeared naturally with increasing age. Hall dispatched his questionnaires to schoolteachers and interested amateurs by the tens of thousands. His voluminous and motley returns became a statistician's nightmare and an a-priorist's delight. As Hall admitted:

Most returns are not made by experts, but by young people with little knowledge of psychology or of the dangers of loose and inaccurate statement, and who are peculiarly prone to exaggeration in describing their feelings. Some returns are seen to be of no value, and are rejected from the start. Many of the floridly described fears are flimsy and no doubt far less real than the language would indicate. Some, too, no doubt are almost purely imagined. (1897, p. 239)

Nonetheless, Hall pushed forward with recapitulatory interpretations of his favorite subjects: fear and play. Of childhood fears, the English leader of child study, James Sully, had written: "Fear appears early in the life of the child as it seems to appear low down in the zoological scale" (1896, p. 92). Hall listed nearly 50 common fears as phyletic vestiges. Some are inheritances from a recent past—big eyes and teeth "must owe some of their terrors to ancestral reverberations from the long ages during which man struggled for existence with animals with big or strange eyes and teeth" (1897, p. 212). Others are remembrances of truly ancient times. Hall attributes the childhood fear of water to a reptilian ancestor recently freed from its pond and not anxious to return:

It would be well for psychologists to postulate purely instinctive vestiges, which originated somewhere since the time when our remote ancestors left the sea, ceased to be amphibious and made the land their home. Do we not dishonor the soul by thinking it less complex or less freighted with mementoes of its earlier stages of development than the body which, in the amniotic fluid medium, unfolds its earlier prenatal stages like a fish. (p. 169).

(Nonetheless, Hall was happy to have it both ways. In other works, he cites the love of many children for water as a recapitulated vestige of our piscine past [1904, 2: 192–195]. Of involuntary swaying motions in children, he writes: "This suggests the slow oscillatory movements used by fish in swimming or maintaining their position in currents of water" [p. 192]. He attributes a fascination for the seashore to our amphibian past: "The shore where these forms first emerged and became amphibian . . . is no less than a passion to children . . . It accounts for a large proportion of all truancies. To paddle, splash, swim, and sun sometimes constitutes almost a hydroneurosis, and children pine all winter and live only for the next summer at the sea" [p. 194].)

Hall based his phyletic interpretation on two arguments—an a priori faith in the biogenetic law, and a conviction that childhood fears reflect the problems of ancestral adults, not the environments of modern children:

> Their relative intensity fits past conditions far better than it does present ones. Night is now the safest time, serpents are no longer among our most fatal foes, and most of the animal fears do not fit the present conditions of civilized life; strangers are not usually dangerous, nor are big eyes and teeth; celestial fears fit the heavens of ancient superstition and not the heavens of modern science. The weather fears and the incessant talk about weather fit a condition of life in trees, caves or tents. (1897, pp. 246–247)

In childhood play, recapitulationists found their primary evidence for the phyletic determination of youthful behavior: "What was once the serious occupation of men becomes in more advanced stages of culture the play of children"(Jastrow, 1892, p. 352). G. Stanley Hall devoted most of his studies to play—"the vestigial organs of the soul." In ascribing play to inheritance from ancestral adults, Hall used the same strategy he had pursued in studying fear. Play is not adaptive as practice or preparation for adult activity. He criticizes Groos's theory of utility as "very partial, superficial, and perverse. It ignores the past where lie the keys to all play activities" (1904, 1:202). Play is the repetition of ancestral patterns. The activities of play may reflect past needs more than present realities, but play must not be curtailed lest the soul's development be distorted; all stages must be passed in their proper sequence. Hall illustrates this doctrine of "catharsis" with an embryological analogy:

> [Play] exercises many atavistic and rudimentary functions, a number of which will abort before maturity, but which live themselves out in play like the tadpole's tail, that must be both developed and used as a stimulus to the growth of the legs which will otherwise never mature . . . I regard play as the motor habits and spirit of the past of the race, persisting in the present, as rudimentary functions sometimes of and always akin to rudimentary organs. The best index and guide to the stated activities of adults in past ages is found in the instinctive, untaught, and non-imitative plays of children . . . Thus we rehearse the activities of our ancestors, back we know not how far, and repeat their life work in summative and adumbrated ways. (1904, 1: 202)

Hall's students and followers pursued the phyletic study of play to a degree that must have seemed extreme even to contemporaries (see, for example, Acher, 1910, on blocks, sand and earth, stones, snow, points and edges, adornment of the body, clothing, striking, and the "psychic stringward tendency" [p. 137] that drives a child to play cat's cradle). Bovet (1923, p. 152) sought a phyletic order in childhood fighting. Babies don't fight and real aggression scarcely commences before age three. This must reflect our bucolic origin in an uncrowded countryside with abundant food and space and no need for fighting (the limited motor skills of a baby do not figure in Bovet's analysis at all). Children develop fighting skills in evolutionary order: scratching and biting, followed by kicking, punching, and weapons. Scratching and biting are remnants of a quadrupedal ancestry.

From these studies, Hall developed his general phyletic theory of childhood. "The child," he wrote, "is vastly more ancient than the man . . . Adulthood is comparatively a novel structure built upon very ancient foundations" (in Strickland, 1963, p. 216).

Hall believed that ages 8–12 (youth) represent a coherent, long stable, and preconscious phase of ancestral development. Teeth are fully developed; the brain has nearly reached its adult size; health is good and activity intense and varied; endurance and resistance to fatigue are great. But creative intelligence and ethical judgment are nearly absent. Hence:

> Reason, true morality, religion, sympathy, love and esthetic enjoyment are but very slightly developed. Everything, in short, suggests the culmination of one stage of life as if it represented what was once, and for a very protracted and relatively stationary period, the age of maturity in some remote, perhaps pigmoid stage of human evolution, when in a warm climate the young of our species once shifted for themselves independently of further parental aid. The qualities now developed are phyletically vastly older than all the neoatavistic traits of body and soul, later to be superposed like a new and higher story built on to our primal nature. (1904, 1: ix–x)

He even believed that he could discern a few physical vestiges of an ancestral sexual maturation at about age six—"as if . . . we could still detect the ripple-marks of an ancient pubic beach now lifted high above the tides of a receding shore-line as human infancy has been prolonged" (p. x).

Hall argued that youths should be left alone to pursue the "savage" rites of their phyletic stage:

> Rousseau would leave prepubescent years to nature and to these primal hereditary impulsions and allow the fundamental traits of savagery their fling till twelve . . . The child revels in savagery, and if its tribal, predatory, hunting, fishing, fighting, roving, idle, playing proclivities could be indulged

in the country and under conditions that now, alas! seem hopelessly ideal, they could conceivably be so organized and directed as to be far more truly humanistic and liberal than all the best modern school can provide. (p. x)

In fact, any attempt to suppress the natural urges of savagery can lead to disaster. Phyletic stages must be passed and expressed in their proper order, lest their repressed effects crop out later in life at inappropriate times. (It is no wonder that Hall befriended Sigmund Freud at the height of his unpopularity in America.)

Rudimentary organs of the soul now suppressed, perverted or delayed, to crop out in menacing forms later, would be developed in their season so that we should be immune to them in maturer years, on the principle of the Aristotelian catharsis for which I have tried to suggest a far broader interpretation than the Stagirite could see in his day. (pp. x–xi)

Hall would have preferred to keep children in nature and out of school until adolescence; but even this recapitulatory romantic bowed reluctantly to the demands of complex urban life:

Another remove from nature seems to be made necessary by the manifold knowledges and skills of our highly complex civilization. We should transplant the human sapling, I concede reluctantly, as early as eight, but not before, to the schoolhouse with its imperfect lighting, ventillation, temperature. We must shut out nature and open books. The child must sit on unhygienic benches and work the tiny muscles that wag the tongue and pen, and let all the others, which constitute nearly half its weight, decay. (pp. xi–xii).

Having accepted this sad necessity, Hall proposes to make the best of it. Since youth (8–12) represents a prereflective phase of our ancestry, it is especially suited to drill, inculcation, and rote learning:

The senses are keen and alert, reactions immediate and vigorous, and the memory is quick, sure and lasting . . . Never again will there be such susceptibility to drill and discipline, such plasticity to habituation, or such ready adjustment to new conditions . . . The method should be mechanical, repetitive, authoritative, dogmatic. The automatic powers are now at their very apex, and they can do and bear more than our degenerate pedagogy knows or dreams of. (p. xii)

Since discipline cannot be enforced by internalized ethics, it must be imposed from without: "Dermal pain is far from being the pitiful evil that sentimental and neurasthenic adults regard it, and to flog wisely should not become a lost art" (p. 402).[8]

But at adolescence (about age 13) the child passes rapidly to phyletic consciousness, ethical discernment, and creative thought. The entire strategy of previous schooling must be reversed and the "liberal" goals of self-discovery, free discussion, and moral argument substituted for the pedantry of drill and inculcation:

> Adolescence is a new birth, for the higher and more completely human traits are now born. The qualities of body and soul that now emerge are far newer . . . the later acquisitions of the race slowly become prepotent. Development is less gradual and more saltatory, suggestive of some ancient period of storm and stress when old moorings were broken and a higher level attained. (p. xiii)

Hall entitled his massive treatise "Adolescence" (1904). It is still widely read and studied, but few modern scholars appreciate the central role of recapitulation in defining both title and subject. Adolescence is not just an exciting and stressful time of rapid change; it represents the phyletic transition from preconscious animality to conscious humanity.

In Hall's work, recapitulation reached the acme of its influence outside biology. By the second decade of our century, recapitulation was collapsing on its home front of embryology and anatomy, and the message leaked out to some perceptive child psychologists: "More recently, disquieting rumors from the source of its origin have been heard to the effect that the principle was formulated without sufficient warrant, and that it cannot be depended upon even as a helpful hypothesis" (Davidson, 1914, p. 2). Davidson translates for child psychologists the standard biological refutations of recapitulation. First, early stages evolve to meet their own necessities (prevalence of cenogenesis over palingenesis). All stages are altered; no principle of terminal addition may be maintained: "Ancestral life-history . . . has been altered with each step in descent rather than extended. In keeping with this view infancy probably has had its own evolution, having been evolved when it was needed and having been altered from age to age from germinal mutation, and by selection as the necessities of its circumstances required" (p. 80).

Second, if early stages have any phyletic information at all, they repeat the early stages of ancestors, not the adult forms (von Baer's laws vs. Haeckelian recapitulation): "Their ancestral reference would be first to ancestral infancies and only indirectly to adult characters for which the infantile condition was preparing" (p. 81). In addition, child psychology can provide some refutations of its own. Traits of human behavior are not genetically fixed like anatomical markers of ancestry. If people in civilized societies behave in a more "advanced" way than their "primitive" cousins, we need invoke only a different education for common material, not a genetic transcendence of ancestral conditions: "The civilized societies work differently with the same human nature and by contrast the fundamental inborn traits seem ancestral when they are only less trained" (p. 93).

By 1928 (pp. 46–51), the Gestalt psychologist Koffka was offering three alternatives for the explication of individual development: (1) recapitulation, which he rejects categorically; (2) Thorndike's "utility" theory—a Darwinian argument that the sequential stages of childhood are selected for their adaptive value in situations encountered by modern children (a cenogenetic theory in Haeckelian terms)—this Koffka also rejects because he regards Darwinism as dead; (3) a "correspondence" theory, according to which ontogeny seems to parallel phylogeny because external constraints impose a similar order on both processes. There are, for example, only so many ways to move from simplicity to complexity, from homogeneity to heterogeneity, from instinct to consciousness. Phylogeny and ontogeny have no direct influence upon each other (as Haeckelian recapitulation requires); each follows a roughly similar path because it is the only path available. (Mudcracks, basalt pillars, soap bubbles, bee cells, and echinoid plates are all hexagonal because only a few regular forms can fill space completely. The external constraints are identical, but no result has any direct influence upon another.) Koffka favors this theory. And our leading scholar of child development today, one of the most respected intellectuals in the world, follows it as well—Jean Piaget.

Piaget was trained as a paleontologist during the heyday of Haeckelian recapitulation (he wrote his dissertation on Jurassic gastropods from France). His pronouncements seem to have a Haeckelian ring—though precision and clarity are not Piagetian hallmarks. It would be reasonable to assert that Piaget studies the ontogenesis of concepts in children because, prompted by his paleontological training in the Haeckelian mode, he believes that children provide the only access to a more interesting question with no direct answer: how, historically, did we learn to think and reason? But this assertion would be at least half wrong. Piaget believes in parallels between ontogeny and phylogeny, but he denies Haeckelian recapitulation as their mechanism.

In a recent article, Piaget expressed his general belief in the importance of parallels:

> The fundamental hypothesis of genetic epistemology [Piaget's name for his school of thought] is that there is a parallelism between the progress made in the logical and rational organization of knowledge and the corresponding formative psychological processes. With this hypothesis, the most fruitful, most obvious field of study would be the reconstituting of human history—the history of human thinking in prehistoric man. Unfortunately, we

are not very well informed in the psychology of primitive man, but there are children all around us, and it is in studying children that we have the best chance of studying the development of logical knowledge, mathematical knowledge, physical knowledge, and so forth. (1969, p. 4)

Piaget has often made more specific claims that seem to support the primitive-as-child argument of classic recapitulation. Speaking, for example, of the child's dualistic belief in both an external reality and an imposition of his own being upon all objects: "There is dualism everywhere—realism on the one hand, subjective adherences on the other . . . The situation is closely analogous to . . . the mentality of primitives . . . We would therefore seem to be in the presence of a very general feature of thought" (1960, pp. 781–782). Or, of the child's closeness to immediate observation, despite an absence of logical consistency:

Compared with ourselves, the child is both closer to immediate observation and further removed from reality. For, on the one hand, he is often content to adopt in his mind the crude forms of actuality as they are presented in observation: one boat will float because it is light, another, because it is heavy, etc. Logical coherence is entirely sacrificed in such cases to fidelity to fact. The causality which results from phenomenism of this kind is not unlike that which is to be found in primitive races. (p. 253)

Piaget is also fond of drawing parallels between the child's acquisition of logico-mathematical knowledge and the history of Western science:

It may very well be that the psychological laws arrived at by means of our restricted method can be extended into epistemological laws arrived at by the analysis of the history of the sciences: the elimination of realism, of substantialism, of dynamism, the growth of relativism, etc. all these are evolutionary laws which appear to be common both to the development of the child and to that of scientific thought. (p. 240; see also Fiske, 1975)

Children, for example, tend at first to classify objects into rigid categories, and only later to develop any notion of relativism. With this sequence, Piaget compares the transition from Linnaean staticism to modern notions of evolutionary continuity (p. 298). An eight year old, asked why a marble falls to the ground, responded in Aristotelian terms: it was moving to its natural place (in Fiske, 1975). Children spontaneously develop an impetus theory much like that of medieval physics. Euclid limited his geometry to relations within a single figure; ask a young child to draw a chimney on a house and he places it perpendicular to the sloping roof—for he can only think of it in relation to the adjoining part of the house, not the "external" ground.

But, Piaget argues with obvious reason, the "Aristotelian" physics of modern children cannot possibly represent a genetic inheritance from adults living a mere two thousand years ago (1971, p. 84). Hence, Haeckel's causal explanation for the parallel must be incorrect. Piaget, by the way, is not unfriendly to the biogenetic law as a general proposition (p. 83); he merely denies its relevance to human psychology. When I wrote to ask what he thought of Haeckel's doctrine, Piaget responded: "I have done very little work in psychology on the relationships between ontogenesis and phylogenesis because, psychologically, the child explains the adult more than the reverse" (letter of Feb. 21, 1972).[9]

Piaget's general theory of conceptual development in children lies midway between two extremes: (1) Chomskyan neo-preformationism, with its claim that our faculties of intelligence are endowed innately with a formal mechanism of logic (though the content of intelligence—knowledge—is gradually acquired through ontogeny); (2) the older empiricism, with its assertion that the mind, at birth, is a blank slate. There are, Piaget avows, inborn components of reason, but they are not static; they themselves evolve in a definite way during ontogeny as the child assimilates external reality to its changing internal structures. In other words, reason itself evolves in response to increasing experience with the external world: "The truth, in short, lies half-way between empiricism and apriorism: intellectual evolution requires that both mind and environment should make their contribution. This combination has, during the primitive stages, the semblance of confusion, but as time goes on, the mind adapts itself to the world, and transforms it in such a way that the world can adapt itself to the mind" (1960, p. 258).

The young child, for example, has very different ideas of causality than the adult. He confuses the self with the external world and with other people. His ego obscures both empirical and formal truth (1960, pp. 301–302). He subscribes to notions of magic, finalism, animism, and dynamism (p. 272). But these "subjective adherences" disappear during ontogeny as the child "becomes conscious of his subjectivity" (p. 246). He comes to notice the existence and mechanism of his own thought, to separate signs from the things signified, to cease believing that names belong to objects, to recognize that dreams are not caused by emanations from the objects that appear in them.

These stages in the ontogeny of reasoning may arise whenever a mind endowed with basically human capacity moves from a preconscious union with the world to the kind of differentiation that logical reasoning as we know it implies.[10] Every child does this in the course of his own growth. But did not humanity do it as well in our phyletic history of successive adults: "Does the human child, during its period of mental growth, only manifest characteristics that are transmitted to it by language, its family, and its school, or does the child itself provide spontaneous productions which may have had some influence, if generalized, on more primitive societies than our own" (1971, p. 83). The sequences run in parallel, but neither causes the other. They both follow similar paths because a common object (the preconscious mind) is pursuing a common history of development (successive assimilations of external reality to produce a sequence in modes of reasoning).

Ironically, Piaget's contemporary explanation of parallels between ontogeny and phylogeny harks back to the earliest theory of all—the Meckel-Serres law of early nineteenth-century Naturphilosophie and transcendental morphology. The Naturphilosophen attributed parallels not to any interaction of one sequence with another (as Haeckel was later to require), but to a common constraint—the single direction of all development—acting separately on two independent sequences. One hundred years later, Hertwig (see Chapter 6, note 30) proposed a different correspondence theory, refuting Haeckel's biogenetic law while affirming the parallels between ontogeny and phylogeny. Hertwig saw the laws of physics and chemistry as a common external constraint. Given a small and simple starting point (the phyletic amoeba or the ontogenetic zygote), nature can only build complexity in a limited number of ways. Piaget uses the same style of argument, however different the content. The parallels are real, but phylogeny does not cause ontogeny. Again, two independent sequences follow similar paths under the influence of a common constraint—the structure of the human mind itself.

Primary Education

In an outrageously mixed metaphor with agricultural and geological components, G. Stanley Hall proclaimed the educational potential of recapitulation:

> Children thus in their incomplete stage of development are nearer the animals in some respects than they are to adults, and there is in this direction a rich but undiscovered silo of educational possibilities which heredity has stored up like the coal-measures, which when explored and utilized to its full extent will reveal pedagogic possibilities now undreamed of. (1904, 2: 221–222)

Hall would have preferred a "school" of nature for young children—a real opportunity to relive the phyletic past as our ancestors did. But he accepted the practical necessity of professional teachers and school buildings. Still, if repetition of phylogeny were repressed on hard school benches, the child would grow up as a psychological cripple. Therefore, curricula must be structured to match the historical sequence of human cultures; the child must relive, if only vicariously, his ancestral past:

> The deep and strong cravings in the individual to revive the ancestral experience and occupations of the race can and must be met, at least in a secondary and vicarious way, by tales of the heroic virtues the child can appreciate, and these proxy experiences should make up by variety and extent what they lack in intensity . . . Thus we not only rescue from the dangers of loss, but utilize for further psychic growth the results of the higher heredity, which are the most precious and potential things on earth. (1904, 1: ix)

Herbert Spencer had urged much the same thing in 1861: "If there be an order in which the human race has mastered its various kinds of knowledge, there will arise in every child an aptitude to acquire these kinds of knowledge in the same order . . . Education should be a repetition of civilization in little" (1861, p. 76).

This argument did not arise directly from Haeckel's biology (indeed, Spencer was a supporter of von Baer). The idea that education should follow some sequential order is scarcely avoidable, but what shall the principle of that ordering be? One might, for example, argue that faculties of the mind unfold in some succession: for example, perception, imagination, memory, reason. Or that some logical principle—increasing complexity of numerical relationships for example—establishes a sequence for learning (De Garmo, 1895, pp. 108–109). But, in the late eighteenth century, the same intellectual ferment that inspired Naturphilosophie and the first flowering of recapitulation inevitably suggested that nature, rather than reason or logic, might supply the key. If all the world is in upward flux along a single path of development, then instruction must follow nature as a child mounts through the stages of lower creatures and primitive civilizations towards a higher humanity. Pestalozzi, Froebel, and Herbart, the great triumverate of German-speaking educational reformers of the late eighteenth and early nineteenth centuries, all supported a vaguely defined notion of recapitulation in this pre-Haeckelian mode.[11] Pestalozzi wrote in his popular work of 1801, *Wie Gertrud ihre Kinder lehrt:* "The child masters the principles of cultivated speech in exactly the same slow order as Nature has followed with the race" (in Strickland, 1963, p. 63). Herbart wrote in 1806: "If they would, however, continue the work of their forefathers, they must have travelled the same way" (in 1895, p. 165). And Froebel proclaimed in 1826: "Inasmuch as he would understand the past and present, [the individual] must pass through all preceding phases of human development and culture, and this should not be done in the way of dead imitation or mere copying, but in the way of living spontaneous self-activity" (in 1887, p. 18).

Nonetheless, recapitulation did not become the basis for curricula in primary schools until the triumph of evolutionary theory and the introduction of Haeckel's powerful arguments. In Haeckel's version of recapitulation, a child literally was a small savage—not merely a lower stage independently generated in all developmental series (as the Naturphilosophen believed). If modern society would not allow him to behave like one, it could at least inspire his interest in school by teaching him the tales of ancestral stages appropriate to his age. Nicholas Murray Butler, educator and pundit, wrote at the turn of the century about the influence of evolutionary theory upon teaching:

> Every conception of this 19th century, has been cross-fertilized by the doctrine of evolution . . . But much remains to be done in applying the teachings of evolution in actual plans and methods of instruction. The logical order is so simple, so coherent, and so attractive, that it seems a pity to surrender it for the less trim and less precise order of development, but this will have to be done if teaching efficiency according to evolution is to be had. The course of evolution in the race and in the individual furnishes us also with the clue of the natural order and the real relationships of studies. (1900, pp. 320–321)

More specifically, the supporters of recapitulatory curricula always invoked the embryological analogue to make their point:

> Just as the embryo of one of the higher animals shows unmistakable evidence of passing through all the essential stages of development manifested by lower orders, so the child in his mental evolution passes through, in little, all the great culture epochs that have marked the development of the race . . . We are fond of thinking of education as the process of realizing in each individual the experience of the race, but we have not emphasized the idea that the child can best get this experience in the same order that the race obtained it. (De Garmo, 1895, pp. 109–110)

The idea of basing primary school curricula upon recapitulation arose within an educational movement that invoked the name of Johann Friedrich Herbart (through the master—who might not have approved[12]—was long dead). In the 1870s and 1880s, German universities maintained only two professorships in pedagogy, and both were held by Herbartians. Tuiskon Ziller (1817–1883) held sway in Leipzig, where he formulated the basic theory of recapitulatory curricula—the *Kulturhistorischenstufen,* or theory of culture epochs. W. Stoy, a conservative Herbartian, reigned at Haeckel's university of Jena and thought very little of Ziller's theory; but Wilhelm Rein succeeded Stoy in 1885 and proceeded to put Ziller's theory to a favorable test. Ziller was a dreamer, Rein a practitioner. Rein and his collaborators published eight painstaking volumes corresponding to the eight years of German primary schooling and demonstrating how Ziller's plan for a recapitulatory curriculum could work in practice. Of these works, an American admirer wrote: "These eight volumes are a monument of patient labor, such as only Germans are capable of executing" (De Garmo, 1895, p. 142).

During these decades, graduate study at German universities was *de rigueur* for aspiring American academics. America's leading Herbartians all studied in Germany and brought Ziller's theory home with them.

Ziller stated the basic principle of his culture-epochs theory in the following way:

> The mental development of the child corresponds in general to the chief phases in the development of his people or of mankind. The mind-development of the child, therefore, cannot be better furthered than when he receives his mental nourishment from the general development of culture as it is laid down in literature and history. Every pupil should, accordingly, pass successively through each of the chief epochs of the general mental development of mankind suitable to his stage of advancement. (in Seeley, 1906, pp. 75–76)

Ziller's recapitulatory plan involved two procedures: selection and concentration. Each of the eight years of primary education must have a central focus. This focus shall be the period of cultural and literary history that the child is recapitulating during the given year. To put it bluntly, let him read about savages when he is a savage himself. Ziller referred to this choice of sequential core material as selection; for the eight years of a German *Volksschule,* he recommended:

1. Stories from epic folklore.
2. Robinson Crusoe (redomiciled in Germany).
3. Biblical patriarchs.
4. Judges of Israel.
5. Kings of Israel.
6. The life of Christ.
7. Apostolic history.
8. The Reformation.

(Our secular age may miss the radical nature of a curriculum that did not mention Jesus Christ until the sixth grade; after all, the Bible itself had previously been used as the basis for moral and religious instruction in all grades.)

The recapitulatory model worked reasonably well for history, literature, and moral instruction, but its application to science and mathematics raised obvious dilemmas. One cannot follow the same historical plan and seriously teach alchemy before modern chemistry or the Ptolemaic before the Copernican system. The culture epochs, Ziller argued, form a single grand lesson in ethics, but science has no moral content and cannot be treated in the same sequential manner. If a culture epoch must form the core of all instruction in a given year, how then shall science be taught? Ziller responded with his notion of concentration: science shall always be treated in terms of modern understanding, but all material for instruction must be concentrated about a culture epoch. The content of science should be modern, but the chosen subjects must match interests inspired by the culture epoch: study tropical faunas while reading Robinson Crusoe, or geology and meteorology while hearing the tales of Noah.

I cannot judge how popular these recapitulatory curricula became in America; I do not think that the Zillerian Herbartians ever achieved a majority position. But they were certainly no fringe movement and the lives of millions of school children were directly influenced by their practices. I have tried to survey the primers and instructional manuals of 1880–1915 and have discovered a strong influence for recapitulation in the establishment of curricula. The leading pundits at traditional universities said little (indeed, most of these schools provided no formal training in education). But the principals of normal schools and the heads of boards of education told a different tale—and these were the people who did the practical job of educating, even though their writings are ignored by traditional historians of ideas. Supporters of recapitulation in education included W. E. Chancellor (1907), superintendent of public instruction in the District of Columbia; A. J. Smith (1899), superintendent of schools in St. Paul, Minnesota; R. N. Roark (1895), dean of the Department of Pedagogy at Kentucky State College, Lexington; W. W. Charters (1913), dean of the School of Education at the University of Missouri; C. A. Phillips, dean of the faculty at the State Normal School of Warrensberg, Missouri; and M. V. O'Shea (1906), professor of the science and art of education at the University of Wisconsin. John Dewey offered several criticisms of the culture-epochs theory, but he also said this in its praise: "It must first be heartily acknowledged that it makes practically the first attempt to treat the curriculum, especially in its sequence, upon other than conventional, or formal and logical grounds. Educational theory is indebted to the doctrine for the first systematic attempts to base a course of study upon the actual unfolding of the psychology of child nature" (1911, p. 241).

Many educators tried to translate Ziller's curriculum into an appropriate American equivalent. H. M. Scott (1897), principal of the Detroit Normal Training School, carried out an extensive classroom experiment to establish an ideal curriculum. She claimed that the notion of culture epochs was primarily poetic, though much reinforced by biology; her own adherence, she argued, was strictly an empirical matter: it worked for young children. She writes:

> The fundamental instincts of the majority of the first-grade children upon entering school were found to be a restless curiosity, a naive sort of imaginativeness, and tendencies toward contrivance of a crude order, in short, such instincts as characterize the Nomadic period in civilization. Stories about Hiawatha suggested themselves as answering the interests of these children, and were successfully used. In the second grade, the Greek myths were found to appeal most strongly to the pupils, as embodying their own instinctive attitude toward life; and after a while in another grade stories of chivalry were demanded by the children in response to the dawnings of chivalric impulse only half recognized in themselves. From such suggestions on the part of the children the entire system has little by little arisen, without any idea at the outset of its being a "system" at all. (1897, p. 5)

Frank McMurray, professor of pedagogy at the University of Illinois, tried to apply Ziller's theory of concentration to his own selection of culture epochs (de Garmo, 1895). It is hard, I confess, not to laugh at some of his choices, but they do illustrate the principle that scientific and mathematical instruction should be clothed in an appropriate literary and historical culture epoch. I suspect that his first graders were mighty tired of conifers by the time they finished the following year:

Literature—Anderson's *The Fir Tree* (the primary culture epoch for concentration).

Science—White Pine (more common around Illinois than any other evergreen).

Number—"Number of needles in a bundle of white, Scotch, or Austrian Pine; in two bundles of White pine; in two, four, five of Scotch or Austrian Pine."

Music—*High in the Top of an Old Pine Tree.*

Third graders, immersed in Robinson Crusoe, would be drawing a stalk of wheat, counting the number of grains in a head of wheat, and happily singing "When the Corn Begins to Sprout." Fifth graders, while delighting in the adventures of John Smith, would be suffering the following exercises in higher arithmetic: "Quantity of tobacco chewed by one person per year, in a lifetime; quantity smoked; its value. Weight of ashes of cigar compared with weight of cigar; one is what percent of the other, etc."(de Garmo, 1895, p. 128).

Explicitly recapitulatory curricula did not survive long into the twentieth century. Some educators knew about the decline of recapitulation in biological circles and urged caution on this account: "The recapitulation theory has been subjected to a lively attack in the realm of biology, and those who base their work upon it in the mental sciences may well ask themselves whether they are not building on a shifting soil" (Bovet, 1923, p. 150). Others knew the status of recapitulation in biology, but deemed the whole subject irrelevant to matters educational. Starch (1927, pp. 26–27) argued that the culture-epochs theory had "built pedagogical mountains out of biological molehills," and that any comparison between anatomical recapitulation in an embryo and mental recapitulation in a child rested on the flimsiest of analogies.

But most of the dissatisfaction arose within pedagogical and psychological circles. Teachers did not like the culture epochs because they wrenched children from a contemporary context. A child might be a savage in some biological sense, but he lived in a world of trains and urban apartments, and these modern surroundings shaped his interests (Judd, 1903, p. 197; Raymont, 1906, p. 173; Klapper, 1912, p. 104; Norsworthy and Whitley, 1918, p. 377.) Monroe et al. write that the culture-epochs theory

> becomes absurd when tested by common sense. Although the development of the child may parallel the development of the race in certain respects, it does not follow that the curriculum should parallel the cultural development of the race. Obviously a child living in the 20th century would pursue a 20th century curriculum. There is no justification for delaying the study of current events and our present community, state, and national life until the child has completed his study of the preceding periods of racial development. (1930, pp. 408–409)

Many educators continued their critique beyond a denial of relevance to an outright rejection of recapitulation itself. Children may pass through a coherent sequence of developmental stages, but the parade of ancestral adults is not its cause. E. L. Thorndike (1919), for example, championed his "utility theory"—a Darwinian proposal that childhood behaviors arise by selection for immediate benefit (and tend to be expressed by inheritance at the same developmental stage in descendants, unless modified by new selection pressures acting upon these future generations). In Haeckelian terms, an ontogenetic sequence of behaviors becomes a set of cenogeneses. Moreover, many supposed recapitulatory behaviors may have nothing to do with inheritance at all, but may be reasonable and spontaneous responses to immediate stimuli: "The infant's dislike of, and fright at touching his mother's fur stole or the family cat may be explained not by any inherited memory of unfortunate racial contact with a mastodon, but by the unusualness of the skin stimulus, the odor of the fur, or the uninterpreted expression of pussy's eyes, whiskers, and tail, let alone the feel of her nose or claws" (Norsworthy and Whitley, 1918, p. 36).

Nonetheless, I believe that recapitulation had a lasting influence in American primary education. First of all, several prominent educators retained the central idea of culture epochs within an expanded curriculum that integrated the ancient and modern. John Dewey, for example, had tried a modified system of culture epochs in his Chicago experimental school during the early 1890s. But he could not interest children in all aspects of supposedly appropriate material for their age: they loved the Roman heroes, for example, but yawned through the study of Roman laws. When Dewey opened his laboratory school at the University of Chicago in 1896, he abandoned any strict adherence to culture epochs. He continued to treat the past in chronological sequence, but he did so selectively and always sought explicit links with an immediate present. Still, Dewey was, at the time, a general supporter of recapitulation: "There is a sort of natural recurrence of the child mind to the typical activities of primitive people; witness the hut which the boy likes to build in the yard, playing hunt, with bows, arrows, spears and so on" (in Strickland, 1963, p. 311).

Dewey was happy to use the past, even to use it in proper recapitulatory sequence, but only if it could directly illuminate the present: "The child is not, educationally speaking, to be led *through* the epochs of the past, but is to be led *by* them to resolve present complex culture into simpler factors, and to understand the *forces* which have produced the present" (1911, p. 241). For a rigid and sterile reliance on a recapitulated past, Dewey had only contempt (1916, p. 88).

Furthermore, recapitulation was a major weapon in the liberalization of education and the increasing freedom of children. It was the bulwark of a naturalistic argument: we must not force children to learn in a pre-set logical pattern; we must, instead, mold education to the child by following the course of his natural development. We must not expect adult behavior and ethical judgment from young children.

Happy results often arise from defective reasons. G. Stanley Hall had a noble goal for recapitulation in education—"to reconstruct the grammar-school course: scientifically, so that school-hours, curricula, exercise, buildings, etc., shall all be . . . in accordance with child-nature, the true norm" (1893, in Strickland, 1963, p. 91). He railed against the stiff formalism of current practice on evolutionary grounds:

Not only has the daily theme spread as an infection, but the daily lesson is now extracted through the point of a pencil instead of from the mouth. The tongue rests and the curve of writer's cramp takes a sharp turn upward, as if we were making scribes, reporters, and proof-readers. In some schools teachers seem to be conducting correspondence classes with their own pupils. It all makes excellent busy work, keeps the pupils quiet and orderly, and allows the school output to be quantified, and some of it gives time for more care in the choice of words. But is it a gain to substitute a letter for a visit, to try to give written precedence over spoken forms? Here again we violate the great law that the child repeats the history of the race, and that, from the larger historic standpoint, writing as a mode of utterance is only the latest fashion. (1904, 2: 462)

Recapitulation, in short, became the strongest argument for child-centered education:

Since it is the order of nature that the new organism should pass through certain developmental stages, it behooves us to study nature's plan and seek rather to aid than to thwart it. For nature must be right; there is no higher criterion . . . The parallelism of phylogeny and ontogeny enforces the argument in favor of natural development . . . It furnishes a double support to the view that education should be a process of orderly and gradual unfolding, without precocity and without interference, from low to ever higher stages; that forcing is unnatural and that the mental pabulum should be suited to the stage of development reached. (Guillet, 1900, in Thorndike, 1919, pp. 104–105)

We have rejected the rationale today (and some of its implications—the dangers of precocity, for example). But much of the little that is good about modern American education follows an ideal that triumphed with the strong aid of recapitulation.

Freudian Psychoanalysis

W. M. Wheeler, student of social insects and one of the most perceptive and widely educated biologists of our century, had little use for Victorian psychology. He rejoiced in Freud, Jung, Adler, Jones, and Ferenczi and expressed his pleasure in a 1917 address, "On Instincts." Of the older school, he wrote:

After perusing during the past twenty years a small library of rose-water psychologies of the academic type and noticing how their authors ignore or merely hint at the existence of such stupendous and fundamental biological phenomena as those of hunger, sex, and fear, I should not disagree with, let us say, an imaginary critic recently arrived from Mars, who should express the opinion that many of these works read as if they had been composed by beings that had been born and bred in a belfry, castrated in early infancy, and fed continually for fifty years through a tube with a stream of liquid nutriment of constant chemical composition. (in Evans and Evans, 1970, pp. 226–227)

Yet, amidst his praises for Freud, we read one note of slightly constraining caution—one area where Wheeler felt that Freud might be asking too much of evolutionary biology: "In nothing is the courage of the psychoanalysts better seen than in their use of the biogenetic law. They certainly employ that great biological slogan of the nineteenth century with a fearlessness that makes the timid twentieth century biologist gasp" (p. 226).

Sigmund Freud had two strong reasons for a favorable predisposition towards Haeckel's doctrine. He was, first of all, trained as a biologist during the era of its domination. Secondly, he was a devout Lamarckian and remained so throughout his long life (see pp. 80–88 for why recapitulation finds an almost automatic justification under Lamarckian notions of inheritance). In his last work, *Moses and Monotheism* (1939), Freud held fast even though evolutionary biology had abandond his favored belief: "This state of affairs is made more difficult, it is true, by the present attitude of biological science, which rejects the idea of acquired qualities being transmitted to descendants. I admit, in all modesty, that in spite of this I cannot picture biological development proceeding without taking this factor into account" (pp. 127–128).

Freud was a devout recapitulationist—and he said so clearly and often: "Each individual somehow recapitulates in an abbreviated form the entire development of the human race" (from the 1916 *Introductory Lectures on Psychoanalysis,* p. 199); or "ontogenesis may be regarded as a recapitulation of phylogenesis, in so far as the latter has not been modified by more recent experience. The phylogenetic disposition can be seen at work behind the ontogenetic process" (from the 1914 preface to the third edition of the *Three Essays on the Theory of Sexuality,* 1905, p. xvi). Statements like these have been cited before as isolated testimonies to Freud's conviction. But the central role of recapitulation in his entire system has rarely been noted. (I thank Frank Sulloway and Robert McCormick of Harvard's History of Science Department for guiding me through this literature and for clearly identifying the role of recapitulation in Freud's thought. See Sulloway [in press] for more details and for a general assessment of biological influences upon Freud.)

In an 1897 letter to Fliess, before he had formalized his theory of psychosexual stages, Freud argued that repression during later ontogeny of olfactory stimuli in infant sexuality had a phyletic basis:

I have often suspected that something organic played a part in repression: I have told you before that it is a question of the attitude adopted to a former sexual zone . . . ; in my case the suggestion was linked to the changed part played by sensations of smell: upright carriage was adopted, the nose was raised from the ground, and at the same time a number of what had formerly been interesting sensations connected with the earth became repellant. (in McCormick, 1973, p. 7)

Freud later linked the infant's oral and anal sexuality to a quadrupedal ancestry before vision became a dominant sense and eclipsed a previous reliance upon smells and tastes. In the *Three Essays* of 1905, Freud wrote that oral and anal stages "almost seem as though they were harking back to early animal forms of life" (1962 ed., p. 96; the idea is not completely defunct in modern psychoanalytic circles—see Yazmajian, 1967, p. 219). Still later, he had this to say on the ontogenetic development of libido and ego: "Both of them are at bottom heritages, abbreviated recapitulations of the development which all mankind has passed through from its primaeval days over long periods of time" (1916, in McCormick, 1973, p. 8).

Freud recognized an essential difference between this mental recapitulation of ideas and behaviors and the Haeckelian physical recapitulation of ancestral morphologies. The difference became an essential argument in his theory of neuroses. Physical recapitulations are transient stages; they are replaced by subsequent forms (indeed their material is remolded to make the later stages). But the stages of mind can coexist. To be sure, they appear in proper phyletic order during ontogeny, but an ancient stage does not vanish to make way for a later one. The earlier stages are characteristically repressed in the healthy adult, but they need not disappear. The repressed, primitive core continues to "reside" in the adult brain. Freud provides a graphic metaphor of this concept in *Civilization and Its Discontents* (1930). Imagine modern Rome with all its buildings perfectly preserved from the days of Romulus to now. Impossible of course, for no two material objects can occupy the same spot. But mental phenomena may correspond to this vision of a truly eternal city: "The earlier phases of development are in no sense still preserved; they have been absorbed into the later phases for which they have supplied material. The embryo cannot be discovered in the adult . . . The fact remains that only in the mind is such a preservation of all the earlier stages alongside of the final form possible, and that we are not in a position to represent this phenomenon in pictorial terms" (1930, 1961 ed., p. 18).

Freud's general theory of neurosis and psychoanalysis relies upon this view of mental recapitulation. Sexual energy (libido) is limited in quantity. It can be compulsively fixed at levels of development prior to maturity by traumatic events of early childhood: "It has long since become common knowledge that the experiences of the first five years of childhood exert a decisive influence on our life, one which later events oppose in vain" (1939, p. 161). "The genesis of neurosis always goes back to very early impressions in childhood" (1939, p. 91). Neuroses, therefore, are expressions of sexual energy appropriate to children but normally repressed and superseded in adults. They arise only because early mental stages survive in adults (though normally in a repressed state). "We were thus led to regard any established aberration from normal sexuality as an instance of developmental inhibition and infantilism" (1905, 1962 ed., p. 136). Neuroses are not only the abnormal retention of stages appropriate to children; they also represent the expression of ancestral tendencies—an atavism to be shunned in any progressivist reading of evolution. Psychoanalysis aims to relieve neurosis by reconstructing and understanding its childhood causes: "You may regard the psychoanalytic treatment only as a continued education for the overcoming of childhood-remnants" (1910, p. 213).

In a particularly graphic image, Freud evoked the ancestral character of neurotic behavior "With neurotics it is as though we were in a prehistoric landscape—for instance, in the Jurassic. The great saurians are still running around; the horsetails grow as high as palms" (notes written in 1938, reprinted in 1963, p. 299). Freud once even argued that differences among mental abnormalities might reflect the different ancestral stages (= periods of childhood) at which libido became fixed. We should be able to arrange the neuroses themselves in phyletic order. In 1915, he wrote to Ferenczi: "Anxiety hysteria—conversion hysteria—obsessional neurosis—dementia praecox—paranoia—melancholia—mania . . . This series seems to repeat phylogenetically an historical origin. What are now neuroses were once phases in human conditions" (in McCormick, 1973, p. 17).

Indeed, Freud did not shrink from completing the recapitulatory system of his beliefs. In the extraordinary closing words to his report on the Schreber case, Freud rediscovers the fourfold parallelism of classical recapitulation: the child, the modern savage, our primitive ancestor, and the adult neurotic all represent the same phyletic stage—the primitive as true ancestor, the savage as a modern survivor, the child as a recapitulated adult ancestor in Haeckelian terms, and the neurotic as a fixated child (= primitive):

> I am of the opinion that the time will soon be ripe for us to make an extension of a principle of which the truth has long been recognized by psychoanalysts, and to complete what has hitherto had only an individual and ontogenetic application by the addition of its anthropological and phylogenetically conceived counterpart. "In dreams and neuroses," so our principle has run, "we come once more upon the *child* and the peculiarities which characterize his modes of thought and his emotional life." "And we come upon the *savage* too," thus we may complete our proposition, "upon the *primitive* man, as he

stands revealed to us in the light of the researches of archaeology and of ethnology." (1911, in 1963, p. 186, italics original)

From this conviction, Freud embarked upon his most ambitious project for recapitulation: nothing less than the reconstruction of human history from psychological data on the development of children and neurotics. Freud often argued that the general libidinal development of individuals recapitulates a sequence of stages in the history of civilization. He compared the narcissism of young children with a primitive belief in the personification and power of thought (animism), the sexual attachment to parents (oedipal complex) with the development of monotheistic religion,[13] and the mature dominance of the reality principle with the later scientific phase of civilization.

> If we may regard the existence among primitive races of the omnipotence of thoughts as evidence in favor of narcissism, we are encouraged to attempt a comparison between the phases in the development of men's view of the universe and the stage of an individual's libidinal development. The animistic phase would correspond to narcissism both chronologically and in its content; the religious phase would correspond to the stage of object-choice of which the characteristic is a child's attachment to his parents; while the scientific phase would have an exact counterpart in the stage at which an individual has reached maturity, has renounced the pleasure principle, adjusted himself to reality and turned to the external world for the object of his desires. (1913, 1950 ed., p. 90)

But Freud had something far more specific in mind for recapitulation as a guide to the reconstruction of human history. From the existence of two coordinated phenomena in different series of the threefold parallelism—the oedipal complex of children (with its preservation in neurotics) and the totemism of savages—Freud made a bold foray into psychological anthropology.

Totem and Taboo (1913) bears the subtitle, "Some points of agreement between the mental life of savages and neurotics." "A boy's earliest choice of objects for his love is incestuous and those objects are forbidden ones"—his mother and sister (1950 ed., p. 17). Although the normal boy liberates himself naturally from these wishes as he matures, the neurotic with his "psychical infantilism" does not. Freud compares this neurotic behavior with a normal pattern in "savages"—where incestuous wishes among normal adult males remain so strong that taboos must be established to prevent fulfillment (that is, adult savages retain the transient juvenile stage of civilized white children): "It is therefore of no small importance that we are able to show that these same incestuous wishes, which are later destined to become unconscious, are still regarded by savage peoples as immediate perils against which the most severe measures of defense must be enforced" (p. 17).

Now, savages are living primitives and should behave as our ancestors did: "We can recognize in their psychic life a well-preserved picture of an early stage of our own development" (p. 1). What, then, can we infer about human history from the existence among savages of incest taboos and the associated doctrines of totemism (identification of the clan with a sacred animal that must be protected and revered throughout the year, save for one solemn holiday when it may be eaten; strict taboos upon males against sexual relations with women in the totemic clan), and from the reoccurrence of totemism and taboos in the oedipal complex of our children. Indeed, Freud states his recapitulatory aim in the preface to *Totem and Taboo:* "to deduce the original meaning of totemism from the vestiges remaining of it in childhood—from the hints of it which emerge in the course of the growth of our own children" (p. x).

In short (and for all its absurdities), Freud argues that the original human social group was a patriarchal horde, dominated by a ruling male, the father. The father dominated all the women and retained exclusive sexual rights to them. One day, his excluded sons banded together to kill and devour him. But they were so consumed with guilt for this deed of parricide that they renounced sexual contact with the women of their clan and identified their slain father with an animal that must be worshipped and not harmed. Yet once a year, they celebrated their deed of liberation in the totemic feast; for on that day the animal representing their father may be killed and consumed:

> If the totem animal is the father, then the two principal ordinances of totemism, the two taboo prohibitions which constitute its core—not to kill the totem and not to have sexual relations with a woman of the same totem—coincide in their content with the two crimes of Oedipus, who killed his father and married his mother, as well as with the two primal wishes of children, the insufficient repression or the reawakening of which forms the nucleus of perhaps every psychoneurosis. (p. 132)

Lest anyone imagine that the oedipal complex recalls only an ancient fear or longing among subjugated sons and not the deed itself, Freud ends his book by insisting that the primal act of parricide had occurred. Citing Faust's counter-comment to St. John, he writes: *Im Anfang war die Tat* ("In the beginning was the deed").

> I should like to insist that its [the book's] outcome shows that the beginnings of religion, morals, society and art converge in the Oedipus complex. This is in complete agreement with the psychoanalytic finding that the same complex

constitutes the nucleus of all neuroses, so far as our present knowledge goes. It seems to me a most surprising discovery that the problems of social psychology, too, should prove soluble on the basis of one single concrete point—man's relation to his father. (pp. 156–157)

His last book, *Moses and Monotheism,* is but a more specific rendering of the same scenario. Moses was an Egyptian by birth who cast his lot with the Jews and attempted to lead them out of captivity. But he was slain in rebellion by his adopted people who, in their crushing guilt, made him the prophet of a single omnipotent God and created the high ethical ideals that still motivate our "Judeo-Christian" civilization.

If recapitulation permits us to rediscover an unobservable past, might it not guide us in predicting an unexperienced future? In his 1930 essay *Civilization and Its Discontents,* Freud draws some gloomy analogies between the maturing of individuals and the human condition in increasingly complex modern societies—for "the development of civilization is a special process comparable to the normal maturation of the individual" (1961 ed., p. 45). Just as a mature man must sublimate his early urges for aggression and domination, so too must all members of society repress an increasingly larger set of basic biological instincts in order to live harmoniously in a more crowded, urbanized, and socially cohesive world: "If the development of civilization has such a far-reaching similarity to the development of the individual and if it employs the same methods, may we not be justified in reaching the diagnosis that, under the influence of cultural urges, some civilizations, or some epochs of civilization—possibly the whole of mankind—have become 'neurotic'" (p. 91).

Freud's early supporters and later rivals accepted his basic belief in recapitulation, but put it to different uses. C. G. Jung, for example, strongly supported recapitulation throughout his long career. He wrote in 1912:

All this experience suggests to us that we draw a parallel between the phantastical, mythological thinking of antiquity and the similar thinking of children, between the lower human races and dreams. This train of thought is not a strange one for us, but quite familiar through our knowledge of comparative anatomy and the history of development, which show us how the structure and function of the human body are the results of a series of embryonic changes which correspond to similar changes in the history of the race. Therefore, the supposition is justified that ontogenesis corresponds in psychology to phylogenesis. Consequently, it would be true, as well, that the state of infantile thinking in the child's psychic life, as well as in dreams, is nothing but a re-echo of the prehistoric and ancient. (1916 ed. pp. 27–28)

Thirty years later, long after biologists had abandoned the biogenetic law, Jung reaffirmed his support with some very wise words on the uses of the past in education:

> Childhood, however, is a state of the past. Just as the developing embryo recapitulates, in a sense, our phylogenetic history, so the child-psyche relives "the lesson of earlier humanity" as Nietzsche called it. The child lives in a pre-rational and above all in a pre-scientific world, the world of men who existed before us. Our roots lie in that world and every child grows from those roots. Maturity bears him away from his roots and immaturity binds him to them. Knowledge of the universal origins builds the bridge between the lost and abandoned world of the past and the still largely inconceivable world of the future. How should we lay hold of the future, how should we assimilate it, unless we are in possession of the human experience which the past has bequeathed to us? Dispossessed of this we are without root and without perspective, defenseless dupes of whatever novelties the future may bring. (1943, in 1954, pp. 134–135)

Yet, even though Jung spoke more elegantly than Freud of his belief in recapitulation, Jung's approach to psychoanalysis guaranteed that he would not make much use of Haeckel's doctrine within his system. The child, to be sure, recapitulates his past in proper phyletic sequence—but this has little relevance to the study and cure of adult neuroses. A child usually develops few psychological problems during the period of his recapitulation. He is dominated by instincts and does not understand the significance of the archetypes he is experiencing. Libido is not exclusively sexual and the causes of neurosis do not lie in the events of childhood; neuroses do not represent a fixation of sexual energy at an infantile (= ancestral) stage that should be repressed and superseded in normal development.

Return, then, to Freud's metaphor for the mind—Rome with all its buildings intact. For Jung, only this adult arrangement matters. The "buildings" do appear in chronological sequence during ontogeny, but this is not important. The adult mind contains an entire history of its past as racial memory in a collective unconscious. Jung's concept is static: knowing the ontogenetic order of racial memories does not facilitate the study and cure of neuroses. For neuroses develop in adults when development to wholeness (and adaptation to society) falters and libido is directed "backwards" into the primitive unconscious—there to animate the archetypes and place an individual under the domination of primitive ways. Neuroses are not infantile stages representing a definite time of ancestral history, but events of the moment that call forth images from a collective unconscious. Jung's appeal is not to recapitulation (an ontogenetically ordered series of ancestral stages), but to a general notion of racial memory (the static possession by adults of a complete racial history). As McCormick states:

"For Freud, the later problems of life arise during the early period of recapitulation when stages of advance are blocked. But for Jung the important stage is long after this period . . . Recapitulation ceases to be a question of research for Jung because the archetypes exist independently of any individual's development" (1973, p. 34).

If Jung found little use for recapitulation in practice, another of Freud's early supporters and later apostates carried Haeckel's doctrine to previously unimagined heights of folly and application—Sandor Ferenczi, in his *Thalassa, a Theory of Genitality* (1924). Ferenczi states explicitly his desire to import biological conclusions into psychology, particularly Haeckel's version of evolutionary theory. By his own admission, Ferenczi wrote *Thalassa* "as an adherent of Haeckel's recapitulation theory" (1968 ed., p. 3).

Today, Ferenczi is known, largely in ridicule, as Mr. Back-to-The-Womb—"where there is no such painful disharmony between ego and environment that characterizes existence in the external world" (p. 18). (I have no wish to stifle the ridicule, but merely to identify the recapitulatory basis of Ferenczi's theory.) Ferenczi saw sexual intercourse as a longing for return to ancestral conditions of repose in a timeless ocean—the "thalassal regressive trend . . . striving towards the aquatic mode of existence abandoned in primeval time" (p. 52). The sex act fulfills this primal urge in three ways: (1) post-ejaculatory repose symbolizes oceanic tranquility; (2) the penis (a symbolic fish, so to speak) reaches towards the womb (though only its secretion actually makes it)—women simply lose out here; and (3) the product of union passes its embryonic life in an amniotic fluid representing the ancestral ocean.

Haeckel had admitted the placenta as a primary example of cenogenesis—an exception to recapitulation. After all, no adult ancestor could have lived in an artificial pond created by its own skin. But Ferenczi argues that the female womb is a recapitulated ocean (This is, of course, utter nonsense in any but a symbolic context—though Ferenczi seems to support a literal interpretation). He compares contractions of the amnion during pregnancy to tidal cycles in the sea; he even claims that the erotic genital secretion of females has an oceanic basis: "the odor of the vagina comes from the same substance (trimethylamine) as the decomposition of fish gives rise to" (p. 57):

> If Professor Haeckel had the courage to lay down the basic biogenetic law of the recapitulation in the stages of embryonal development of the evolutionary history of the species (palingenesis), why should one not go further and assume that likewise in the development of the means of protection of the embryos (which up to this time has been regarded as the paradigm of cenogenesis) there is contained a bit of the history of the species . . . The arrangements for the protection of the germ cells are not new creations, and so do not

belong to cenogenesis, but on the contrary they too represent a kind of recapitulation—the recapitulation, namely, of the environmental situations which have been experienced during the development of the species. (pp. 45–46)

But Ferenczi does not stop with a recapitulatory comparison of womb and ocean. If sexual intercourse expresses a longing for return to a tranquil ocean, its symbolic striving may not only be towards a piscine past, but further back towards the ultimate tranquility of a Precambrian world without life. The death wish is itself a memory of our inorganic ancestors: "We have represented in the sensation of orgasm not only the repose of the intrauterine state, the tranquil existence in a more friendly environment, but also the repose of the era before life originated, in other words, the deathlike repose of the inorganic world" (p. 63).

Thus, the recapitulatory cycle begins with coitus (= striving for death = the earth before life) followed by impregnation (= the dawn of life). The fetus then begins its embryonic life by repeating the earliest stages of an amoeboid past. Birth represents the colonization of land by tetrapods (even though any Haeckelian biologist would have argued that amphibian or reptilian stages were long superseded by this time). Believe it or not, the latency period following infant sexuality recapitulates the ice ages of our phyletic past (p. 70). (Though lest one wonder why we didn't do ourselves in by declining to copulate during cold times, Ferenczi assures us that the ice ages only redirected *some* of our genital drives to the development of "higher" intellectual and moral activity.)

Few intellectual movements have had as much influence (from national consciousness to cocktail party conversation) as twentieth-century psychoanalytic theory. I have tried to argue that these theories cannot be properly assessed or even understood without recognizing their links to the biogenetic law. Yet these links have rarely been mentioned because so few psychologists and historians have any inkling of Haeckel's doctrine and its impact.[14] Millions of lives have been influenced or molded by theories shaped in the light of a basic tool for any "enlightened" late nineteenth-century thinker—recapitulation. I can offer no greater testimony to Haeckel's influence and no better demonstration of why it behooves us to study and to understand this abandoned doctrine.

Epilogue

If I may practice the historian's sin of judging the past in a current context, I find a tension throughout this chapter between two uses of recapitulation in nonbiological fields. On the one hand, recapitulation is cited in the name of greater individual freedom and liberation from ancient constraints—mold education to the child's nature, for he is repeating his ancestry and it must be so; do not impose adult criteria for discipline and morality upon a savage child. On the other hand, it is used to deny freedom by consigning certain individuals to biological inferiority—criminals and "lower" races.

The common theme is biological determinism. All supporters of recapitulation have used it to make biological claims about human nature and to defend a notion of inevitability for selected aspects of behavior and social status. Lombroso, for example, regretted that nature made criminals, but defended a treatment of them as incorrigible: "We are governed by silent laws which never cease to operate and which rule society with more authority than the laws inscribed on our statute books. Crime . . . appears to be a natural phenomenon and, if we may borrow from the language of philosophy, a necessary phenomenon, as are birth, death, conception and mental illness" (1887, p. 667).

Statements about innate biology might seem, at first, to be an equally good strategy for liberals—what better argument can anyone advance against a restrictive practice than the claim that it prevents the expression of our genetic nature. But the argument can always be turned around. Equating innate with inevitable almost always tips the balance in favor of the status quo—for how could it arise in opposition to our nature. Consider, for example, this defense of capitalism as a biological necessity:

> Do we believe that the child recapitulates the history of the race? If so we may not be surprised to find the passion for property-getting a natural one. Selfishness is the cornerstone of the struggle for existence, deception is at its very foundation, while the acquiring of property has been the most dominant factor in the history of men and nations. These passions of the child are but the pent up forces of the greed of thousands of years. (Kline and France, in Thayer, 1928, p. 64)

One man—and a very acute one—saw through the liberal usage to recognize the profoundly conservative nature of recapitulatory arguments as a species of biological determinism: John Dewey. Dewey began his career as an educational reformer by supporting the culture-epochs theory. But in 1916, he rejected it decisively in a book with an appropriate title, *Democracy and Education.* He had recognized the philosophical context of the theory and had learned to appreciate the general dangers of deterministic arguments. He finally rejected the culture epochs for their implication that "past life has somehow predetermined the main traits of an individual, and that they are so fixed that little serious change can be introduced into them" (p. 86).

Dewey did not deny a role for heredity in setting limits to education. He did not make the dogmatic plea that arguments about innate nature be rejected a priori (he was quite prepared to work with them if the evidence warranted it). He merely stated his opposition to determinism in the absence of persuasive evidence. We now believe that all the arguments cited in this chapter for constraints based upon recapitulation are fundamentally incorrect. Yet they carried great weight in their day because science stood for them and a liberal assertion of unconstrained potential smacked of sentimentality. Antideterminists always have this cross to bear, but the collapse of recapitulatory arguments about human potential suggests that history may be on their side.

—6—

Decline, Fall, and Generalization

A Clever Argument

Ernst Haeckel, consummate as ever in debate, had structured his argument ingeniously. His biogenetic law was an exhaustive taxonomy of all possible results, for its umbrella extended to include the treacherous interpolations of cenogenesis along with the phyletic markers of palingenesis. It could engender no refutation because it included all phenomena. Haeckel's sycophants could ignore this tautological necessity, and rejoice with Schmidt that "the biogenetic law has no exceptions" (1909, p. 125).

Yet a fair moderator of this debate would have admitted a crucial question about relative frequency: the utility of Haeckel's law depended upon the dominance of palingenesis over cenogenesis. But if exceptions are more frequent then the supposed rule, then the theory must fall. It is a common assumption, repeated in almost every textbook, that Haeckel's theory of recapitulation collapsed under the ever-increasing weight of slowly accumulated exceptions—particularly of larval adaptations and paedomorphosis.[1] My aim in this chapter is to argue that nothing could be further from the truth.

Natural history does not refute its theories by cataloguing empirical exceptions to them (while working within a paradigm that engendered the theory in the first place). With millions of potential examples in a discipline second to none for its superabundance of empirical information, how can a catalogue of counter cases ever refute a theory—especially when the theory itself allows a "reasonable" number of exceptions?[2] Proponents can always furnish their lists as well. And since each list must include a ridiculously small percentage of all possible cases, how can a theory of natural history be rejected by simple enumeration? We cannot know whether cenogenesis is "really" more common than palingenesis. What we can test is the validity of proposed laws—terminal addition and condensation—that entail the dominance of palingenesis.

In this chapter, I will treat the unsuccessful attempts of empirical cataloguers to refute Haeckel's theory of recapitulation. I will then argue that the biogenetic law fell only when it became *unfashionable in approach* (due to the rise of experimental embryology) and finally *untenable in theory* (when the establishment of Mendelian genetics converted previous exceptions into new expectations). The biogenetic law was not disproved by a direct scrutiny of its supposed operation;* it fell because research in related fields refuted its necessary mechanism. If these arguments offend some scientists' beliefs about the way science should operate, they reflect, nonetheless, the way it does operate.

An Empirical Critique

Many critics tried to refute recapitulation by using it unsuccessfully—by working within its confines to obtain confusion rather than phylogeny. This empirical approach engendered three main objections to Haeckel's doctrine:

1. Acceleration is not general or equal for all organs. Each ontogenetic stage is an inseparable mixture of organs in different stages of ancestral repetition.

2. Larvae and embryos have evolved many features as adaptations to their own mode of life. New characters can be introduced at any stage of ontogeny, not only as additions to the end of ancestral growth.

3. Development can be retarded as well as accelerated. Embryonic or larval stages of ancestors can become the adult stages of descendants—a phenomenon directly opposite to recapitulation.

* I do not mean to imply that this direct scrutiny only served to buttress recapitulation. In fact, it produced interminable arguments that could not be resolved in the absence of firm criteria for distinguishing palingenetic from cenogenetic features. I am only arguing that there was nothing in these debates to compel believers to question the premises of the theory itself (Marshall, 1891, and MacBride, 1914, for example, express sorrow at the continued wrangling but are confident that further factual inquiry will resolve all the issues). The debates did, however, discourage many young scientists from working with the theory and led them to experimental work, with its promise of true testability. Both Roux and Driesch, after all, had studied with Haeckel.

To each of these arguments, Haeckel's supporters had ready replies:

1. We will redefine recapitulation to apply to individual organs rather than to entire organisms.

2. It doesn't happen often. Or, if it does happen often, it can be recognized and its effects removed.*

3. Any descendant endowed with previously larval features is a degenerate exception to the progressive evolution of lineages.

None of these objections was, or could have been, rejected. The first was handled by accommodation, the last two by a denial of their importance. All three, after all, belonged in the domain of permissible cenogenesis. I shall consider them in order before passing on to the successful attack.

* The problem of recognition produced severe difficulties. How can one tell whether a given larval feature is an interpolated adaptation or an accelerated ancestor? Reliance on embryological data alone involves the danger of circular reasoning, since both the criterion and decision may arise from the same source. The standard response, discussed at great length by Haeckel, invoked Agassiz's threefold parallelism: we seek corroborative evidence from the sequences running parallel to ontogeny—comparative anatomy and paleontological succession. Thus, the parasitic larval stages of fresh-water unionid clams (living as parasites on fish gills) are clearly interpolations because no parasites are found among related but more primitive adults, and because the fossil record of this group includes only a sequence of ordinary bottom-dwelling clams (the appropriate fish had probably not even evolved in time to harbor such a potential parasitic ancestor). In cases lacking such corroboration, careful supporters of the biogenetic law often declined to apply it at all (Mehnert, 1897, p. 5). Gegenbaur, for example, defended the classic comparative anatomy of adults as a necessary handmaiden to the use of ontogenetic data in elucidating phylogeny: "Comparative anatomy is no mere substitute [*Ersatz*] for the gaps that exist in ontogeny. It is no phylogenetic stopgap that will disappear if, one day, the entire domain of ontogenetic knowledge is made manifest and clear . . . Comparative anatomy furnishes the corrections for features that are introduced into ontogeny by cenogenesis" (1889, pp. 8–9).

In practice, the separation of cenogenetic features was not difficult when they were highly specialized characters of very small groups. Comparative embryology alone would have satisfied most scientists for the case of unionid clams—the larval adaptation is unique, evidently useful, and present in the ontogeny of no other taxa in this large class of molluscs. If it is palingenetic, then either unionids have an ancestry separate from all other clams or else all others have lost this ancestral stage. While not illogical, neither of these positions invites belief. An ontogenetic stage held in common by all members of a large group often presented serious problems. The placenta could easily be cast as an embryonic adaptation since its form made no sense as an adult configuration. But the trochophore larva, for example, functioned perfectly well as an independent organism. Its widespread occurrence could be interpreted as the sign of its distant ancestral status, its development as a larval adaptation in a common ancestor, or its evolution as a convergent feature of several groups. Cases like this were the bugbear of recapitulation; they inspired the endless and fruitless arguments that could never disprove recapitulation, but could easily drive young scientists away from comparative embryology in utter frustration.

Organs or Ancestors: The Transformation of Haeckel's Heterochrony

Many have traced the demise of Haeckel's theory to the inadequacy of his gastraea—that "lean animal-specter"[3]—as a satisfactory ancestor for all metazoans; or rather to the general inadequacy—or absurdity—of the almost countless creatures that recapitulationists elevated from their fleeting appearance in modern embryos to hypothetical adult ancestors of great antiquity.

The epitome of Haeckel's *Natürliche Schöpfungsgeschichte* (1868) is a chart of twenty-two human ancestors based upon embryology, comparative anatomy, and paleontology. Later stages rested heavily on the evidence of fossils and the anatomy of modern animals, but the primeval forms were constructed solely from earliest ontogeny. Though their existence relied upon no other evidence, they were given names and placed at the base of Haeckel's famous tree (Fig. 20). The following account of five stages appears in later editions of *Natürliche Schöpfungsgeschichte* (1892, for example). Gastraea makes no appearance in the first edition of 1868, since Haeckel did not publish his gastraea theory until 1874.

Stage 1—Monera (the primordial, anucleate ancestors of all animals).[4] Evidence—the monerula (the fertilized ovum after disappearance of the germinal vesicle). As Russell writes: "It was still believed by many that the egg-nucleus disappeared on fertilization. The true nature of the process was not fully made out till 1875, when O. Hertwig observed the fusion of egg- and sperm-nuclei in *Toxopneustes,* a sea urchin" (1916, p. 291).

Stage 2—Amoeba. Evidence—the cytula (the ovum after reformation of its nucleus). As evidence, Haeckel cited the amoeboid motions observed in some egg cells (Fig. 21). With characteristic assurance, he argued: "An irrefutable proof that such single-celled primeval animals really existed as the direct ancestors of Man, is furnished according to the fundamental law of biogeny by the fact that the human egg is nothing more than a simple cell" (1892, 2:381).

Stage 3—Synamoeba (the first association of undifferentiated amoeboid cells to form the earliest multicellular organism). Evidence—the morula (the mass of cells produced by initial cleavages of the fertilized egg).

Stage 4—Blastaea (an ancestral free-swimming form). Evidence—the blastula (hollow sphere of cells formed after the initial cleavages).

Stage 5—Gastraea (a two-layered differentiated form; the common ancestor of all Metazoa). Evidence—the gastrula (two-layered sac

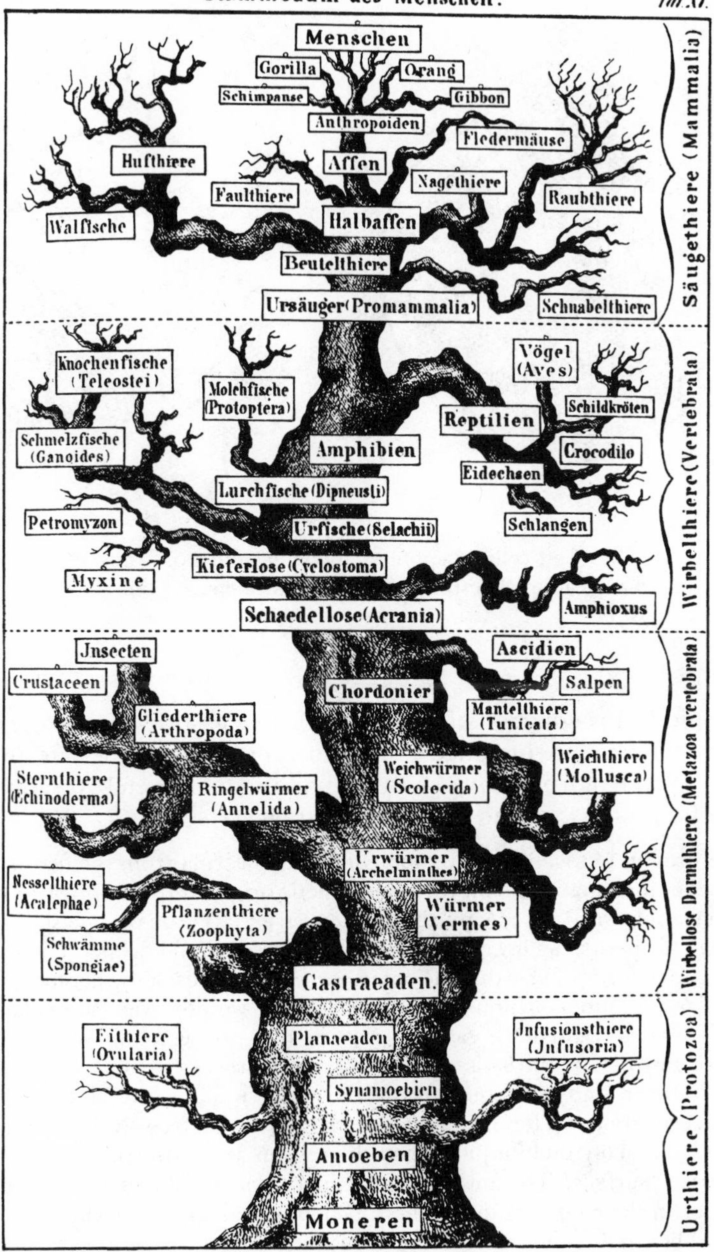

Fig. 20. Haeckel's evolutionary tree. The first five stages (monera, amoeba, synamoeba, planaea, and gastrea) are reconstructed almost entirely from ontogeny of higher forms. (From Haeckel, 1874.)

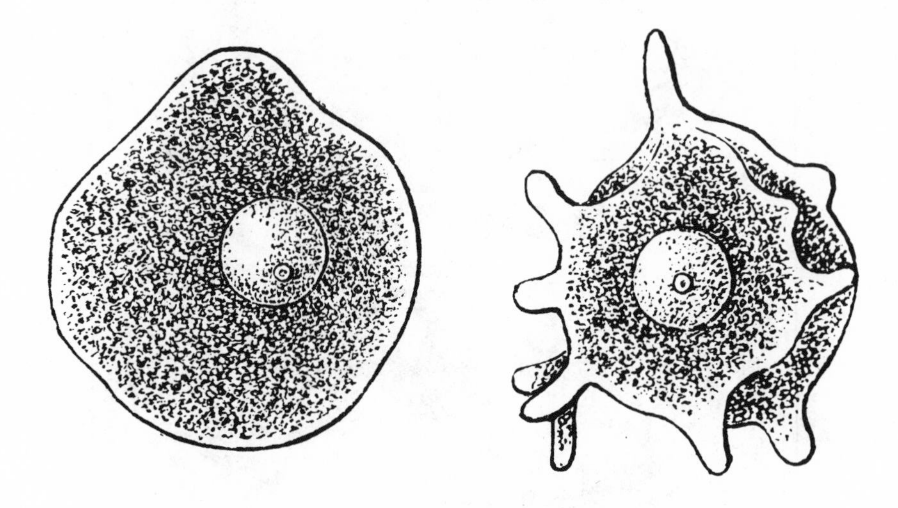

Fig. 21. Egg cells of sponges showing amoeboid movements and indicating ancestry as free-living amoeba. (From Haeckel, 1874.)

formed by invagination of the blastula). The gastraea, Haeckel's most famous invention, provided an explanation in phylogenetic terms not only for the morphology of gastrulation in modern embryos, but also for the products of the resulting germ layers:

The present ontogenetic development of the gastrula from the blastula still provides information about the phylogenetic origin of the gastraea from the planaea [blastaea] . . . A pit-shaped depression appears on one side of the spherical blastula, an invagination which gets deeper and deeper. Finally, this invagination goes so far that the outer, invaginated part of the blastoderm lies right on the inner or non-invaginated part. If we now want to explain the phylogenetic origin of the gastraea (repeated, according to the biogenetic law, by the gastrula) on the basis of this ontogenetic process, we must imagine that the single-layered cell-community of the sphaerical planaea [blastaea] began to take in food preferentially at one part of its surface. Natural selection would gradually build a pit-shaped depression at this nutritive spot on the spherical surface. The pit, originally quite flat, would grow deeper and deeper in the course of time. The functions of taking in and digesting food would be confined to the cells lining this pit; while the other cells would take over the functions of locomotion and protection. Thus, the first division of labor arose among originally similar cells of the planaea [blastaea]. This earliest histological differentiation had, as a consequence, the separation of two different kinds of cells—nutritive cells in the pit and locomotory cells on the outer surface. (1874a, pp. 392–393)

These hypothetical creatures are well known; not so celebrated is the twenty-first stage, Haeckel's hypothetical *Pithecanthropus alalus,* the speechless ape-man. This genus is listed first in the classification of *Generelle Morphologie* (1866, 2:clx) and justified later in *Natürliche Schöpfungsgeschichte* (1868). Haeckel drew upon all aspects of the threefold parallelism to construct our immediate ancestor, but opted for a speechless state primarily because babies do not talk: "The evolution of language also teaches us (as well, to be sure, from its ontogeny in each child as from its phylogeny in each group [*Volk*]) that uniquely human conceptual speech was developed gradually only after the rest of the body had attained its characteristic human form" (1874a, p. 496).

Pithecanthropus may be the only zoological genus that received its name before it was found. When Eugene Dubois (1896) uncovered the bones of an ancestral human in Java, he happily applied Haeckel's name, though he changed the specific designation to reflect another of Haeckel's correct predictions (1868, p. 508)—that this ancestor would be fully erect though small brained—thus, *Pithecanthropus erectus* (now generally classified as *Homo erectus*).

Despite its few successes, Haeckel's method of creating hypothetical ancestors from the stages of ontogeny plunged the conclusions of recapitulation into endless and fruitless controversy. If organs were accelerated at different rates, and if cenogenetic features were interpolated into early ontogeny, how could any distinct ontogenetic stage represent, in toto, an extinct ancestor? When opponents of recapitulation needed a whipping boy, they fastened eagerly upon these inventions of facile, if uncritical, minds. Consider two statements: one published in the midst of battle in 1886, the other in a postmortem of 1928:

> Courageous hypotheses—daring conclusions—these almost always are of service to Science. But Schemata injure her if they bring existing knowledge into an empty and warped pattern, and claim thereby to give deeper understanding. Unfortunately, the gastraea was not fertile, but it was strongly infectious; it has propagated itself as Neuraea, Nephridaea, etc. and is guilty of all the Original-animals, the Trochosphaera, the Trochophora, the Original-insect, and I know not what besides. (Kleinenberg, 1886, in Oppenheimer, 1967, p. 270)

> As a result of such speculations multitudes of phylogenetic trees sprang up in the thin soil of embryological fact and developed a capacity of branching and producing hypothetical ancestors which was in inverse proportion to their hold on solid ground. (Conklin, 1928, p. 71)

How did recapitulationists react to the telling criticism that no ontogenetic stage could represent a complete ancestor. Some, like

Haeckel, pressed on regardless. Most simply altered their method of application. No committed believer abandoned the theory on this account.

The obvious alteration involved a simple shift of perspective. If organs are accelerated at different rates, then each organ must be considered separately, for each still repeats the stages of its own evolution. Heterochrony—Haeckel's term for cenogenesis involving the unequal acceleration of organs—destroyed the attempt to invent entire ancestors, but it retreated to irrelevance when recapitulationists decided to analyze each organ separately.

We have already seen how Cope and Weismann developed this alteration; we have also argued that it quickly became the methodological focus of nearly all serious work in comparative embryology. Recapitulation was preserved—indeed, it was strengthened—by accommodation.

In the 1890s, a group of German morphologists sought to study directly the unequal acceleration of organs (Oppel, 1891; Keibel, 1895, 1898; Mehnert, 1891, 1895, 1897, 1898). They developed a series of *Normentafeln* (standard tables) to depict the relative development of organs in the ontogeny of several common vertebrates. By comparing tables, the peculiarities of individual species could be identified and a reason for especially slow or rapid acceleration sought in the form and function of adult organs. Heterochrony, they believed, should not merely be treated as an exception to palingenesis; its own operation should be codified to yield laws explaining why some organs develop at such uncommon rates.

Mehnert (1891) began by trying to link an organ's time of appearance in ontogeny to its size or relative importance in the adult. Later, both he (1897, 1898) and Keibel argued strongly that rapid acceleration always accompanied early and persistent function: "I seek," wrote Keibel, "a connection between the function of an organ and its time of appearance in ontogeny. I hold that an organ appears early in ontogeny if it begins to function early" (1898, p. 786). Massart examined heterochrony in plants and reached similar conclusions:

1. Organs which must function first develop first.
2. Organs which must function at the same time develop in the order of their [final] size. (1894, p. 236)

Although Keibel and Mehnert reached the same conclusions and lavished the highest praise on each other's work (Keibel, 1898, p. 787), they held completely different opinions of its effect on the validity of recapitulation. Mehnert fully accepted the decision of Cope and

Weismann to view each organ as a separate case of recapitulation. Thus, differential acceleration is only "cenogenetically modified palingenesis" (1895, p. 436) and each organ "repeats its phylogeny in the most minute manner" (1897, p. 106). Mehnert wrote:

> The biogenetic law has not been shaken by the attacks of its opponents. The development of each organ is entirely and exclusively dependent upon phylogeny. But we must not expect that all the stages evolving together in a phylogenetic series will appear *at the same time* in the ontogeny of descendants because the development of each organ follows its own specific rate . . . The embryo does not repeat in detail one and the same phyletic stage; it consists rather of an assemblage of organs, some at a phyletically early stage of development, some at a phyletically older stage. (1898, pp. 148–149)

Mehnert viewed his entire effort as a positive contribution to the biogenetic law. Heretofore, heterochrony had been classed as an exception and falsification of the phyletic record. But if he could establish a law relating heterochrony to evolutionary changes in the time and intensity of an organ's function, then it would have phyletic significance: "Insofar as embryonic cenogenesis is only a direct and lawful result of phyletic developmental energies [*Entfaltungsenergien*], we are completely unjustified if we—as has sometimes happened—designate it as a falsification."[5]

But Keibel refused to accept a redefinition of recapitulation in terms of individual organs. Using the same data, he attacked Mehnert's interpretations:

> Mehnert stands up very strongly for the biogenetic law. Yet it is perfectly obvious that Mehnert's biogenetic law is something quite different from Haeckel's. I have spoken out against the biogenetic law because the temporal shifts that occur in the organogenesis of higher animals make the appearance of ancestral stages completely impossible. Yet the doctrine of ancestral stages is, in Haeckel's opinion, one of the most important components of the so-called biogenetic law. (1898, p. 790)

Despite Keibel's disclaimer, the biogenetic law had been rescued from the strictures of its inventor. Haeckel's reading fell from favor and his law, refractory as ever to empirical criticism, retained its popularity through a redefinition in terms of individual organs. This redefinition has persisted ever since (Yezhikov, 1933, p. 75; Lebedkin, 1937, p. 393). The authors of the most recent and complete treatment of recapitulation write: "The ties that bind ontogeny and phylogeny must not be studied animal by animal, but rather organ by organ" (Delsol and Tintant, in press).

Interpolations into Juvenile Stages

Early studies of marine plankton by Johannes Müller and others had revealed an astonishing diversity of larval forms, exquisitely designed for the conditions of their juvenile existence. These adaptations for dispersion and protection were well known to Haeckel, himself no stranger to marine exploration (Haeckel wrote the volumes on siphonophores and radiolarians for the *Challenger* reports). Even Haeckel could not attribute ancestral significance to all these structures, for they were clearly the transitory adaptations of floating or swimming larvae belonging to lineages whose adults had been sessile throughout their geologic history. Thus, Haeckel gathered them into his first and largest category of cenogenesis—larval adaptations interpolated into the palingenetic record, but not altering it in any other way.

Critics of the biogenetic law often cited the frequency of such interpolations as an argument for abandoning the law as useless in practice. Giard (1905), for example, gathered under the name of *poecilogonie* all cases known to him in which similar adults developed from strikingly different larvae. The biogenetic law, he argued, would attribute the adult similarities to convergent evolution and view the divergent larvae as signs of distinct ancestry; but such a position would be absurd for the many cases of nearly identical adults bearing all the characteristic features of their phylum. The larval differences are adaptations: "The larvae have become divergent by adapting themselves to different environments. Heredity has maintained the similarity of adults" (p. 154). Giard argued by enumeration to ridicule the biogenetic law, and affirmed that he had adopted his position only as his counter-cases accumulated: "When, about fifteen years ago, I reported the first known cases of poecilogony, the facts appeared rare and exceptional. Since then, they have been observed very often and in almost all groups of animals" (p. 155).

The standard response of recapitulationists was generous and frustrating. They simply admitted every case and remained supremely confident that none of them could render illegible a primarily palingenetic record. F. M. Balfour, England's staunchest recapitulationist, had written: "There is, so far as I see, no possible reason why an indefinite number of organs should not be developed in larvae to protect them from their enemies and to enable them to compete with larvae of other species, and so on. The only limit to such development appears to be the shortness of larval life" (1880, 2:364).

As a stronger tactic, some opponents tried to make a quantitative assessment, arguing that cenogenesis was far more common than a supposedly primary palingenesis. Rabaud wrote: "In summary, the 'falsifications' are so pervasive that they completely mask the homologies upon which naturalists have tried to establish the filiation of organisms . . . They are sufficient to render inapplicable in practice the principle of Fritz Müller" (1916, p. 275).[6]

The response of recapitulationists was equally generous and even more frustrating. They were usually quite willing to classify most juvenile features as cenogenetic—while they stoutly maintained that such features could be recognized and separated from the palingenetic markers of ancestry. Carl Gegenbaur, Haeckel's dear friend and firm though critical supporter, wrote: "We meet on each path [of ontogenetic development] many, I would even say more, features that are never realized as permanent stages of lower creatures. They often cover the palingenetic features with a thick veil" (1888, p. 4). Yet recapitulationists retained their faith that the veil could always be lifted:

> But we must not on that account "empty out the child with the bath" and conclude that there is no such thing as a "biogenetic law" or recapitulation of the phylogeny in the ontogeny. Not only is there such a recapitulation but—as F. Müller and Haeckel have already said —ontogeny is nothing but a recapitulation of the phylogeny, only with innumerable subtractions and interpolations, additions and displacements of the organ-stages both in time and place. (Weismann, 1904, pp. 174–175)

Introduction of Juvenile Features into the Adults of Descendants

Recapitulation requires that adult features of ancestors appear in the juvenile stages of descendants. Nothing, therefore, can be more contrary to its operation than the incorporation of previously juvenile features into the adult stages of descendants. In the 1920s Walter Garstang emphasized this contradictory process in a series of articles (1922, 1928, 1946) and a delightful book of posthumously published poetry (1951). Many have supposed that Garstang's elucidation of "paedomorphosis" disproved recapitulation.[7] In fact, recapitulationists had recognized this phenomenon from the beginning; they had discussed it at length, and had catalogued as many cases as Garstang ever knew (Kollmann, 1885; Boas, 1896).

Every subject has its *Drosophila;* in this case, the exemplar's role is held by axolotl, a Mexican salamander that usually reproduces as an aquatic larva (see Chapters 8 and 9 for a summary of its biology). Garstang devoted one of his poems to this reluctant metamorphoser; it explains the situation better than any tedious paragraph from my pen.

The Axolotl and the Ammocoete

Amblystoma's[8] a giant newt who rears in swampy waters,
As other newts are wont to do, a lot of fishy daughters:
These Axolotls, having gills, pursue a life aquatic,
But, when they should transform to newts, are naughty and erratic.

They change upon compulsion, if the water grows too foul,
For then they have to use their lungs, and go ashore to prowl:
But when a lake's attractive, nicely aired, and full of food,
They cling to youth perpetual, and rear a tadpole brood.

And newts Perennibranchiate[9] have gone from bad to worse:
They think aquatic life is bliss, terrestrial a curse.
They do not even contemplate a change to suit the weather,
But live as tadpoles, breed as tadpoles, tadpoles altogether!

(Garstang, 1951, p. 62)

Although the axolotl, a favored Aztec delicacy, had been known to western scientists since the Spanish conquest, its ontogenetic status had remained something of a mystery. Some accepted it as a "finished" form and classed it among the perennibranchiate amphibians; others, like Cuvier (1828, p. 416), insisted that it must be the larva of an unknown salamander. Milne-Edwards (1844, p. 77), foreshadowing the coming resolution, accepted it as an adult but refused to accord it a "primitive" status among amphibians. He explained its persistent larval features as an arrest of development within the idealized urodelian type (see also Leuckart, 1821, p. 262).

The observations of A. Duméril at the menagerie of Paris's *Muséum d'Histoire Naturelle* finally resolved the issue. In January of 1864, he received six axolotls. One year later, they reached sexual maturity in the larval stage and mated successfully; in September of 1865, the eggs hatched. The axolotl had matured and bred successfully as a tadpole and its status seemed assured (1865a). Later that September, to Duméril's great surprise (1865b), two of the offspring metamorphosed to adult salamanders of the genus *Ambystoma* (Fig. 22). By the beginning of 1866, eleven offspring had metamorphosed, although the original parents remained in their larval state. Duméril's attempts to induce metamorphosis by excising the external gills and forcing the tadpoles to a more terrestrial existence were only marginally successful (1867, 1870).

Earlier, in 1861, de Filippi had discovered sexually mature larvae of *Triturus alpestris* among a collection of normally metamorphosed newts. These two cases proved that highly developed forms could revert to simpler stages by becoming sexually mature as juveniles. This discovery also forced a reassessment of the perennibranchiate amphibians, with their permanent larval features. Previously, the perennibranchiates had generally been regarded as ancestral salamanders (in evolving metamorphosis, the higher salamanders simply added a stage to the end of ontogeny in orthodox compliance with recapitulatory expectations). But if axolotl represented a juvenile stage of higher salamanders, then the perennibranchiates had merely gone a step further and committed themselves to permanent youth by dispensing entirely with their adult form—an undoubted exception to recapitulation. In 1885,[10] Kollmann designated this retention of larval features as "neoteny" (p. 391).

Meanwhile, further cases—observed and inferred—began to accumulate in nearly all phyla and at an ever-increasing rate. Anton Dohrn, for example, upheld the annelid theory of vertebrate origins. To this belief, the existence of *Amphioxus* and larval tunicates posed a threat as potential ancestors having nothing to do with annelids. Dohrn circumvented this problem by casting them as degenerated fish. He speculated that if the ammocoetes larva of cyclostome fishes became sexually mature, a form representing *Amphioxus* would quickly arise.

> Now look at *Ammocoetes* there, reclining in the mud,
> preparing thyroid-extract to secure his tiny food:
> If just a touch of sunshine more should make his gonads grow,
> The Lancelet's claims to ancestry would get a nasty blow![11]
>
> (Garstang, 1951, p. 62)

Bolk's theory of fetalization (1926c) was not, as many suppose, the first invocation of paedomorphosis* as an agent for human evolution (Moreover, Bolk—a strong supporter of recapitulation [1926b][12]—introduced it as an exception, not a threat, to Haeckel's theory.) Cope, America's foremost recapitulationist, ascribed many human physical traits to the retention of ancestral, embryonic features through retardation. Speaking of the human face, he wrote: "As these characters result from a fuller course of growth from the infant, it is evident that in these respects the apes are more fully developed than man. Man stops short in the development of the face, and is in so far more embryonic. The prominent forehead and reduced jaws of man are characters of 'retardation'" (1883, in 1887, p. 286).

* Paedomorphosis is a general term designating the retention of youthful ancestral features by adult descendants. We now restrict "neoteny" to cases involving the retardation of somatic development and designate as "progenesis" those cases of paedomorphosis produced by accelerated maturation and the precocious truncation of ontogeny. See Chapter 7 for definitions. Nineteenth-century writers often used neoteny in the broad meaning that paedomorphosis bears today.

Fig. 22. The first successful transformation of the Mexican axolotl into an adult *Ambystoma.* (From Duméril, 1867.)

B

A

A, Axolotl non transformé. — B, Axolotl transformé.

More direct examples continued to accumulate from ontogeny's partners in the threefold parallelism. In paleontology, agnostid trilobites (Jaekel, 1909) and several genera of terebratellid brachiopods (Beecher, 1893) were ascribed to embryonic retention:

> They are true fixed genera. This stability can even be favored by the fact that development of genital glands can occur at the first stages [of ontogeny] and that the individual, capable of reproduction before attaining its definitive form, can therefore give rise to a series of descendants having a tendency to stop definitely at an [early ontogenetic] stage of their ancestors. This fact permits us to explain the irregularity that we encounter in searching to establish a rigorous parallelism between ontogenetic and phylogenetic development. (Fischer and Oehlert, 1892, p. 3)

The ammonites, Hyatt's bulwark of recapitulation, began to yield their opposite examples (Pavlov, 1901; Smith, 1914). Even Hyatt himself (1870) once admitted that several dwarfed species had retained the juvenile features of their ancestors.

From comparative anatomy, Chun (1880, 1892) discovered that larval ctenophores of most species are sexually mature.[13] And, like a Phoenix from the ashes of Naturphilosophie, old von Baer spoke from his Russian outpost to report several cases of parthenogenesis in insect larvae structurally unable to copulate (1866). Von Baer named this phenomenon "paedogenesis." This term was widened by Hamann (1891)[14] to include both sexually reproducing larvae and the possibility of phyletic fixation in the larval state ("phylo-pädogenie").

Axolotl was one of the causes célèbres of late nineteenth-century morphology; every recapitulationist had to deal with it, and the growing number of similar examples, in some way. Haeckel acknowledged it as an "exceedingly curious case" (1868, p. 192), but managed somehow to convince himself that axolotl might represent a persistent ancestral type. Eimer concurred: "In the spotted Salamander and its nearest allies the metamophosis into an exclusively air-breathing animal has become the rule, while in *Siredon* [the generic name given to axolotl before its metamorphosis was discovered] it is only commencing, and occurs at first in rare individual cases" (1890, p. 47).

Few others dared to deny that axolotl was an exception to recapitulation. But no one abandoned the biogenetic law on its account, for there was an easy and eminently respectable way out. Recapitulation was intimately tied to the idea of progressive evolution, since it required that phylogeny proceed by *adding* stages to ancestral ontogenies. But paedomorphosis implies a *subtraction* of stages, a truncation of development—a phyletic reversion to a previous ancestral state. Paedomorphic forms are degenerate exceptions to life's history of pervasive, progressive development.

Weismann treated the axolotl in a long article and wrote of its impact upon recapitulation:

According to this [biogenetic] law, each step in phyletic development, when superseded by a later one, must be preserved in ontogeny, and must therefore appear at the present time as an ontogenetic stage in the development of each individual. Now my interpretation of the transformation of the axolotl seems to stand in opposition to it [recapitulation], for the axolotl, which was *Amblystoma* [sic] in earlier times, preserves nothing of *Amblystoma* in its ontogeny. The contradiction is, however, only apparent. As long as we deal with a true advance in development and, therefore, with the attainment of a new stage never reached before, then we will find ancestral stages in ontogeny. But this is not so when the new stage is not truly new, but had once been the final stage of an earlier ontogeny; or, in other words, when we deal with a reversion to the previous phyletic stage . . . In this case, the previous final step of ontogeny is simply eliminated . . . [The axolotl] has reverted to the perennibranchiate stage, and the only trace that remains of its former developmental status is the tendency, more or less retained in each individual, again to ascend to the salamander stage under favorable conditions. (1875, p. 328)

Moreover, lest anyone think that the biogenetic law had been in any way compromised by this peculiar larva, Weismann hastened to add: "Its general accuracy and validity can be shown, in an inductive way, to be so highly probable [*in so hohem Grad wahrscheinlich gemacht werden*] that, today, it is doubted by very few scientists who deal with embryology and comparative morphology" (p. 328).

Recapitulationists engaged in a good deal of semantic quibbling about the evolutionary role of axolotl (atavism, developmental arrest, simple truncation, or true reversion), but the general line of Weismann's argument was followed by all (Cope, 1869, in 1887, p. 88; Kollmann, 1885, p. 392; Wiedersheim, 1879; and Balfour, 1880, 2: 143). The only original notion came as a speculation from E. Ray Lankester, one of England's staunchest supporters of the biogenetic law. Had its implications not been ignored, it might have provided a serious challenge to Weismann's resolution. Lankester noted that paedomorphosis need not result in simple reversion to a former adult stage, for the features transferred from larva to adult need not be the accelerated adult stages of former ontogenies. They may have arisen as adaptations cenogenetically interpolated into larval life. The paedomorphic adult would then display features never before seen in its adult ancestors. Paedomorphosis can produce something new and progressive—not only the recall of a distant past. "By super-larvation [Lankester's term for paedomorphosis] it would be possible for an embryonic form developed in relation to special embryonic conditions and not recapitulative of an ancestry, to become the adult form of the race, and thus to give to the subsequent evolution of that race a totally and otherwise improbable direction" (1880, in Coleman, 1967, p. 129). Yet, the appendix to a treatise on *Degeneration* was not an auspicious place to launch such a suggestion. There it languished awaiting resuscitation under the aegis of an evolutionary theory more willing to include paedomorphosis among its orthodox phenomena. And then it re-emerged as de Beer's "clandestine evolution" (1930, p. 30).

Meanwhile, Weismann's argument was pursued by recapitulationists who dismissed paedomorphosis as an anomalous and unimportant exception (Schmidt, 1909, p. 46). Mehnert, for example, declared "that, on the one side, acceleration of ontogeny and progressive evolution are bound together; on the other, retardation of development and regressive tendencies" (1897, p. 92). The argument was pursued by paleontologists who gave further evidence for degeneration by linking the occurrence of paedomorphosis to the imminent extinction of lineages. Smith (1914), for example, traced a racial life cycle of acceleration, stasis, ontogenetic stretching (ontogeny is retarded, but the adult stage is reached eventually), arrest of development, and, finally, reversion (and extinction). Of a lineage of Silurian gastropods, Ulrich and Scofield wrote:

> This and the two preceding genera, *Euphemus* and *Warthia*, are of unusual interest because we believe they show that in the decline of a family it actually retraced its steps by the adoption of primitive characteristics. In other words we regard them as atavistic types in which the progressive development of the individual was arrested in the embryo, and in which, because of the failure to develop the adult features of their immediate ancestors, certain characters that under previous conditions were larval only became permanent. (1897, p. 856)

What then did Walter Garstang do? Why did paedomorphosis become so popular under his urging? Why was it hailed as original and important if Garstang added little in new documentation? The answer is simply that Garstang spoke in the context of an evolutionary theory revised by Mendelian genetics—and such a revised theory regarded paedomorphosis as orthodox. This revision of evolutionary theory, not the empirical study of ontogeny, assured the downfall of recapitulation (see pp. 202–206).

But until that revision occurred, recapitulationists could meet any empirical challenges to their doctrine simply by treating them as exceptional or irrelevant. Keibel posed the challenge of unequal acceleration; Mehnert incorporated it. Giard spoke of larval adaptations, but so did Balfour and Haeckel. Rabaud said that larval adaptations were more common than ancestral repetitions; Gegenbaur replied that the repetitions could still be recognized. The axolotl brashly flaunted its permanent youth and Weismann branded it degenerate.

What Had Become of von Baer's Critique?

I have argued that critics of the biogenetic law could not disprove it by cataloguing exceptions within its theoretical structure; I have also indicated that its downfall came (and could come) only when that structure was replaced. Yet this presents a historical problem that I can resolve only imperfectly: there already was another theory of development readily available—von Baer's laws, with their trenchant critique of recapitulation.[15] Why didn't the evolutionary critics attack recapitulation from von Baer's standpoint and argue that supposed repetitions of ancestral *adults* were only the common *embryonic* stages of primitive and advanced forms alike? This was, after all, the position of Darwin himself.

Toward the end of his life von Baer wrote wistfully: "Agassiz says that when a new theory is brought forth, it must go through three stages. First men say that it is not true, then that it is against religion, and, in the third stage, that it has long been known" (1866, p. 91). But von Baer neglected the most terrible fate of all—Mendel's dilemma: that it may be true, but utterly ignored.[16] Coleman noted: "While von Baer's attack may today appear to have been decisive, in the midnineteenth century it seems at best to have taught a few embryologists to be somewhat more cautious in their utterances and to have discouraged even fewer from speculating hopefully along the older lines" (1967, p. xiv).

One reason for this neglect lies in the descriptive methodology (and ethic) of so much nineteenth-century morphology. Many criticisms of the biogenetic law were entered by-the-way in descriptive treatises when a particular finding seemed contrary to its expectations. The authors of these superb but tedious monographs felt strongly that they should play the mason's role in adding a few bricks to the temple of science. They did not study morphology to illustrate or test any theory. If an observation seemed contrary to accepted dogma, they simply recorded it; they did not seek to encompass it within a different theory—for that would have placed theory before fact, and fact was both primary and unsullied.[17]

A second reason for the neglect of von Baer's alternative is simply that most late nineteenth-century evolutionists had not read it and did not have an accurate notion of what it contained. Haeckel, in a scandalous but not uncharacteristic trick of debate, had managed to assimilate von Baer among his supporters! Vialleton described what he charitably called "this omission" as "unfortunate and passing strange" ("malencontreuse et passablement singulière"—1916, p. 101). And it engendered a monumental confusion that persists to this day (see p. 2). Kohlbrugge enumerated the many authors who had cited von Baer as a recapitulationist and labeled their misinterpretation as "certainly remarkable, one might even say tragic" ("gewiss merkwürdig, fast möchte man sagen tragisch"—1911, p. 452).

But the third and most important point is that recapitulationists had an easy, though quite inconclusive, rebuttal to von Baer's theory. This is best illustrated in the interesting exchange between Hurst and Bather. Maintaining that "the two views—von Baer's and the Recapitulation Theory—are irreconcilable," Hurst (1893, p. 365) defended the former:

> So it is in all the alleged cases of recapitulation. The gill-arches and clefts, the blood-vessels of an embryo bird or mammal, present that striking resemblance to the corresponding parts of the *embryo* of a fish which is expressed in von Baer's law . . . The ontogeny is not an epitome of the phylogeny, [it] is not even a modified or 'falsified' epitome, [it] is not a record, either perfect or imperfect, of past history, it is not a recapitulation of the course of evolution. (1893, pp. 197–199)

F. A. Bather, keeper of fossil invertebrates at the British Museum, replied that von Baer's law would apply only to "fraternal" relationships between two species descended from a common ancestor (not from one another), for in such a case the two ontogenies would be identical up to the recapitulated adult stage of the most recent common ancestor; thereafter, they would diverge, producing a result deceptively similar to von Baer's predictions. Only paleontologists encountered true evolutionary sequences, or "filial" relationships. The relationship of two contemporary species will almost always be fraternal; one will not repeat the adult stage of the other in its ontogeny. But in filial sequences, the adult ancestor will become the juvenile descendant because evolution proceeds by the addition of stages to the end of ontogeny, and previous adult stages are accelerated in development to appear as juvenile stages in descendants.

In other words, Bather defended the biogenetic law by reasserting the principles necessary to its operation. He accused Hurst of ignoring the principle of acceleration and misconstruing the idea of terminal addition (1893, pp. 276-277). He then explained Haeckel's theory by invoking these processes: "Variation, or change from parent to offspring, takes place by the addition of features at the end of the ontogeny; and these features are, by subsequent successive additions, gradually pushed back to earlier stages of ontogeny, so that what is the ultimate stage of one form is the penultimate of the next" (p. 277). To this argument, Hurst could muster only a feeble, slightly apologetic, and rhetorical reply.

Recapitulation would fall only when the two laws necessary to its operation—terminal addition and condensation—became untenable in theory. In the meantime, neither a catalog of exceptions, nor the assertion that undisputed facts fit another theory better, would budge it. Only when Mendelian genetics made terminal addition and condensation untenable in theory, could T. H. Morgan gain acceptance for his reincarnation of von Baer's position as "the repetition theory." Only then could Shumway sound his battle cry—"let us return to the law of von Baer" (1932, p. 98). Only then could Garstang proclaim in his gentler way: "Ontogeny is not a lengthening trail of dwarfed and outworn gerontic stages. Youth is perennially youth and not precocious age" (1922, p. 90).

Benign Neglect: Recapitulation and the Rise of Experimental Embryology

The empirical critiques of the first thirty years had not shaken Haeckel's theory. In 1890, Marshall devoted his presidential address before the British Association to the biogenetic law, "which forms the basis of the science of Embryology, and which alone justifies the extraordinary attention this science has received" (1891, p. 827). And, in 1893, Bather praised a journal for daring to print Hurst's support of von Baer's principles against the biogenetic law: "*Natural Science* is to be congratulated on the publication of an article so opposed to current belief . . . for it has thereby shown that it will not burke views simply because they are unfashionable, but rather that it is ready to afford a free field to all genuine knights-errant who dare to smite the shields of authority."

By 1914, MacBride was lamenting: "In these days this law is regarded with disfavour by many zoologists, so that to rank oneself as a supporter of it is to be regarded as out-of-date. The newest theory is, however, not necessarily the truest" (p. 49). What had happened in the twenty years separating Bather and MacBride?

The Prior Assumptions of Recapitulation

As late nineteenth-century science became more mechanistic and experimental,[18] the basic explanatory structure that nurtured the biogenetic law began to ring with a progressively more archaic sound.

This structure included a complex set of ethical and methodological beliefs imposed by evolutionary morphologists upon their data. Since data were analyzed in its light, data could hardly refute it. When the structure of experimental science replaced it, the biogenetic law was quietly forgotten—not labeled incorrect (for its structure had been set aside, not overthrown), but simply deemed irrelevant.

To the four basic questions of developmental biology Haeckel's school responded in the following way:

1. What is an embryonic stage? To recapitulationists, it was a *sign* of ancestry, not a *design* for immediate existence. Haeckel's morphology was pure and formal, not functional. Wilhelm His complained of its causal emptiness and characterized its results as "rigid morphological diagrams, abstracted by merely logical operations" (1888, p. 295). The attitudes of formal and functional morphologists were bound to clash in interpreting the data of early ontogenetic stages. The functional morphologist wants to know how things work; he is committed to an explanation of structure in terms of its immediate use by the organism. An embryonic adaptation is an adaptive design, not an annoying exception to the repetition of ancestry. The formal morphologist of Haeckel's time saw in embryonic shapes only the marks of ancestry. Embryonic stages were often labeled as nonfunctional and adaptive interpolations were regarded as confounding nuisances. Only from this standpoint could Balfour[19] (1880, p. 702) exhaust the explanatory value of a tadpole's gill by attributing it to the frog's permanently aquatic ancestral state (without even acknowledging that it also keeps the tadpole alive by extracting oxygen from water). An early statement in the career of an excellent functional biologist, Walter Garstang, expresses this frustration with the formal school (it also reflects a basic attitude that undoubtedly prompted Garstang's much later attack upon the biogenetic law—an attack rooted in Garstang's conviction that larval morphology must exist to serve larvae).

> A good deal of skepticism has been expressed in recent years by various writers as to the utility of the more trivial features which distinguish the genera and species of animals from one another. I do not think that such skepticism can excite much surprise if one remembers that the vast majority of "biologists" are almost exclusively engaged in the study of comparative anatomy and embryology. The amount of attention paid to these branches of biology has long been utterly out of proportion to the scant attention devoted to the scientific study of the habits of animals and of the function of the organs and parts composing their bodies . . . The subject is invested with so much intrinsic interest, as well as with such important bearings on the problems of evolution. (1898, pp. 211-212)

2. What is the "cause" of an embryonic stage? Movements in science often appropriate only one of the legitimate meanings of causality and treat it as the *only* determinant of phenomena. Debates between different movements are, on this account, often as devoid of substance as they are vital to professional prestige and position. Aristotle delineated four aspects of causation, and many pseudo-debates arise when movements acclaim one of them as a fully satisfactory mode of explanation. Take, for example, any adaptive structure. A modern developmental biologist will ask what built it and invoke the results of regulated gene action and mitosis (efficient cause). An evolutionary biologist will ask what it is for and seek to understand its role in the successful design of an organism (final cause). Both explanations are legitimate and complementary, yet the tendency of each discipline to encompass all biology within its favored mode sparked one of the most acrimonious debates of the 1950s and 1960s.

There are often several legitimate levels within the Aristotelian notion of "efficient" cause. We can, in Romanes' terms (1896, p. 98), seek a proximate cause in the enzymes, hormones, or mechanical pressures that actually mold a structure, or an ultimate cause in the process of natural selection that superintended its evolutionary development. To recapitulationists, *the* cause of an embryonic stage is ultimate and efficient: the stage appears today because it was the adult stage of an ancestor, now transferred to early ontogeny according to principles of terminal addition and condensation. Thus Balfour posed a dilemma and answered it:

> [Why do animals] undergo in the course of their growth a series of complicated changes, during which they acquire organs which have no function, and which, after remaining visible for a short time, disappear without leaving a trace . . . The explanation of such facts is obvious. The stage when the tadpole breathes by gills is a repetition of the stage when the ancestors of the frog had not advanced in the scale of development beyond a fish. (1880, p. 702)

To the new school of experimental embryologists, *the* cause of an embryonic stage was proximate and efficient. The debate centered upon the nature of explanation, not the content of development.

3. How do we study embryonic stages? Haeckel's school answered: by *observation* of normal development and by *comparison* with similar stages of related organisms.

4. Why should we study embryonic stages? What can we learn from them? Adherents to the biogenetic law responded: "to form a basis for Phylogeny, and to form a basis for Organogeny or the origin and evolution of organs" (Balfour, 1880, p. 4). Evolutionary morphologists sought to establish the genealogy of life as their ultimate goal

(Russell, 1916, p. 268), yet the pitifully imperfect record of fossils precluded any attempt to do so directly. How fortunate then that the immediately accessible ontogenies of modern forms could reveal life's history even more faithfully. It is largely for this reason that Haeckelians insisted so strongly on the repetition of *adult* ancestors by descendant embryos. Early in his career, William Bateson sought phylogeny in embryology; recalling, in a sadly skeptical old age, his study of *Balanoglossus* under W. K. Brooks at the Johns Hopkins summer laboratories, he wrote "Morphology was studied because it was the material believed to be most favorable for the elucidation of the problems of evolution, and we all thought that in embryology the quintessence of morphological truth was most palpably presented. Therefore, every aspiring zoologist was an embryologist, and the one topic of professional conversation was evolution" (1922, p. 56).

Wilhelm His and His Physiological Embryology: A Preliminary Skirmish

Among early opponents of recapitulation, Wilhelm His, professor of anatomy at Leipzig, was surely the most effective. He did not achieve this status by marshaling the most telling rebuttals to recapitulation; his specific arguments were, in fact, fairly weak (1874, pp. 165–176). Rather, he challenged Haeckel's methodology and asserted that the most important causes of embryological shapes were proximate and efficient. He sought to explain the complexity of developing form by displaying it as the automatic result of simple mechanical pressures produced by local inequalities of growth. He compared the embryonic layers of the chick to elastic sheets and tubes, and "constructed" the principal organs by cutting, bending, pinching and folding. In his great work of 1874, *Unsere Körperform und das physiologische Problem ihrer Entstehung,* His noted the extraordinary resemblance between embryonic organs and simple manipulations upon rubber tubes (Figs. 23–25). "We must start," His wrote, "from the fact that the brain, at its beginning stages, is a tube with moderately elastic walls" (p. 96). With a strong thread, His attached one end of his rubber tube to a fixed point and bent the tube toward that point (Fig. 23). He compared this with the initial attachment of the medullary tube to the foregut and invoked the same simple force as a proximate cause of morphogenesis:

The foregut plays the role of the fixed thread, and the form assumed by the anterior end of the brain [*vordere Gehirnende*] corresponds exactly to the paradigm. In fact, you need only compare [Figs. 23 and 24] in order to find the

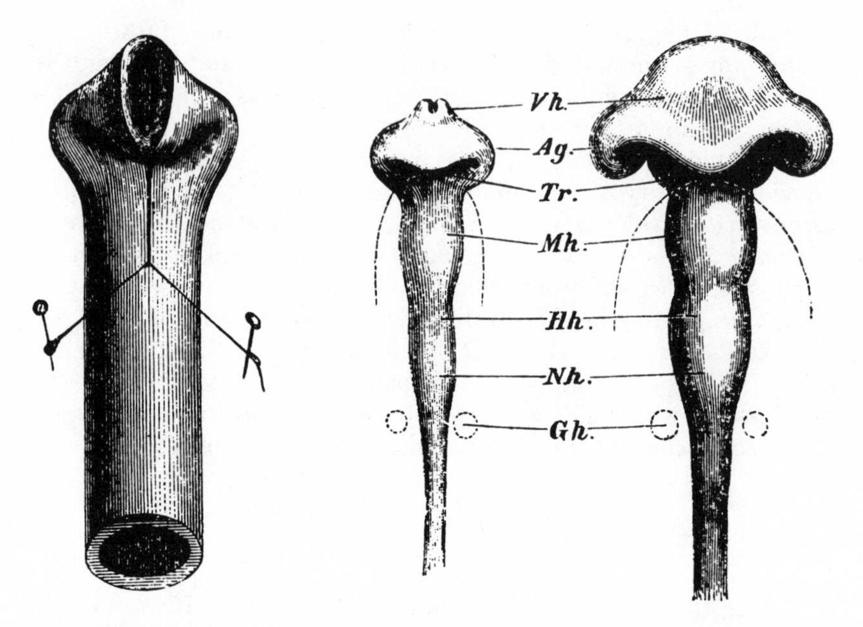

Fig. 23. A rubber tube with a string attached at one end and bent back upon itself compared with the developing chick brain. *Ag.* is the Anlage of the optic lobes (corresponding with the lateral projections of the bent tube). *Tr* is the stalk of the hypophysis (corresponding with the point of attachment for the fixed thread of the tube). (From His, 1874.)

greatest possible agreement [*grösstmöglichste Übereinstimmung*] in all important points. You will find in the stalk of the hypophysis the fixed point of the bent tube, in the *Anlagen* of the two optic lobes its two lateral projections.[20] (p. 100)

Later, His summarized his method and conclusions:

These examples, which could easily be multiplied, may be sufficient to prove the general importance of elementary mechanical considerations in treating morphological questions. They show at the same time how the means that nature uses in forming her organisms may be very simple. The segmented germ divides itself into the primitive embryonic organs by a few systems of foldings . . . Even the most complicated of our organic systems, the nervous system, follows a course of the most astonishing simplicity. (1888, p. 297)

His took great pains to point out that his preferred explanations did not exhaust the content of causality (1874, pp. 172–176). He admitted often and gladly that the proximate forces responsible for bending and folding had an ultimate phyletic origin passed down to

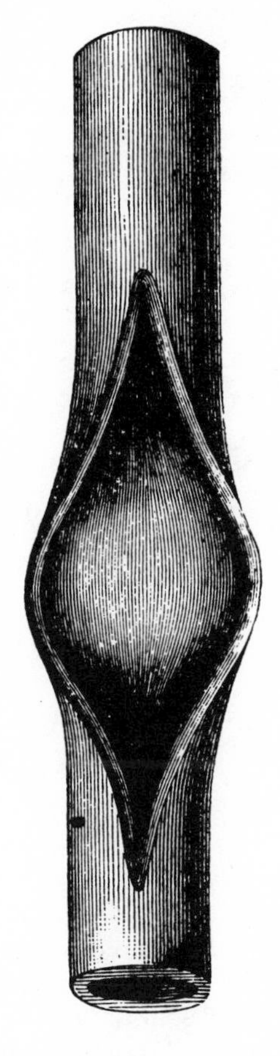

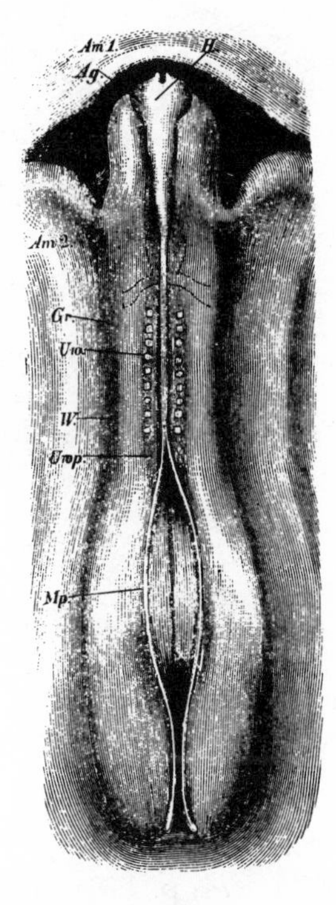

Fig. 24. Comparison between a slit rubber tube with convex bending and the early chick embryo. (From His, 1874.)

the embryo through heredity: "The mechanics of development and heredity are facts of a different order" (1894, p. 2).[21] Yet he clearly preferred his own, admittedly partial set of causes as both more significant and more modern in its conceptual link to the mechanistic physiology of his time: "I should be the last to discard the law of organic heredity . . . but the single word 'heredity' cannot dispense science from the duty of making every possible inquiry into the mechanism of organic growth and of organic formation. To think that heredity will build organic beings without mechanical means is a piece of unscientific mysticism" (1888, pp. 174–175). His's attack upon the biogenetic law was far more telling and fundamental than that of Haeckel's empirical critics, for His questioned the basic method of evolutionary morphology and suggested that embryologists do something quite different: "An array of forms, following one after the

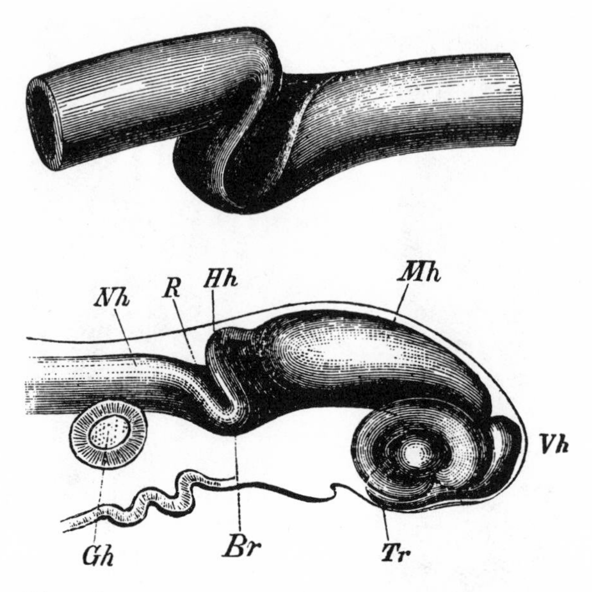

Fig. 25. A slit rubber tube bent back upon itself and the developing brain of a chick embryo. (From His, 1874.)

other is really, and this must be emphasized again and again, no explanation" (1874, p. 176).

No other explanation of living forms is allowed than heredity, and any which is founded on another basis must be rejected.The present fashion requires that even the smallest and most indifferent inquiry must be dressed in phylogenetic costume, and whilst in former centuries authors professed to read in every natural detail some intention of the *creator mundi,* modern scientists have the aspiration to pick out from every occasional observation a fragment of the ancestral history of the living world. (1888, p. 294)

Haeckel obviously grasped the scope of His's challenge, for he lavished upon it his most withering rhetoric. He labelled His's argument as the "rag-bag" or "rubber-tube" theory (*Gummi-Schlauch Theorie*) and called it

one of the curiosities of the embryological literature. The author imagines that he can build a "mechanical theory of embryonic development" by merely giving an exact description of the embryology of the chick, without any regard to comparative anatomy and phylogeny, and thus falls into an error that is almost without parallel in the history of biological literature . . . He imagines constructive Nature to be a sort of skillful tailor. The ingenious operator succeeds in bringing into existence . . . all the various forms of living things by cutting up in different ways the germinal layers, bending and folding, tugging and splitting . . . a sartorial theory of embryology.[22] (1905, pp. 49–50)

Yet Haeckel's scorn for the bending of rubber tubes rested on no claim of *inaccuracy,* for he freely admitted that correspondences between bent tubes and developing brains might specify the proximate cause of embryonic stages. He railed only against the inadequacy of such explanation. Proximate causes specify how heredity operates in building an animal. But heredity is established and changed by phylogeny. Phylogeny determines the sequence of embryonic stages; and phylogeny will explain their succession in a deeper and more interesting way than any statement about the direct construction of a particular embryo.

When this "descriptive embryology" rises in spite of its restriction, to an explanation of the facts it describes, it assumes the proud title of "physiological embryology." It fancies it has found the real *mechanical* causes of the facts of embryology when it has traced them to simple *physical* processes, such as the bending and folding of elastic plates . . . The chief defect of this "exact" or physiological . . . method in embryology is seen in its attempt to reduce most complex *historical* processes to simple *physical* phenomena. When, for instance, the spinal cord of the vertebrate embryo severs itself from the general envelope, or when the 5 cerebral vesicles are formed by transverse folds at its bulbous upper extremity, it might seem to a superficial observer that these are simple physical processes. But we do not really understand them until we trace them to their true phylogenetic causes, and see that each of these apparently simple processes is the recapitulation of a long series of *historical* changes. (1905, p. xix)

Haeckel sensed correctly that His was a far more serious competitor than his empirical critics. The empirical critics worked with his methods and his modes of explanation, but His would have substituted a drastically different approach and relegated the biogenetic law to irrelevancy—a fate far worse and far more irrevocable than any odor of inaccuracy. Therefore, when his defense rose above the polemical, Haeckel counterattacked with statements about the limits of proximate causation in historical sciences:

I am one of those scientists who believe in a real "natural history," and who think as much of an historical knowledge of the past as of an exact investigation of the present. The incalculable value of the historical consciousness cannot be sufficiently emphasized at a time when historical research is ignored and neglected, and when an "exact" school, as dogmatic as it is narrow, would substitute for it physical experiments and mathematical formulae.[23] Historical knowledge cannot be replaced by any other branch of science. (1905, p. 881)

Roux's Entwicklungsmechanik and the Biogenetic Law

Embryology is an historical science only in part. Could that part possibly maintain its popularity against the aggressive supporters of experimental methods and mechanistic outlooks? The experimental method had triumphed in physiology and promised—despite Du Bois-Reymond's *ignorabimus* ("we will never know")—to reduce organic function to the exact laws of physics and chemistry (Hertwig, 1901). Even Haeckel paid lip service to the ideal of reduction.[24]

His's attack had come about ten years too early. In its time it was an isolated incident; but by the late 1880s and early 1890s, two of Haeckel's apostate students—Wilhelm Roux and Hans Driesch—were advancing experimental methods in embryology and relegating the biogenetic law to a backshelf of outmoded methods.[25] Before the century's end, T. H. Morgan (1899, p. 195) could write: "If I mistake not, there is a tendency at present, that is slowly gaining ground, to give up as unprofitable the interpretation of . . . embryological phenomena in terms of speculative phylogeny." This time, proximate causation triumphed and set the fashion for the next half-century, one of the most exciting and fruitful periods in the history of embryology.[26]

Experimental embryologists rejected all aspects of Haeckel's methodology (see p. 187). They were interested in how the structures of juvenile stages worked; they experimented by disturbing the normal course of development; they studied embryonic stages to discover their proximate causes in previous conditions and to assess their influence upon following ones. In attempting to reduce the complexities of development to laws of physics and chemistry, they focused upon the earliest stages, which recapitulationists usually ignored (patterns of cleavage might yield to mechanical analysis, though the morphogenesis of complex organs seemed intractable). But the greatest clash between the two approaches took place on the battlefield of causality. Experimental embryologists relentlessly asserted that their kind of cause (proximate and efficient) exhausted the legitimate domain of causality. All that had come before them was merely descriptive; they had established the first causal science of embryology. Developmental mechanics [*Entwicklungsmechanik*] would solve the riddles of ontogeny that had, heretofore, only been recorded in their proper sequence. Thus, Wilhelm Roux began the prolegomenon to his new journal —*Archiv für Entwicklungsmechanik*—with these words: "Developmental mechanics . . . is the doctrine [*Lehre*] of the causes of organic forms . . . We may designate as the general goal of developmental mechanics the ascertainment of formative forces or energies" (1894, p. 1). Papers in the Haeckelian tradition were simply ruled out of the journal as having nothing to do with the discovery of cause. Roux dubbed them "preliminary analyses" (*erstere Analyse*), and excluded "papers in comparative anatomy which reduce the forms of organisms exclusively to the factors of variation and heredity, without striving for any further analysis of these 'inconstant' complex components" (1894, pp. 36–37).

He pleaded specifically for a causal analysis of phenomena described by the biogenetic law: "Both heredity and adaptation are urgently in need of causal explanation, i.e., of analysis into their uniformly operating components; this analysis is a task of developmental mechanics. The same is true for the so-called 'biogenetic law' and for cenogenesis" (p. 26). As an example, Roux discussed the development of the mammalian liver (pp. 6–7). In ontogeny, the liver is transformed from a tubular structure to a reticular gland with the narrowest possible meshes. Haeckel would have seized upon the gross morphology of this transformation and asked the phyletic question: what ancestor possessed a tubular liver in its adult state? Roux sought the efficient cause of this transformation in cellular processes. The cells of all tubular glands have a bipolar differentiation, with a secreting surface at one end, and, at the other, a basal surface that takes up nutriment from capillaries (other surfaces merely function as areas of contact with adjacent cells). The cells of reticulated glands have a multipolar differentiation with several surfaces for absorption and secretion. An explanation for the ontogenetic transformation of the liver must be sought in physical and chemical factors that determine this change in differentiation, for "the multipolar differentiation of the liver cells stipulates or causes the transformation of these cells from the tubular [*Schlauchtypus*] to the framework type [*Fachwerktypus*]" (p. 7). This example illustrates many aspects of Roux's approach: the concern with how embryonic organs work, the explanation of whole organs in terms of the cells that build them, and the further reduction to physical or chemical forces shaping the cells and setting their function.

Yet, though Roux's methods led him to ignore the biogenetic law, nothing in his doctrine led him to oppose it. Not wishing to incite the influential Haeckel any further, he carefully staked out a different area of research and avoided any direct attack upon recapitulation. In fact, in a classic instance of damning with faint praise, he ranked himself among its supporters:

> To be sure, we agree with Haeckel's dictum: "phylogeny is the mechanical cause of ontogeny." . . . But the "biogenetic law" merely designates the fact of repetition and its general necessity, and therefore expresses the causal con-

nection only in the most general way. It teaches us nothing . . . about the operating causes and their intensities. The experimental study of developmental mechanics can bring us this knowledge. (1905, p. 253)

The ultimate goal of Roux's quest for reduction was to find explanations based on the "simple components" of physics and chemistry. But the extreme complexity of morphogenesis converted this goal into a devout wish for a distant future. In the meantime, "complex components," still in the organic realm but simpler than the phenomena they explained, could be isolated, categorized, and manipulated. The mode of cell differentiation, to return to our previous example, is a complex component that determines the gross morphology of organs.* Likewise, the biogenetic law is a complex component that describes the operation of heredity in preserving ancestral structures. But the biogenetic law is an uninteresting complex component because it resides at such a "high" level of complexity itself. It should be reduced toward its own simple components, not utilized in Haeckel's manner.

Although recapitulation could coexist peacefully with developmental mechanics in a purely intellectual realm, it could never do so in the domain of human beings. Both schools had to compete for a limited number of academic positions and the status they entailed. To establish themselves, experimental embryologists had to displace a generation of Haeckelian morphologists. As Fleming states so well:

Ernst Haeckel . . . infuriated Roux by his insistence upon the so-called "fundamental biogenetic law." . . . On the face of it, this was a matter for empirical resolution; and even if true perfectly compatible with the study of *Entwicklungsmechanik.* Roux, however, with some justification attributed to Haeckel the more ambitious design of establishing this kind of "description" not merely as valid but as exhausting the content of embryology and precluding the necessity for any mechanical analysis of development. By the same token, Haeckel was undercutting Roux's endeavor to elevate biology to the estate of Newtonian physics. More concretely, Haeckel was pointing away from the experimental embryology of Roux to the speculative imposition of evolutionary schemes upon the embryo; from the microscope slide to the hypothetical evolutionary tree. (in Loeb, 1964, p. xx)

* A similar distinction is found in D'Arcy Thompson's (1917) attempt to explain organic form as the result of physical forces acting upon responsive matter. For such simple configurations as the external form of protozoans, Thompson advocates a complete physical explanation—they are shapes assumed under the influence of surface tension. For the complex shapes of crabs or fishes, one has to accept the basic form as given and try to explain *differences* among species by simple physical forces displayed in Cartesian transformations.

The younger generation of experimentalists extended their challenge. Speaking at the Marine Biological Laboratory in Woods Hole, E. B. Wilson asserted the birthright of a new movement:

[It is] a just ground of reproach to morphologists that their science should be burdened with such a mass of phylogenetic speculations and hypotheses, many of them mutually exclusive, in the absence of any well-defined standard of value by which to estimate their relative probability. The truth is that the search after suggestive working hypotheses in embryological morphology has too often led to a wild speculation unworthy of the name of science; and it would be small wonder if the modern student, especially after a training in the methods of more exact sciences, should regard the whole phylogenetic aspect of morphology as a kind of speculative pedantry unworthy of serious attention. There can be no doubt, I think, that this state of things is leading to a distaste for morphological investigation of the type represented, for instance, by Balfour and his school, while the brilliant discoveries of the cytologists and experimentalists . . . have set up a new tendency that gathers in force from day to day. (1894, pp. 103–104)

An ancedote cited by Oppenheimer (1967, pp. 74–75) illustrates the acrimony inspired by such assertions. In 1891, Hans Driesch sent to Haeckel a book in which he tried to explain the orientation of cleavage planes by mathematical formulae and physical principles: "He knew his book would not be looked upon with favor by Haeckel, but he sent him a copy, together with a letter asking whether development of the individual might not be considered from this new point of view. Neither the letter nor the book was acknowledged, but in due time Haeckel sent him an unwritten message, through a mutual friend, suggesting that Driesch take off some time in a mental hospital."

He did not (though his experimental colleagues might have made the same suggestion many years later after he converted to vitalism and shifted to philosophy). And Haeckel eventually retired to the empyrean, harmless height of elder statesman. What his empirical critics could not achieve by direct attack, his methodological opponents won by benign neglect.[27] One can search through volumes of Roux's *Archiv,* scan the longest textbooks of experimental embryology, and not find a single reference to recapitulation.

Recapitulation and Substantive Issues in Experimental Embryology: The New Preformationism

The first major controversy within experimental embryology so strongly recalled the attitudes of a previous debate that its names were reincarnated. The new epigeneticists, like Driesch, spoke of a "harmonious equipotential system" among the first cleavage cells. Each cell contains the latent potential to produce a complete organism. Differentiation occurs because forces surrounding the blastomeres (the first embryonic cells) vary according to differences in spatial position and time of origin for these cells; these forces impress different characters upon an initially undetermined cell and eventually fix its fate. In the classic experiment of this school, Driesch obtained complete larvae from blastomeres separated from a sea-urchin embryo at the four-cell stage.

The new preformationists believed that the fate of an embryo is fixed in the fertilized ovum. In this "mosaic" theory of development, the egg is as structured as the adult. It is divided into regions (*Keimbezirke*) destined to produce specified parts and organs of the completed animal. Cleavage is merely a process by which these determinants are sorted into different cells and, finally, into tissues and organs. Ontogeny is a true "evolution"—an unfolding of predetermined structure—in the eighteenth-century sense of that term. As Jenkinson wrote in his famous text: "The factors on which the differentiation of the whole and of each part depend are essentially internal, and all that happens is that by a continued process of cell division the parts are separated from one another and the structure thus made palpable and manifest" (1909, p. 158). Thus Roux and Weismann held that the nuclear divisions of cleavage were "unequal," since the determinants of later structures were partitioned into separate blastomeres. When Roux obtained only a half-embryo after destroying one blastomere of a frog at its two-celled stage, he provided the standard empirical support for this doctrine. (Although Roux had destroyed one blastomere, he had not severed it from the one remaining. When the two are separated, whole embryos can develop in some situations.)

Conklin contrasted the two schools and correctly maintained that their link to the older controversy lay more in similar attitudes than in the content of belief:

> But while this modern controversy recalls the ancient one between the adherents of evolution and those of epigenesis, it does so chiefly because it proceeds from the same temper of mind, and not because anyone today is ready to defend the views of either the evolutionists or the epigenesists of a century ago. No one now expects to find in the egg or sperm a predelineated germ with all adult parts present in miniature, neither can anyone now maintain that the egg is composed of unorganized and non-living material. Everyone now admits that the truth is somewhere between these two extremes; the real problem is how much or how little of organization is present, and not whether the germ is organized at all.[28] (1905, p. 5)

I have presented what Oscar Hertwig called "*the* biological problem of today" in its most extreme contrast (Hertwig, 1894). As Conklin noted, everyone soon acknowledged that the egg is organized in some sense. This basic tenet of the new preformationism had a strong effect, mostly negative, upon the biogenetic law. When they bothered to comment upon the biogenetic law at all, most supporters of experimental embryology relied upon this tenet to dismiss it. The egg, with localized regions delineating adult organs, is as much a terminal product of evolution as the complex form that develops from it. How can it represent the primeval amoeba, a completed form bearing in its architecture no potential for anything higher. The egg is as organized as the adult; it can only be the precursor of itself: "The hen's egg is no more the equivalent of the first link in the phylogenetic chain than is the hen itself" (Hertwig, 1906, p. 160).[29] Our two-celled ancestor, Hertwig argued, was a loose federation of two independent entities; the two-celled mammalian embryo is an intimately unified precursor. Of the fertilized ovum, he wrote: "Its daughter cells no longer become independent of each other as a result of their cleavage; rather, they are bound together into a higher organic unity . . . The [egg] cell . . . becomes ever richer in new *Anlagen* and, by this means, becomes more and more different from the primitive ancestral cell . . . The more complicated the end product of an ontogeny, the more complicated the corresponding *Anlagen*" (1906, pp. 158–159). The same argument was advanced by Conklin (1905, p. 110), Goodrich (1924, p. 147), Montgomery (1906, p. 191), and Roux (1881, p. 57).

Yet, despite its frequent use, I fail to see in this contention any more than a debating point. The egg's complexity could scarcely be denied. It was as strongly supported in Haeckel's notion of perigenesis as in Roux's idea of preformation; for Haeckel's mammalian ovum contained an immense concentration of wave-energies—a remembrance of all the transformations experienced through millions of years by the original amoeba and its heirs. Haeckel was concerned with *visual appearance,* not latent potential.[30] The egg carries a complete set of determinants, but it faithfully assumes the form of an amoeboid progenitor; the human embryo is destined to be man, but it still grows the tail of our quadrupedal ancestors.

Although the fact of the egg's organization tended to foster an attitude unfriendly to recapitulation, the phyletic problem of how that complexity had evolved was another matter. One could argue that evolutionary changes arose in the germ and became manifest at various times during the course of development. This was the usual solution and it offered little comfort for recapitulationists. But there was another possibility: that complexity had arisen late in development and had been pressed back upon the germ according to Hyatt's principle of acceleration, the motor of the biogenetic law. Thus recapitulation managed to insinuate itself into the new preformationism just as Bonnet had introduced it into the old (Chapter 2).

This notion was introduced by Lankester under the name of "precocious segregation." Lankester believed that the separation of ectoderm and endoderm first arose in evolution by delamination of the blastula. Yet he thought he detected this distinction in the two-celled stage of many modern animals. The separation now occurs before the stage at which selection could have introduced it; therefore, it must have been shunted back to first cleavage by the principle of acceleration:

This hypothesis may be called that of precocious segregation: "precocious" since it is the acquirement of a condition in the developing organism, in virtue of heredity, at an earlier period of development than that at which such acquirement was attained by its forefathers through adaptation. The tendency to precocity in this sense, in regard to important structural arrangements, has been insisted on by Haeckel in discussing what he terms "heterochrony in the palingenetic phenomena of ontogeny." (1877, p. 411)

Although proposed from the heartland of speculative phylogeny,[31] Lankester's suggestion was later adopted by many of the new preformationists. E. B. Wilson contrasted spiral with radial cleavage. He identified the spiral type as primary in evolution and provided a mechanical explanation for it (alternation of cells as the result of mutual pressure—the general argument for hexagonal closest packing in three-dimensional structures). How then does bilateral cleavage arise from it? Wilson answers that this secondary pattern of cleavage is a reflection of later symmetry pressed back into earlier stages originally conditioned by laws of physics:

The characteristics of the spiral period are, in their broadest outlines, the result of mechanical conditions which have no relation to the adult structure. What, then, is the origin of bilateral forms of cleavage? It appears to me that they must be the result of a throwing back or reflection of the adult bilaterality upon the early stages. In some cases this influence has extended to the very beginning, as in the Cephalopod or in the ascidian, or even to the unsegmented ovum itself . . . In some cases, of which *Nereis* is a beautiful example, it has not extended so far; the early stages are still dominated by the mechanical conditions peculiar to them, and the bilateral form only appears when these conditions have been in a measure overcome. (1892, pp. 453–454)

Lillie generalized the argument to maintain that early stages of cleavage were useful in establishing phyletic relationships *only* when they had been affected by precocious segregation: "The fundamental forms of cleavage are primarily due to mechanical conditions, and are only significant morphologically in so far as they have been secondarily remodelled by processes of precocious segregation" (1895, p. 38). Wilson (1904) extended his earlier suggestion and attributed many localizations in the egg itself to precocious segregation.

In a superb illustration of the methodological clash between Haeckelian and experimental embryology, Conklin chided his colleagues for lapsing into old habits of thought. Even if it were true (which it is not), precocious segregation said nothing about cause and could be no more than a description from the wrong perspective:

> The early appearance of differentiations is usually explained as a "throwing back of adult characters upon the egg." The whole life cycle is viewed from the standpoint of the adult; the embryo and germ exist for the purpose of producing a certain end; the adult is primary, the germ secondary. But do not all such ideas put the cart before the horse? What is the evidence that any inherited modification of an adult structure can arise without an antecedent modification of the germ? We know that the adult is moulded upon the egg, that specific modifications of the germ do, in some cases, produce specific modifications of the adult, but the converse proposition is certainly not established. "Precocious segregation" represents the backward rather than the forward look; it is a teleological rather than a causal explanation. (1905, p. 110)

Precocious segregation may have supplied some comfort to recapitulationists confronted with the organization of the egg. Yet the beliefs that led to a recognition of this organization clearly heralded the death of the biogenetic law. The attention of embryologists and students of heredity was directed away from the developing sequence of ontogenetic stages and focused upon the germ itself. When Conklin (1903) attributed the reverse symmetry of some gastropods to an inverse organization of the egg, and then linked most important changes in phylogeny to the "alteration of germinal organization" (1905, p. 111), he left little room for recapitulation and its insistence upon the terminal addition of new features.

Recapitulation could not survive the basic attitude that underlay the new preformationism—the mechano-structuralist bias that had been the kernel of eighteenth-century "evolution" as well. If preformationism had one cardinal tenet, it was a desire to trace structure to pre-existing structure—to see in the complexities of adult form only an elaboration of structures present at the very beginning of ontogeny. How can the stages of ontogeny be a parade of ancestral forms if all the essential features of the highest stage have structural precursors in the first (even if these be invisible)? Recapitulationists could refute a specific argument about the egg's internal organization by invoking a criterion of visual appearance against the fact of latent potential, but they could not survive an attitude that came close to denying process altogether.

Mendel's Resurrection, Haeckel's Fall, and the Generalization of Recapitulation

The preformationist attitude had not destroyed recapitulation. After all, Weismann was an enthusiastic supporter of the biogenetic law and Roux, in public, was at least indifferent. Yet the shift of attention to the germ and its structure redirected the study of heredity and led ultimately to the Mendelian synthesis. Experimental embryology had abandoned recapitulation as unfashionable, but genetics would render it untenable.

Early Mendelians had the same general reasons as students of *Entwicklungsmechanik* for neglecting recapitulation. As a putative "structure" in the fertilized ovum (and elsewhere), the gene was a perfect particle to support the biases of preformationist thought about early organization. "Benign neglect" again played its role. Several famous Mendelians began their careers as recapitulationists in the tradition of speculative phylogeny. Bateson (1886), Morgan (1891), and Castle (1896) had sought the origin of vertebrates in the embryology of primitive chordates; Davenport (1890) had tried to unravel the history of bryozoans from the ontogeny of modern forms. After their conversion to Mendelism, most of these men never mentioned the biogenetic law in print.

But the main impact of Mendelism was much more specific: it ultimately disproved the two "laws" of evolution that recapitulation required for its general occurrence—terminal addition and condensation. The more astute recapitulationists had long recognized that a catalog of cases could provide no ultimate justification for their beliefs. As long as the mechanism of heredity lay shrouded in mystery, recapitulationists could always postulate a convenient and purely hypothetical set of laws to yield their preferred results. The laws of terminal addition and condensation were of this type. Beyond the vague analogy to memory pursued by some Lamarckians, they had never had any justification beyond the argument: (1) recapitulation is true; (2) these laws yield recapitulation as a general result; (3) these laws must be true. This argument is neither illogical nor circular (one can, in theory, test the first statement by accumulating more and more cases, and then argue that no other laws fit the second statement). It is unsatisfying because it can be vindicated only by induction from its results. While the mechanism of heredity remained unknown, there was no other way to argue. Yet without a mechanism, terminal addition and condensation could only have a tentative status as empirical laws; no one would be fully satisfied until they could be displayed as deductive consequences of a satisfactory theory of heredity.[32] As Lebedkin wrote: "the chief task in studying the recapitulation problem is to ascertain the Laws of Heredity" (1937, p. 561). August Weismann, perhaps the most astute of Haeckel's supporters, recognized this very well when he wrote: "If we could see the determinants, and recognize directly their arrangement in the germ-plasm and their importance in ontogeny, we should doubtless understand many of the phenomena of ontogeny and their relation to phylogeny which must otherwise remain a riddle" (1904, p. 189).

In the first years of this century, the determinants Weismann sought were identified as Mendelian genes located on chromosomes. The laws of terminal addition and condensation could finally be tested. Like Daniel's king, they were weighed and found wanting.

Throughout Chapter 4, I emphasized that the laws of terminal addition and condensation fit naturally and comfortably into the Lamarckian theory of inheritance (but stood as unjustified, ad hoc assumptions within other theories). When Mendelians discarded the inheritance of acquired characters, they also rejected the most promising theoretical basis for the biogenetic law. In the 1910s and 1920s, the strongest support for the collapsing theory of universal recapitulation came from a few unrepentant Lamarckians. Moreover, they tied the fate of recapitulation explicitly to their hopes for a Lamarckian resurrection. In his textbook of embryology, E. W. MacBride defended the law of acceleration, by the analogy to memory and habit: "But, the reader will exclaim with horror, does not this explanation postulate the acceptance of that Lamarckian heresy, the inheritance of acquired characters? . . . The answer to this question is twofold: first, *the difficulty of framing any other theory of recapitulation seems to be insuperable;* and, second, the experiments which have been held to disprove the inheritance of acquired characters are far from conclusive" (1914, pp. 650–651, my italics). And Paul Kammerer, defending his toads and salamanders against Bateson's onslaught, wrote: "Without an inheritance of acquired characteristics, these 'biogenetic repetitions' would seem impossible" (1924, p. 218; see also Koestler, 1972; Gould, 1972b).

The rejection of Lamarckian inheritance only provided an indirect argument against recapitulation; the direct assault was mounted upon the laws of terminal addition and condensation.

Terminal addition had to be discarded because the genes that control characters are present from conception, and evolutionary change occurs by mutational substitution. With these propositions, what possible justification can be offered for a belief that new features must be *added terminally*. "A house," Garstang wrote, "is not a cottage with an extra storey on top" (1922, p. 84). T. H. Morgan attacked the biogenetic law by arguing that these substitutions can be expressed at any point in ontogeny.[33]

> Genetics has contributed two facts that have a bearing on the recapitulation theory . . . In the first place, there is abundant evidence that a new gene may bring in a change at any stage of development. In the second place, it is well known that a new gene may change the final stage of development—not by adding something to the end stage (although this, too, may be possible), but by replacing it by substitution. (1932, p. 185)

> This idea of germinal variation therefore carried with it the death of the older conception of evolution by superposition. (1916, p. 18; see also 1934, p. 148)

This argument reverberated back almost a hundred years to vindicate the neglected von Baer. From this standpoint, von Baer's attack upon the recapitulation of "lower" adults had been completely justified (as had Darwin's evolutionary transformation of the same argument). The mutations of descendants are expressed at various points in their development. Before these points of expression, the ontogenies of ancestor and descendant are identical; afterwards, they diverge. The gill slits of a human fetus are not those of an adult fish; they represent the common embryonic state of all vertebrates. They specify common descent, but they do not consitute a parallel between the stages of ontogeny and phylogeny. Most cases in the catalog of recapitulation would have to be reinterpreted as "embryonic survivals rather than as phyletic contractions" (Morgan, 1916, p. 21). Of our embryonic gill slits, Morgan wrote: "Is it not then more probable that the mammal and bird possess this stage in their development simply because it has never been lost? Is not this a more reasonable view than to suppose that the gill slits of the embryos of the higher forms represent the adult gill slits of the fish that in some mysterious way have been pushed back into the embryo of the bird" (pp. 20–21).

Recapitulation, in altered form, might have survived the collapse of terminal addition had it been able to retain a law of condensation. Recapitulationists would have had to admit that the final stage of ontogeny had no special claim as a locus of evolutionary novelties, but they could still have maintained that new characters, wherever they arise, are always transferred back to appear earlier in descendant ontogenies. Ancestral features would always appear in more juvenile stages of descendants. This last hope for universal recapitulation was dashed by the discovery that genes act by controlling the *rates* of processes.

E. B. Ford and Julian Huxley spoke of the "brilliant work of the Morgan school," but complained: "so far, however, the genes are known only as the heritable basis whose ultimate effect is the production of one or more visible characters in the adult organism, while the developmental stages by which these are obtained are still for the most part obscure" (1927, p. 112). They set as their aim: "to investigate genetically-controlled rates of development, in an endeavor to obtain further information on the mode of action of genes."

As early as 1918, Richard Goldschmidt had spoken of "rate genes" (see Goldschmidt, 1923, and 1938, pp. 51–78, including discussion of Sewall Wright's [1916] studies of coat color in rabbits). He discovered that "genetic races" of the gypsy moth, *Lymantria dispar,* differed only in genes controlling the depositional rates of pigment in caterpillars. In some races, a pattern of light markings persists until pupation; in others, this pattern is gradually covered by a dark cuticular pigment deposited at definite rates. Goldschmidt found that these rates differed among races and were intermediate in heterozygous hybrids of intermediate color. Ford and Huxley (1927) studied eye coloration in the amphipod *Gammarus chevreuxi.* Red and black are Mendelian alternatives. In this species, all colored eyes (some are uncolored) are red at first and change to black as melanin is deposited at definite rates during development. Ford and Huxley discovered a set of genes that produced a graded series of colors by altering both the rates and times of onset for deposition of melanin. Goldschmidt generalized this theme:

> The mutant gene produces its effect, the difference from the wild type, by changing the rates of partial processes of development. These might be rates of growth or differentiation, rates of production of stuffs necessary for differentiation, rates of reactions leading to definite physical or chemical situations at definite times of development, rates of those processes which are responsible for segregating the embryonic potencies at definite times. (1938, pp. 51–52)

If genes produce enzymes and enzymes control the rates of processes, what possible justification can be offered for universal acceleration in phylogeny? Acceleration, to be sure, does occur, but there is no reason to consider it any more fundamental, or even any more common, than retardation. The retardation of somatic characters usually results in paedomorphosis, while acceleration yields recapitulation. If both acceleration and retardation are equally valid in theory, then these results are equally orthodox. Paedomorphosis can no longer be cast aside as an exception to universal recapitulation. This new condition of "equal orthodoxy" is emphasized in J. B. S. Haldane's article on "The Time of Action of Genes, and Its Bearing On Some Evolutionary Problems":

> There has been a common tendency in evolution for development to accelerate, i.e., for certain characters to appear progressively earlier in the life cycle . . . This presumably means that the time of first action of certain genes has tended to be pushed back . . . Another common tendency has been a retardation of certain characters relative to the life-cycle, so that originally embryonic characters persist in the adult. This is known as neoteny. (1932, pp. 15–16; see also Huxley, 1923, p. 616)

In writing his "critical restatement" of the biogenetic law, Garstang (1922) did not "discover" paedomorphosis or any of the other well-known exceptions to recapitulation. Instead, he recognized that they could no longer be dismissed as exceptional, for they were now the expectations of a new theory, and they demanded equal status with all the phenomena of recapitulation. Paedomorphosis is no "degenerative exception" to universal acceleration. Cenogenesis is not the "secondary falsification" of an essentially palingenetic development. The facts that had lurked so long in the limbo of exception were elevated to orthodoxy by the discoveries of Mendelian genetics.

But recapitulation was not "disproved"; it could not be, for too many well-established cases fit its expectations. It was, instead, abandoned as a universal proposition and displayed as but one possible result of a more general process—evolutionary alteration of times and rates to produce acceleration and retardation in the ontogenetic development of specific characters. "Recapitulation is only responsible for a certain fraction—not even nearly half—of the relationships between juvenile and ancestral [adults] forms . . . [It is] neither a law, nor even a rule, but only one mode among many" (Franz, 1927, p. 36). I shall devote the rest of this book to exploring the consequences of this generalization.

PART TWO

HETEROCHRONY AND PAEDOMORPHOSIS

—7—

Heterochrony and the Parallel of Ontogeny and Phylogeny

Acceleration and Retardation

Confusion in and after Haeckel's Wake

A nice dilemma we have here,
That calls for all our wit:
And at this stage, it don't appear
That we can settle it.

W. S. Gilbert, *Trial by Jury*

Ernst Haeckel might have borrowed a line from his nation's adversary, Louis XV, and exclaimed: "Après moi le déluge." Once recapitulation had lost its universal status and become but one mode among many, numerous authors tried to elaborate more complete taxonomies of the relationship between ontogeny and phylogeny. These new schemes proliferated under so many criteria of such diverse standing that an almost anarchic confusion soon arose. There is no way that I can bring order to the next few pages. I wish only to record the confusion of these complex schemes as a prelude to my subsequent attempt at resolution and simplification.

The elaborations proceeded from each of Haeckel's laws. Some authors expanded the law of terminal addition; as a basis for classification they used the stage of ontogeny at which new features arise in evolution (traits might also be deleted, and this too could occur at any stage). Many elaborate classifications were based entirely upon the addition and subtraction of characters at various times. Franz (1927), for example, distinguished four "biometabolic modes": (1) prolongation, or the addition of characters to the end of ontogeny (producing recapitulation of ancestral adult stages by juvenile descendants); (2) abbreviation, or the subtraction of characters from the end of ontogeny (yielding paedomorphosis, the opposite of recapitulation); (3) ontogenetically increasing deviation: characters introduced at an intermediate stage of ontogeny continue to alter the subsequent course of development (and yield the predictions of von Baer's law: similarity of ancestral and descendant development until the stage of introduction, increasing deviation thereafter); (4) ontogenetically culminating deviation: characters introduced at an intermediate stage of ontogeny influence the intermediate stages only and have no effect upon adults.

Matveiev's classification (1932), an elaboration of Severtzov's "phylembryogenetic modes" (1935), includes six categories in two groups (see also Lebedkin, 1937; Kryzanowsky, 1939; and Delsol and Tintant, 1971 and in press, for other taxonomies based on addition and subtraction of characters). In Matveiev's first group, characters may be *added* in phylogeny: (1) at terminal stages (Severtzov's "anaboly"), producing recapitulation; (2) at intermediate stages (deviation), yielding the expectations of von Baer's laws; (3) at the earliest stages (Severtzov's "archallaxis"), producing neither recapitulation (Haeckel) nor the repetition of embryonic stages (von Baer); we gain no clues to ancestry from the descendant's ontogeny.

In the second group, characters may be *deleted* in phylogeny: (4) from the end of ancestral ontogenies (abbreviation), yielding paedomorphosis; (5) from intermediate stages (Matveiev calls this "acceleration" since a deletion of intermediate stages will cause terminal stages of ancestors to appear earlier in descendants. Most other authors use "acceleration" for the speeding-up of developmental rates, not for simple deletion. The result is recapitulation in either case); (6) from the earliest stages ("negative archallaxis" of Severtzov).

Other authors expanded and generalized the law of condensation: A feature already present in the ontogeny of ancestors may be either *accelerated* or *retarded* to appear, respectively, *earlier* or *later* in the ontogeny of descendants. Matveiev (1932) included these changes in rates of development as a seventh and minor category in his classification.

Still other authors generalized both of Haeckel's laws (Smith, 1956). Zimmermann's elaborate chart (1967, p. 126) specifies both when a character originates and what happens subsequently to its developmental rate. De Beer (1930, 1958) mixes both criteria in his eight "morphogenetic modes"—though he claims to base his classification only on alterations in developmental rate. Since this seminal work is the foundation for so many modern studies, I shall present an explicit critique later (see pp. 221–228).

As a further elaboration, several authors tried to extend the concept of recapitulation itself. They reasoned that phylogeny is not the disconnected array of adult stages that Haeckel had envisaged, but a sequence of ontogenies (Garstang, 1922; Severtzov, 1927; Schindewolf, 1946). Recapitulation is the repetition of phylogeny during ontogeny. If phylogeny is construed as a sequence of complete ontogenies, then any stage held in common by ancestors and descendants is a "recapitulation." This theme was stressed by the Russian School of A. N. Severtzov and his followers (Severtzov, 1927, Matveiev, 1932; Yezhikov, 1933, 1937; Lebedkin, 1937; Kryzanowsky, 1939). This remarkable redefinition turned many accepted terms into their opposites. Thus, Schindewolf (1946) actually classified cenogenesis as a subcategory of palingenesis! For, in this new definition, palingenesis applies to any feature present in both ancestor and descendant, including common adaptations of juvenile stages.

We encounter the ultimate confusion in Peter's revision of the terms "palingenesis" and "cenogenesis." To him, any character held in common by ancestor and descendant is palingenetic (since it produces "recapitulation" under this widened and distorted definition). Any character newly evolved by a descendant is cenogenetic: "A cenogenetic feature reaches no further back in phylogeny than the species or group for which it is characteristic; a palingenetic [feature] is already present in evolutionary history" (1955, p. 68). Peter's concepts are totally divorced from the meanings they carried through 80 years of debate, since there is no longer any reference to the ontogenetic stage at which characters appear. The only distinction is between new ("cenogenetic") and old ("palingenetic"). This revision carries the further disadvantage (as Peter readily admits) that the terms are now purely relative in application. The malleus and incus of our middle ear are palingenetic with respect to our position among mammals, and cenogenetic when we are contrasted with reptiles.

Guidelines for a Resolution

Any reader who has worked his way through the labyrinth of the last few pages should now understand why ontogeny and phylogeny has become such an unpopular subject of late. When the well of new concepts has dried up, one can always argue about terms, permute their meanings, and arrange them in ever more complex classifications. I think that the time has come to make some distinctions and divisions. We should cease trying to gather into one uncomfortable scheme such a heterogeneous collection of processes and results—for the common property of these phenomena is only that they describe some way of extracting phyletic information from ontogeny (where no one would deny it resides). There must be a simpler way to present the essence of this great historical theme in modern guise. To do this, I shall pose four questions before reducing de Beer's complex classification of results to two simple processes.

1. How shall we depict phylogeny? Obviously, phylogeny unfolds historically as the sequence of ontogenies for all organisms making up a lineage. It does not follow, however, that the appropriate display of phylogeny is a complete motion picture, from egg to adult, of all these ontogenies. If asked to depict a phyletic sequence of ten successive species, I would not draw the myriad steps of each ontogeny—just as I would not display complete life cycles if I were asked to portray all the monarchs of England. I would seek some *criterion of standardization,* to render each form at a *comparable stage* of development. Traditionally, this stage has been the adult, but I could as well use the eggs of each, the hatchlings, or the corpses—and the more stages we can compare, the more we will learn, as Bonner (1965, 1974) has argued so forcefully. The justification for depicting phylogeny as a sequence of adults does not arise from a claim that only this stage is important in evolution, but merely from the mundane need to consider a sequence of processes at comparable points.

This notion must be emphasized because defenders of the sequence of adults tend to apologize when faced with the argument of Garstang (1922), Kryzanowsky (1939), and Schindewolf (1946) that complete phylogenies are sequences of ontogenies. The presumed issue is a red herring. The sequence of adults is no provisional picture of knowledge that will one day be perfected; it is a standardized sequence that allows us to view each step of a lineage at a comparable point. Phylogeny is a sequence of ontogenies; it is depicted by presenting comparable stages at chosen standardized points (usually, by tradition and for convenience, the adult).[1]

2. How can ontogeny be related to phylogeny? As I emphasized in the introduction, the subject of this book is the ancient contention that a parallel exists between the stages of ontogeny and phylogeny. This is not a side issue extracted from a larger subject. The notion of a parallel has been among the most important themes in the history of biology since Aristotle's time. I have tried to justify this contention by discussing the theories used throughout history to support or attack the kind of parallel implicit in the notion of *recapitulation*—that individuals in the course of their own ontogenetic development pass through

stages representing the *adults* of their ancestors. I have a faith that the most formidable intellects of the past cannot have been so deluded that they persistently centered their discussion on a trivial part of a larger subject. I will therefore assume that it is still important to discuss what constitutes a parallel between the stages of ontogeny and phylogeny, and to distinguish the processes producing such parallels from other relationships between embryology and evolution.

The discovery of parallels between ontogeny and phylogeny does not exhaust the evolutionary information to be gained from the study of individual development. When, for example, the ontogenies of ancestor and descendant display the early identity and later deviation predicted by von Baer's laws, we derive important evolutionary information, though we encounter no parallel between ontogeny and phylogeny. The embryonic features that we share with all vertebrates represent no previous adult state, only the unaltered identity of early development. Though they do not allow us to trace the actual course of our descent in any way, they are full of evolutionary significance nonetheless; for, as Darwin argued, community of embryonic structure reveals community of descent.

3. How are parallels between ontogeny and phylogeny produced? We have seen that the relations between embryology and evolution can be classified on the basis of two contentions well summarized by Garstang:

> I shall simply assume . . . (1) that, instead of new characters tending to arise only towards the end of the ontogeny, they may arise at any stage in the ontogenetic sequence; and (2) that, instead of new characters always tending to push their way backwards in the ontogeny, they may extend into adjoining stages in either direction, either backwards from the adult towards the larva and the embryo (tachygenesis) or forwards to the adult from the embryo and the larva (paedomorphosis). (1928, p. 62; see also de Beer, 1958, p. 170)

Which of these phenomena produces a parallel between ontogeny and phylogeny? Not the first one, for the introduction of an evolutionary novelty at an early or intermediate stage of ontogeny only makes a descendant different from its ancestor. Such an introduction produces the deviation of von Baer's law if its effects persist (and increase) throughout ontogeny, or a juvenile adaptation if they do not persist. Since the effects of such novelties vary according to their time of introduction in ontogeny (and since we must know the ontogenies of ancestor and descendant in order to assess them properly), we may say that they "relate" the study of growth and evolution. But, to repeat, *they produce no parallel between the stages of ontogeny and phylogeny.*

The situation is quite different for Garstang's second contention—features appearing at one stage of an ancestral ontogeny may be shifted to earlier or later stages in descendants. Figure 26 shows how these displacements produce parallels between ontogeny and phylogeny. If an adult feature of an ancestor is progressively shifted back, then the stages of a descendant's ontogeny repeat an evolutionary sequence of adult forms (Fig. 26A); if a juvenile feature of an ancestor is progressively shifted forward, then the evolutionary sequence of adults repeats this ancestral ontogeny in reverse order (Fig. 26B).

All parallels between ontogeny and phylogeny fall into these two categories: If a feature appearing at a standardized point of ancestral ontogeny arises earlier and earlier in descendants, we encounter a direct parallel producing *recapitulation* (the descendant repeats in its own ontogeny a sequence of stages that characterized ancestors at their standardized point—Fig. 26A). If a feature appearing at a standardized point of ancestral ontogeny arises later and later in descendants, we encounter an inverse parallel (Fig. 26B) producing *paedomorphosis* (early features of an ancestral ontogeny are carried forward to appear at the standardized point of a descendant).

Although this simple distinction includes all parallels, it is subject to two provisos that I shall explore throughout this chapter. First, we encounter ambiguities if we do not distinguish the various criteria for standardization—size, age, and developmental stage. I try to resolve these ambiguities with a "clock model." Second, recapitulation and paedomorphosis are *results,* and they can be produced by several processes. Previous studies have often classified evolutionary changes only by these results, thus mixing together distinct processes with differing evolutionary significances. I shall focus on the processes.

I must emphasize that classifications based upon addition and displacement completely exhaust the morphological description of how evolution can occur. Evolutionary changes must appear in ontogeny, and they can arise only by the introduction of new features or by the displacement of features already present. The second process produces parallels between ontogeny and phylogeny; the first does not. Together, they describe the course of morphological evolution. *The continued relevance to modern biology of the great historical theme of parallels between ontogeny and phylogeny rests entirely upon the relative frequency of evolution by displacement rather than by introduction.*

Moreover, since displacement involves no more than a change of timing for developmental stages already present in ancestors, its genetic basis probably resides in the regulatory system. The frequency of evolution by displacement rather than by introduction might provide a minimum estimate for the relative frequency of regu-

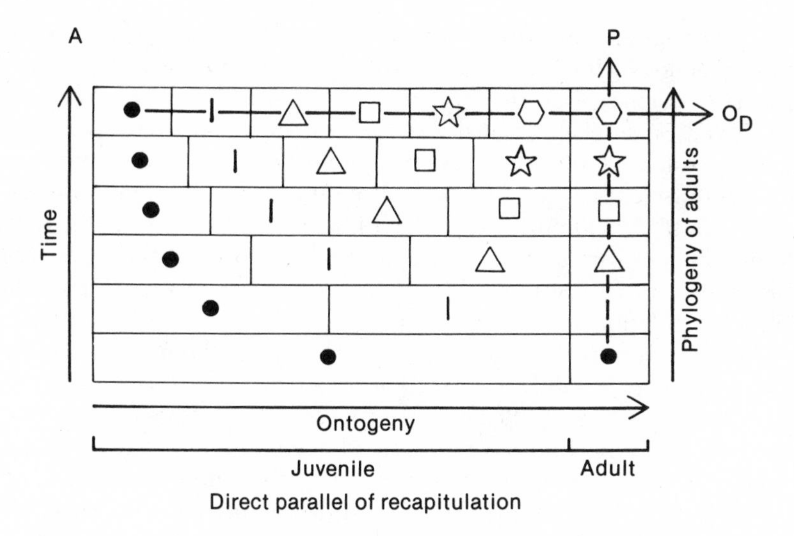

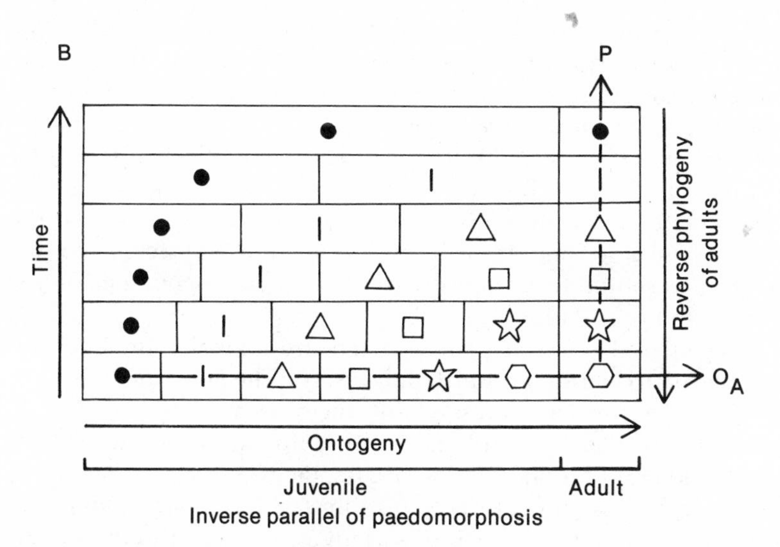

Fig. 26. Types of parallels between ontogeny and phylogeny (P = phylogeny; O_A = ancestral ontogeny; O_D = descendant ontogeny). (*A*) The direct parallel of recapitulation—ontogeny of the most advanced descendant repeats the adult stages of phyletic series of ancestors. (*B*) The inverse parallel of paedomorphosis—ontogeny of the most remote ancestor goes through the same stages as a phylogeny of adult stages read in reverse order (since progressively more juvenile stages of ancestors become the adult stages of successive descendants).

latory changes in evolution—a current and elusive issue in evolutionary biology. (It is a minimum estimate because an accumulation of quantitative changes in timing can also lead to the evolution of apparent novelties; shall we regard a supernumary molar tooth as a novel feature coded by specific structural genes or a result of earlier and more intense expression of a molarization field?) This minimum estimate should provide invaluable data for any hypothesis that favors a dominant role for regulation—since it allows one to argue: "at least this many and maybe more." Hypotheses that seek to underplay the role of regulation are not tested by the estimate because we cannot know how many regulatory changes lurk behind the introduction of apparent novelties. In any case, I merely wish to assert that karyotypes and molecular studies do not provide the only tests for hypotheses about the importance of regulation in evolutionary change. Classical morphology may have an important contribution to make.

4. Which is more important in producing parallels between ontogeny and phylogeny: the shifting of stages by acceleration and retardation or by prolongation and truncation? There are two ways to shift a feature from one stage of development to another. It may be shifted absolutely if ontogeny does not change its length in descendants and the feature is *accelerated* or *retarded* to appear at earlier or later stages of ontogeny. Or it may be shifted relatively if it continues to appear at the same time in development while the descendant's ontogeny is progressively *prolonged* or *truncated* in time.[2] Which of these processes is more important in evolution? Two arguments convince me that we should focus our attention upon the phenomena of acceleration and retardation.

(i) The paradox of infinite length. The traditional argument for the primacy of acceleration-retardation rests on the absurdity of denying it. Any organism has thousands of ancestors in its lineage; if phylogeny proceeded by simple addition, ontogeny would grow inconceivably long. "If recapitulation occurs, tachygenesis [acceleration] is inevitable, unless ontogeny is to be a process prolonged indefinitely" (George, 1933, p. 134). The only serious attack upon this paradox was made by the great Russian morphologist A. N. Severtzov (1927, 1935). Severtzov distinguished two phases in ontogeny: an early and very short period of *morphogenesis* and a much longer period of *growth*. He argued that almost all features (their existence *and* their proportions) are established during the period of morphogenesis. At the close of this short stage, the tiny animal is essentially a miniature adult; it will complete its ontogeny only by increasing in size (and, perhaps, by altering proportions in a minor way). Severtzov's theory of "phylembryogenesis" proclaims that evolution proceeds in two major modes: (1) by substitution of one step in morphogenesis for another (archallaxis—this produces no parallel between ontogeny and phylogeny), and (2) by additions to the end of morphogenesis (anaboly), lengthening this phase of development "at the expense of" the subsequent period of growth (1927, p. 175). Severtzov believed that anaboly is by far the more common of these two phenomena. As an example, he cites the needle-jawed teleost *Belone* (Figs. 27 and 28). At first, both jaws of this fish are equally short; then the lower jaw increases while the upper remains short; finally, the upper jaw lengthens as well. The lengthening of the upper jaw occurs fairly early in ontogeny, and the subsequent course of development involves only an increase in size. Severtzov assumes that *Belone's* adult ancestor ended its period of morphogenesis with short and equal jaws; these were maintained throughout the period of growth. The lengthening of the lower jaw, and, later, the corresponding increase of the upper jaw are anabolies added to the end of morphogenesis and shortening the subsequent period of growth (Fig. 28). They produce a parallel between ontogeny and phylogeny since the intermediate stages of *Belone's* morphogenesis were once the adult stages of its ancestors; but they do so by simple prolongation without any acceleration of developmental rates. "Accordingly, the law of recapitulation of Müller and Haeckel is the immediate and necessary consequence of evolution by addition to the final stages of morphogenesis (anaboly)" (1927, p. 175).

Severtzov's concept of anaboly avoids the paradox of infinite length by allowing prolongation only at the end of an early period of morphogenesis, and only by shortening a much longer period of growth. The total length of ontogeny remains constant: "The duration of the period of morphogenesis is very short compared with that of growth, so that the lengthening of morphogenesis shortens the period of growth by only a small fraction" (p. 178). Anabolies can be added almost without bound in the history of a lineage since they can never completely consume the subsequent period of growth.

If parallels between ontogeny and phylogeny are produced by the prolongation of morphogenesis, what role is left for acceleration and retardation? Severtzov acknowledges the existence of these displacements, but he claims that they "do not occur very often" (p. 123), and that acceleration produces only a "secondary recapitulation" (p. 178). I cannot accept Severtzov's argument because I see no clear distinction between periods of morphogenesis and growth. No one would deny that, in higher animals at least, the greatest changes in form occur early in ontogeny. But these changes do not cease abruptly at some fixed point corresponding to an underlying genetic or bio-

Fig. 27. The ontogeny of the needle-jawed teleost, *Belone.* Lower jaw (*u*) lengthens first, upper jaw (*o*) later. (From Severtzov, 1927.)

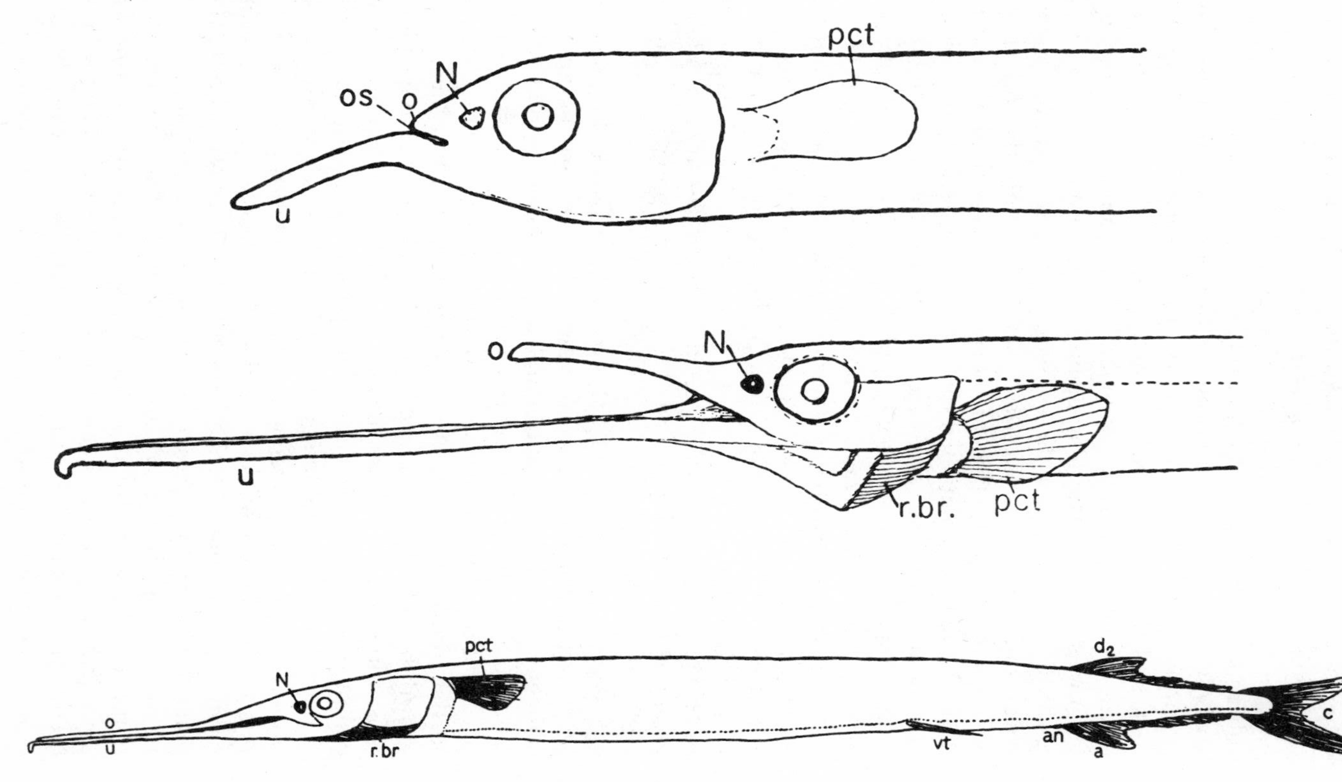

Stadien der Phylogenese.	Periode der Morphogenese.			Wachstumsperiode.						Beispiele:
x_1	a	b	c	c_1	c_2	c_3	c_4	c_5	C . . .	Brustflosse.
	f	g	h	h_1	h_2	h_3	h_4	h_5	H . . .	Oberkiefer.
	m	n	o	o_1	o_2	o_3	o_4	o_5	O . . .	Unterkiefer.
x_n	a	b	c	c_1	c_2	c_3	c_4	c_5	C . . .	Brustflosse (konst.).
	f	g	h	h_1	h_2	h_3	h_4	h_5	H . . .	Oberkiefer (konst.).
	m	n	o	p	p_1	p_2	p_3	p_4	P . . .	U.kiefer verlängert.
x_q	a	b	c	c_1	c_2	c_3	c_4	c_5	C . . .	Brustflosse (konst.).
	f	g	h	i	i_1	i_2	i_3	i_4	I . . .	O.kiefer verlängert.
	m	n	o	p	q	q_1	q_2	q_3	Q . . .	U.kiefer sehr lang.
x_z	a	b	c	c_1	c_2	c_3	c_4	c_5	C . . .	Brustflosse (konst.).
	f	g	h	i	k	k_1	k_2	k_3	K . . .	O.kiefer sehr lang.
	m	n	o	p	q	q_1	q_2	q_3	Q . . .	U.kiefer sehr lang.
	Periode der Morphogenese.			Wachstumsperiode.						

Fig. 28. Severtzov's phyletic interpretation of Fig. 27. New features are added within ontogeny to the end of a period of morphogenesis and at the expense of a much longer and subsequent period of growth (*Wachstumsperiode*); they are called anabolies. First the lower jaw lengthens (anaboly p at stage x_n and q at x_q); then the upper jaw (anaboly i at x_q and k at x_z). The pectoral fin (*Brustflosse*) does not change during phylogeny. (From Severtzov, 1927.)

chemical transition; they simply decrease their rates gradually throughout ontogeny. (In fact, the *rates* of change are often maintained throughout growth since a simple power function accurately describes the complete postnatal ontogeny of so many organisms.) Severtzov developed his views before the work of Huxley (1932) and hundreds of colleagues on allometric growth. We have tabulated so many thousands of cases involving significant postnatal changes of proportions (see Gould, 1966) that the general existence of a long and discrete period of growth in geometric similarity cannot be maintained. Therefore, Severtzov's anabolies, common though they be, are not additions to end stages; they are interpolations into intermediate stages of ontogeny, and they affect the subsequent course of development. As such, they are equivalent to the deviations that affirm von Baer's laws. The two short jaws of *Belone's* early ontogeny are a feature held in common by ancestor and descendant; they do not mark the miniaturized adult stage of a short-jawed ancestor. Anabolies produce no parallel between ontogeny and phylogeny.

I conclude that the paradox of infinite length is valid. Recapitulation and paedomorphosis may, for a time, result from a simple prolongation or truncation of ontogeny; but this cannot be long extended in phyletic lineages, lest ontogeny become absurdly long or disadvantageously short. Acceleration and retardation must be the cause of most extended parallels between ontogeny and phylogeny.

(ii) Most cases of prolonged or truncated ontogeny are caused by the acceleration and retardation of maturation. Although limited by the paradox of infinite length, prolongation does certainly occur. In fact, Fritz Müller based his original evolutionary formulation of recapitulation upon it: "Descendants reach a new end stage . . . by passing along this course of previous ontogeny without deviation, but then, instead of standing still, advancing still farther . . . Descendants pass through the entire development of the progenitors and . . . the historical development of a species will be mirrored in its developmental history" (1864, in 1915, p. 250). De Beer (1930, p. 80) included prolongation among his modes of evolution, but denied it any great importance. Paleontologists might reject this denial, because evolutionary size increase is such a common event in phyletic lineages—often deemed of sufficient generality to be canonized as "Cope's law." When Cope's law applies, phylogeny may proceed by a simple extrapolation to larger sizes of the allometric tendencies already present in ancestral ontogenies. This phenomenon is called "allomorphosis" (Gould, 1966, p. 601); some authors, Rensch in particular (1959), grant it an important role in evolution: "In many cases, the characters of grown individuals [of a phyletic series] are related to each other as if the different species were earlier or later growth stages of a single species; as if, therefore, the largest species was developed by additions to its end stages" (Rensch, 1971, p. 14). The opposite phenomenon of truncation encounters fewer theoretical difficulties and has been known (by such names as paedogenesis and progenesis) for more than a century.

But how do prolongation and truncation occur? Truncation arises when the gonads are *accelerated* in development. Maturation occurs early, development often stops, affected individuals reproduce while retaining their larval form (paedomorphosis). De Beer cites the classic case: "Another example is the worm *Polystomum integerrimum,* which is parasitic on a frog, usually in its bladder, where it takes three years to reach maturity. But should it infect an early tadpole stage of the frog, it remains in the gill chamber and becomes sexually mature in five weeks" (1930, p. 27). Though truncation is only facultative in this case, evolutionary fixation has arisen in others. *Polystomum ocellatum* is permanently paedomorphic and resembles the truncated form of *Polystomum integerrimum* (de Beer, 1930, p. 27).

Any major prolongation will be correlated with a *retardation* of maturation. This retardation is not a secondary consequence or an ancillary correlate of prolongation; on the contrary, selection for delayed maturation may be the major determinant of phyletic size increase (Chapter 9).

A small degree of prolongation or truncation may develop without any advance or delay in maturation, but major evolutionary changes in the length of ontogeny involve selection for *acceleration* or *retardation* of sexual maturity. Thus, prolongation and truncation are but another aspect of acceleration and retardation (of sexual organs rather than somatic characters). We can trace almost all parallels between ontogeny and phylogeny to acceleration and retardation.

DeBeer made a major contribution by forcefully linking acceleration and retardation in the single phenomenon of temporal displacement. But he introduced a regrettable confusion by applying Haeckel's term "heterochrony" to this phenomenon. To Haeckel, heterochrony was the displacement of one feature relative to other features of the same organism. For de Beer, it became the displacement of a feature relative to the time that this same feature appeared in an ancestral form.* This transmutation has been so widely adopted, however, that I have no choice but to use "heterochrony" in de Beer's sense throughout the rest of this book.

Heterochrony is the mechanism that produces recapitulation and paedomorphosis as its result. If parallels between ontogeny and phylogeny are of major significance in the history of life, it is because heterochrony—the temporal displacement of characters—is a pervasive phenomenon among evolutionary processes.

The Reduction of de Beer's Categories of Heterochrony to Acceleration and Retardation

In a series of remarkable books that established the synthetic theory of evolution, Gavin de Beer's *Embryology and Evolution* was the first

* The odyssey of heterochrony is exceedingly curious. In its original use, it defined a class of *exceptions* to recapitulation since it precluded the complete transference of an entire ancestral adult to juvenile stages of descendants. In Mehnert's reinterpretation, it became consistent with recapitulation because its prevalence required that recapitulation be redefined for single characters only. In de Beer's altered reading, it becomes the *mechanism* of recapitulation.

and the shortest (1930; expanded and retitled *Embryos and Ancestors,* 1940; 3rd edition, 1958). In 116 pages, de Beer brought embryology into the developing orthodoxy by attacking Haeckel's theory of recapitulation as inconsistent with modern evolutionary theory. For more than forty years, his book has dominated English thought on the relationship between ontogeny and phylogeny. I believe, however, that it has become deficient in several important respects. Most of these deficiencies have arisen naturally and inevitably as modern work in ecology and biometrics has suggested different evolutionary significances for heterochrony (Chapters 8–9). But one problem was present from the start, as a confusion in de Beer's classification of eight types of heterochrony. I shall try to show that: (1) De Beer mixed two criteria in distinguishing his "morphological modes"; (2) by focusing on results, he created a cumbersome classification that obscured the simplicity of the underlying processes; and, (3) his eight "morphological modes" can be reduced to the two aspects of a single process—acceleration and retardation.

(For historical reasons I will base my quotations and discussion primarily on de Beer's first edition of 1930. Lest this seem unjust to a much-revised book, I note that the latest edition of 1958 does not differ in any way on these points. In the introduction to this latest edition, de Beer writes: "During the intervening years a great deal of new evidence has become available, and these fresh data have fitted into place in my scheme like pieces of a puzzle, for I have no reason to alter the plan of my former book in the slightest degree. The present book is my previous one brought up to date and enlarged" [p. 5].)

In stating his three principles, de Beer clearly distinguishes between the introduction of new characters and their displacement in time by heterochrony.

1. Qualitative evolutionary novelties can and do appear at all stages in ontogeny, and not solely in the adult.
2. Characters can and do change the time and order of their appearance in the ontogeny of the descendant as compared with that of the ancestor.
3. Quantitative differences between characters, resulting in heterochrony, play a part in phylogeny in addition to the introduction of qualitative novelties. (1930, p. 88)

De Beer then presents his classification of "the eight possible types of relation between ontogeny and phylogeny" (1958, p. 173). It appears in a chapter entitled: "Heterochrony and Its Effect in Phylogeny" (1930, pp. 34–40). If heterochrony is the basis of classification, then we must assume that the eight modes refer to quantitative, temporal displacements, not to the qualitative introduction of evolutionary novelties. Starting with the 1940 edition, de Beer presents his famous chart of the eight modes (Fig. 29); the verbal descriptions are identical in all editions. Of the eight modes, four have nothing to do with heterochrony; and of these four, three are identical.

1. "A character which is present or makes its appearance in the young stage of an ancestral animal may in the ontogeny of a descendant appear in the young stage only, producing youthful adaptations or *cenogenesis*, and not affecting phylogeny" (1930, p. 37). Haeckel's definition of cenogenesis had been much broader, but de Beer restricts it to embryonic and larval adaptations that have *no effect* upon adult organization. Severtzov (1927, pp. 147–148) had introduced this restriction by arguing (quite justly) that many of Haeckel's cenogeneses are primarily adaptations of the adult stage, even though they may affect early stages of ontogeny as well. He wanted to make a clear distinction between truly larval adaptations and phylogenetic alterations of the adult that begin early in ontogeny. Severtzov's restriction has spread via de Beer through almost all literature in English on the subject. Much as I deplore the confusion that attends such major shifts of meaning, "cenogenesis" is now so widely known only in this restricted sense, that I shall so use it hereafter in this book.

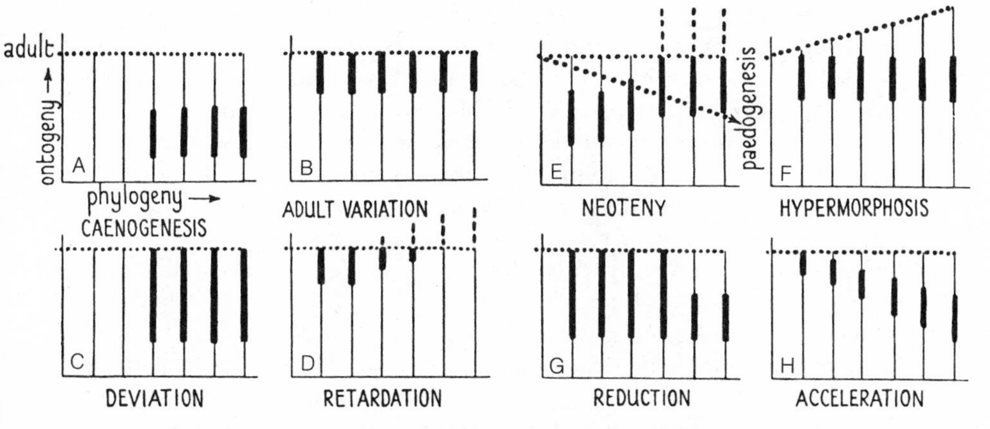

Fig. 29. De Beer's eight categories of "heterochrony." See text for explication and critique. Only four are actually modes of heterochrony and these reduce to the two processes of acceleration (paedogenesis = progenesis, acceleration) and retardation (neoteny, retardation, hypermorphosis). (From de Beer, 1940.)

As de Beer's own figure (Fig. 29A) shows, there is no heterochrony in cenogenesis. The character, once it arises, does not alter its time of appearance in ontogeny. Evolutionary change occurs only through the introduction of the character itself; but this is not heterochrony, and it produces no parallel between ontogeny and phylogeny. De Beer has confused two criteria by improperly including the addition of a new feature as a category of heterochrony.

2. "A character which is present or makes its appearance in the young stage of an ancestral animal may in the ontogeny of a descendant appear in the young and adult stage, producing a substitution of a new adult condition for the old, resulting in progressive *deviation* in the ontogeny of the descendant from that of the ancestor" (1930, p. 37—Fig. 29C). The newly introduced character now extends its influence (not itself) to all later stages. This produces a deviation between ancestor and descendant, not an incorporation of youthful ancestral features into adult descendants by heterochrony. There is no displacement of any character in time and no heterochrony—only the addition of a character and its subsequent effects. The results of deviation are those predicted by von Baer's law—embryonic identity and subsequent divergence between ancestral and descendant ontogenies. As I have demonstrated before, and as von Baer himself so vigorously argued in pre-evolutionary terms (1828, pp. 192–262), these cases provide no parallels between ontogeny and phylogeny. De Beer himself seems to acknowledge this in one passage: "These are the cases which led von Baer to propose his principle of the greater resemblance between young stages . . . These cases are of the utmost importance because they have been used erroneously to support the biogenetic law" (1930, pp. 45–46). Moreover, as I shall argue later in this chapter, most deviations may develop when the first effect of a character introduced late in ontogeny is transferred to earlier stages of descendants. The static fact that two ontogenies deviate does not permit an evolutionary reconstruction of the event. A new character may be introduced early with effects gradually extending to the adult (the only method de Beer allows); it may be introduced late with initial effects propagated gradually backward; or it may alter the entire course of subsequent development right from the start. We must avoid the fallacy of assuming that a historical process can be inferred unambiguously from a modern result.

3. "A character which is present in the young and adult stages of an ancestral animal may in the ontogeny of a descendant appear in the young stage only, resulting in the *reduction* of the character to a vestige" (1930, p. 37). But this is just deviation viewed from the stand-

point of the character replaced rather than the one added (Fig. 29G). As the effects of a new character are propagated towards the adult, previously adult features gradually disappear (Figs. 29C and 29G fit together like pieces of a jigsaw puzzle). Reduction is the same process as deviation, and it is not heterochrony. De Beer acknowledges this in writing that, "These cases really constitute the other side of the picture presented by deviation" (p. 71).

4. "A character which is present or makes its appearance in the adult stage of an ancestor may in the ontogeny of a descendant appear in the adult stage, resulting in those differences which distinguish individuals, varieties, and races: *adult variation*" (1930, pp. 37–38). A new adult feature is merely a deviation that has little morphological effect because it appears so late in ontogeny (Fig. 29B). Adult variation falls in the same category as deviation and reduction. As de Beer himself notes: "Under this heading we include what is really only a special case of the phenomenon which was described under the term deviation . . . We cannot draw any hard and fast line between the characters which, substituting themselves for others in phylogeny, appear early in ontogeny, and those which appear late. However, the later a character appears in ontogeny, the smaller as a rule is the change which it produces by its presence" (pp. 73–74). De Beer even admits that adult variation does not involve heterochrony: "The substitution of one character for another in the adult does not usually involve heterochrony, and produces only small phylogenetic effects" (p. 74). Yet de Beer presents all eight morphological modes as categories of heterochrony!

We may isolate and remove these four of de Beer's eight categories from the domain of heterochrony. They deal with the introduction of new features, not the displacement of characters in time. They are important in evolution, but they produce no parallels between ontogeny and phylogeny. Moreover, they reduce to two phenomena: cenogenesis and deviation (including reduction and adult variation). De Beer subdivides deviation according to where in ontogeny a new character appears and whether we shall consider its effect or the feature it replaces; this confusion and proliferation illustrates the unnecessary complexities that we engender in producing *taxonomies of results* rather than *explications of processes.*

We are left with four categories that treat some aspect of heterochrony:

5. "A character which is present or makes its appearance in the adult stage of an ancestor may in the ontogeny of a descendant appear in the young stage producing precocious appearance of the an-

cestral character and *acceleration*" (1930, pp. 37–38). This is the classic category of heterochrony; it produces Haeckelian recapitulation (Fig. 29H).

6. "A character which is present or makes its appearance in the young stage of an ancestral animal may in the ontogeny of a descendant appear in the adult, by a relative retardation of the development of the bodily structures as compared with the reproductive organs, resulting in *paedogenesis* and *neoteny*" (1930, p. 37). Here de Beer mixes two processes because their results are the same—youthful features of ancestors become adult features of descendants (paedomorphosis). "This state of affairs," de Beer admits (p. 57), "may of course be brought about in many ways." We sense this confusion in Figure 29E, where the two processes are clearly superposed. The problem lies in the concept of "relative retardation," for this term hides two distinct phenomena with differing evolutionary significances (Chapters 8–9).

De Beer describes paedogenesis as follows: "with the rate of development of the soma remaining constant, this effect will be produced if the germen is accelerated" (1930, p. 57). Here the result of paedomorphosis arises from the *acceleration* of maturation.

I retain all terms in their original meanings—and invent no new ones[3]—unless a later but prevailing usage clearly requires the acceptance of change (as with cenogenesis and heterochrony). I must therefore reject de Beer's "paedogenesis" as a designation for the acceleration of maturation. Von Baer coined paedogenesis to define a type of reproduction that happens to yield truncated development as a by-product. He restricted it to cases of parthenogenesis in insect larvae structurally unable to copulate: "If fully developed females lay eggs that develop without being fertilized . . . we call this mode of reproduction parthenogenesis; while we propose that such asexual reproduction be called paedogenesis if it occurs in early stages of development" (1866, p. 121). Hamann (1891) extended von Baer's term to include all precocious maturation (though Chun, 1892, attacked him vigorously and urged a return to von Baer's concept). Some authors, including de Beer, have adopted this generalization, but it is not widespread and entomologists still use von Baer's term in its original meaning (Wyatt, 1963). For accelerated maturation I shall use Giard's term *progenesis* (1887, p. 23): "We say that an animal exhibits progenesis when sexual reproduction occurs in a more or less precocious fashion, that is to say when the sexual products (eggs or spermatozoa) form and mature before the animal has attained its complete development."

De Beer's second paedomorphic process involves a retardation of somatic development. For these retardations, Giard (1887) used Kollmann's term *neoteny*. This is faithful to Kollmann's original usage in describing the giant larvae of common European amphibians (where larval retention results from delayed somatic development rather than accelerated reproduction).

Neoteny must not be confused with what we call progenesis. We have neoteny when an animal, in becoming adult, retains certain infantile characters. An adult man who has kept his milk teeth (we know of a case), is an example of partial neoteny. In neotenic animals, growth continues but with more or less numerous arrests of development. On the contrary, we have progenesis when, in normal development with normal growth, the genital organs develop prematurely and allow the animal to reproduce before it has assumed adult characters. The appearance of signs of puberty and functioning of genital organs in children of either sex is a well known case of progenesis in humans. (Giard and Bonnier, 1887, p. 195)

De Beer uses "neoteny" to describe both acceleration of gonads and retardation of somatic characters: "Since there can be no hard and fast distinction between paedogenesis [= progenesis] and neoteny, and paedogenesis is of no significance in progressive evolution, all these cases are included here under the term neoteny" (1930, p. 64). I reject this extended usage because I believe that the two phenomena play very different evolutionary roles (Chapters 8–9) and that progenesis is widespread and important; "progress," whatever it may be, is not the only phenomenon of life's history.

I use Garstang's term "paedomorphosis" to specify the common *result* of progenesis and neoteny—the appearance of youthful characters of ancestors in later ontogenetic stages of descendants (Clark, 1962, p. 300). This is faithful to Garstang's meaning: "New characters . . . may extend . . . forwards to the adult from the embryo and larva (paedomorphosis)" (1928, p. 62). De Beer (1930, pp. 59 and 90), however, uses paedomorphosis in a wider sense to include both neoteny *and* deviation (since he construes deviation as the increasing influence of originally juvenile characters upon adult stages). I reject this extension for two reasons: first, deviation has nothing to do with heterochrony since it involves no displacement in time of any feature; second, deviation may also arise by the propagation backwards of the first effects of originally adult characters.

I summarize this regrettably necessary terminological excursion by stating that my choices all have sensible etymologies, are faithful to their original meanings, and are widely understood:

Paedomorphosis (Garstang, 1922—"shaped like a child"): the displacement of ancestral features to later stages of the ontogeny of descendants. This morphological phenomenon is a common *result* of two distinct *processes:*

Progenesis (Giard, 1887—"generation before [the ancestral time]"): paedomorphosis produced by an *acceleration* of maturation.

Neoteny (Kollmann, 1885—"retention of young features"): paedomorphosis produced by a *retardation* of somatic development.

Thus, in this second category of heterochrony, de Beer mixed an aspect of acceleration with one of retardation. The problems of a taxonomy of results are again evident.

7. "A character which is present or makes its appearance in the adult stage of an ancestor may in the ontogeny of a descendant appear in the late adult stage, i.e. too late, resulting in the reduction of the character to a vestige by *retardation*" (1930, pp. 37–38—Fig. 29D). Here, de Beer merely presents the other side of neoteny (just as he recorded the obverse of deviation in the separate category of reduction). If youthful characters are retarded, then previously adult characters must be gradually "pushed off" the end of ontogeny and eliminated. "Retardation of structures to vestiges is therefore the other side of the picture presented by the phenomenon of neoteny" (p. 75).

8. "A character which is present or makes its appearance in the adult stage of an ancestor may in the ontogeny of a descendant appear in the same stage, which is no longer adult, the new adult stage being relatively delayed, resulting in 'overstepping' the previous ontogenies or *hypermorphosis*" (1930, pp. 37–38—Fig. 29F). I argued previously that this prolongation of ontogeny results from the retardation of maturation. De Beer (1930, p. 77) agrees: "This additional development or hypermorphosis, may then be expected in cases where the rate of development of the reproductive glands is delayed relatively to that of the body characters."

De Beer's four admissible categories of heterochrony reduce to two processes with a common basis: acceleration and retardation. These processes affect reproductive organs and somatic features differently. We simplify de Beer's complexities by distinguishing true heterochrony from the introduction of new features, and by focusing on processes instead of results. I summarize my reduction of de Beer's categories of heterochrony in Table 3.

A Historical Paradox: The Supposed Dominance of Recapitulation

My argument has arrived at the following point: Parallels between ontogeny and phylogeny are produced by heterochrony. Heterochrony proceeds by acceleration or retardation. We have no a

Table 3. The categories of heterochrony.

Timing			
Somatic features	Reproductive organs	Name in de Beer's system	Morphological result
Accelerated	—	Acceleration	Recapitulation (by acceleration)
—	Accelerated	Paedogenesis [= progenesis]	Paedomorphosis (by truncation)
Retarded	—	Neoteny	Paedomorphosis (by retardation)
—	Retarded	Hypermorphosis	Recapitulation (by prolongation)

priori basis for assuming that one of these processes is more frequent than the other; their attendant results—recapitulation and paedomorphosis—should be equally common. At this point we encounter a paradox. The great comparative embryologists and anatomists of the late nineteenth century had an encyclopedic knowledge of the facts of their subject. We shall never see their like again unless science returns, as it surely cannot, to an ethic of observation and description. Almost to a man, they upheld a principle of universal recapitulation, under which paedomorphosis included only a minor class of exceptional cases. Their impression was, of course, formed in the light of a theory that expected these results. The "cloven hoofprint" of theory (Hanson, 1969) marked every perception and must have been the major reason why these scientists always "observed" the dominance of recapitulation. Only recapitulation, they believed, permits progressive evolution. Only recapitulation allows the tracing of evolutionary lineages through their preservation as ontogenies.

But I doubt that the bias of expectations will resolve this paradox completely. We must explain why the expectations arose in the first place (for there was a large empirical input, even though general perspectives of the time favored recapitulation a priori). And we must explain the continued confirmation of recapitulation, recalling the argument of Chapter 6 that the eventual downfall of the biogenetic law had little to do with empirical refutation. Surely the *impression* of recapitulation's dominance has some basis in nature. This dominance poses a problem because it *seems* to imply a far greater frequency of acceleration over retardation; for acceleration, as Hyatt so amply demonstrated, is the motor of recapitulation. I shall try to resolve this paradox by arguing that the dominance of acceleration is but one, and probably the weakest, of three reasons why recapitulation seems to be far more common then paedomorphosis.

1. Von Baer argued strenuously against recapitulation. Yet, ironically, the evolutionary transformation of his laws guaranteed that recapitulation would seem to prevail where it does not exist at all. To take the classic case: our fetal gill slits are features held in common by all vertebrates. Their appearance has nothing to do with heterochrony; they appear in ontogeny now as they have appeared for 400 million years. Yet it is characteristic of evolution in "progressive" lineages that the adults of primitive forms diverge less from the common embryonic plan than do the groups that evolved from them (Darwin's argument for the relationship of embryology and evolution). Our gill slits are those of embryonic fishes; but adult fishes retain these organs in elaborated form. Although it seems that we recapitulate the adult stage of an ancestor, we only repeat an embryonic stage that ancestral adults modify very little. Moreover, if we interpret our gill slits as ancestral adult structures, then we are driven to a principle of acceleration because they appear when the human fetus is so much smaller than our ancestral fish. I have no doubt that most supposed "recapitulations" of adult ancestors have nothing to do with heterochrony and only reflect the evolutionary transformation of von Baer's law—that development proceeds from the general to the special. The acceleration that they seem to require does not exist. They are embryonic repetitions that represent, as Garstang said, "merely the static aspect of inheritance" (1922, p. 86). Thus, Garstang interpolated his biting comments to undercut MacBride's (1917, p. 428) defense of the biogenetic law:

"The young Hermit-Crab swims freely about in the water and has a symmetrical abdomen like that of Shrimps and Prawns" (but so have the *young* stages of these creatures!); "the young Flatfish swims with its ventral edge down and its dorsal edge up, and has an eye on each side of the head" (but so have the young of all Teleosti!!); "the young Comatulid is fixed to the bottom by a stalk like other Crinoids" (and their young too, in all probability!); and "the young American Oyster possesses a foot like that of other bivalves by which it crawls about" (and, I may add, as the *young* of nearly all other Lamellibranchs crawl about!) . . . It is the adult Oyster which has lost its foot, not the young Oyster which has acquired it. It is the adult *Portunion*[4] which has lost its legs, not the young *Portunion* which has acquired them by tachy—(or any other kind of) genesis from its adult ancestors! . . . His [MacBride's] province was to show that by virtue of Haeckel's Biogenetic Law he could reconstruct the prominent features of an *adult* ancestor from a developmental stage. All he has done is imperfectly to confirm Von Baer's prae-Haeckelian doctrine, that animals resemble one another more closely in their young stages than in their adult stages. (1922, pp. 89–90)

The same argument formed the basis of T. H. Morgan's rebuttal of universal recapitulation: "But how, it may be asked, can we explain the apparent resemblance between the embryo of the higher form and the adult of lower groups. The answer is that this resemblance is deceptive, and in so far as there is a resemblance it depends on the resemblance of the adult of the lower form to its own embryonic stages" (1903, p. 75; see also de Beer, 1930, pp. 101–102; 1958, p. 62).

Two further corollaries of von Baer's laws enhance the temptation to render embryonic repetition in recapitulatory language:

First, the earliest stages of ontogeny are particularly refractory to major change since they control the induction and differentiation of all later structures. Since alterations tend to have cumulative effects in ontogeny, early changes usually disrupt the ancestral course of development. Modifications of late stages are more easily incorporated as advantageous variants. Thus, deviations rarely begin very early in a descendant's ontogeny; a major period of identity almost always persists in the early ontogenies of ancestor and descendant, leaving much scope for a deceptive recapitulatory reading. Even if successful mutations are distributed randomly in their ontogenetic time of expression, later stages will be more profoundly modified than earlier ones simply because most mutations have a cumulative effect. A leading modern textbook cites this argument as the primary reason for apparent "recapitulation" (Balinsky, 1970, p. 590).

Second, although Severtzov erred in drawing an absolute distinction between morphogenesis and growth, he was certainly right in affirming that the major transformations of ontogeny usually occur very early. A rather young embryo may be very similar to its adult in all but size. Even if a deviation occurs early in ontogeny, it will usually follow these major transformations. The last step common to ancestor and descendant will be close in form to the ancestral adult. On this basis, Severtzov tried to "synthesize" the fundamentally irreconcilable laws of Haeckel and von Baer: "The final stages of morphogenesis in an ancestor are preserved in the ontogeny of descendants; therefore, we have true recapitulation of characters of adult ancestors because these final stages [of morphogenesis] differ from the corresponding parts of adult ancestors only in their smaller size" (1927, p. 173). "Recapitulated characters," wrote Yezhikov (1933, p. 74) "are embryonic characters and, at the same time, characters of the adult condition."

2. As a second reason for the apparent dominance of recapitulation, true recapitulation can occur without acceleration. If ontogeny is simply extended without being compressed, the adult characters of ancestors appear at the same time in descendants, but as intermediate stages of ontogeny. As Table 3 shows, evolution by prolongation is usually accompanied by *retardation* of sexual development—not by acceleration of anything. This is probably responsible for the remarkable persistence of recapitulatory beliefs among paleontologists. Since paleontologists deal so often with phyletic size increase (Cope's Law), they most often encounter cases of recapitulation by simple prolongation. Tilley (1973, p. 15), for example, has shown that interpopulational variation in body size reflects differing ages of maturation in the salamander *Desmognathus ochrophaeus*.

3. We arrive finally at the classic argument: recapitulation dominates because acceleration is more common than retardation. Zimmermann (1967, p. 126) has estimated the dominance of recapitulation over paedomorphosis at 80 percent in plants, while Remane (1962, p. 574) estimates 80–90 percent for organs of higher animals in later stages of embryonic development, after differentiation from the primary germ layers. But these are mere guesses. Empirical tabulation will not solve the problem; there are simply too many cases to count and prior attitudes always dictate the selection. Moreover, even if we accept these figures, Zimmermann and Remane do not separate pseudorecapitulatory deviations and prolongations from true acceleration of somatic characters. Can we cite any theoretical argument to justify the dominance of acceleration over retardation?

Stebbins (1974) has recently based such an argument upon the fact that early stages of ontogeny are highly refractory to immediate change. As Zuckerkandl states:

> Processes of terminal program additions imply that contemporary larval stages should be closely related to ancestral adult stages . . . and present adult stages, strictly speaking, should have no equivalent in the distant past. Many evolutionists are opposed to such a view. It seems however difficult to escape it altogether. Terminal addition of gene programs should be particularly "easy," in that they can hardly interfere with the programs for preceding developmental stages. (in press)

Stebbins agrees that the canalization of development virtually precludes the introduction of major innovations at points other than at or near the end of ontogeny. If innovations are advantageous at the end points of development, they might be equally valuable at earlier times, but they cannot arise there. Thus, innovations move preferentially in one direction—back into earlier stages of ontogeny, by acceleration.

If they had been introduced early and suddenly, they would have produced a disruption too great to bear. But they can be introduced later and gradually insinuate themselves backwards through the action of newly evolved modifiers. Stebbins refers to this process as the "increasing precocity of gene action" (1974, p. 115). It represents another mode of regulatory change, with important potential for macroevolution.

Stebbins' proposal can be tested against the traditional Haeckelian pattern of recapitulation. The conventional mechanism requires an addition of characters to the end of ancestral ontogenies and a consequent shifting back into earlier development of previously adult features. For increasing precocity, this shifting back may occur with no further alteration of the adult stage: an advantageous modification accelerates its time of first appearance without undergoing any change itself.

If one is inclined to feel friendly toward speculative phylogeny in the old tradition, Jagersten's (1972) interesting book offers powerful support to Stebbins. Jagersten believes that an original, holopelagic, radially symmetrical Blastaea descended to spend the latter part of its life as a benthic organism. This life cycle of pelagic juvenile and nonsedentary benthic adult is ancestral for all the higher Metazoa. As the adult stage evolved and acquired new adaptations for its benthic existence (often for sedentary life), it became more and more unlike the original holopelagic progenitor. A true metamorphosis in development gradually evolved, as pelagic and benthic phases of the life cycle diverged. But if this rapid and profound restructuring must occur all at once when the larva descends from the plankton, then ontogeny provides little margin for error. Jagersten believes that an advantageous margin evolved by the acceleration of certain adult features into the pelagic phase in order to prepare the larva, so to speak, for its coming transformation. Jagersten refers to this process as "adultation," and recognizes it as a category of acceleration (1972, p. 6). He cites the shell and foot of veliger larvae and states that the shell, or at least the shell gland, can already be distinguished in the gastrula stage of some molluscan embryos. These adultations cannot be examples of von Baer's laws; they must represent a true acceleration because the characters are advantageous only to adults and the original life cycle was holopelagic. The characters must have been accelerated into pelagic larvae from an adult state that evolved much later. But their acceleration has not been caused by any subsequent additions to adult ontogeny. Once an adult specialization evolves, it need develop no further in adults in order to extend its influence progressively back into pelagic juvenile stages. It is accelerated because an earlier appearance will greatly aid the mechanics of metamorphosis. The increasing precocity of gene action can explain adultation most effectively.

In conclusion, I do not know if recapitulation is a more common result than paedomorphosis. Stebbins' concept of increasing precocity and the predominance of increase in phyletic size argue, respectively, for a greater frequency of acceleration over retardation and of prolongation over truncation. Both would justify a belief in the dominance of recapitulation. But I am also convinced that the impression of an overwhelming dominance rests, ironically, on cases that have nothing to do with recapitulation, but only mimic it in the workings of von Baer's laws.

Dissociability and Heterochrony

Correlation and Dissociability

When we eliminate the numerous cases that only reflect von Baer's laws, we are still left with thousands of documented heterochronies and the conclusion that this process has dominated the evolution of many important lineages (Chapter 8). Is there any general observation about the nature of ontogeny that will help us to understand why acceleration and retardation are so common in phylogeny? I believe that the notion of *dissociability* provides such a key.

There is an old prejudice in biology, expressed as much in ancient notions of perfect harmony as in Cuvier's correlation of parts. It dictates that organisms be treated as sublimely integrated systems, always changing in perfect coordination. In phylogenetics, this prejudice surfaced as the "harmonious development of the type." Fossils were often denied an ancestral status if they showed a mixture of primitive and advanced characters, rather than perfect intermediacy between an earlier ancestor and its modern descendant. Ernst Mayr (personal communication) tells of an argument he had in the 1940s with Franz Weidenreich on the status of *Australopithecus.* Weidenreich would not allow this small-brained but fully erect hominid into our lineage because it was an inharmonious type. The concept of "mosaic evolution," developed by Louis Dollo and others, refuted the notion of harmonious development by affirming that individual organs could have independent phyletic histories, despite the evident correlation of parts within any organism.[5] Correlations are no more immutable than species themselves.

The notion of primary harmony motivated Haeckel's view of recapitulation, while the recognition of dissociability prompted the first major alteration of his system. Haeckel expected to find entire ancestors evenly accelerated into early phases of a descendant's ontogeny; he regarded any mixture of stages as an exceptional sign that development had been thrown out of phase. Cope and Mehnert argued that all parts are dissociable in phylogeny: each develops at its own rate and every stage of a descendant's ontogeny must display different ancestral stages of the various organs. Recapitulation must be considered in terms of dissociated parts rather than entire organisms.[6]

Joseph Needham defined a dissociability far more basic than that of primary organs: "the dissociability of the fundamental processes in ontogenesis." We can distinguish three "fundamental processes" that are crucial to the study of parallels between ontogeny and phylogeny; dissociation between any two must result in heterochrony.[7] The three are (1) growth, or "increase in spatial dimensions and in weight" (Needham, 1933, p. 181); (2) maturation; and (3) development. I include under "development" both differentiation, or "increase in complexity and organization" (p. 181), and any change in shape during ontogeny. Bonner (1974, p. 20) has urged an extension of the term "development" to include allometries of late ontogeny as well as fetal and juvenile differentiation. I construe growth in the purest sense, as size increase with geometric similarity; any differentiation of parts or change in their proportions is an aspect of development. I am aware that my union of differentiation and allometry will seem both unconventional and unbiological to many, for allometry is often regarded as an aspect of simple growth. Many authors would deny any meaningful distinction between growth and development (Falkner, 1966, p. xv). I make the distinction because it allows me to formulate a compellingly simple clock model of heterochrony. The model, as an abstraction for measuring the extent of heterochrony, will, in any case, find its primary justification in its utility. If its distinction of growth and development should find some causal basis in cellular processes, it would acquire a gratifying reality as well.

Novak has proposed a causal basis for separating pure growth from development in hemimetabolous insects. He states that "the growth in each larval instar may be interpreted as consisting of two parts: the isometric growth of the whole body (both larval and imaginal) and the disproportionate increase of the imaginal parts by allometric growth" (1966, p. 115). The allometric (development) and isometric (growth) phases are causally separate. The isometric phase occurs only in the presence of an effective concentration of juvenile hormone. After a molt, and before this concentration is attained, mitosis occurs only in imaginal parts of the epidermis, and allometric development progresses; when effective concentrations are reached, mitosis proceeds equally over the entire body. Since the titer of juvenile hormone decreases steadily throughout larval ontogeny, the effective concentration is reached later and later in each successive molt, thus permitting a longer allometric phase in each instar and the eventual production of an adult (ontogeny following the last larval molt is entirely allometric, corresponding to the first phase of other instars). Novak (1966, p. 116) emphasizes that the normal relation of growth and development can be dissociated, to produce, for example, giant supernumerary larvae geometrically similar to smaller, normal forms.

W. M. Krogman, the leading American student of child growth, makes the same distinctions between growth, development, and maturation. He defines growth as "*proportionate* changes in size," and adds:

> At the risk of oversimplification it is possible to set up three aspects of human growth and development.
> 1. We grow. This is size . . .
> 2. We grow up. This is proportion.
> 3. We grow older. This is maturation. Every tissue in the body bears an indelible register of the passage of biological time. (1972, p. 3)

Dissociation of the Three Processes

Attempts to separate simple growth from changes in shape date at least from Aristotle's speculation that the embryo receives two kinds of food from its mother, one to generate form, the other to increase size: "Everywhere the nutriment may be divided into two kinds, the first and the second; the former is 'nutritious,' being that which gives its essence both to the world and to the parts; the latter is concerned with growth, being that which causes quantitative increase" (*Historia animalium,* 744b, 32–36). Needham (1933) presents a fascinating array of separations between growth and development, ranging from "anidian" embryos that never develop a primitive streak and continue to cleave without differentiating, to the experimental production of dwarfs geometrically similar to their larger forebears. Polyploid larvae of salamanders have fewer and larger cells than normal forms, but they build the same internal structures with the same size and shape (Goss, 1964, pp. 31–33). Whatever controls the development of form is at least partly independent of the size and number of building blocks. Wolpert points out that this ability of a system to maintain its form as parts are added or removed (in fact, to exhibit general size invariance) is a classic issue in pattern recognition. He refers to it as the "French Flag Problem"—what are "the necessary properties and communications between units arranged in a line, each with three possibilities for molecular differentiation—blue, white, and red—such that the system always forms a French Flag irrespective of the number of units or which parts are removed" (1969, p. 5). A fragment of hydra, for example, can make an almost complete animal only one-hundredth its normal volume. The proportions of mesenchyme, endoderm, and ectoderm in sea urchin embryos can remain constant over an eightfold range in size. Wolpert develops models based on his concept of "positional information" to interpret such invariance.

The further dissociability of maturation from growth and development adds to the variety of heterochronic change. Delsol and Tintant provide an impressive list of dissociations in sexual development, particularly in amphibians, and draw the general conclusion that

> Independence between the development of gonads and the rest of the organism . . . [is] an absolutely general phenomenon . . . We must consider that the genital glands of batrachians develop according to their own rhythm and independently of the embryological development of the other parts of the body. Without doubt, this independence explains why embryological perturbations and the phenomena of heterochrony, whatever they may be, are observed so frequently in batrachians. It also explains the role that neoteny can play in evolution. (Delsol and Tintant, in press; see also de Beer on Slijper's work, 1958, pp. 73–74, and Uhlenhuth, 1919, on general dissociability in amphibian development)

Needham presented his notion of dissociability by analogizing development as the engagement of gears to the primary shaft of metabolism.

> In the development of an animal embryo, proceeding normally under optimum conditions, the fundamental processes are seen as constituting a perfectly integrated whole. They fit in with each other in such a way that the final product comes into being by means of a precise co-operation of reactions and events. But it seems to be a very important, if perhaps insufficiently appreciated, fact, that these fundamental processes are not separable only in thought; that on the contrary they can be dissociated experimentally or thrown out of gear with one another. The conception of out-of-gearishness still lacks a satisfactory name, but in the absence of better words, dissociability or disengagement will be used in what follows . . . There are many instances where growth and differentiation are separable. It is as if either of these processes can be thrown out of gear at will, so that, although the mechanisms are

still intact, one or the other of them is acting as "layshaft" or, in engineering terms, is "idling."* (1933, pp. 180–181)

If growth and development can be dissociated experimentally, often by such simple stimuli as changes in temperature (Needham, 1933, p. 183), then heterochrony can be easily exploited as a pathway to evolutionary change. Bonner has stressed the importance of disengagement in permitting the flexibility that complex organisms require if they are to change at all. Bonner includes both the independence of organs from each other and the separation of stages in ontogenies of individual structures:

> If the steps did not occur in blocks or units that can be shifted or altered in toto without seriously affecting the rest of the organism, evolutionary change in complex organisms might have been virtually impossible. It would have meant that a slight alteration in one small link might have completely disrupted the whole chain, for each step of each part would be coupled into all the steps of neighboring parts. But if there can be separate autonomous units, then all that is presumably needed is to alter the initial cue or stimulus which initiates it; if the cue is moved forward in time relative to the other processes, then the unit as a whole will move forward. (Bonner, 1965, p. 123; see also 1974, p. 164)

And, we might add, produce evolutionary change by heterochrony.

A Metric for Dissociation

The primary desideratum for a study of dissociability would be a satisfactory explanation for it at the cellular level. Such, if I understand an alien field correctly, is not now available. I can only offer in its place, from an area more familiar to me, some questions for a metric to gauge its quantitative effect.

As an illustration, let us consider the simplest dissociation of size and shape. If we can measure the correlation of size and shape in an ancestor, the change of this correlation in descendants will indicate whether heterochrony has occurred and, if it has, will measure both its direction and its magnitude.

Allometry is the study of relationships between size and shape

* Needham was evidently rather pleased with his insight, for he ended his article thus: "The fundamental processes have been envisaged as so many secondary gears engaging or disengaging, as it were, with the primary shaft of basal metabolism. Whatever the value of this analogy may be, there can be no doubt about that of the facts which have been brought together in this review. It might indeed have been entitled, had the author possessed sufficient Kantian audacity, 'Prolegomena to any future theory of the integration of the developing organism'" (1933, pp. 219–220). I will not argue that the idea of dissociability is synthetic a priori.

(Gould, 1966). For more than 50 years, since Julian Huxley generalized its earlier use in the study of brain-body relationships, allometric work has relied largely upon the power function:

$$y = bx^a \quad (1)$$

From this common formula, we can extract a simple measure of heterochrony. Take a standard logarithmic plot of organ size against body size for an ancestral ontogeny and designate the best estimate of shape at a standard point (usually the adult) as *C* (Fig. 30).[8] (I am completely bypassing the evident statistical problem and presenting a purely geometric argument.) Pass through point *C* a line of slope $a = 1$. A line of slope $a = 1$ is a *line of isometry* because shape (the y/x ratio) is constant at all points upon it. Size and shape are dissociated if the standard shape *C* appears at a different size in descendants. If shape *C* appears at smaller sizes in descendants, we have an acceleration of shape relative to size (line segment *A-C* of Fig. 30); if *C* is delayed to larger sizes, we have retardation (segment *C-R* of Fig. 30). We

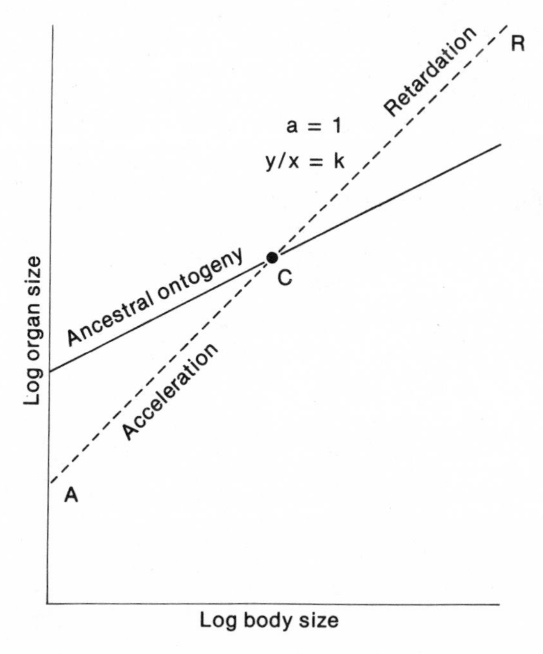

Fig. 30. Bivariate representation of heterochrony. If shape at standard size *C* in ancestors occurs in descendants at smaller sizes, we have acceleration (line *AC*); if at larger sizes, retardation (*CR*).

simply plot the descendant's ontogeny on the graph and read the effects of heterochrony directly. The logarithmic plots of ontogeny do not even have to be straight lines (as long as they are monotonic). We simply want to compare the relative sizes at which a characteristic value of y/x appears in ancestor and descendant.

A particularly interesting and rather common case involves ancestral and descendant ontogenies of the same slope (White and Gould, 1965; Gould, 1971). Since the relative difference in size along any slope $a = 1$ is identical when ontogenetic plots of ancestor and descendant are parallel, heterochrony affects all stages of ontogeny equally and we need not designate a standard point. Moreover, we can extract a measure of heterochrony directly from the coefficients of power functions for ancestors and descendants, according to the following derivation from Gould, 1972:

Consider two regressions of constant $a \neq 1$. For any point on regression 2, there is one and only one point of the same shape on regression 1, such that:

$$\frac{y_1}{x_1} = \frac{y_2}{x_2} \qquad (2)$$

and

$$\frac{y_1}{y_2} = \frac{x_1}{x_2} \qquad (3)$$

At these points, the two regression equations are

$$y_1 = b_1 x_1^{\,a} \qquad (4)$$

and

$$y_2 = b_2 x_2^{\,a} \qquad (5)$$

Therefore

$$\frac{b_1}{b_2} = \frac{y_1/x_1^{\,a}}{y_2/x_2^{\,a}} \qquad (6)$$

and from (3)

$$\frac{b_1}{b_2} = \frac{x_1/x_1^{\,a}}{x_2/x_2^{\,a}} \qquad (7)$$

or

$$\frac{b_1}{b_2} = \left(\frac{x_1}{x_2}\right)^{1-a} \qquad (8)$$

Now x_1/x_2 is the desired quantity—the relative difference in size at which shape on the two regressions is the same. Let us call this quantity s. Finally:

$$s = \left(\frac{b_1}{b_2}\right)^{\frac{1}{1-a}} \tag{9}$$

s is our required measure of heterochrony since it expresses the relative difference in size at which ancestors and descendants have the same shape. I have named s "White's criterion of geometric similarity" (Gould, 1972; see also White and Gould, 1965). If, for example, $s = 3$, then descendants have the same shape as ancestors when they are three times as large, and shape has been *retarded* with respect to size.

The use of White's criterion as a measure of heterochrony can be illustrated with paleontological examples of recapitulation and paedomorphosis:

1. Acceleration in ammonoids. Newell (1949) studied the relationship of suture length to shell size in an evolutionary sequence of Paleozoic ammonoids. Parallel lines move progressively to the left, illustrating an acceleration in development of the suture as previously adult stages are reached at smaller and smaller sizes in descendants (Fig. 31). For the comparison of ancestral *Uddenites* with descendant *Medlicottia*, $s = 0.18$. The descendant attains the same sutural complexity as its ancestor when it is only one-fifth as large; thus, White's criterion has measured the acceleration of shape with respect to size.

2. Retardation in *Gryphaea.* This Jurassic oyster begins its postlarval life by secreting a flat shell cemented to a hard substrate. It soon breaks the attachment and coils for the remainder of its life (Fig. 32). In Liassic rocks of England, Hallam (1969) documented a lineage that decreased in coiling as it increased in size. (Reduced coiling afforded greater stability, as Hallam showed in flow-channel experiments.) For coiling in ancestral *Gryphaea arcuata* from the *Angulata* zone and descendant *Gryphaea gigantea* from the *Spinatum* zone, $s = 7.42$. Descendants would reach the same shape as ancestors at 7.42 times the ancestral size. In fact, adult descendants are only 1.76 times as large as adult ancestors. Thus, adult descendants have the same shape as juvenile ancestors and paedomorphic evolution proceeded by the retardation of coiling relative to size.

3. Retardation and apparent paedomorphosis in *Gryphaea*—the problem of standardization. Hallam's *Gryphaea* are the descendants of populations in the classical sequence studied by Trueman (1922). In these more ancient *Gryphaea,* it had been long assumed that coiling increased through time (though it did not—see Hallam, 1959, and

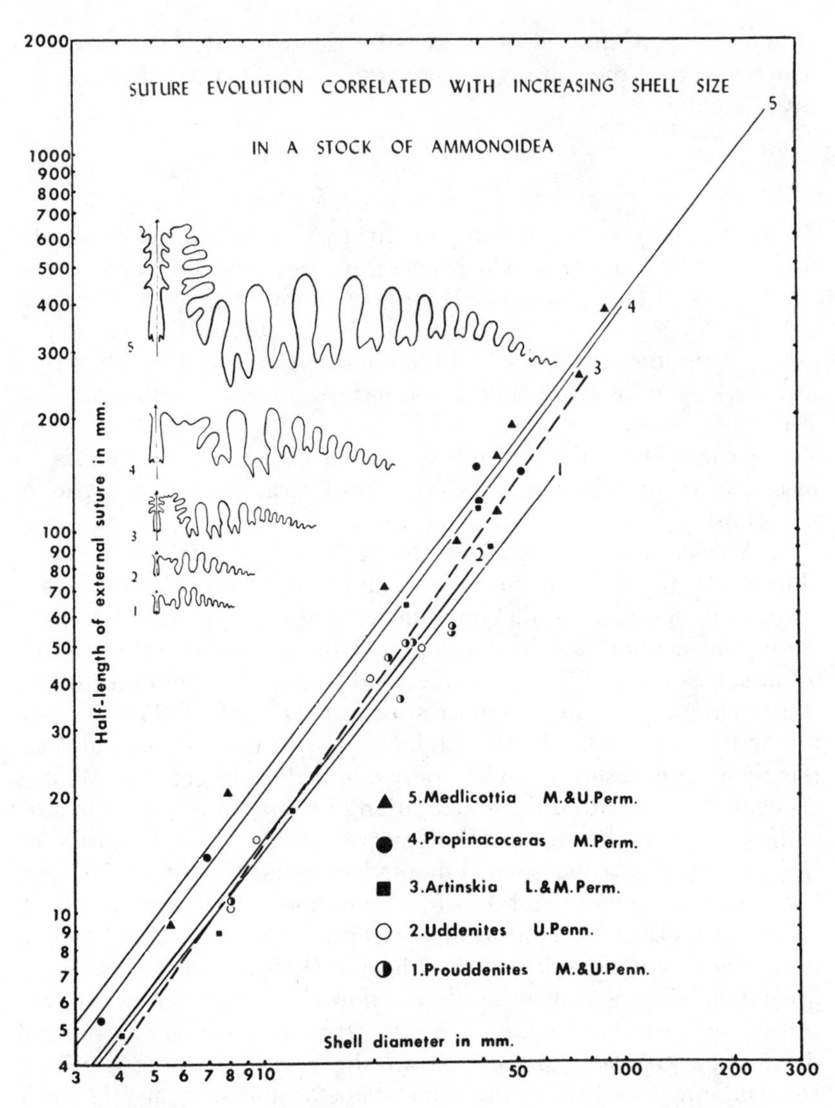

Fig. 31. Acceleration in ammonites as given suture length is reached at smaller sizes in descendants than in ancestors. (From Newell, 1949.)

Gould, 1972). Burnaby (1965) startled paleontologists by announcing that descendants in this classical sequence were *less* coiled than ancestors of comparable size. He designated this as a case of paedomorphosis and concluded that coiling had decreased through time. In-

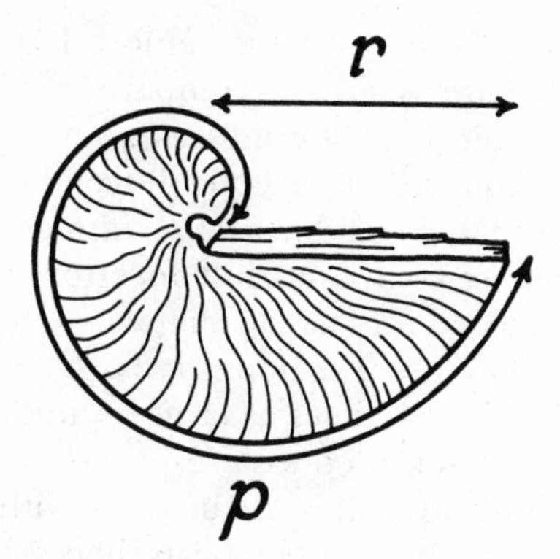

Fig. 32. *Gryphaea incurva* from lower Jurassic rocks of Britain, showing measure of coiling—the ratio of coiled valve (p) to flat valve (r) lengths.

deed, White's criterion would seem to confirm this conclusion, for $s = 1.2$ and descendants do not attain the coiling of their ancestors until they are 1.2 times as large: shape is retarded *with respect to size.* Here we encounter a problem of standardization that can best be illustrated with another example.

Walton and Hammond (1938) plotted leg length versus body length of Shetland ponies and Shire horses. They obtained parallel lines with a slope of approximately 0.8, illustrating the well-appreciated fact that relative leg length declines during ontogeny. Since the line for Shires lies above that for Shetlands, they concluded that Shetlands are shorter-legged than Shires. Indeed, this is true for animals considered at the same size; but standardization by size is biologically inappropriate since it forces the comparison of a juvenile Shire with an adult Shetland. Since relative leg length declines in ontogeny, it is scarcely surprising that juvenile Shires are longer legged than adult Shetlands. In fact, when corresponding adults are considered, the leg/body ratio is identical for both forms (though the Shetland is, of course, much smaller at this corresponding developmental stage). The downward, parallel transposition of the leg length–body length regression for Shetlands represents the only way that this smaller breed can maintain the *same proportions* as Shires at correspondingly smaller sizes (see Gould, 1971, on the significance of allometric transpositions in maintaining shape at increased or decreased sizes).

We have a similar situation in *Gryphaea.* Shape is retarded with respect to size, but standardization by size is biologically inappropriate because adult descendants are considerably larger than adult ancestors. If we standardize by size, we compare an adult ancestor with a juvenile descendant. Since coiling increases during ontogeny, ancestral adults are more coiled than juvenile descendants with incomplete growth. The criterion for standardization should be developmental stage, not size. When we compare adult ancestors with adult descendants, we note that descendants, as calculated by *s,* have the same shape as ancestors when they are 1.2 times as large. Furthermore, adult descendants are actually 1.2 times as large as adult ancestors (Gould, 1972, p. 108). Coiling is retarded with respect to size but the biological result is not paedomorphosis. It is size increase with *geometric similarity.* Large descendants have the same shape as their smaller ancestors; retardation occurred in order to transpose the same shape to larger sizes attained in evolution. (This proviso does not disturb the conclusions of our previous example, *Gryphaea gigantea.* Here we had true paedomorphosis since the rate of phyletic size increase lagged far behind the value of *s.* They must be equal for evolution to proceed in geometric similarity.) This example illustrates two important points: First, paedomorphosis and recapitulation are general *results* that depend upon a criterion of standardization and bear no ineluctable relation to any heterochronic *process.* The retardation that often leads to paedomorphosis can yield an equality of shape at corresponding developmental stages of ancestor and descendant. Second, all results must be considered at a biologically appropriate criterion of standardization. This may be size, age, or developmental stage.

Temporal Shift as a Mechanism of Dissociation

In the examples of the previous section, ancestral and descendant ontogenies followed the same slope and differed only in *y*-intercept. In this common situation, a temporal shift in development may often be identified as the mechanism of dissociation: an allometric feature grows at the same rate in descendants, but it begins to grow at smaller or larger body sizes. Consider, again, *Gryphaea*: Burnaby (1965) gives these equations for the periphery of the coiled valve versus the length of the flat valve (the ratio coiled/flat measures the amount of coiling—our measure of shape): $C = .237F^{1.766}$ for ancestors, and $C = 0.210F^{1.766}$ for descendants. Coiling begins when the coiled/flat ratio exceeds 1.0 (before that, the valves are of equal length and the shell is flat like an ordinary oyster—the equations apply only to the coiled portion of ontogeny). According to Burnaby's equations, the key ratio of 1.0 is reached at a size (flat valve length) of 6.5 mm in ancestors and 7.7 mm in descendants. The parallel regressions demonstrate that coiling proceeds at the same rate in ancestors and descendants; it merely begins at a larger body size in descendants. We predict that descendants simply reach a larger size as attached juveniles (either by growing longer or by growing faster). In other words, heterochrony here involves a change in the relative timing of ontogenetic events, not an alteration in growth rates of characters used to define shape. This prediction can be checked against Hallam's (1968) data for attachment scars (the mark of flat juvenile growth preserved in adult specimens). Indeed, scars for descendants are 1.25 times as large as ancestral scars (Gould, 1972, p. 110). This compares favorably with the size displacement calculated from Burnaby's equations for coiling ($s = 1.17$). The entire ontogeny—not just coiling—has been shifted to larger sizes in descendants.

In other cases, the genetic basis for temporal shifts can be identified. The creeper gene in fowl is an autosomal semidominant; in homozygotes it is lethal and in heterozygotes it is chondrodystrophic (causing markedly shortened limb bones). Cock (1966) plotted tarsometatarsal length against body weight for prenatal growth from seven days until hatching. Creepers and normals have equal slopes, but the y-intercept of creepers is significantly lower. The developmental basis of creeper is not known, but parallel regressions imply that the tarsometatarsus forms later in the limb bud of creepers and grows at the normal rate once it appears.

Although temporal shifting is the simplest explanation for parallel regressions, we may not assume that the more common case of alteration in allometric slopes must indicate a change in growth rate rather than a shift in timing. Laird, Barton, and Tyler (1968) and Barton and Laird (1969) have shown that the slope of a power function may record a relative displacement in *time* between two identical Gompertz functions for the absolute growth of y and x. The difference in allometric slope between ancestor and descendant would then record a change in the amount of temporal shift between y and x; moreover, the change in shift could be estimated directly from the difference in slope.

This emphasis on time of onset follows a long tradition in quantitative studies of heterochrony. Much of the early work on "rate genes" focused not only upon changes in the speed of reactions, but also on shifts in time of onset for developmental events. Goldschmidt argued that "small temporal displacements [*kleine zeitliche Verschiebungen*] in the *Anlagen* of individual parts . . . are sufficient to explain complicated changes" (1927, p. 220). Schmalhausen (1927) also traced several allometries in the growth of chickens to different times of origination for the *Anlagen* of parts under comparison.

A Clock Model of Heterochrony

It is often impossible to decide whether we deal with speeding up or slowing down in evolution unless we have a standard criterion for measuring time [*ein Orthochronisches Vergleichungsobject*]. To lay hold of such a criterion is the most important thing we can do.

E. Mehnert, 1897

I find it remarkable that so little attention has been directed toward a synthesis of the two great literatures on size and shape: the quantitative measurement of allometry, long treated as bivariate in Huxley's (1932) formulation, but now attaining a multivariate generalization (Teissier, 1955; Jolicoeur, 1963; Gould, 1966; Hopkins, 1966; Mosimann, 1970; Sprent, 1972), and the study of heterochrony, a subject that has doggedly maintained a purely qualitative and descriptive approach.

The standard techniques of allometry do not provide an optimal metric for heterochrony because they subtly reinforce a prejudice directed against the dissociability upon which heterochrony depends. Mosimann (1970, p. 943) argues persuasively that the "functional relationships mold" of bivariate plotting places undue emphasis upon the functional association of size and shape. The form of a regression comes to be viewed as a primary feature. The abstracted straight line becomes a key character. Mosimann writes: "I do feel strongly that in many cases the use of functional relations in allometry has been a rebirth of the 'type' concepts of taxonomy" (personal communication, March 3, 1970). Any phyletic change is regarded as a "break" or "disruption" of this primary correlation. Association is primary, disassociation exceptional. The plotting of size and shape as a functional relation is inherently uncongenial to the notion of dissociability.

But the functional relation is but one method among many. Mosimann (1970) prefers a nonfunctional approach that considers the vectors of size and shape separately. In this model, isometry is no longer a rigid correlation of two variables with a slope of $a = 1$, but an expression of the "stochastic independence of some shape vector from some size variable" (1970, p. 931). The nonfunctional approach is rooted in the concept of dissociability. Heterochrony is no longer the disruption of a primary correlation, but rather the simple expression of differential changes in the independent vectors of size and shape.

The bivariate regression is not "truth"; it is a valid picture that directs thought in ways that are rarely appreciated because alternatives are not presented. A proper attention to dissociability requires a new picture in which the ordinate and abscissa of bivariate plots acquire a potentially independent status. In deriving a new model, I thought of independent vectors moving at their own rates, and devised a rather unconventional clock. I quickly saw that my clock had the happy feature of making explicit the bugbear of studies in heterochrony: the concept of standardization—for it allows us to depict all the possible standardizations of size, age, and developmental stage in a single framework. It also quantifies the temporal shifting that I identified in the last section as the principal cause of dissociation. This shifting is only implicit in allometric regressions that bypass time to plot the sizes of parts against each other.

1. THE BASIC IDEA. We want to plot size and shape as two potentially independent vectors operating during the lifetime of an organism. To do this, we set up a semicircular clock with two hands. One hand represents our best statistic for a measure of size. It may be body length, body weight, the projection of a specimen on the first principal component for ontogenetic data within a population, or some other measure of size. The other hand measures shape, which may be "disconnected" from its ancestral relationship to size during evolution.[9] It is generally depicted as a dimensionless ratio or angle.

I shall, in this discussion, use the *Gryphaea* example as an illustration (Fig. 32). As a measure of size, we may use the length of the flat valve (Hallam, 1959). For shape, we determine the ratio of periphery (coiled valve) and length (flat valve) to yield a measure of coiling—the dimensionless property that sparked one of the greatest debates in the paleontological literature (Trueman, 1922; Hallam, 1959; Burnaby, 1965, summarized in Gould, 1972). The hands move forward during a lifetime, beginning from the left margin of Figure 33.

2. SETTING THE SCALES. The clock has three scales corresponding to size, shape, and age (Medawar, 1945). Developmental stage dictates the placement of the scales relative to each other. Let us compare a descendant with its ancestor at the same developmental stage—at, say, the attainment of adulthood (defined in some unambiguous way).

(i) Setting the scale of age. We calibrate the scale to place the age of the ancestor's chosen developmental stage at the midline (Fig. 34). The initial age is recorded at the origin; intervening ages are evenly interpolated within and extrapolated, if necessary, beyond.

(ii) Setting the scales of size and shape. If we wish to see how the vectors of size and shape have become dissociated in a descendant, we must calibrate them to move together in the ancestor. To do this, we tabulate our best estimates of ancestral size and shape (Table 4). The

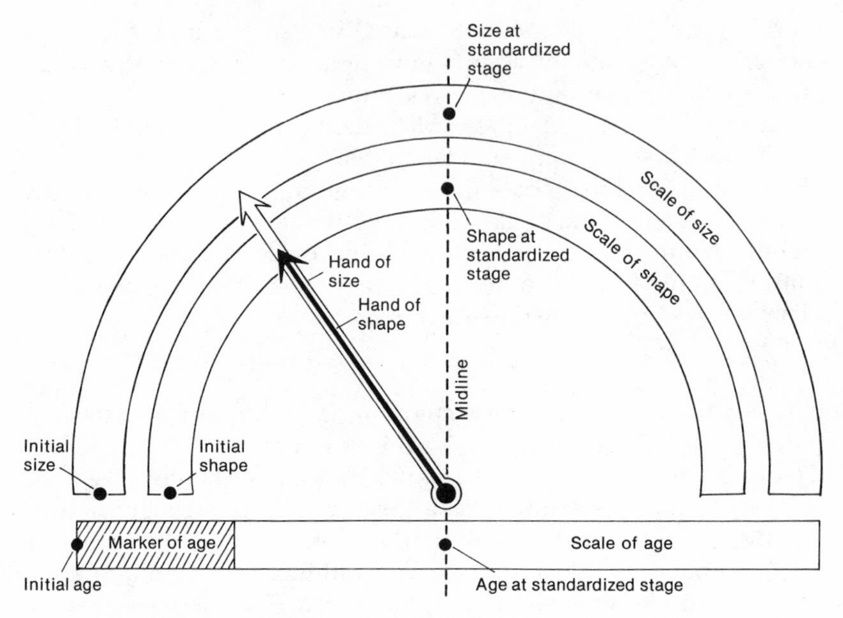

Fig. 33. Framework for clock model of heterochrony, showing scales of size, shape, and age.

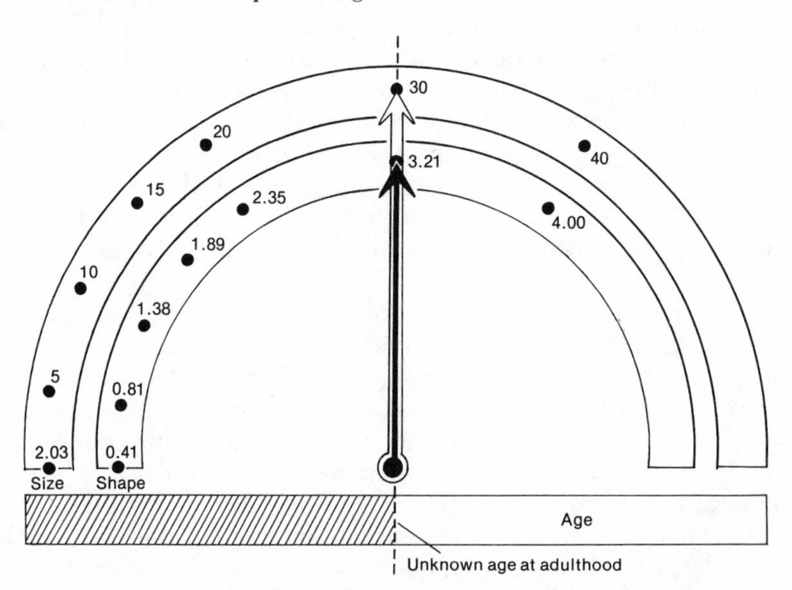

Fig. 34. Scales of the clock model calibrated for *Gryphaea* problem of Table 4. Adulthood is chosen for the standardized stage of midline values. All ancestral markers lie on the midline for ancestral adulthood.

Table 4. Relationship of size (length of flat valve) and shape (coiling as expressed in ratio of valve lengths) in ancestral *Gryphaea angulata.* (From Gould, 1972, equation 18, p. 102.)

Size	Shape
2.03 (average size at beginning of coiling)	0.41
5	0.81
10	1.38
15	1.89
20	2.35
30 (average adult size of ancestor)	3.21
38 (maximum size of ancestor)	3.84
40	4.00
46 (maximum size of descendant)	4.45

scales on our clock will be calibrated to reflect this correspondence and guarantee that the hands of size and shape remain together during growth on the *ancestral* clock. Since evolution often involves phyletic increase in size, the scales must include values never reached in ancestral ontogenies. We estimate these values by the dubious procedure of extrapolation from the ancestral bivariate curve.[10] (We shall obviously be in difficulty if the relationship of size and shape is nonmonotonic. If the same shape is achieved at different sizes, the scale of shape will reverse itself and we may not know where to place the hand of shape for a descendant ontogeny if it displays one of these "repeated" shapes at the developmental stage selected for comparison with an ancestor.)

We have calibrated the scales of size and shape relative to each other; we must now fix them along the periphery of the semicircle. To do this, we simply place at the midline the pair of values characterizing ancestors at the developmental stage marked on the midline of the age scale. Thus, if an ancestral *Gryphaea* became sexually mature at two years, the midpoint of its age scale would be so set, and the midpoints of its size and shape scales would depict the values of these variables at that age and stage. (Values of size and shape are fixed at the origin and at the midline. We may use any scale we wish to interpolate values between and extrapolate them beyond. In practice, I lay out size on a linear scale and enter the designated values for shape at the appropriate sizes. Unless size and shape have a very regular relation, the scale for shape will follow no simple pattern. It may behave like the wriggling eel that Bertrand Russell once compared with a ruler in trying to explain relativity. He also reminded us that the ruler wriggles in the eel's framework.)

3. RUNNING THE CLOCK—ANCESTRAL ONTOGENY. At the outset of ontogeny,[11] all markers are set at their initial positions and the scales register values for size, shape, and age at this developmental stage. In *Gryphaea,* for example, the "outset" of ontogeny is the developmental stage at which the juvenile shell breaks its attachment to the substratum and begins to coil as a free-living organism. The values recorded at the initial positions of the scales are the length of the flat valve (size), the ratio of periphery of the coiled valve to length of the flat valve (shape),[12] and the absolute age (since hatching or since metamorphosis) at which the shell breaks its attachment (Fig. 34).

We may now run the clock for ancestral ontogeny. As the marker on the scale of age advances in good Newtonian fashion (equal linear increments in equal intervals of time), the hands of size and shape advance together at the rate dictated by their scales. Since increase in size (and rates of change in shape) are usually most rapid early in life, the hands will generally move quickly at first and then slow down. (If this seems undesirable, the scale of size can be recast in some appropriate transformation. The important point is only that scales of size and shape be so calibrated with respect to each other that the hands move together in ancestral ontogeny.)

The clock can now be read at any moment in ancestral ontogeny. For the developmental stage at which we wish to compare ancestor and descendant, the age marker and the hands of size and shape will all lie on the midline for ancestral ontogeny (Fig. 34).

We do not really need to run the clock for ancestral ontogeny since its positions are fixed by our methods. Quantitative data for ancestral ontogeny serve to set the scales of the clock for comparison with the descendant—for we will plot the descendant's ontogeny on the same set of scales.

4. DESCENDANT ONTOGENY—HETEROCHRONY AND THE REALMS OF DISSOCIATION. Since we are testing for heterochrony with common developmental stage as a criterion of standardization, we want to know where the markers of descendant ontogeny lie on the ancestral scales when the descendant has reached the developmental stage selected for comparison. At this point, we may divide each of the scales into two realms separated at the midline (Fig. 35):

(i) The domains of shape. If heterochrony has occurred, the hand for descendant shape will not lie on the midline at the common developmental stage. If it lies between the origin and the midline, paedo-

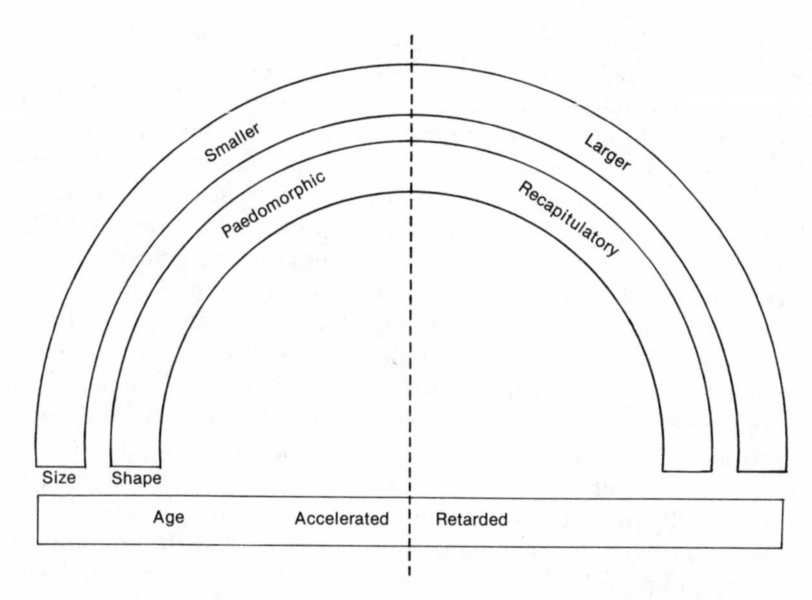

Fig. 35. The domains of heterochrony—accelerated and retarded on the age scale, paedomorphic and recapitulatory on the shape scale, and smaller or larger on the size scale.

morphosis has occurred because the descendant displays, at this developmental stage, a shape attained at a more youthful stage in ancestors. If it lies beyond the midline, we have a case of recapitulation because the descendant has "gone beyond" the ancestral shape for this developmental stage—that is, the ancestral shape has already occurred at a more youthful stage in the descendant.

(ii) The domains of age. The descendant may reach the selected developmental stage when it is younger or older than the ancestor at that stage. If it reaches the common developmental stage at a younger age, maturation has been accelerated; if it reaches this stage when it is older, maturation has been retarded. If developmental stages are retarded or accelerated while size and shape remain in their ancestral relationship, we observe heterochrony in evolution. When we standardize by developmental stage, the hand of shape will still be in the domain of paedomorphosis if the standard stage is reached more rapidly in descendants than in ancestors; if the standard stage is reached more slowly, the hand for shape will have moved beyond its ancestral value into the domain of recapitulation.

(iii) The domains of size. The descendant, at the common developmental stage, may be either smaller or larger than the ancestor. Heterochrony is defined as the evolutionary displacement of a specific feature (a "shape") relative to a common standard of size, age, or developmental stage. Therefore, when size alone is altered from its ancestral condition, we do not generally speak of heterochrony (since the hand of shape remains in its ancestral position at each developmental stage); nonetheless, the process that alters size alone and produces proportioned dwarfs and giants is an aspect of the same phenomenon of dissociation that yields all types of heterochrony.

Descendant *Gryphaea incurva* are larger, but have the same shape as ancestors at the common developmental stage (adulthood). Unfortunately, we cannot set the scales of absolute age for ancestors (though techniques are potentially available for aging fossil molluscs via astronomical periodicities reflected in growth lines—Wells, 1963; Scrutton, 1965; Clark, 1968). Using the clock, we can easily spot the evolutionary significance of the altered relationship between size and shape that Burnaby (1965) misinterpreted (Fig. 36). Phyletic size increase has occurred with geometric similarity. Burnaby thought he had a case of paedomorphosis because shape is, indeed, retarded with

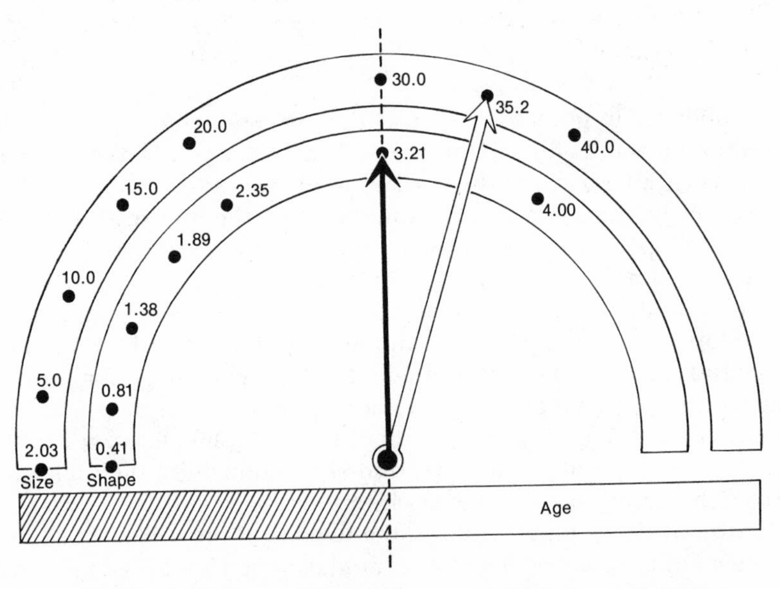

Fig. 36. Markers for descendant *Gryphaea* upon scales calibrated for ancestors (see Table 4). The nature of heterochrony is now apparent; ironically, we have size increase but no change of shape. Dissociation of size and shape yields phyletic increase in size with the maintenance of geometric similarity in coiling.

respect to size. But the *processes* of *retardation* and *acceleration* are one thing, and the *results* of *paedomorphosis* and *recapitulation* are another. In this case, shape is retarded with respect to size in order to maintain the same shape at the same developmental stage reached by descendants at larger sizes. (This could not have been achieved by a simple extrapolation of the ancestral ontogenetic curve to larger sizes. The ancestral curve is strongly allometric, yielding increased coiling at larger sizes.)

Retardation of shape with respect to size often does lead to paedomorphosis as a result. During the late Pleistocene, the Bermudian pulmonate *Poecilozonites bermudensis* produced several paedomorphic derivates (see pp. 275–279 for a fuller account). These animals are (at adulthood) about the same size as their nonpaedomorphic adult ancestors, but *all* measures of shape are strongly retarded to the values of very juvenile ancestors (Gould, 1968, 1969). Here (Table 5 and Fig. 37), the result (and presumably the selective significance) of retardation is paedomorphosis.

Yet, to reinforce my contention that retardation is a process not in-

Table 5. Size and shape in ancestral land snail *Poecilozonites bermudensis zonatus*, Pleistocene of Bermuda. (From Gould, 1969, pp. 520–521.)

Width of spire (mm)	Size (width plus height)	Shape (width/height)
3.31	4.54[a]	2.68
3.5	4.84	2.61
4.0	5.66	2.41
5.0	7.45	2.04
6.0	9.46	1.73
6.35	10.22	1.64[b]
7.00	11.70	1.49
8.0	14.16	1.30
8.94	16.68[c]	1.15
10.0	19.80	1.02
12.0	26.42	0.83

I use an average sample of nonpaedomorphic *P. bermudensis zonatus*—no. 73 in Gould, 1969—as a model for an ancestor; sample 21b in Gould, 1969, is my model for a paedomorphic descendant. Values differ slightly from those in Gould, 1969, because here I estimate shape from the allometric regression of height versus width (reduced major axis), not from the actual average of specimens. I use whorl number as a criterion of relative age and assume (for the sake of illustration only) that adulthood is reached at the end of the fifth whorl in both ancestor and descendant.

[a] Initial size at onset of allometric spire growth.

[b] Shape of descendant at adulthood.

[c] Size at end of fifth whorl (taken as adulthood).

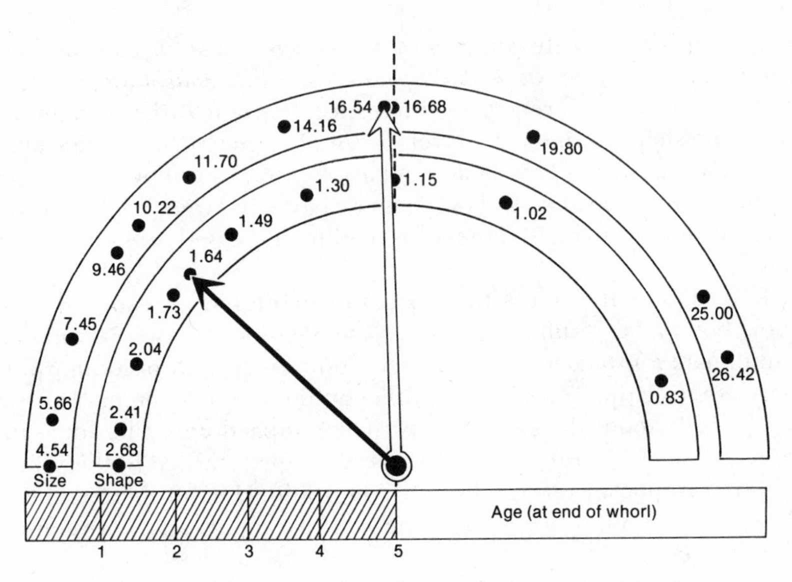

Fig. 37. Paedomorphosis in Pleistocene land snails from Bermuda (see Table 5 and Gould, 1968, 1969). Scales are calibrated for ancestors. Descendants at the same age (number of whorls) are the same size, but are strongly retarded (juvenilized) in shape.

exorably linked with its common result of paedomorphosis, I point out that retardation of shape with respect to size could even yield the result of recapitulation. Suppose, for example, that *Gryphaea* had continued its evolution by growing further along the descendant's ontogenetic curve to attain adulthood at still larger sizes (Fig. 38). Shape is still retarded with respect to size on the ancestral scales, but size increase has proceeded so far that the ancestral adult shape is now reached at a juvenile stage of descendants—even though the descendant, at this juvenile stage, is larger than the ancestor was at its adult stage.

5. TYPES OF HETEROCHRONY ON THE CLOCK MODEL. In Table 3, the types of heterochrony are presented in their "pure" form. All are the products of dissociation and disruption of ancestral correlations. In contrast with de Beer's complex system (1930, 1958), they reflect but two processes—retardation and acceleration—and yield but two results—paedomorphosis and recapitulation. There is, however, no simple one-to-one correspondence either between retardation and paedomorphosis on the one hand, or between acceleration and reca-

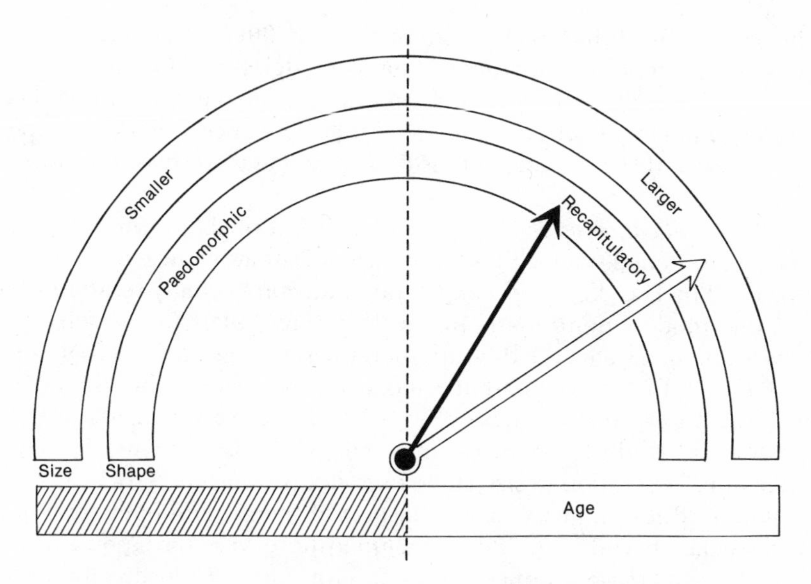

Fig. 38. Hypothetical recapitulation in *Gryphaea* to show lack of necessary correspondence between the process of retardation and the result of paedomorphosis. Shape is retarded with respect to size, but phyletic size increase is so pronounced that this retarded shape is still in the realm of recapitulation at the standardized developmental stage.

pitulation on the other. Four basic types of heterochrony appear on the clock model:

(i) *Paedomorphosis by progenesis* (acceleration of maturation with respect to somatic development). Here paedomorphosis is not attained via the "classic" route of retarded shape (Fig. 39A). Descendant ontogeny is simply truncated by the early attainment of sexual maturity (reflected by placement of the age marker in the domain of acceleration at the developmental stage—in this case sexual maturity—common to ancestor and descendant). In somatic development, the descendant is both smaller and paedomorphic—in other words, it is a sexually mature juvenile. The correlation of size and shape is unchanged.

(ii) *Paedomorphosis by neoteny* (retardation of shape with respect to developmental stage). Here, the vector of shape is retarded while size and developmental stage remain unchanged from the ancestral condition (Fig. 39B).

(iii) *Recapitulation by hypermorphosis* (retardation of maturation with respect to somatic development). The correlation of size and shape is unchanged from the ancestral condition (Fig. 39C). Ontogeny is simply prolonged because maturation has been delayed. The age marker shows that the selected developmental stage—sexual maturity in this case—occurs at a later age in descendants. The ancestral adult shape is attained at the same size, but the descendant at this size is still a juvenile.

(iv) *Recapitulation by acceleration* (acceleration of shape with respect to developmental stage). This is the "classic" situation, urged as nearly universal by Haeckel, Cope, and Hyatt, and convincingly rendered as but one mode among many by de Beer (Fig. 39D). At the selected developmental stage, the descendant is the same age and size as its ancestor. But the vector of shape has been dissociated and "speeded up"; the ancestral shape has been pushed back to a younger age and smaller size, and the vector of shape has progressed further.

A "pure" alteration in size alone does not produce heterochrony in evolution. But change in size is an aspect of the general phenomenon of dissociation and it should be included here. A retardation in the rate of size increase produces a dwarfed form geometrically similar to its ancestor (Fig. 39E). If the rate of size increase is accelerated, a proportioned giant evolves (Fig. 39F).

Of course, these idealized situations are abstract "end members" in a continuum of complex change. One factor is rarely altered alone; nonetheless, actual examples yield the same interpretations. Figure 40, for example, is a qualitative account of neoteny in humans. Shape might represent the ratio of facial to cranial length, with body weight marking size and sexual maturity chosen as the developmental stage for comparison. All vectors are altered from their ancestral positions. Sexual maturity is attained later and at larger size; shape is clearly retarded, since in many respects an adult human skull resembles the standard juvenile condition of most primates (Chapter 10). Moreover, the position of the three vectors probably reflects a common cause—prolongation of rapid juvenile growth with a delay of sexual development, leading both to larger final size and to the retention of proportions marking juvenile stages.

Nothing in my statements on dissociation should be construed as a denial of common correlations between size, shape, and developmental stage. The most rapid rates of change in size and shape generally occur together in early stages of development (see Needham's discussion [1968, pp. 39–41] of Murray's "admirably ingenious attempt" to quantify change in shape in order to establish this point). The onset of sexual maturity may mark both a great slowdown (or cessation) of size increase and a series of changes in shape. Correlations of size and shape are often tenaciously maintained or quickly

Fig. 39. Types of heterochrony in their "pure" form. (*A*) Progenesis by truncation of ontogeny with early sexual maturation. (*B*) Neoteny by retardation in somatic development. (*C*) Hypermorphosis by delay in maturation and simple extension of growth. (*D*) Acceleration by speeding up of somatic development. (*E*) Proportioned dwarfism by slower growth with constant rate of development. (*F*) Proportioned giantism by more rapid growth.

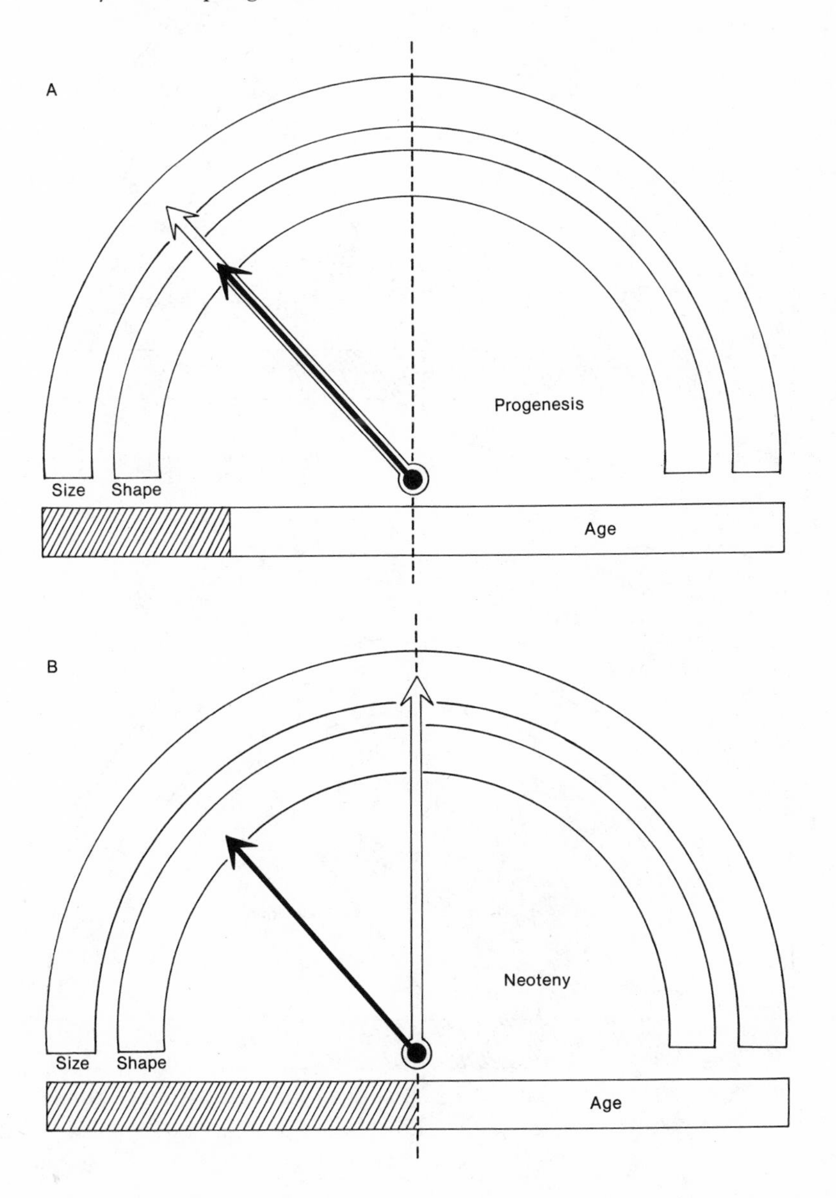

Fig. 39. Continued

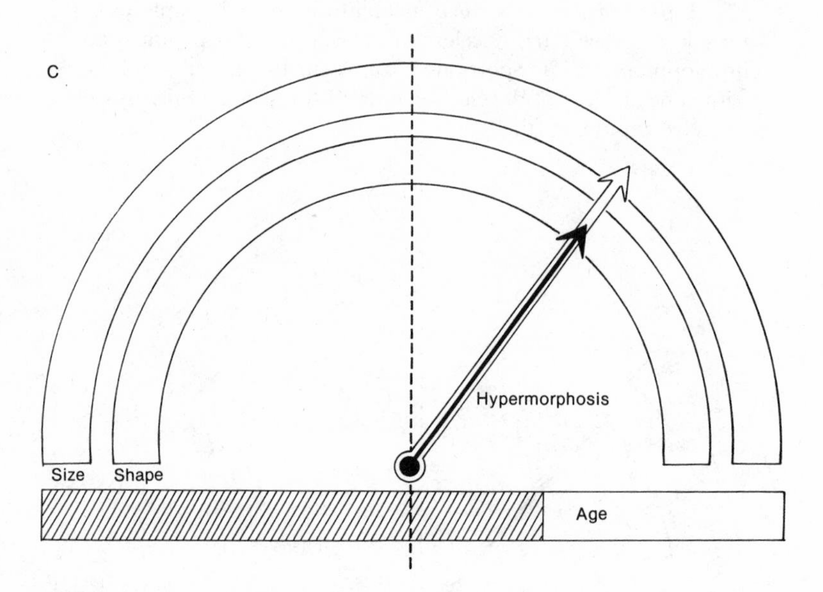

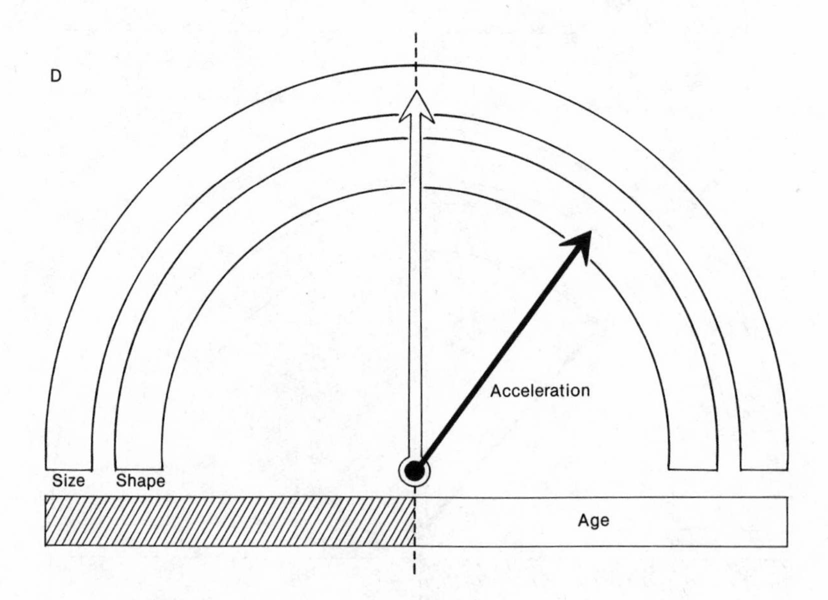

Fig. 39. Continued

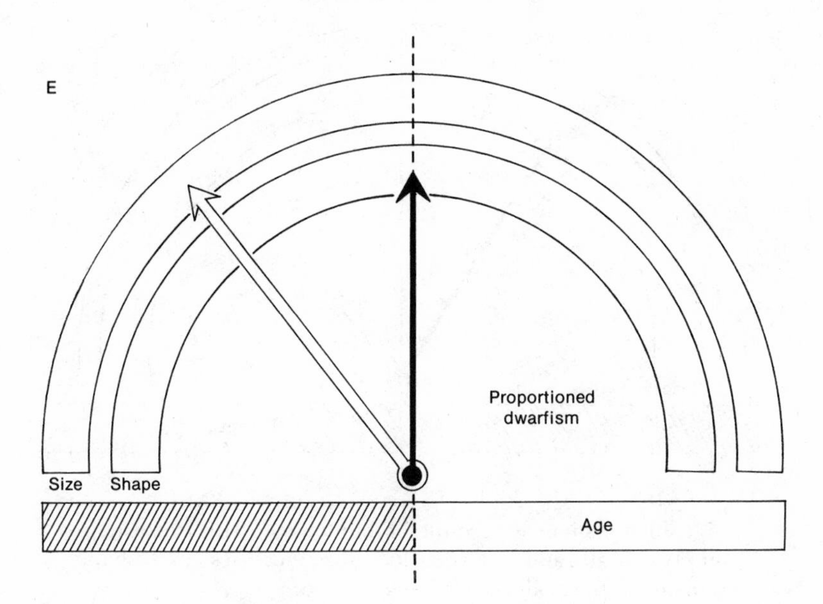

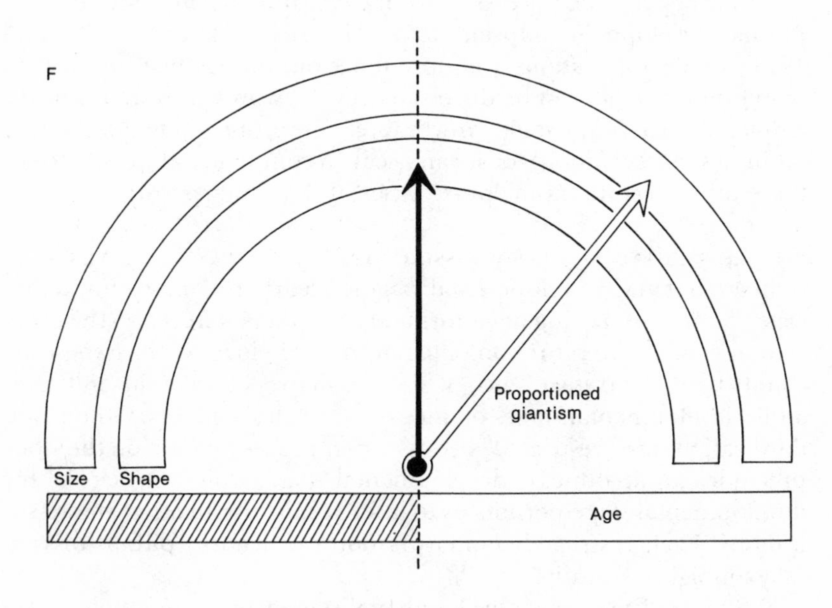

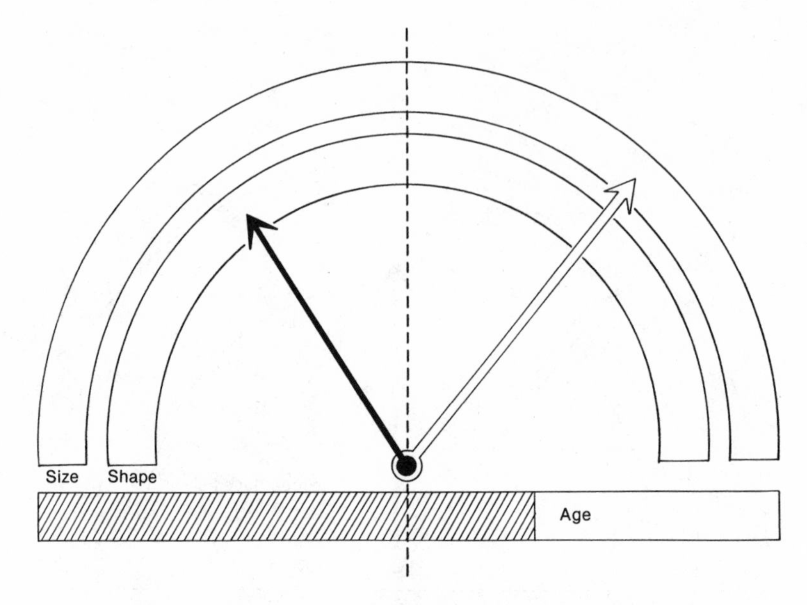

Fig. 40. Qualitative account of human neoteny (see Chapter 10). Maturation is retarded, size increases, and shape remains in the realm of juvenile ancestors.

reestablished in the face of extreme disruptions in conditions of normal development (Mosier, 1972; Mosier and Jansons, 1969 and 1971). Some correlations are more tenacious and difficult to disrupt than others; yet all can be dissociated. A delay in maturation, for example, can "carry" juvenile growth rates and proportions along with it (as in human evolution), or it can occur in complete independence of these other vectors (as in "pure" cases of hypermorphosis).

6. RESTRICTED MODELS—MISSING DATA AND STANDARDIZATION BY OTHER CRITERIA. Developmental stage is clearly the most satisfactory criterion for comparing ancestors and descendants. It is also the criterion of classical literature on ontogeny and phylogeny, since these accounts treat the transference of specific shapes between juvenile and adult developmental stages of ancestors and descendants. Other standardizations are "restricted" because their representation on the clock precludes an account by developmental stage, while the "clock" for developmental stage permits us to infer results for all other standardizations. Each restricted standardization can lead to pitfalls or causally ambiguous results.

(i) Standardization by developmental stage when absolute ages are not known. This is a standard source of frustration for paleontologists and museum taxonomists. Morphology clearly records the attainment of adulthood, but not absolute age. In such cases, we cannot set the scale of age. The *Gryphaea* story is incomplete and partly unexplained for this reason. We know that descendant *Gryphaea* increased in size but maintained the same shape at adulthood; the dissociation of size and shape yielded geometric similarity with phyletic size increase. But how did this increase in size occur? To resolve this issue, we must have data on absolute age. Did ancestral and descendant *Gryphaea* reach adulthood at the same age (as I have guessed on Fig. 36)? In such a case, we could assert that rates of growth speeded up while rates of change in shape remained in their ancestral state. As an alternative, descendant *Gryphaea* may have reached adulthood at a greater age. Phyletic size increase would reflect the simple prolongation of ancestral growth rates over a longer period of time. If descendants were older as adults, then rates of change in shape slowed down to yield the ancestral shape at the descendant's greater age.

(ii) Standardization by age when developmental stage is not known. We lose the data of acceleration or retardation in developmental stage, because the age marker of descendants always rests on the midline. Suppose, for example, that sexual maturity were accelerated while the relationship of size, shape, and age continued as in ancestors.[13] If we standardized by age, the clock for descendant ontogeny would be identical with that for ancestors. An event of potentially great evolutionary importance (as a regulator of generation time) would be completely hidden.

(iii) Standardization by size when neither age nor developmental stage are known. When we standardize by size, no age scale can be recorded on the clock, for data on age and developmental stage are completely lacking. We compare ancestor and descendant at a common size by placing the "hand" for size on the midline in descendants. We can only tell whether shape has been accelerated or retarded with respect to size. On this criterion of standardization, we must refer to any retardation as paedomorphosis and to any acceleration as recapitulation. This can lead to pitfalls of the kind previously illustrated by Burnaby's conclusions for *Gryphaea* (Fig. 36). As I argued previously, Burnaby (1965) compared ancestral and descendant *Gryphaea* at a common size, thus contrasting an adult ancestor with a juvenile descendant. Since shape is retarded with respect to size, he judged his result to be a case of paedomorphosis. Indeed, the hand for shape does lie in the paedomorphic domain on this criterion of standardization (Fig. 41). Yet a more complete comparison at common developmental stages shows that shape was retarded with respect to size in order to yield the same shape in larger descendant adults.

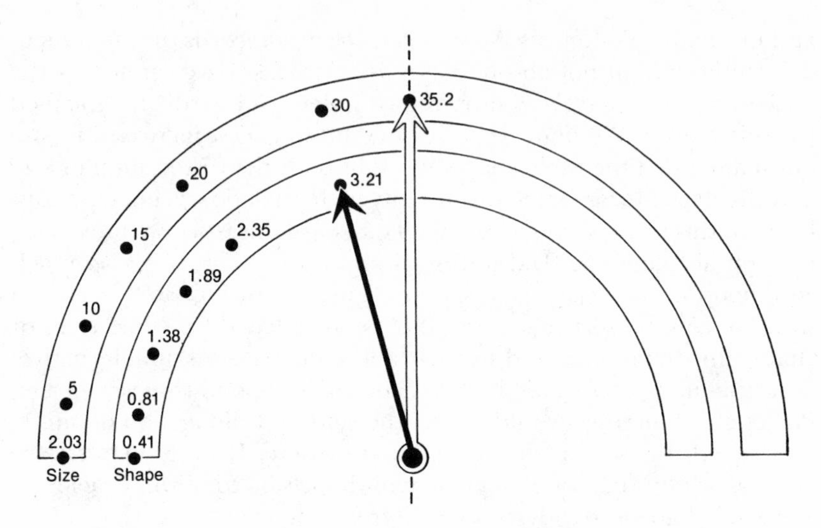

Fig. 41. Restricted model for *Gryphaea* considering only size and shape (see Table 4). Shape is retarded with respect to size and we are forced to make a false judgment of paedomorphosis.

While the complete model of standardization by developmental stage permits us to assess the interactions of all four relevant properties—size, shape, age, and developmental stage—each of the restricted models eliminates one or more of these properties from consideration.

7. CONCLUDING STATEMENT. Paedomorphosis and recapitulation are defined by the location of the shape vector in the left (paedomorphic) or right (recapitulatory) quadrant of the descendant's clock at the selected point of standardization with ancestors. The developmental stage of adulthood, or attainment of sexual maturity, is both the traditional and the generally preferred criterion of standardization. Paedomorphosis and recapitulation are results produced by a multitude of processes with differing evolutionary significance. They should not be used as a basis for classifying the relationships between ontogeny and phylogeny. We should classify by process and its evolutionary significance. The fundamental processes are acceleration and retardation. They bear no simple relationship to the results of paedomorphosis and recapitulation. The retardation of shape with respect to size, for example, can produce paedomorphosis, evolution in geometric similarity, or even recapitulation.

Appendix: A Note on the Multivariate Representation of Dissociation

Experimental studies of individual characters have provided most recorded cases of dissociability, but through the multivariate study of form separate patterns of variation in size and shape have been detected as an almost universal property of covariation within natural populations. The loadings of original variates upon orthogonal factor axes usually cluster very strongly. Measures of size tend to associate with the first principal component, while loadings upon subsequent axes reflect differences in shape.* Since axes are orthogonal, the clusters of size and shape are "independent" in the most technical sense. This is a constraint of the method; it must not be equated with biological reality.

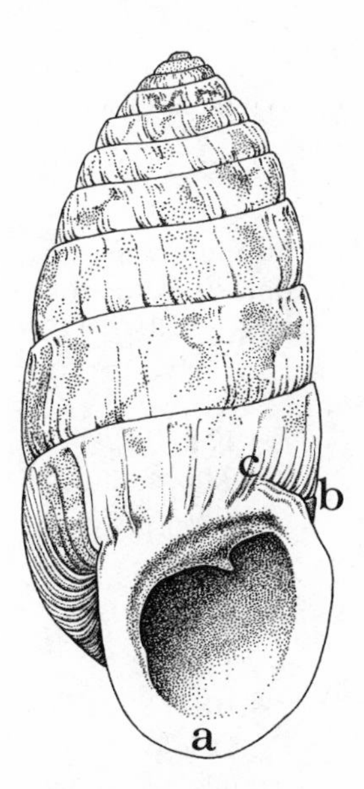

Fig. 42. A *Cerion* shell from Bimini. Markers of development are indicated: (*a*) the adult lip; (*b*) the suture of the last pre-adult whorl and (*c*) the overlap of the adult aperture onto the last pre-adult whorl (the extent of overlap, *c* minus *b* measured parallel to the axis of coiling, is a measure of development).

* This arises from the unenlightening fact that most parts are larger in bigger animals. This primary correlation nearly always subsumes the greatest amount of variance and provides the first principal component. Jolicoeur (1963) has derived a general test for isometry using this axis; this is important for my definition of size differences as geometrically similar.

The greatest misuse of factor analysis arises from a temptation to interpret any association of high loadings as a functional complex. It need be no more than a mathematical artifact, particularly for axes that encompass very little information. (I have an intuitive feeling that Lomax and Berkowitz [1972, p. 232] have been driven to absurdity by their assumption that clustering in loadings must have biological significance. They find, on the very small 13th axis of their 19-axis solution, an association of milk production with florid vocalization, and conclude: "Factor 13, in which the availability of milk products is linked to vocal dynamics, suggests that this extra source of protein accounts for many cases of energetic vocalizing.") Still, the hope of factor analysis remains that functional associations may be identified by reducing large matrices to fewer generating factors. The persistent separation of variation in size and shape is by far the strongest indica-

Table 6. Factor scores of 16 variables in Q-mode factor analysis of the Cuban land snail *Cerion moralesi.*

	Axis 1: Strong development without large shell	Axis 2: Width	Axis 3: Strong development and large shell
Variable			
1. Adult whorls	−.174	.105	*.598*
2. Ribs on 1st whorl	−.028	−.195	.029
3. Protoconch width	.329	.047	.088
4. Protoconch height	.376	.088	.104
5. Height at 4th whorl	.465	.085	−.038
6. Height at 7th whorl	.318	−.268	−.291
7. Width at 4th whorl	.336	−.265	−.252
8. Adult height	−.161	−.169	*.270*
9. Adult width	−.125	−.464	.074
10. Aperture height	.004	−.355	.147
11. Aperture tilt	*.299*	.111	*.355*
12. Aperture width	−.020	−.419	.112
13. Aperture protrusion	*.314*	−.079	*.265*
14. Umbilical width	.029	−.334	.110
15. Width of lip	*.152*	−.192	*.280*
16. Thickness of lip	*.176*	.107	*.270*

Four measures of strong development are italicized for their coordinated high scores on axes 1 and 3 (measures of adult size also italicized on axis 3). Wide, thick lips on protruding and highly eccentric apertures represent greatest departure from juvenile growth patterns and serve as markers of intense development.

Table 7. Factor loadings of 16 variables in R-mode factor analysis of 52 samples of *Cerion* from Little Bahama Bank.

Variable	Axis 1: Large Shell and (to a lesser extent) strong development	Axis 2: Strong development alone (no other high loadings)
1. Adult whorls	.329	.074
2. Ribs on 1st whorl	−.144	.087
3. Protoconch width	.789	−.021
4. Protoconch height	.200	.043
5. Height at 4th whorl	.305	.272
6. Height at 7th whorl	.541	.102
7. Width at 4th whorl	.931	.013
8. Adult height	.900	.189
9. Adult width	.941	.033
10. Aperture height	.909	.328
11. Aperture tilt	*.090*	*.809*
12. Aperture width	.910	.300
13. Aperture protrusion	*.689*	*.648*
14. Umbilical width	.876	.150
15. Width of lip	*.485*	*.593*
16. Thickness of lip	*.565*	*.407*

Measures of strong development are italicized.

tion we have that this hope is well grounded. The separation of size and shape on orthogonal axes does not mean that they are independent and dissociable in organic development. It does not guarantee that the associations reflect any simple and distinct control over size and shape. Yet this is certainly a tempting interpretation for many cases.

Cerion moralesi, a Cuban land snail, ceases growth at a definite point by strongly altering the direction of its aperture and secreting a thick lip at the final position of the aperture (Fig. 42). In *Cerion,* a thick lip almost always accompanies a strongly altered orientation, forming an association marking intense development. Is this association correlated with the dominant association of all measures that have large values in large animals (the "size" cluster)? We might predict such a correlation by arguing that snails ceasing to grow when large should also have more intense development if size and shape are correlated. This prediction is affirmed by scores on the third *Q*-mode varimax axis (Table 6): these show high values for measures of size *and* development. But, the same measures of development are also associated as high scores on axis 1. Since the axes are orthogonal, growth and development are dissociable if the scores reflect real associations. Large size implies intense development, and this is scarcely surprising; but intense development is also dissociable from size. If selection can affect the measures of development as a unit, it can easily produce heterochrony by accelerating or retarding their expression with respect to size. This independence of size and development may be quite general for several levels of variation in *Cerion.* It occurs again and again for the major determinants of within-sample variation, as in Table 6. I have also found it (Table 7), expressed just as clearly and strongly, in the largest among-sample study I have attempted: a factor analysis of mean vectors for 52 samples of *Cerion* from Little Bahama Bank (Abaco and Grand Bahama Islands). In a massive study of cockroaches (446 measurements in 37 species), Huber (1974) separated, on orthogonal axes, characters correlated with body size from those associated with maturation.

—8—

The Ecological and Evolutionary Significance of Heterochrony

The Argument from Frequency

The Importance of Recapitulation

How common is evolution by heterochrony? We can no more answer this question than we could resolve the issue of relative frequencies for recapitulation and paedomorphosis. We can be certain, however, that its effects have been catalogued thousands of times and that it is the dominant mode of evolution in many important lineages. If Haeckel's law of recapitulation is "a vague adumbration of the truth" (Julian Huxley, personal communication, August 1971), the truth must be that heterochrony is extremely important.

Although I put little stock in the tradition of argument by enumeration in natural history, I begin this chapter with selected examples of the dominant role of heterochrony in lineages ranging from phylum to endemic genus. Here, as in all the following sections of this book, I emphasize paedomorphosis (this chapter takes as its general theme the importance of distinguishing progenesis from neoteny as distinct phenomena both in ecological significance and evolutionary importance). In so doing, I do not wish to cast my lot with de Beer (1930, 1958) in claiming that recapitulation is rare and unimportant in evolution. I believe that it is every bit as significant as paedomorphosis. Recapitulation, like paedomorphosis, is a consequence of two processes with different evolutionary meaning (Table 3). The first process is hypermorphosis, the extension of ancestral ontogeny. I shall discuss hypermorphosis within the framework of retarded maturation. Since hypermorphosis is the usual correlate of extensive phyletic size increase, it is a major phenomenon in evolution insofar as Cope's rule holds (Stanley, 1973). The second process is acceleration in the development of somatic characters. We can disregard with confidence the mechanism invoked by Haeckel and his followers—the literal pushing-back of old characters with addition of new features to the end of ancestral ontogenies. Nonetheless, there is a germ of truth in F. Müller's insight that recapitulation occurs because advantageous characters will often augment their selective value by appearing earlier in ontogeny. Since early ontogenetic stages are so refractory to change, evolutionary novelties (if they are to develop gradually) will almost always make their first appearance in late stages. As Stebbins argues in his principle of "increasing precocity of gene action" (Chapter 7), these features may slowly work their way back into early ontogeny by selection of modifiers to permit earlier expression. The result is recapitulation—the character appears earlier and earlier in descendant ontogeny. But it is not pushed back by new features added terminally; it merely extends its selective advantage over a greater portion of ontogeny. I agree with Stebbins (1974, pp. 115–119) that increasing precocity may be the primary source of recapitulatory effects. If it is as general a principle of development as Stebbins believes, then the high frequency of recapitulation cannot be doubted.

Does recapitulation ever occur by the classic route of pushing back with terminal addition of new features? I believe that it does, but in a very different and more mundane manner than that invoked by the biogenetic law. For Haeckel and his followers, new features were often unrelated to previous adult stages; they appeared as independent advantages and established themselves because a second and separate process accelerated the previously adult features into earlier ontogenetic stages. But in most recapitulations by addition, the pushing back and the terminal novelty must be related as simple quantitative expressions of selection for more rapid development of specific features. Take, for example, the most widely cited case of recapitulation during its heyday: the development of greater complexity in ammonite sutures, with acceleration of simpler patterns to earlier ontogenetic stages. Suppose that an adult ammonite of a certain size obtains a selective advantage by evolving a more complex suture. Since ontogeny (with many exceptions) proceeds towards increasing elaboration of suture pattern, this adaptation arises most economically when the developmental rate for sutures is increased relative to the general rate of growth. There is no pushing back by a new character added terminally, merely a disruption of the ancestral correlation between size increase and sutural development. The only precondition for such an event is ancestral allometry, a nearly universal property of growth (Gould, 1966). (Isometric ontogenies possess no potential for simple heterochronic change. Allometric ontogenies can be retarded to retain juvenile states or extrapolated into previously unexpressed morphologies.) Moreover, the ammonite suture, for all its apparent complexity, may be the mechanical result of very few generating factors; an increase in complexity may involve no more than an earlier inception or more rapid change in one of these factors—a potentially simple change in gene regulation. When the following three conditions are met, evolution by classic acceleration is almost inevitable:

1. Evolutionary change involves the increasing complexity of a feature already present in ancestors.
2. Ancestral ontogeny is positively allometric for increasing complexity in the feature during growth.
3. Evolution proceeds without lengthening either the ancestral time of maturation or the life span.

Since these conditions are common, evolution by acceleration must be an important path to the development of morphological complexity in specific features.

In summary, three situations generally lead to recapitulation; their widespread occurrence guarantees a high frequency of recapitulation among evolutionary events: (1) the prolonged extension of ancestral ontogeny in size and time—hypermorphosis; (2) a selective advantage for earlier appearance of novelties originally introduced near the end of growth—Stebbins' principle of increasing precocity of gene action; (3) a selective advantage for increased complexity in characters already developing with positive allometry in ancestors.

The Importance of Heterochronic Change: Selected Cases

A PHYLUM: THECAL EVOLUTION IN GRAPTOLITES. Silurian monograptids are colonial animals; their astogeny[1] begins with a sicula (Fig. 43) and continues by the progressive budding of individual thecae along a single stipe. Each theca is a complete animal; those nearest the sicula (proximal) are oldest, those furthest away (distal) are youngest. Each theca has its own ontogeny; it begins as a straight tube (simple) which may later bend and acquire a complex aperture (such thecae are termed "elaborate"). Elles (1922, 1923) argued that two recurrent trends marked the evolution of *Monograptus* (Fig. 44): either elaborate thecae arise proximally and spread distally (the "progressive" sequence), or thecae become simplified distally and spread this alter-

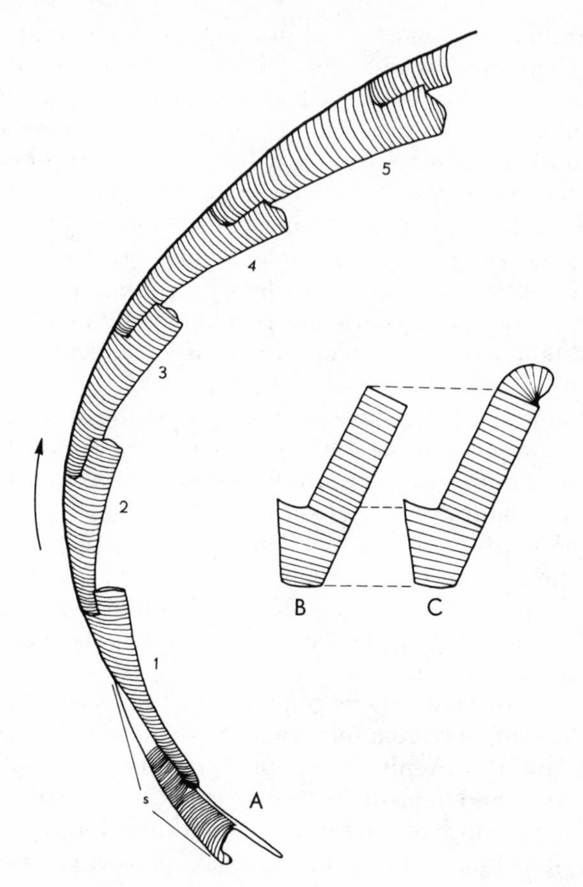

Fig. 43. (*A*) Sicula (*s*) and proximal thecae of a monograptid colony. (*B*) Simple theca. (*C*) Elaborate theca. (From Urbanek, 1973.)

ation proximally (the "regressive" sequence). These trends not only occurred again and again in monograptids, but also characterized the evolution of two-stiped *Didymograptus* in the Llandeilo (Ordovician) (Elles, 1923). Why should they be so common and other events (like the simultaneous introduction of novelties in all thecae) so rare or unknown? Elles' trends represent the primary modes of heterochrony; they reflect the evolutionary limits of simple alterations in developmental rates. The progressive sequence is an example of recapitulation. Proximal thecae are the oldest. Features of their late ontogeny are, at first, not attained by the younger, distal thecae. By a progressive acceleration in development, even the youngest

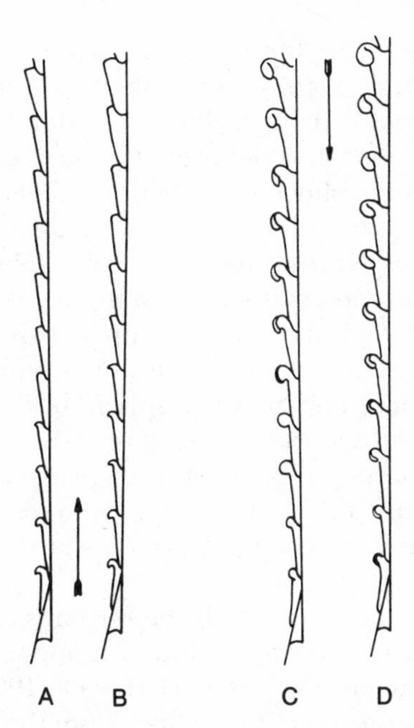

Fig. 44. Evolution by heterochrony in monograptid colonies. (*A* and *B*) Proximal complexity spreading distally (*A* is ancestral, *B* descendant)—recapitulation by acceleration. (*C* and *D*) Distal complexity spreading proximally. (From Urbanek, 1973.)

thecae of descendant species go through the entire potential ontogeny.[2] The regressive sequence represents paedomorphosis. Youngest thecae are the first to lose adult characters and to retain permanently their juvenile form. This tendency spreads to older thecae until all are paedomorphic.

Urbanek (1960, 1966, 1973) has confirmed most of this scheme, but with some reservations. He acknowledges that phyletic sequences in monograptids usually proceed by the distal spread of proximally introduced complexity (1973, p. 488). But he also shows that some lineages begin with distal introductions that spread proximally. He doubts the validity of Elles' "regressive" sequences, but affirms (1960) that several trends to simplification begin in distal thecae and spread proximally (reduction and disappearance of periderm in the Archiretiolitinae and simplification of thecae in triangulate monograptids). But Urbanek's main contribution to graptolite studies lies in his elegant explanation of all these trends as simple results of evolutionary changes in a morphogenetic substance produced by the sicula. All individuals are united along the stipe and all retain their connection to the primary sicula. Natural experiments in regeneration indicate that sections of stipes without siculae never develop elaborate thecae no matter how large or how old these thecae may become. If sicular substances control morphogenesis, then proximal thecae will become elaborate first because they have been in longer and closer contact with such substances. (If these morphogenetic substances act as inhibitors rather than promoters, then distal thecae will be the first to become elaborate.) Recapitulation in the "progressive" sequence reflects a strengthening or acceleration in rates of production for this morphogenetic substance. The proximal spread of distal novelties marks a weakened effect or decreased rate by retardation. (If the substance is a promoter, distal simplification spreads proximally; if an inhibitor, distal complexity spreads down the stipe—the result is paedomorphic in any case as features of young thecae appear in progressively older thecae.)

Heterochrony is the cause of all these trends; it is, therefore, the primary determinant of thecal evolution in monograptids. New features, "in the vast majority of cases" (Urbanek, 1973), are introduced into a few thecae at one end of the stipe; from there they may spread gradually over the entire colony. This spread reflects a change in rate or timing of ancestral processes—a disruption of established correlations between growth and differentiation.

A CLASS: ADAPTIVE THEMES IN BIVALVE EVOLUTION. Stanley has written that "an immense amount of evolutionary potential is stored within the postlarval bivalve" (1972, p. 182). Nearly all postlarval bivalves develop two features absent in most adults—byssal threads for attachment and an active foot for crawling. Both aid the small postlarva in finding and establishing a benthic site for later life. Of the byssus, Stanley writes:

> A postlarval byssus is apparently present in virtually all living species of burrowing bivalves. Attachment following settlement is critical because a tiny clam, being the same general size as a grain of sand and even less dense, is highly vulnerable to transport and destruction by bottom currents. It seems likely that the juvenile byssus greatly increased the survival rate of ancestral burrowing bivalve species and triggered their diversification. (1975, p. 379)

Ancestral bivalves were shallow burrowers. Stanley (1968, 1972) has traced an increasing evolutionary diversity to two morphological "inventions" permitting the invasion of new habitats: mantle fusion with evolution of siphons for effective deep burrowing and byssal attachment for epifaunal life on hard substrates.

Yonge (1962) has suggested that the adult byssus arose by paedomorphic retention of the postlarval attachment threads. A roster of epifaunal byssate bivalves illustrates the significance of this paedomorphosis: Mytilacea (mussels), Pteriacea (pearl oysters), Pectinacea (scallops—swimmers show the morphological signs of prior phyletic attachment), Anomiacea ("jingle shells"), Ostracea (oysters—cement for the attached valve is secreted by the byssal glands), Dreissenacea (zebra mussels), some Arcacea (ark shells), and Tridacnidae (giant clams). Yonge has also suggested that the modern Erycinacea may represent a structural stage in the evolution of an adult byssus; for these tiny, progenetic clams are modified in habit, but not otherwise in form by their byssal retention. The erycinids average 3 to 6 mm in length (with a range of 1 to 17mm); their shells are generally thin and unornamented. In size and overall appearance, they are like the postlarvae of many ordinary bivalves (Kautsky, 1939). Their accelerated maturation leaves the mobile foot and byssus in its postlarval state; with these organs, they locate and secure their substrate, typically small holes and cracks in rock, though they often lead a commensal or parasitic existence on worm tubes, the abdominal segments of crabs, the esophagus of holothurians, or the anal spines of sea urchins. Stanley (1972) favors an infaunal to epibyssate transition at such small sizes because adaptive zones merge, so to speak, at the sand grain. When particle size equals animal size, there is no discontinuity between life in soft sediment and on a hard substrate—the tiny progenetic ancestor of an epibyssate taxon can attach to a single grain.

The evolutionary radiation of the bivalves is one of the success stories of invertebrate evolution; few other marine taxa have shown such a steady and consistent increase in diversity. The paedomorphic retention of a postlarval byssus provided one of the two major contributions to this success; it opened to bivalves a large ecospace formerly unavailable to them.

> No group of animals is better fitted for life in soft substrate than the Bivalvia with complete enclosure of the body in the laterally compressed, hinged shell, insinuating and terminally dilating foot, and most usually with both inhalent and exhalent apertures, i.e. all contacts with the water above, confined to the posterior end. It is initially surprising to find such animals, and often with supreme success, exploiting the possibilities of life on hard, sometimes highly exposed and surf-beaten strata. (Yonge, 1962, p. 113)

Moreover, Stanley (1972, p. 182) has traced several evolutionary reversions from epibyssate to infaunal life (among the arcoids, for example). These he also attributes to paedomorphosis since the active foot for infaunal reversion most plausibly arises by retention of postlarval mobility.

Finally, to mention specific adaptations rather than general trends, the external ornament of bivalve shells usually develops toward increasing complexity during growth. If selection favors either a weaker or stronger ornament in descendant adults, the obvious paths to adaptation are paedomorphosis (with retention of juvenile ornament) or recapitulation (with acceleration and hyperdevelopment of an allometry already present). Nevesskaya (1967) cites an excellent case (Fig. 45): the Lower Sarmatian cockle *Cardium plicatum* bears 13 to 18 squamate costae as an adult. The juvenile shell, at a length of 5 mm, has fewer costae covered with spines. During ontogeny, the spines wear away and costae increase in number. A Middle Sarmatian descendant, *Cardium fittoni,* retains both the juvenile number of costae and their characteristic spines—a clear case of paedomorphosis by neoteny (Fig. 46).

A FAMILY: EVOLUTIONARY PATTERNS IN PLETHODONTID SALAMANDERS. Wake (1966) has traced the dominant role of paedomorphosis in the evolution of plethodontid salamanders. He divides the subfamily Plethodontinae (with 18 of the 21 plethodontid genera) into three tribes. Paedomorphosis has largely determined the evolutionary radiation of the two most diverse tribes, the Hemidactyliini and the Bolitoglossini. In the hemidactylines, 10 of 20 species (in four of the six genera) are permanently larval in their morphology. They are, for the most part, inhabitants of subterranean waters at the peripheries of their ancestral ranges. Several are eyeless cave dwellers. Paedomorphosis plays a different role in the terrestrial bolitoglossines with direct development. The aquatic larva has been eliminated from the life cycle, but selected juvenile features are retarded in development and remain in the adult stage. The 125 species of bolitoglossines in-

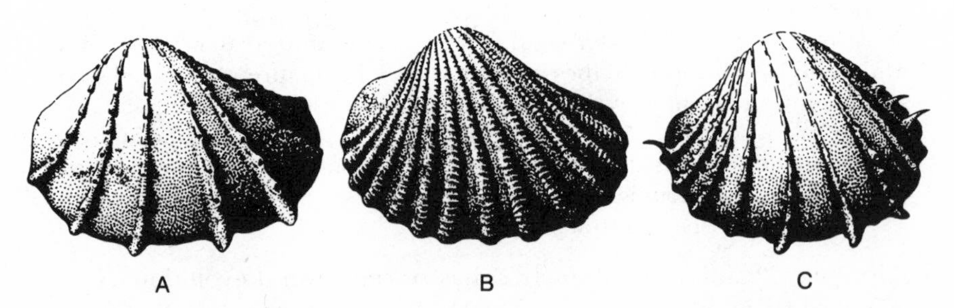

Fig. 45. Neoteny in Sarmatian cockles. (*A*) Young specimen of ancestral *Cardium plicatum,* 5 mm in length. (*B*) Adult at 17 mm in length. (*C*) Descendant *C. fittoni,* 35 mm long but with ribbing of the ancestral juvenile (*A*). (From Nevesskaya, 1967.)

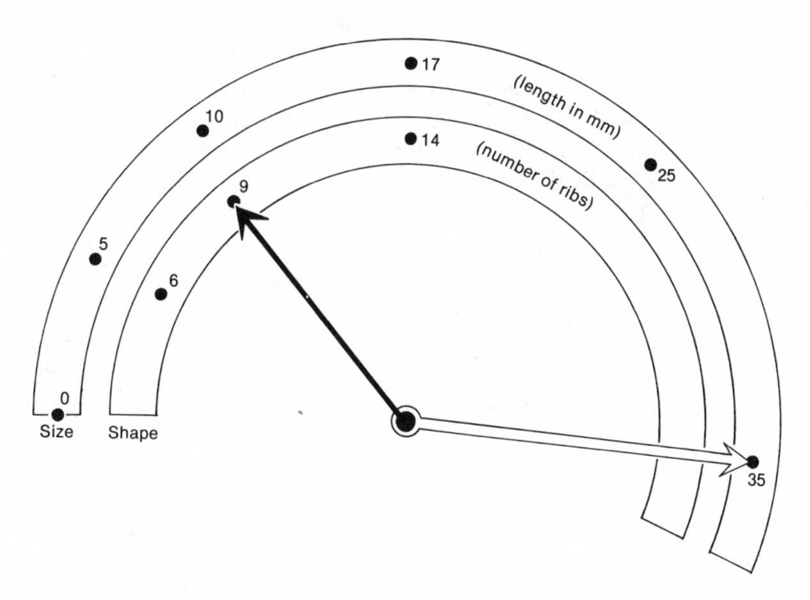

Fig. 46. Clock for neoteny of Fig. 45. No information for age; size is greatly increased while shape (number of ribs) remains in juvenile stage of ancestor.

clude a large chunk of the entire order Caudata, which contains fewer than 300 species (Wake, 1966, p. 2). They are also the only salamanders that have successfully invaded the tropics (where the vast majority of bolitoglossine species dwell). Wake believes that extensive paedomorphosis has provided the primary impetus to tropical success. The most paedomorphic bolitoglossines are the most tropical, and several genera with extensive latitudinal ranges display an impressive and gradual increase of paedomorphic characters in the lower latitudes (Wake, 1966, p. 81). Wake concludes:

> The highly adaptive tropical radiation has resulted in the evolution of several genera and over 100 species of salamanders. Paedomorphosis seems to have been a means of releasing the evolutionary potential of the group . . . The fact that all of the more advanced species of the various genera are increasingly affected by paedomorphosis demonstrates its dynamic, progressive influence in the group. The active role of paedomorphosis has been the dominant feature in the evolution and adaptive radiation of the tropical salamanders. (1966, p. 83)

A GENUS: ITERATIVE NEOTENY IN BERMUDIAN LAND SNAILS. Paedomorphosis dominates the evolutionary history of the land snail *Poecilozonites bermudensis* in the Pleistocene of Bermuda (Gould, 1968, 1969, 1970). During the major climatic fluctuations induced by advance and retreat of continental glaciers, the main lineage displayed only the most trifling changes. Since these changes are comparable in direction and magnitude to the purely phenotypic effects inflicted by climate upon the shells of modern land snails, they need have no genetic basis (Gould, 1969, pp. 483–492). But on at least four separate occasions (Fig. 47), peripheral populations became strongly paedomorphic (see Gould, 1969, pp. 473–479, for justification of their independent derivation from the central stock). All characters were affected; the paedomorphic shells are scaled-up replicas of early ontogenetic stages in the main lineage (Fig. 48). Figure 49 displays factor loadings for three groups of snails upon the first two axes of a Q-mode factor

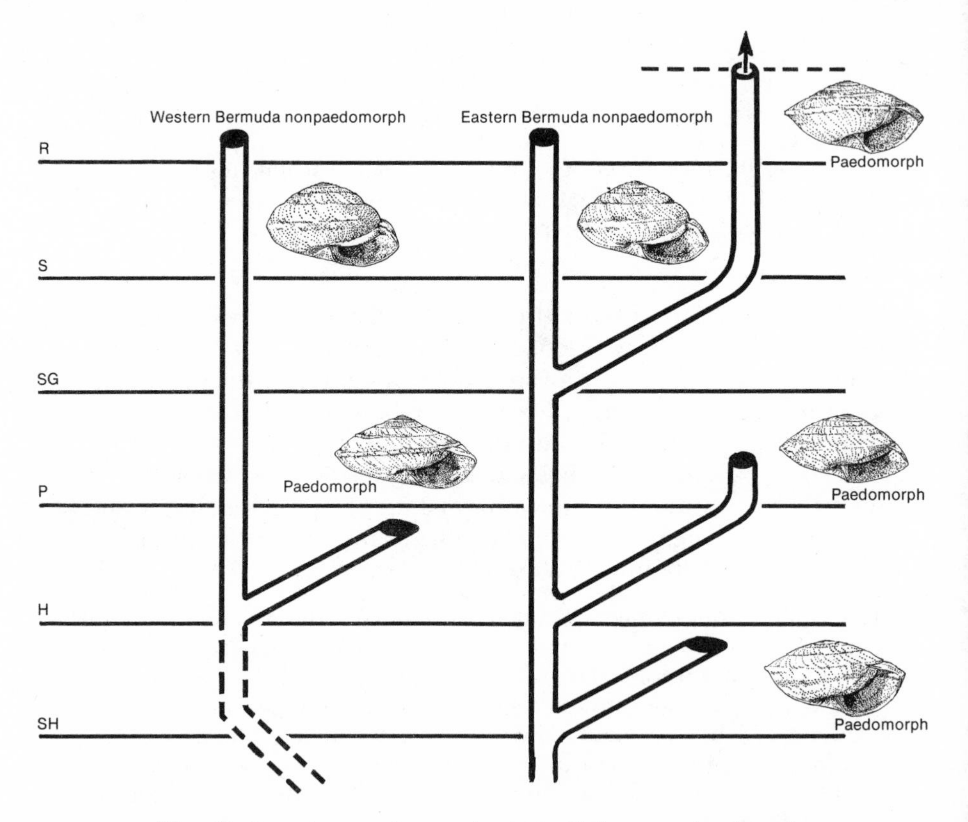

Fig. 47. Iterative paedomorphosis in Pleistocene land snails from Bermuda. On at least four separate occasions, the central stock of *Poecilozonites bermudensis zonatus* gave rise to strongly paedomorphic descendants bearing shells that are little more than scaled-up replicas of ancestral juveniles (see Fig. 48). (From Gould, 1969.)

analysis (these axes explain more than 95 percent of the total information). The snails are defined by seven ratio measures of shape; shell size does not enter the analysis at all. The three groups include: tiny juvenile shells (*Y*) of a nonpaedomorphic ancestor, adult shells (*A*) from the same nonpaedomorphic sample, and adult shells (*P*) of a paedomorphic descendant (same size as the nonpaedomorphic adults). Ancestral juveniles and adult paedomorphs group together; adult nonpaedomorphs are separate. The paedomorphs are truly scaled up replicas—in shell form, shell thickness, and color—of ancestral juveniles.

This iterative paedomorphosis is the only evolutionary event in the recorded history of *P. bermudensis,* and it happens again and again. Moreover, its adaptive trigger can be identified in the almost totally lime-free soils that served as substrate for the most paedomorphic forms. (Most nonpaedomorphs lived in carbonate dunes.) Land snails build their shells from ingested lime and the correlation of abundant snails with limestone substrates is among the best documented phenomena of animal ecology. When faced with a great scarcity of lime, land snails either die, move, or grow very thin shells. The juvenile shell of *Poecilozonites* is paper thin, the adult quite thick. Paedomorphosis must have been the easiest path to the large and rapid reduction of shell thickness that changing environments required.

Frequency of Paedomorphosis in the Origin of Higher Taxa

It is unfortunate that most literature on paedomorphosis is cast in the same mold that bolstered recapitulation during the previous half century—speculative phylogeny of higher taxa. There is scarcely a major group of animals that has not inspired a paedomorphic theory for its origin. And when one considers that insects (de Beer, 1958, derives them from myriapod larvae), copepods (Gurney, 1942, pp. 22–26; Noodt, 1971; from a larva like the decapod protozoea), and vertebrates (Garstang, 1928; Berrill, 1955; Whitear, 1957; Bone, 1960; from hemichordate larvae, tunicate tadpoles, and echinoderm larvae) have been among the most popular candidates, the number of modern paedomorphic derivates could be overwhelming. All phyla have been implicated, from protists (Hadzi, 1952) to higher plants (Carlquist, 1962; Takhtajan, 1969). General supporters of paedomorphosis have supplied long lists: de Beer (1958) favors a paedomorphic origin for, among others: ctenophores, siphonophores, cladocerans, copepods, pteropods, insects, chordates, appendicularian tunicates, ratite birds, hominids, hexacorals, proparian trilobites, and grapto-

1
2
3
4
5
6
7
8
9

(Facing page)
Fig. 48. Paedomorphosis in Bermudian land snails. (*1*) Adult of the normal nonpaedomorph (actual width, 19.7 mm). (*2*) Juvenile shell of the normal form enlarged to the size of adults for comparison with the paedomorphs (actual width, 11.5 mm). (*3–6*) The four independently evolved paedomorphs, adult shells comparable in size with *1* (22.2 mm, 21.5 mm, 20.2 mm, and 23.0 mm, respectively). (*7*) Coloration of a paedomorph (juvenile "flames" are preserved; they never coalesce into bands). (*8* and *9*) Typical coloration of nonpaedomorphs; flames coalesce into bands. (From Gould, 1968.)

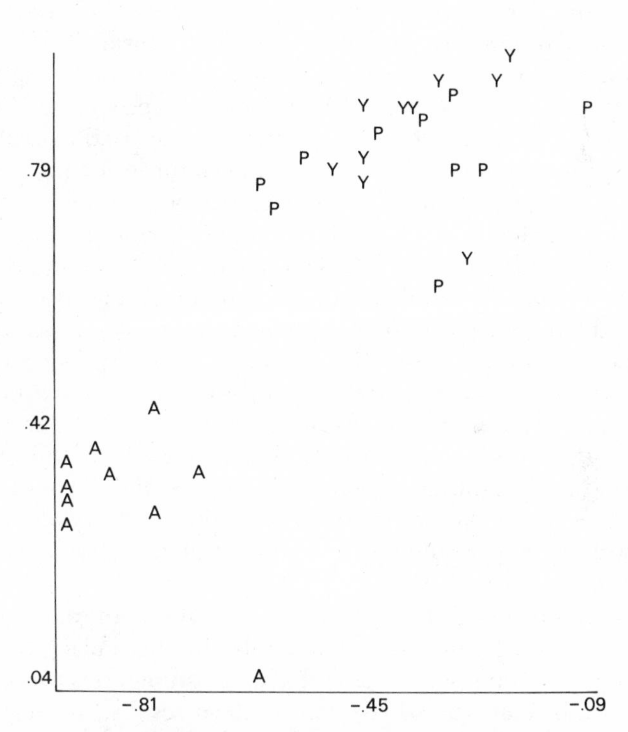

Fig. 49. Multivariate comparison of adults (*A*) and juveniles (*Y*) of a nonpaedomorphic population, with adults (*P*) of a paedomorphic population. The variables are seven ratio measures of shape; the two varimax axes encompass 91.8 percent of the total information. Juvenile nonpaedomorphs cluster with adult paedomorphs and plot far from their own adults. (From Gould, 1968.)

lites. Clark (1964, p. 221) is convinced of the paedomorphic origin of all pseudocoelomate phyla; since he views the pseudocoelomates as polyphyletic, paedomorphosis had to occur several times.

Even general opponents are always willing to grant a few phyla: Hadzi (1963) favors a paedomorphic origin for chaetognaths from brachiopods and for ctenophores from something like Müller's larva of turbellarians (Heath, 1928, discovered a sexually mature turbellarian looking much like Müller's larva). Even Jägersten (1972) will allow ctenophores, rotifers, and entoprocts into the paedomorphic fold. Garstang's poetry (1951) remains the easiest introduction to the literature of speculative paedomorphosis.

Frankly, I have never known what to make of this literature. The cases cannot all be wrong, but they are all unproved. And they present a problem common to most phylogenies inferred entirely from modern organisms. The fossil record is notoriously uninformative on the origin of new morphological designs; usually, higher taxa simply appear in the record without clear antecedents or incipient stages (indeed, this abruptness has inspired most paedomorphic postulates, since promotion of a larva to adult status is one way to generate a new design without intermediates). Since an ancestor with a larva ripe for paedomorphosis is almost invariably hypothetical, it need not have existed at all. The "larval progenitor" could just as well have been an ancestral adult which later augmented its own ontogeny to evolve the modern animal now pressed into service as a model for the supposed ancestor. Thus, I have never been impressed with the paedomorphic theory of vertebrate ancestry because we need not view the tunicate tadpole (for example) as an ancestral larval form. The real ancestor is lost in Precambrian darkness. Something like the tunicate tadpole might have been the definitive *adult* form of our ancestor. Later, it may have evolved a sessile adult, the modern tunicate, while another branch elaborated the ancestral adult into a more efficient swimmer.

This is but one example of a general trap often entered in phyletic arguments based upon modern animals. In this fallacy of unique inference, the assumption is made that an unambiguous historical hypothesis can be constructed from its modern results. But several historical reconstructions are usually consistent with a single set of results. Davis and Birdsong, for example, support Fryer's (1959) notion that water-column foragers among fish are often paedomorphic:

> During the ontogeny of species which forage benthically, the morphological changes involve greater modification of the jaws and "pharyngeal mills" in cichlids, centrarchids, pomadasyids, sparids, and numerous other perciform families . . . Among the water-column foragers, however, the basic adaptive

modifications for feeding remain virtually unchanged during ontogeny. Consequently, the morphological hard parts, and associated musculature which ultimately represents the taxonomic distinction among various benthic forms, simply does not differentiate comparably in adults of water-column foraging species. (Davis and Birdsong, 1973, p. 303)

But the paedomorphic argument needs a subsidiary hypothesis not advanced by Davis and Birdsong: one must demonstrate the phyletic priority of benthic adaptations, for only then can the reduced differentiation of water-column foragers be viewed as a paedomorphic retention. Otherwise, the lesser differentiation of water-column foragers may be ancestral, with the benthic adaptations added later. The best hypotheses of paedomorphosis provide direct evidence or strong inference about ancestors. For hominid neoteny, for example (Chapter 10), we have australopithecines as well as the general patterns of primate and mammalian ontogeny—humans are juvenilized with respect to all three, and we cannot propose our ontogeny as ancestral. Such additional evidence may be available for fishes, but Davis and Birdsong do not supply it.

Although this argument will surely reduce the number of valid paedomorphic higher taxa, the balance may be redressed by another bias that commonly precludes the recognition of paedomorphic groups. To many biologists, "simple" morphology still means "primitive" in ancestry—generalized structure is confused with temporal priority. Since larval morphology is usually quite simple, the adults that retain it are treated as ancestral groups rather than recent paedomorphic derivates. I have been particularly struck by how often the "primitive" label is attached to tiny forms in unusual environments; I would prefer to view these groups as progenetic adaptations to the requirements for small size and rapid reproduction in their specific environments (see pp. 332–336). Thus, Swedmark and Teissier (1966, p. 131) choose to view the interstitial actinulid coelenterates as "archaic" despite their larval features. Pennak and Zinn (1943, p. 4) consider the interstitial mystacocarids as "undoubtedly the most primitive living Crustacea" while admitting that their heads contained persistent larval features. Madsen (1961), however, attacks the tendency of many colleagues to view the deep sea fauna as archaic; instead, he attributes several taxa to paedomorphosis (porcellanasterid starfish and elasipod holothurians, for example).

Finally, we are left with the usual frustration of enumerative arguments in natural history: we cannot judge relative frequency, for there are too many unrecorded cases. I am, nonetheless, inclined to agree with Rensch's assessment: "by far, most phylogenetic changes in form arise by heterochrony" (1971, p. 12). Still, I would rather rest my case on fulfilled expectations drawn from ecological and evolutionary theory. To these, I now turn.

A Critique of the Classical Significance of Heterochrony

The Classical Arguments

Traditional views die hard, and their influence extends long past the time of their public abandonment. In its heyday, recapitulation was the motor of evolutionary progress, a mechanism to build increasing complexity by adding new stages to previous, simpler ontogenies. Paedomorphosis, in this light, is degenerative; it merely lops stages off a complex ontogeny and returns an organism to a simpler, ancestral state (literally, in fact, to the adult stage of a long-extinct forebear).

Few modern authors would state the traditional view as baldly as Ruzhentsev has: "Ontogenetic recapitulation is one of the basic laws of movement in the organic world, a replacement of one form of life by another, more perfect one" (1963, p. 943). Yet the influence of recapitulation lingers, if only in setting the stage for criticism.

It has lingered more directly in the persistent statement that one mode of paedomorphosis is degenerate and nonprogressive—progenesis, or juvenilization by the acceleration of maturation and the truncation of ontogeny. Simplification of form, shortening of the life span, and reversion to a largely unaltered larval or juvenile morphology have been contrasted unfavorably with the prolongation and potential for complexity associated with evolutionary progress. To Cloud (1948, p. 333), progenesis is "strictly regressive"; to de Beer (1958, p. 64) it is "of no significance in progressive evolution." Wake writes: "Paedogenesis [= progenesis] is genetically fixed, leads to specialization and degeneration, and is of little significance as far as future phylogenetic progress is concerned" (1966, p. 79).

Neoteny, on the other hand, has been granted a progressive role, for here ontogeny is maintained at its ancestral length (or even extended) while specific somatic features are delayed in appearance. De Beer writes: "If [paedomorphosis] is achieved by accelerating the rate of development of the reproductive glands and hastening the time of maturity, the phylogeny usually results in a simplification often associated with parasitism. But if [it] is brought about by a slowing down of the rate of development of the body relatively to that of the reproductive glands, then changes are brought about which have considerable importance in phylogeny" (1930, pp. 70–71).

But this positive role is still anchored to traditional thinking, for neoteny is treated as a mode of rescue from a progressive trend gone wrong. Phylogeny usually proceeds by addition, by further differentiation, or by increasing complexity in specialization. Often, however, this trend leads to a cul-de-sac or evolutionary dead end. Animals become too committed to the peculiarities of their environments. By evolving a complex, fine-tuned design for a highly specific mode of life, they sacrifice plasticity for future change. Neoteny can now come to the rescue and provide an escape from specialization. Animals can slough off their highly specialized adult forms, return to the lability of youth, and prepare themselves for new evolutionary directions. As Hardy writes: "The chance comes earlier, before it is too late, and such lines are switched by selection to new pathways with fresh possibilities of adaptive radiation" (1954, p. 167). "The adult stage," claims Hilzheimer, "is a rigid, unchangeable endform, adapted to specific conditions of life. New directions of development can arise only from the still changeable and adaptable youthful stages" (1926, p. 110). In his original formulation, de Beer made the point in cosmic fashion, complete with analogies to thermodynamics:

> We see, then, that evolution by gerontomorphosis produces relatively small changes which sacrifice the power of changing further, and that on the other hand paedomorphosis produces large changes which do not sacrifice that power. If gerontomorphosis were the only possible method of evolution, as Haeckel's theory of recapitulation would suggest, phylogeny would gradually slow down and become stationary. The race would not be able to evolve any further, and would be in a condition to which the term "racial senescence" has been applied. It would be difficult to see how evolution was able to produce as much phylogenetic change in the animal kingdom as it has, and it would lead to the dismal conclusion that the evolutionary clock is running down. In fact, such a state of affairs would present a dilemma analogous to that which follows from the view that in the universe energy is always degraded. If this were true, we should have to conclude that the universe had been wound up once and that its store of free energy was irremedially becoming exhausted. We do not know how energy is built up again in the physical universe although it must happen somehow; but the analogous process in the domain of organic evolution would seem to be paedomorphosis. A race may become rejuvenated by pushing the adult stage of its individuals off from the end of their ontogenies, and such a race may then radiate out in all directions by specializing any of the stages in the ontogenies of its individuals until racial senescence due to gerontomorphosis sets in again. (1930, p. 95)

The plasticity invoked is usually portrayed as morphological—the more general and adaptable forms of a pluripotent youth versus the committed complexity of a highly specialized adult. But a genetic argument has also been advanced in terms of "unemployed genes" (stressed by Takhtajan, 1969, p. 29). If neoteny occurs rapidly, many genes for adult structures remain in the genome, yet are not transcribed; hence, they become available for modification.

Neoteny has also been favored because it provides one of the few mechanisms for rapid and profound evolutionary change in a Darwinian fashion without the specter of macromutation. A descendant with a mixture of ancestral juvenile and adult characters (or in extreme cases a sexually mature giant larva) may immediately enter a new adaptive zone; yet the genetic input need involve no more than some changes in regulatory genes—for the larval features are already coded in the ancestral genome. "The significance of neoteny," writes Takhtajan "lies in the attainment of maximum phenotypic effects by means of minimal genotypic change" (1969, p. 30).

Many paleontologists have been attracted to neoteny because it seems to provide an orthodox justification for gaps in the fossil record—the traditional barrier to acceptance of Darwinism by paleontologists. Thus, Cloud argues: "This concept is perhaps one of the most important in evolutionary theory, for paedomorphosis may well be the mechanism that produced or contributed to the production of the seemingly bridgeless gaps between some major systematic categories" (1948, p. 331). In discussing giant crustacean larvae with developing gonads, Gurney writes: "If this precocious appearance of sexual characters were so accentuated that sexual maturity occurred before metamorphosis, we should get at one step a new decapod group the relationship of which would be most obscure" (1942, p. 73). And Takhtajan argues that "the evolution of the angiosperms was therefore not only rapid . . . but also discontinuous as a result of neoteny" (1969, p. 33).

A third argument for the importance of neoteny is embodied in de Beer's notion of "clandestine evolution" and Schindewolf's theory of proterogenesis. The plasticity argument involves no change in juvenile stages; it merely invokes their unspecialized nature in the service of future modification. But juvenile stages may be changing by cenogenesis while adults remain in their ancestral condition. Since phylogenies of the fossil record are largely sequences of adults, this juvenile evolution would pass undetected—de Beer's clandestine evolution. If neoteny then promoted these cenogenetic changes to adult status, they would appear as sudden transitions in the fossil record.

The most ambitious theory of macroevolution by neoteny was advanced in a related argument by the German paleontologist Otto Schindewolf (1936, 1950): all (or virtually all) major evolutionary novelties are introduced suddenly by some essentially non-Darwinian process into the juvenile stages of ancestors. Thence they move slowly forward, gradually invading and displacing ancestral adult stages by a process that Schindewolf called proterogenesis.

In summary, the classical significance of paedomorphosis has been portrayed in the following way: Progenesis can be dismissed as a factor in macroevolution because it leads only to degeneration by truncation. Neoteny is an essential process in macroevolution for two reasons: First, it provides an escape from specialization in the replacement of inflexible adult structures by generalized juvenile forms (for instance, Clark, 1964, p. 247, on the "rescue" of vertebrates from a deuterostome stock whose adults had become "acutely specialized" by a sessile mode of life). Second, it supplies one of the very few Darwinian justifications for large and rapid evolutionary transitions, by permitting major changes in morphology without extensive genetic reorganization. In short, the classical arguments are entirely *morphological* and *macroevolutionary.*

Retrospective and Immediate Significance

With one exception—the denigration of progenesis—I do not disagree with the classical arguments. I do, however, regard them as woefully incomplete, for a fundamental theoretical reason: they are entirely retrospective in design. They look upon a case of neoteny after its descendants have evolved and attribute meaning in terms of the aggregate success. But what of the actual species that experienced neoteny? It did not realize that it would be a herald of future diversity because it had sloughed off some ancestral specializations. It was not impressed by the fact that it had gained so much in morphology for so little genetic effort. It became neotenic for its own immediate reasons—its own ecologic strategy in its own particular environment.

Retrospective significance is a legitimate inquiry in macroevolutionary studies. The ancestral species of successful modern groups formed a tiny fraction of their contemporary biotas; more than 99.9 percent of all species are not sources of great future diversity. It is therefore important to seek in retrospect the common features of these exceedingly rare sources of later success. Traditional inquiry in paleontology has done just this. We learn that successful ancestors have usually been smaller than their later descendants (Cope's law of phyletic size increase) and of fairly generalized structure (law of the unspecialized). But these "laws" say nothing about the immediate significance of these features for the ancestors themselves. The ancestors were not small to provide a potential for future change, but to gain some selective advantage in their own environment.

This distinction between immediate and retrospective significance has been blurred by an uncritical acceptance of the continuationist doctrine in macroevolution—namely, that events at the species level can encompass by extrapolation all the characteristics of higher taxa and their evolution. When Simpson popularized this position (1944, 1953), he was opposing a long tradition for separate, non-Darwinian explanations of major features in the evolution of life—a tradition that threatened the generality and success of modern evolutionary theory. His emphasis on continuity and extrapolation from the species level was legitimate and well placed. But it has been extended and uncritically rigidified. It is no denial of Darwinism to distinguish between immediate and retrospective significance in evolution. To cite an example: evolutionary trends are usually explained in the continuationist tradition as a result of orthoselection. The ancestral species responds to directional selection for certain adaptations. Subsequent species respond to the same selection and accentuate the trend. Retrospective and immediate significances are the same. The first hypsodont horse evolved high-crowned teeth for grazing; the subsequent trend towards increasing hypsodonty permitted better grazing in larger animals. Eldredge and Gould (1972) have presented a different interpretation of evolutionary trends: Wright (1967) suggested that speciation might be random with respect to the direction of evolutionary trends, just as mutation is random with respect to the direction of selection within populations. Following Wright's analogy, trends represent the differential preservation of certain speciation events, but the adaptive significance of each species must be sought in the immediate environments that nurtured it, not in the net result of the developing trend. Still, the net result or retrospective significance is a perfectly legitimate concern; it is just not the same thing as immediate significance.[3]

I shall be emphasizing the immediate significance of heterochrony throughout the rest of this book, primarily because it has been so widely ignored. In so doing, I am neither attacking traditional arguments based on retrospective significance, nor trying to undermine the concept of retrospective significance in general. In fact, my ulterior motive as a paleontologist is to prove the importance of my profession by demonstrating that the study of macroevolution, with its emphasis on retrospective significance, cannot be subsumed in the study of living populations, with its necessary concern for immediate significance alone.

I wish first to illustrate with four examples how immediate significance has been ignored in conventional explanations of paedomorphosis:

1. Rudwick (1960) argues that brachiopods can build a spiral lophophore in only two mechanically sound ways. It is hard to imagine the evolutionary transformation of one design into the other, but Rudwick believes that such a transition may have occurred and he invokes paedomorphosis to explain it. Both types of spiral go through a plane (schizolophous) stage in their early ontogenies; from the schizolophe, either form of spiral can be constructed: "The intermediate stage would have been represented by a form that became adult in the schizolophous stage, from which either type of spirolophe could have been derived subsequently by the operation of different growth patterns in the lophophore" (pp. 381–382). Paedomorphosis has the retrospective significance of permitting an evolutionary transition impossible at adult complexity (the traditional "escape from specialization"). But why did the paedomorphic intermediate retain a schizolophe as an adult? What was its immediate adaptive significance (for it knew nothing of the new spirolophe to come)? Was it an instance of progenesis with accelerated maturation and truncation of ontogeny at juvenile sizes and shapes? If so, does immediate significance lie in the advantages of small size or rapid reproduction rather than in the design of the lophophore itself?

2. Baluk and Radwanski (1967) argue that creusiine barnacles evolved from paedomorphic balanids. They cite plasticity for environmental change as the significance of paedomorphosis; adult balanids could not have made the transition. Creusiines live within the corolla of anthozoans—an odd spot for a barnacle. How are they able to live there anyway? Did paedomorphosis provide any specific feature for adaptation to this peculiar environment (rather than just the abstract "plasticity" to leave the balanid life style)?

3. Williams and Wright (1961) believe that terebratuloid brachiopods evolved paedomorphically from spiriferoids. They cite the potential for sexual maturation at very early stages in most living brachiopods. *Terebratulina septentrionalis,* for example, may become mature at one-fourth adult size with its lophophore still in the early plectolophe stage. Retrospective significance may lie in the potential for forming a new order, but what did the tiny, progenetic spiriferoid gain from its small size and simple structure?

4. Stanley (1972) presents an elegant argument for the progenetic transition from byssate to free-burrowing taxa (Fig. 50). Endobyssate taxa like *Cardita floridiana* (*A*) grow allometrically in ontogeny: "The juvenile is equant, and the elongate adult shape arises from it by more rapid growth in a posterior direction than in an antero-ventral direction. The juvenile has a well-developed foot and byssus. The adult retains the byssus, but has a relatively smaller foot" (pp. 195–197).

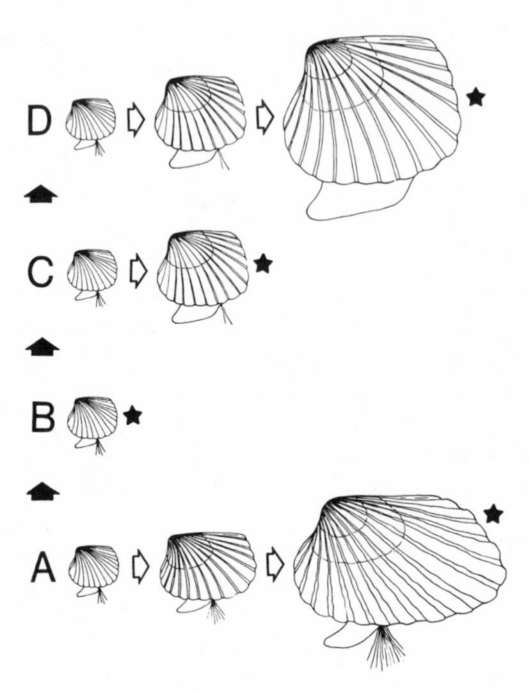

Fig. 50. Hypothetical derivation of a free-burrowing species (*D*) by progenesis (*B*) from a byssate ancestor (*A*). Horizontal sequences on each line represent ontogeny; adults are starred. (From Stanley, 1972.)

Stanley postulates that an adult free-burrower (*D*) might evolve from *A* via a progenetic intermediate (*B*) like the juvenile of *A*. This progenetic ancestor would then evolve isometrically towards larger size while the juvenile foot would be emphasized at the expense of the byssus. Stanley favors a small progenetic intermediate because transitions between adaptive zones are hard to accomplish at large size (see p. 336). The retrospective significance of progenesis might lie in its facilitation of a transition within habitats not available to the larger sizes of either ancestor or descendant. But what did *B* gain from its own progenesis? Did immediate adaptive significance lie in small size itself, or in accelerated maturation?

In a search for immediate significance, we descend from the sweep of macroevolution into a shorter temporal framework. The slow changes of a million years in evolutionary time must yield to the rapidly fluctuating requirements of ecological time. We must consider the demands of immediate environments rather than the tendencies of entire lineages. In setting this ecological theater for the evolutionary play, different reasons for heterochrony become apparent. In particular, the timing of maturation (rather than the morphology obtained by speeding up or slowing down) seems to demand special emphasis.

Heterochrony, Ecology, and Life-History Strategies

In a few cases, proposals have been advanced for the immediate significance of paedomorphosis. Here, we can usually detect the influence of another bias in explanation. Adaptive significance has been sought in terms of *morphological* advantages alone. Thus, when Stubblefield (1936, p. 430) discussed the significance of progenesis in minute agnostid and eodiscid trilobites, he could offer only the possibility of superior enrollment in forms with very few thoracic segments. Maybe so, though larger trilobites with many thoracic segments rolled up tolerably well. Stubblefield considered neither the direct significance of marked reduction in size nor, especially, the consequences of small size for population structure and dynamics—the probability of more individuals, reduced longevity, and earlier reproduction.

Classical evolutionary theory portrayed adaptation in terms of morphology, physiology, and, perhaps, behavior. The size, structure, and dynamics of populations were very rarely considered, though I expect that most evolutionists would have admitted their potential importance if pressed. No biases are more insidious than those leading to the neglect of things everyone knows about in principle. In this case, morphology, physiology, and behavior all reinforced a typological prejudice for considering individuals as efficient machines. The more efficient replace the less efficient and, in this sense, a population exists. But such an attitude does not invite attention to the individual advantages most readily inferred from population size, age structure, and turnover rates. As Cole wrote in his pioneering paper: "Comparative studies of life histories appear to be fully as meaningful as studies of comparative morphology, comparative psychology or comparative physiology. The former type of study has, however, been neglected from the evolutionary point of view, apparently because the adaptive values of life-history differences are almost entirely quantitative" (1954, pp. 134–135).

I regard the rise of theoretical population ecology as one of the most significant events in evolutionary theory during the past twenty years. For, in focusing on immediate significance in the short run of ecological time, it has proved that the components of life history strategies—timing of reproduction, fecundity, and longevity, for example—are adaptations in themselves, not merely the consequences of evolving structure and function. Moreover, they are adaptations to components of the environment not considered in previous theories—among them, patchiness, grain, and the intensity, periodicity, and predictability of fluctuation. In short, theoretical population ecology has given us a new set of parameters for assessing adaptation.

Heterochrony seems to cry out for a reinterpretation in this light. It has been portrayed in terms of morphological results, yet the process that produces it—displacement in time by acceleration and retardation—is among the quantitative measures identified by Cole as a subject of previous neglect. I argued in Chapter 7 that most heterochronic effects result from a change in the timing of maturation. Progenesis arises from the truncation of ontogeny by accelerated maturation. Retarded maturation with no change in rates of size increase or allometry will yield hypermorphosis; if the common correlation of maturation and development is not disrupted, then a retardation in somatic development will accompany delay in maturation, and neoteny will result—as in human evolution (Chapter 10). The timing of maturation is a primary variable in setting life history strategies. We have a prima facie case for ascribing direct significance to the change in developmental timing itself, not only to its morphological consequences. The immediate significance of heterochrony might consist more often than not in this change of timing: "One of the most striking points revealed by this study is the fact that the age at which reproduction begins is one of the most significant characteristics of a species, although it is a datum which is all too frequently not recorded in the literature of natural history" (Cole, 1954).

The most widely discussed topic in studies of life history strategies, the theory of *r* and *K* selection, offers particular promise as a framework for understanding the immediate significance of heterochrony in ecological terms. (See Cole, 1954; Margalef, 1959; and Cody, 1966 for harbingers; MacArthur and Wilson, 1967, for a short formulation; Gadgil and Bossert, 1970; Gadgil and Solbrig, 1972; King and Anderson, 1971; McNaughton, 1975; Pianka, 1970 and 1972; Southwood et al., 1974, for discussion. The generality of the theory has been appropriately criticized by Hairston et al., 1970; Hoagland, 1975; and Demetrius, 1975; but it has proved its utility in such empirical studies as Gadgil and Solbrig, 1972, and McNaughton, 1975.) I do not mean to suggest that the theory of *r* and *K* selection exhausts the content of studies in life-history strategies. It is clearly only a small part of a vastly larger, and exceedingly complex, subject (see excellent review of Stearns, 1976). I choose it to exemplify my concerns because it has generated so much attention, and because it makes clear predictions about the timing of maturation—a key theme in heterochronic change.

In brief, *r* and *K* theory tries to establish which environmental conditions would favor the maximization of *r* (the intrinsic rate of natural increase) and which would lead to the maximization of *K* (the carrying capacity of the environment); both parameters cannot be maximized at the same time. (Obviously, as Pianka [1970, 1972] has emphasized, *r* and *K* strategies are end points of a continuum, not strict alternatives.) It then specifies what an organism can do with its life history and morphology in order to maximize *r* or *K*. Finally, it seeks to discover if these attributes of organisms are preferentially correlated with life in the expected environments. McNaughton writes:

> We expect an *r-K* continuum, the former occurring in a perfect ecological vacuum with no density effects and no competition, the latter occurring in a completely saturated ecosystem where density is high and competition for resources is intense. In the ecological void the optimal adaptive strategy channels all possible resources into progeny, thereby maximizing the rate at which resources are colonized. At ecological saturation, the optimal strategy channels all possible resources into survival and production of a few offspring of extremely high competitive ability. (1975, p. 251)

In simplest terms, *r* selection will predominate when the density-independent component of natural selection is in control—when populations can expand with no negative feedback on growth rate by dwindling resources. *K* selection will prevail when the density-dependent component dominates—when increase in one genotype must be at the expense of another. Situations favoring *r* selection might include large, frequent, and unpredictable environmental fluctuations; frequent catastrophic mortality; superabundant resources; and lack of "crowding." *K* environments tend to be crowded, stable, and benign. Important *r* environments include those available for colonization (superabundant resources) and those subject to large and unpredictable fluctuations imposing catastrophic, density-independent mortality on the few species able to survive in them. We expect, therefore, to encounter *r* selection during early stages of colonization and periods of increase in fluctuating populations.

Some attributes of *r*-selected organisms might include: high fecundity, early maturation, short life span, limited parental care, rapid development, and a greater proportion of available resources committed to reproduction. *K* strategists might employ low reproductive effort with late maturation, longer life, and a tendency to invest a great deal of parental care in small broods of late maturing offspring.

In some empirical confirmations, Gadgil and Solbrig (1972) found that dandelions from disturbed habitats had higher seed outputs, a greater proportion of their biomass in reproduction, and lowered competitive ability. Considering all herbaceous plants on three sites, they found a strong correlation between disturbed habitats and an increased proportion of aerial biomass in reproductive tissue. In several species of the cattail *Typha*, McNaughton (1975) correlated the following *r* attributes with supposedly *r* environments: greater developmental speed, higher fecundity, more offspring with less energy invested in each, reduced selection for traits that increase competitive ability in situations of high density, and increased selection for efficiency in colonization.

Among attributes most relevant to *r* and *K* strategies, the timing of maturation has often been granted special importance. Harper, for example, writes (commenting on work of Cole and Lewontin): "Their studies show that high intrinsic rate of natural increase (*r*) can be obtained by producing a few offspring early in life—*precocious reproduction is all important*" (1967, pp. 255–256, my italics). Lewontin (1965) emphasized the pronounced effect of earlier maturation by demonstrating that, in rapidly expanding populations, small changes in developmental rates (of the order of 10 percent) are approximately equivalent to major increases in fertility (of the order of 100 percent). On this basis, MacArthur and Wilson concluded that "speeding the developmental rate and increasing the span over which reproduction can occur are usually more effective than increasing fecundity" (1967, p. 93).

If change in the timing of maturation is so important in *r* and *K* strategies, then heterochrony should be a common process in ecological adaptation. The immediate significance of heterochrony may well lie directly in its primary mechanism of acceleration and retardation of development—with acceleration tied to *r* regimes and retardation to *K* regimes. I am led, from these considerations, to formulate a specific hypothesis to test my major contention of Chapter 7. I argued there that the *processes* of heterochrony were more fundamental than their *results*. I did this by showing that the classical results of recapitulation and paedomorphosis do not have a one-to-one relationship with the causal processes of acceleration and retardation. Specifically, paedomorphosis can be produced either by acceleration (of the gonads, in progenesis), or by retardation (of somatic features, in neoteny, often accompanied by delayed maturation). If the processes of acceleration and retardation determine the immediate significance of heterochrony, then we should be able to take all the cases usually classed together as paedomorphosis and divide them into two groups with radically differing ecological significance (or, rather, to acknowledge the complexity of natural history, a continuum with radically differing significances at the extremes). I predict that *progenesis will be associated with* r *strategies and neoteny with* K *strategies.* If this correlation can be established, then paedomorphosis has little meaning as a category of immediate adaptation. It is an artificial amalgam of two processes conferring the common result of juvenilized morphology. These processes, as markers of extremes in the *r-K* continuum, could hardly differ more in immediate significance.

Before testing the hypothesis, one proviso must be entered for conventional expectations about *r*-selected organisms. Increased fecundity with decreased parental care has usually been cited as a major criterion (reviewed in McNaughton, 1975, pp. 255–256). This correlation may apply when the *r*-selected organism and a more *K*-selected species chosen for comparison are above a certain minimum size. But, just as classical evolutionary theorists ignored life-history strategies and dwelled on morphology, we can criticize theoretical ecologists for abandoning morphology in their "billiard-ball" models and not recognizing the constraints of design (see Liem, 1973). Organisms that respond to *r* selection by progenesis are likely to be very small (many of the tiniest animals are progenetic), and their contrast in size with ancestral species may be extreme. A positive allometry between body size and egg production is among the best documented phenomena of relative growth (Gould, 1966): it may not hold for intraspecific variation over small ranges of size, but it surely applies to large differences among closely related species, especially when the small species are near the lower limits for their taxon. In her critique of standard criteria for *r* and *K* strategies, Hoagland (1975) has argued that small, marine *r* strategists cannot be more fecund than larger *K*-selected forms. She provides impressive documentation based on her own work with the "slipper snail" *Crepidula* and an extensive survey of the literature. By the allometric rule, tiny animals cannot generate enough energy to produce an absolutely large number of eggs at their size (though they may still devote a greater percentage of their biomass to reproduction).* Constrained by their design to produce relatively few eggs, they opt for what is usually taken as a mark of *K* selection: brood protection of relatively few offspring, since they

* R. Bakker has made the intriguing suggestion (personal communication) that fecundity may be enhanced in tiny progenetic animals if a physiological juvenilization accompanies paedomorphic morphology. Young animals have higher metabolic rates than related (but not progenetic) adult organisms of the same size. If reproductive activity is accelerated into the higher metabolic regime of a "young" animal, more energy might become available for the production of eggs.

cannot afford to liberate such a small number freely into the plankton. In my reading, I have been struck by how many of the smallest progenetic organisms, living in obvious *r* environments and possessing all other classical attributes of an *r* strategist, brood a few relatively large eggs. Neither Hoagland nor I are attacking the concept of *r* and *K* strategies, but merely one of the criteria advanced for larger organisms and then extrapolated to smaller ones without adequate consideration of structural and energetic constraints. We must remember that lowered fecundity does not imply a smaller *r* if maturation is correspondingly accelerated. If, as Lewontin argued (1965), accelerated maturation is more effective than increased fecundity in raising *r*, then a progenetic trade of reduced fecundity for markedly accelerated maturation will provide an excellent *r* strategy for tiny animals. Harper (1967, pp. 255–256), in the quotation cited previously, noted the effect of a *few* precociously produced offspring in raising *r*.

To be effective in ecological time, heterochrony must be a potentially rapid response to changing conditions. Before testing my general hypothesis, I will discuss the most important data concerning potential ease and speed of heterochronic change.

The Potential Ease and Rapidity of Heterochronic Change

The Control of Metamorphosis in Insects

If ontogeny includes a change in form sufficiently abrupt and substantial to warrant the term metamorphosis, then heterochronic effects can be easily diagnosed either by an alteration in timing of metamorphosis itself or by the differential acceleration and retardation of morphological traits with respect to metamorphosis. Moreover, the hormonal basis of several metamorphoses has been established (Jenkin, 1970) and heterochronies can be produced and replicated experimentally (though biologists have rarely discussed this experimental research in the context of relationships between ontogeny and phylogeny).

Although Willis (1974, p. 98) reminds us that "much of this scheme is still subject to controversy," the classical interpretation of hormonal control of metamorphosis in holometabolous insects involves the interaction of two substances. Molting is regulated by ecdysone (molting hormone), secreted by the prothoracic glands. The morphological results of any molt, however, are determined by the juvenile hormone produced by the corpus allatum, an endocrine organ lying just behind the brain. This hormone has been synthesized and found to be relatively nonspecific in its effects; juvenile hormone from one insect is generally effective in other species, as are several chemical analogs and mimics. Its chemistry is reasonably well understood, and it has been discussed extensively as an ecologically benign method of insect control. It engenders as many review articles each year as some popular subjects inspire in the primary literature (see Schneiderman, 1972; Truman and Riddiford, 1974; and Willis, 1974, for example). Metamorphosis depends upon the concentration of juvenile hormone. In the presence of a high titer of juvenile hormone, ecdysone will produce a larval molt; with low titers, the larval-pupal transformation is initiated, while an absence of juvenile hormone allows the pupa to molt into an adult. Thus, juvenile hormone is not an antagonist to the ecdysones, as once believed, but acts with them in normal development. Schneiderman has contrasted the hormonal control of maturation in vertebrates and insects: "Maturation in man and other higher vertebrates is promoted by the secretion of maturation hormones, the gonadotropins of the pituitary. The juvenile condition in man hinges upon the absence of these maturing hormones. The situation in insects depends upon the continued presence of the juvenile hormones, which act on the cells themselves, and prevent them from maturing" (1972, pp. 10–11).

Williams and Kafatos (1972) have proposed a genetic model for the action of juvenile hormone in holometabolous insects. They envision a genome composed of three gene sets, for larval, pupal, and adult characters, respectively. These gene sets are under positive control by juvenile hormone. Juvenile hormone conveys no detailed instructions to particular differentiating cells, but acts in the turning on and off of entire gene sets. Williams and Kafatos propose that juvenile hormone acts as a corepressor by activating the repressors of pupal and adult gene sets. If juvenile hormone binds directly to these repressors, but has a lower affinity for the pupal repressor than the adult repressor, then normal development proceeds as the titer of juvenile hormone drops. At high titers, both pupal and adult gene sets are repressed; at low titers, only the adult set is repressed.

No one, least of all Williams and Kafatos, expect the eventual story to be so simple. But it does seem likely that normal development is controlled by gradually decreasing concentration of a hormone acting primarily at high levels of the regulatory system. This is also an ideal mechanism for the simple and rapid production of heterochronic effects. Any acceleration of adult characters by reduction in the titer of juvenile hormone, or extension of juvenile traits by maintenance of a high titer, represents heterochrony.[4] Since minor alterations in the concentration of a hormone can lead to substantial changes in morphology, heterochrony may play an important role in geographic variation (secretion of juvenile hormone is influenced by temperature and photoperiod, for example), polymorphism (including sex, caste, and phase) and speciation itself.

The experimental manipulation[5] of juvenile hormone levels has generated a complete range of potential heterochronies.[6] In his classical experiments on the bug *Rhodnius,* Wigglesworth (1936, 1940) produced the entire gamut of heterochronies, from precocious miniaturized adults to supernumerary larvae. *Rhodnius* initiates each molting cycle with a single blood meal. Ecdysone is released soon afterwards, and the molt is determined. Subsequent decapitation (with removal of the corpora allata) does not interfere with the progress of the molting cycle. Morphological effects can be regulated by the timing of decapitation. Wigglesworth and subsequent experimenters have obtained the following spectrum of results:

1. Precocious miniaturized adults. If the corpora allata of *Rhodnius* are removed by decapitation soon after the secretion of ecdysone in the fourth larval instar, the next molt produces a miniature adult instead of a fifth instar larva. The source of juvenile hormone has been removed before an effective amount could be released. Bounhiol (1938) removed the corpora allata of silkmoths after their second and third larval molts. He induced complete precocious metamorphosis to nymph and imago, with the suppression of one or several larval stages and a reduction in the duration of larval life by nearly one half.

2. Mixtures of juvenile and adult characters. If *Rhodnius* is decapitated later in the fourth larval instar, the next molt will yield a fifth instar larva with many adult characteristics. Similar mixtures are obtained when a decapitated fourth instar larva is joined in parabiosis to a normal fourth instar larva. The two animals possess a common circulation and they molt at the same time, with their shared concentration of juvenile hormone insufficient for the complete suppression of adult characters in either. Such mixtures are the most common outcome of experiments in the manipulation of juvenile hormone levels. The results can be particularly complex because juvenile hormone is not equally effective at all stages of larval development (it may have no influence when added either too early or too late in a molting cycle); moreover, it affects different organs at different times, rates, and concentrations. Jenkin (1970, p. 213) notes that late application often does not affect the development of wings and genitalia, while it may induce the cuticles of tergites to remain in the larval state. Sehnal and Meyer (1968, see also Sehnal and Novak, 1969) injected purified juvenile hormone from cecropia silkmoths into last instar larvae of the waxmoth *Galleria mellonella.* Injection on the seventh and eighth

days of the instar had no effect; injection one or two days earlier produced "pupal-like intermediate forms"; more larval intermediates arose from treatment on the fourth day, while "morphologically perfect superlarvae" followed injection on the third day. Implantation of extra corpora allata into first and second instar larvae of the silkmoth *Hyalophora cecropia* "can produce a restricted retention of characters of that instar at the next molt" (Staal, 1967, p. 13). Srivastava and Gilbert (1969) applied juvenile hormone to cuticles of young pupae to obtain pupal-adult intermediates in the dipteran *Sarcophaga bullata.* Riddiford (1972) could not induce supernumerary larvae with application of juvenile hormone to fifth instar larvae of cecropia silkmoths, but she did obtain pupae with larval integumentary structures (see also Riddiford, 1975). Willis (1974, p. 109) cites several studies in which adult epidermis, forced to synthesize new cuticle in the presence of juvenile hormone, produced a structure with some features of earlier stages.

3. Supernumerary larvae. If fourth instar larvae of *Rhodnius* are joined in parabiosis to third instar larvae with their higher titer of juvenile hormone, the fourth instar larva molts to a form no more advanced than its previous state. If extra corpora allata are transplanted into fifth instar larvae (which should metamorphose into adults), perfect sixth and even seventh instar supernumerary larvae can be produced. Ilan et al. (1972) applied juvenile hormone to first day pupae of *Tenebrio molitor* and obtained a second pupal molt eight days later. Riddiford (1970) treated eggs of the bugs *Pyrrhocoris apterus* and *Oncopeltus fasciatus* with juvenile hormone analogs four weeks prior to their expected metamorphosis to the adult state. She obtained one or more supernumerary molts to form giant larvae which died without further metamorphosis.

This continuum of effects has special relevance for one classical issue in the study of heterochrony. The distinction between complete and partial paedomorphosis has often been regarded as fundamental (and worthy of separate names, to confuse the literature further). Since the common mechanism of control by juvenile hormones can produce complete paedomorphs (supernumerary larvae) and partial paedomorphs (adults with larval and pupal characters) as graded responses to a continuum in time and intensity of experimental disturbance, I do not regard this distinction as important for discussions of heterochronic mechanisms (though complete and partial paedomorphosis may have different adaptive significances).

The most crucial experiments for heterochrony are those that mimic natural occurrences rather than produce developmental anomalies. Wigglesworth (1954, pp. 70–71) submitted fourth-stage larvae of *Rhodnius* to different temperatures. Those raised at low temperatures molted to fifth-stage larvae more "juvenile" than normal, while larvae grown at high temperatures were more "adult" than normal in their fifth stage. Wigglesworth argues that low temperatures slightly accelerate the action of juvenile hormone, while high temperatures depress it. Long photoperiods also tend to increase activity of the corpora allata. Arctic and high mountain insects often have reduced wings and other juvenile characters as adults. Matsuda (in press) implicates low temperatures and long days as possible causes of this paedomorphosis through their effect on juvenile hormones.

Other experiments point to hormonal control of paedomorphic morphs, castes, and phases in various insects. Apterous (wingless), parthenogenetic aphids are clearly paedomorphic (see pp. 308–310). Hille Ris Lambers (1966) applied juvenile hormone to third instar alatoid female larvae of *Megoura viciae* and obtained apterous-alate intermediates as adults. Solitaria phase locusts are juvenilized as adults compared with conspecifics of the swarming gregaria phase (see pp. 312–319). If extra corpora allata are transplanted into gregaria nymphs, the resulting adults are more like solitaria in appearance. Lebrun (1970) implanted corpora allata of the cockroach *Periplaneta* into nymphs of the termite *Calotermes flavicollis* destined to become adults. Instead, they molted to sexually mature intercastes between soldiers and adults, with reduced wings and other signs of paedomorphosis. Thus, in several nongenetic insect polymorphisms, the paedomorph reflects an enhanced action of juvenile hormones.

The invocation of a similar mechanism for genetic elaboration of species differences is an extrapolation, but perhaps not an unreasonable one. Johnson and Birks (1960) argue that the ancestors of aphids were alates (winged) and that all modern parthenogenetic aphids begin their development as presumptive alatae (Mittler, 1973). If so, then the evolution of genetic potential for apterous development is true progenesis, since apterae are accelerated in speed of ontogeny relative to alatae (see pp. 308–309). Kennedy (1956) regards solitaria locusts as an ancestral form. If so, then gregaria phase locusts are "superadults" evolved by acceleration (not hypermorphosis, since gregaria develop more rapidly and in fewer molts than solitaria). Novak has written:

> [Paedomorphosis] is also the mechanism of origin of wingless, larva-like females in various insect groups such as Lepidoptera (Psychidae, Geometridae, etc.), Coleoptera (Lampyridae), and Strepsiptera. It appears to be one of the commonest and simplest mechanisms of the origin of new species in insects. It may be explained as a small genetically fixed deviation in the quantity of the

juvenile hormone as the result of which different further phylogenetic changes have arisen secondarily. (1966, p. 135)

Finally, Costlow (1968, p. 37) has suggested that gigantism and neoteny in the larvae of some crustaceans may be elucidated by experiments on the extirpation of eyestalks in other species. The hormonal control of development in crustaceans is poorly known, but molting is initiated by a hormone released from the Y-organ following a decrease in concentration of molt-inhibiting hormone secreted by the X-organ–sinus gland complex of the eyestalks. If both eyestalks are removed prior to the third day in third-stage zoeal larvae of the mud crab *Rhithropanopeus harrisii,* one or two supernumerary zoeal stages may be interpolated before metamorphosis to the first crab stage (Costlow, 1966).

Amphibian Paedomorphosis and the Thyroid Gland

The endocrine control of paedomorphosis in salamanders has been recognized ever since Gudernatsch demonstrated the thyroid control of amphibian metamorphosis in 1912. In one early experiment, for example, Hoskins and Hoskins (1919) removed the thyroid anlage before its differentiation began and obtained giant larvae (two to three times the size of controls) that never metamorphosed, in both *Ambystoma punctatum* and *Rana sylvatica.*[7]

Some early and naive hopes for a simple mechanism of paedomorphosis, and a science of rejuvenation for the aged, succumbed to the complexities of thyroid response by paedomorphic salamanders. These salamanders form a graded series, from species with rare neotenics, through paedomorphic populations transformable only under unusual laboratory conditions, to perennibranchiate salamanders that have never been induced to metamorphose. This structural series is matched by decreasing sensitivity to thyroid hormones. The series is purely structural and not evolutionary (the causes of facultative neoteny need not be viewed as first steps in the evolution of perennibranchiates). But the continuity of causes does argue for an evolutionary initiation of paedomorphosis as a rapid (and probably facultative) response to immediate demands of the environment. The series includes:

1. Facultative paedomorphs. Lynn and Wachowski (1951), Brunst (1955), and Kollross (1961) reviewed earlier work on successful induction of metamorphosis by thyroxin and various organic iodines in facultative paedomorphs of several *Ambystoma* species.[8] Neotenic *A. tigrinum* populations from cold Rocky Mountain lakes can be induced

to transform simply by increased temperature in the laboratory (Jenkin, 1970). The correlation holds well in nature, since populations from warmer lakes usually do metamorphose; moreover, in Pleistocene *A. tigrinum* from the Kansas-Oklahoma area, glacial specimens are giant and neotenic, whereas interglacial forms are normal and metamorphosing (Tihen, 1955). Snyder (1956) induced metamorphosis in neotenic *A. gracile* from cold ponds around Mount Rainier. In all these cases, as for the axolotl of *A. mexicanum,* metamorphosis is not suppressed by any deficiency of the thyroid gland itself; for it is normal in morphology and potentially capable of secreting its hormone in amounts sufficient for transformation (Dent, 1968). The thyroid gland can function autonomously at a low level, but higher levels require activation by thyrotropin (also called thyroid-stimulating hormone, or TSH) produced by the anterior lobe of the pituitary (Etkin, 1968). TSH is itself regulated by thyrotropin releasing factor (TRF) produced by the hypothalamus. Jenkin (1970) suggests that the failure of *A. tigrinum* to transform at low temperatures involves a retardation in production of TRF and TSH; these hormones only reach their threshold for action when the temperature is raised. Prahlad and DeLanney (1965) and Norris et al. (1973) showed that axolotls (*A. mexicanum*) do not release sufficient TSH to activate their own fully potent thyroid hormones.

2. Paedomorphic species that can be induced to transform. Kezer (1952) used exogenous thyroxin to induce metamorphosis in two species of *Eurycea* that never transform in nature. The pituitary of *Gyrinophilus palleucus* produces insufficient TSH to stimulate its normal and potentially active thyroid. Transformation is unknown in nature, but Dent et al. (1955) induced it with both thyroxin and TSH treatments. Dent (1968) reports some partial successes (that might have led to full metamorphosis, had not the patient died) in the cavernicolous salamanders *Typhlomolge* and *Haideotriton.*

3. Permanent paedomorphs. In the first flush of thyroxin's success, just after World War I, the scientific world waited impatiently to see the long-lost adult of *Proteus* and *Necturus.* But nature won again and, fifty years later, remains as recalcitrant as ever. Again there seems to be nothing wrong with the thyroid of these "perennibranchiates" (for many years it was believed that, alone among vertebrates, *Typhlomolge rathbuni* lacked a thyroid, but this turned out to be false—Dent, 1968, p. 285). In fact, the thyroids of *Proteus* and *Necturus* produce typical hormones that induce metamorphosis in the larvae of other species (Lynn, 1961). Minor results in the "right" direction can be obtained with very strong solutions of thyroxin, but nothing approaching transformation has ever been achieved. Gills

may atrophy (*Necturus*) or suffer some reduction (*Siren* and *Pseudobranchus*) or resorption (*Proteus*). *Necturus* may shed its skin and protrude its eyes (Lynn and Wachowski, 1951). The skin of *Proteus* does not respond to thyroxin on its own body, but it can be induced to undergo partial transformation when it is transplanted to metamorphosing salamanders of other species (Dent, 1968, p. 281). Still, in all perennibranchiates, most larval features are entirely unaffected by thyroxin. The tissues have apparently lost their capacity to respond to a thyroid stimulus, and paedomorphosis is truly permanent.

Heterochronic effects of thyroid action are by no means confined to the Amphibia; they are merely easier to identify and manipulate when a discontinuous metamorphosis characterizes ontogeny. Scow and Simpson (1945) noted a marked retardation in growth and maturation of rats thyroidectomized at birth. The features most affected included secondary ossification centers, transformation of hair from infant to adult type, opening of eyes, and eruption of teeth. At 89 true days, one rat had a skeletal age of only 18 days.

Khamsi and Eayrs (1966) have established among vertebrates a continuum of heterochronic effects very similar to those associated with juvenile hormone in insects (though the action is opposite, since thyroxin is a stimulator of adult traits, while juvenile hormone is a repressor). Juvenile hormone effects range from miniaturized adults to supernumerary larvae. Very high levels of added thyroxin can accelerate maturation and retard growth by raising the metabolic rate above what available nutrients can supply; dwarfs with perfect adult proportions may be formed. Khamsi and Eayrs obtained intermediate results with lowered levels of supplementary thyroxin. In feeding thyroid hormones to newborn rats, they accelerated maturation at a rate exponentially related to the size of the dose, but did not affect growth except at the highest levels. Scow and Simpson's study represents the intermediate case of thyroid deficiency—an incomplete juvenilization. More severe deficiency may lead to complete juvenilization, as in neotenic *Ambystoma.*

Thus, two divergent systems of control in two very different animal groups have common features that emphasize the potential ease and rapidity of heterochronic change as an ecological response.[9] Marked heterochronic effects can arise by simple experimental alterations in the endocrine system. Mixtures of juvenile and adult characters are intermediate stages in a continuum from precocious miniaturized adults to permanent larvae. The common distinction between total and partial paedomorphosis need not be maintained as a fundamental difference in mechanism; complex partial effects may have a simple cause. (Complete paedomorphosis has often been depicted as a simple genetic, or even nongenetic, response; and partial paedomorphosis as a complex and slowly developed adaptation with many characters under separate genetic control.) Experiments often mimic natural situations and suggest an equally direct control for heterochronies observed in geographic variation and polymorphism. Genetic differences in evolutionary sequences are often elaborated by extrapolation of the basic mechanism underlying more immediate ecological responses (perennibranchiate tissues lose sensitivity to thyroxin; facultative paedomorphs of *Ambystoma* have merely developed a way of preventing thyroxin from reaching their transformable tissue). Since heterochrony can arise rapidly and "easily" by an alteration in endocrine balance,[10] it seems reasonable to consider even large paedomorphic changes in terms of their immediate significance for evolution of life-history strategies in differing ecological circumstances.

—9—

Progenesis and Neoteny

In this chapter, I plan to test the hypothesis that progenesis and neoteny—despite their common consequence of paedomorphosis—evolve as adaptations to strikingly different ecological conditions: progenesis, with its acceleration of maturity, is a life-history strategy for *r* selection; while neoteny tends to evolve in the stable, favorable, and "crowded" situations that specify *K* selection.

For a primary test, I shall contrast the most famous and unambiguous cases of paedomorphosis in modern organisms: progenesis in various insects (including wingless aphids and parthenogenetic larval gall-midges), and neoteny in ambystomatid salamanders. These cases are especially favorable for several reasons:

1. Their fame has led to an extensive, recent literature.
2. Heterochrony can be induced experimentally by disturbing an endocrine mechanism that probably regulates the occurrence of paedomorphosis in nature as well.
3. Both involve a profound metamorphosis in ontogeny, providing good markers for the diagnosis of heterochrony.
4. Paedomorphosis is not genetically fixed in either case. It is one of several ontogenetic paths potentially available to each member of a population. Its regulation by the immediate environment can be easily tested. Moreover, in deciding whether a paedomorphic form is progenetic or neotenic, comparisons can be made with genetically similar, conspecific, normal forms, not with inferred generation times in hypothetical ancestors.

Insect Progenesis

Prothetely and Metathetely

Entomologists generally make a distinction between two types of juvenilization in adult insects. They speak of *prothetely* when the adult develops too "early" (perhaps at a precocious molt), leaving some characters in a larval state. When development proceeds at its normal pace, but certain characters retain their juvenile form in the imago, *metathetely* has occurred (Singh-Pruthi, 1924; Wigglesworth, 1954; Novak, 1966; Matsuda, in press). This classification duplicates the distinction between progenesis (prothetely) and neoteny (metathetely).

Before the elucidation of endocrine control for metamorphosis, entomologists often denied the validity of such a separation, arguing that prothetely and metathetely merely represent the end points of a continuum (which, of course, they do) and that the distinction had no significance either in mechanism or adaptive meaning (just as students of other animal groups often denied a meaningful distinction between progenesis and neoteny).

The demonstration of potentially different mechanisms for the two modes of juvenilization has verified the importance of separating prothetely and metathetely. Southwood (1961) has pointed out that brachyptery (short-wingedness) is often a juvenile trait, especially when it represents part of a polymorphism or spectrum of geographic variability within a species. It may arise in two distinct ways: by excessive influence of the juvenile hormone, leading to juvenile characters in the adult (metathetely); or by depression of juvenile hormone levels, leading to adult characters in the larva (prothetely), perhaps by suppression of a molt (prothoracic glands are often lost in the adult; if a pronounced drop in juvenile hormone induces the adult state at an early molt, later molts are usually suppressed). Southwood discusses

> the problem of distinguishing between the former, production of an adult with larval characters, which could perhaps be called neoteny, and the latter, a larva with adult characters, which is a condition homologous with paedogenesis [= progenesis] . . . Such a differentiation may appear artificial, but it does have the important distinction, in the present theory, that the former results from a lengthening of the influence of the juvenile hormone and the latter from a reduction in this period. (1961, pp. 63–64)

If progenesis and neoteny usually have such different efficient causes, then my hypothesis of their separate status is supported.

Southwood (1961) uses instar number to test his hypothesis of different causes for brachyptery. For geographic variation within species, he finds a correlation between metathetelous (neotenic) brachyptery and cold temperatures and between prothetelic (progenetic) brachyptery and warm temperatures. High-altitude brachyptery in various Heteroptera is metathetelous (same number of instars as normal-winged lowland populations of the same species). In other Heteroptera (see review in Matsuda, in press), brachyptery associated with high temperatures is prothetelic. *Dolichonabis limbatus,* for example, has four instead of the usual five larval instars in brachypterous lowland populations, but a normal-winged morph with the usual number of instars inhabits cold, mountainous regions. Wigglesworth (1952, 1954) implicated temperature as a determinant of experimental prothetely and metathetely in *Rhodnius* (Fig. 51); he also argued that juvenile hormone activity is enhanced by low temperatures. Southwood's natural experiment therefore has support from the laboratory.

Attractive as Southwood's hypothesis may be, it will not explain most of the insect paedomorphs discussed in this section. Solitaria lo-

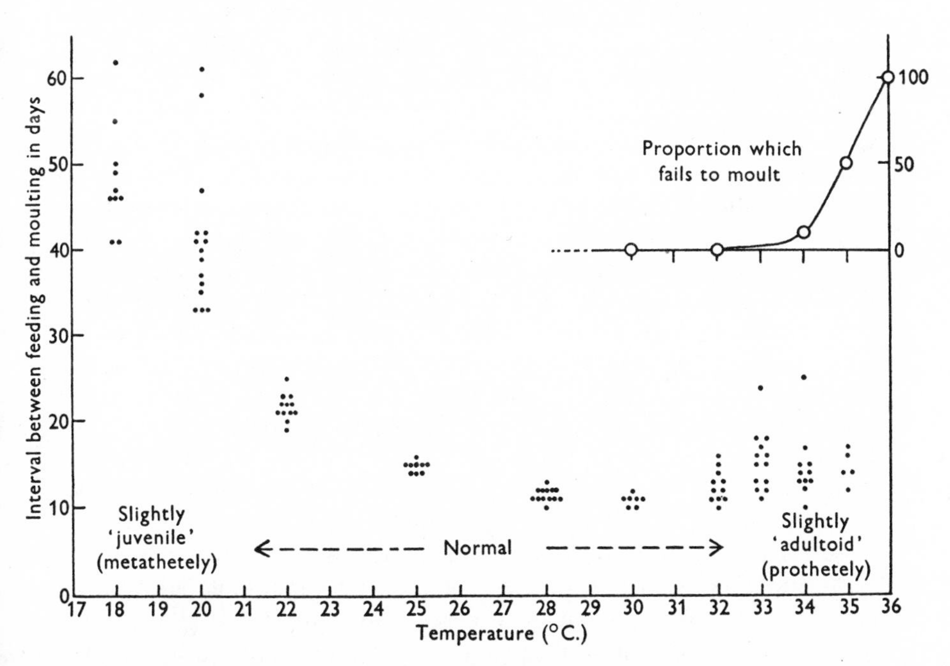

Fig. 51. Effect of temperature on molting of fourth stage larvae of *Rhodnius.* Molting is delayed at low temperatures with resulting metathetely (juvenilization). High temperatures speed development and produce a slight prothetely. (From Wigglesworth, 1954.)

custs may be metathetelous in Southwood's sense (enhanced juvenile hormone for paedomorphosis with an extended period of development). But progenetic gall midges and aphids are not prothetelic; they are not precocious adults that happen to retain some larval characters. They are, at least for gall midges, true larvae with completed sexual maturity. Their precocious maturation is not a standard effect of juvenile hormone, for it involves none of the increased somatic differentiation that accompanies the precocious development of adults by lowered levels of juvenile hormone. As larvae, these gall midges maintain high titers of juvenile hormone. Their parthenogenesis is a true disruption of the normal correlation between somatic differentiation and reproductive maturation.

Paedogenesis (Parthenogenetic Progenesis) in Gall Midges and Beetles

The last of von Baer's contributions to the study of heterochrony (1866) may seem trifling compared with his earlier accomplishments: he gave the name *paedogenesis* to a remarkable method of parthenogenetic reproduction in gall midges (Diptera, family Cecidomyiidae). These insects can reproduce by the normal route, with a full set of larval and pupal molts culminating in the production of winged, ovipositing imagos. But they can also propagate as larvae by viviparous parthenogenesis: larvae of the next generation grow *within* the body of their larval mother and devour her tissues (previously liquified for them by maternal histolysis). When the larvae emerge, already prepared to rear the next generation, their own parent is often no more than a hollow shell. Greater love hath no woman.

Paedogenesis has arisen in at least four distinct groups of the Cecidomyiidae (Wyatt, 1967, p. 95). There can be no doubt that these larval paedomorphs are progenetic rather than neotenic, for their development is markedly accelerated relative to the sexual phase of their life cycle. The paedogenetic cycle of *Miastor metraloas* includes one larval instar with a single molt to the reproductive hemipupa. Up to 60 young may be reared by a single mother in 12 days. The sexual cycle, by contrast, includes two larval and one pupal molt before the emergence of a sexual imago (Wyatt, 1967). *Mycophila speyeri* undergoes only one molt and reproduces as a true larva; up to 38 offspring may emerge in a cycle of only five days (although 19 young in six days is the average). The sexual adults require two weeks to develop (Wyatt, 1964). In laboratory cultures of the same species, Ulrich et al. (1972) obtained up to 47 offspring from a single larval mother in four days.

These progenetic larvae maintain a phenomenal capacity for rapid increases in population (high *r*). Populations of *Mycophila speyeri* introduced to mushroom beds may increase exponentially for five weeks, reaching a density of 20,000 reproductive larvae per square foot (Wyatt, 1964). This is an *r* strategy with a vengeance. The sexual phase is induced either by crowding (Harris, 1925) or by exhaustion of available food. Nikolei (1961) produced 250 consecutive paedogenetic generations by supplying enough food and preventing crowding. Glycogen derived from fungal hyphae is required as a source of energy in larval reproduction (Kaiser, 1969). Larvae destined to become imagos have little glycogen in their fat bodies. The sequence of changes in response to decreasing food has been studied by Ulrich et al. (1972) in *Mycophila speyeri.* Paedogenetic females are of three distinct types: female-producing, male-producing, and both male- and female-producing. Paedogenetic mothers with abundant food generate all-female broods in four to five days. With less food, male-producing and male-female–producing larvae develop as well. If female larvae are given no food, they grow into imagos.

Cecidomyian ecology clearly dictates an *r* strategy, and paedogenesis just as clearly provides it. The paedogenetic species are fungal feeders; they are often very troublesome pests in commercial mushroom beds because they can increase so rapidly following colonization. Their food is ephemeral in time and patchy in distribution. When present, it is usually superabundant (one mushroom can feed quite a clone of these tiny flies). The flying, sexual phase must find the resource, but one can hardly imagine a nutritionally richer environment more free from intraspecific competition than a pristine mushroom. The paedogenetic phase uses the resource to build an enormous population. High *r* is not a result of impressive fecundity (20–40 individuals per mother is trifling compared with many egg-laying forms), but of extremely rapid generation time due to the acceleration of maturation by progenesis. As expanding populations exhaust a resource, the exploratory sexual phase is generated automatically. Since *r* selection usually predominates in the expanding phase of fluctuating populations, paedogenesis must be interpreted as a consummate *r* strategy for life histories.

Little is known of the hormonal control of paedogenesis. Kaiser (1972) reports that the corpora allata are larger in larvae destined to reproduce by paedogenesis than in those destined to form imagos. Given the potential capacity for egg formation within the larva (an adaptation not related to the conventional action of juvenile hormone), the gonadotrophic role of juvenile hormone in the maturation of oocytes may influence whatever mechanism switches a larva toward paedogenesis or toward becoming a sexual imago.

Although paedogenesis seems to be such a highly peculiar and specific adaptation, it nonetheless recurs with very little difference in a group totally unrelated to the Cecidomyiidae, the beetle *Micromalthus debilis* (Scott, 1938, 1941). *Micromalthus* also has three paedogenetic forms (female-producing, male-producing, and male- and female-producing). (In an interesting variation that might intrigue Gilbert's Mikado, the male-producer gives birth to a single offspring, though several may begin to develop. After adhering to his mother's cuticle for four to five days, the larva inserts his head into her genital aperture and devours her.) The paedogenetic forms live in wet, rotting wood. If the wood dries out, they become sexual and seek a new resource.

Progenesis in Wingless, Parthenogenetic Aphids

Charles Bonnet, the first hero of this book, discovered parthenogenesis in aphids in 1746. Female monoecious aphids are either sexual (oviparae) or parthenogenetic (virginoparae). The parthenogenetic forms are either winged (alatae) or wingless (apterae). Wingless virginoparae have been identified as paedomorphs by several authors (Wigglesworth, 1966, p. 205; Johnson, 1959, p. 96; Mittler, 1973, p. 71; Kennedy and Stroyan, 1959). This assessment is based not only on the absence of wings, but on the coordinated retention of a large suite of juvenile characters.

The potential for rapid increase in population size among parthenogenetic forms is legendary. Again, this is not a function of individual fecundity (few mothers produce more than 100 offspring), but of extremely rapid maturation. Aphids have developed an extreme form of viviparity and ovulation is more a juvenile than an adult function (though aphids do not give birth before they become adult). Embryonic development actually begins in a mother's body before her own birth, and two subsequent generations may be telescoped within each "grandmother." Kennedy and Stroyan write: "Individual fecundity of the virginopara is modest . . . It is the telescoping of the generations which gives aphids their unequalled rates of multiplication and the oft-quoted astronomical totals of theoretical progeny from one female in one year—such as 524 billion for *Aphis fabae* Scopoli" (1959, p. 140).

Needless to say, parthenogenetic aphids have more rapid generations than sexual forms. Photoperiod seems to be the dominant influence in switching to sexual cycles. In most temperate aphids, the shortening of days induces the production of oviparae, which lay eggs that may overwinter until the spring flush of vegetation.

A more interesting comparison can be made between winged and wingless virginoparae. The wingless forms are paedomorphic (Fig. 52). Since wingless forms develop more rapidly than winged forms, this paedomorphosis is progenetic rather than neotenic. Kennedy and Stroyan (1959, p. 147) state that wingless virginoparae are generally more fecund, have a shorter generation time, and feed more efficiently than winged virginoparae (see also Mittler, 1973, p. 71). The winged forms, they argue, must develop a complex and efficient sensorimotor apparatus (an adult character); this takes more time and requires a sacrifice of reproductive and feeding efficiency. Johnson (1959, p. 85) studied the comparative development of winged and wingless morphs in *Aphis craccivora* on healthy bean plants at 20°C. Both forms have four nymphal instars, but winged forms spend one more day in the last instar. Wingless morphs reach their fertile adult state in seven days, winged in eight days. Virginoparae of *Drepanosiphum dixoni* develop two winged phases: brachypterous (short-winged) and macropterous. The brachypterous forms correspond morphologically with feeding wingless morphs of other species; macropterous forms arise under conditions of crowding. Dixon (1972,

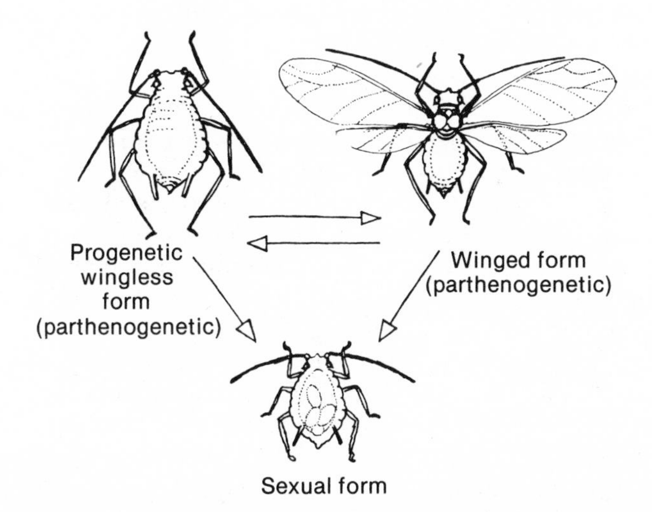

Fig. 52. Polymorphism in aphids. A paedomorphic, apterous (wingless) virginopara (top left), compared with an alate (winged) virginopara (top right). Crowding induces the transformation from wingless to winged, while falling day length induces the sexual phase (bottom). (From Wigglesworth, 1966.)

pp. 463–464) reports a greater reproductive rate for brachypterous forms.

The ecological strategy of wingless versus winged virginoparae seems to mirror precisely the contrast between paedogenetic and sexual gall midges. The progenetic wingless forms exploit superabundant, ephemeral resources and build up enormous populations by maintaining a very high *r* due to rapid turnover of generations rather than high individual fecundity. The paedomorphic morphology is well adapted for feeding and reproduction; energy need not be invested in producing an "expensive" sensorimotor system for mobility. Aphids are "parasitic" sap feeders; they tap a plant's own nutrient stream by attaching to the phloem sieve tubes (Kennedy and Stroyan, 1959).

The uncolonized leaf, like the unexploited mushroom, is a classical environment for *r* selection—a superabundant resource engendering little intraspecific competition for the expanding phase of a fluctuating population. Aphids respond by producing progenetic wingless forms with accelerated maturation. The shift to winged forms is not usually invoked by dwindling food, but by its indirect indicator—tactile stimulation following crowding (Lees, 1966; Johnson, 1974); brief encounters of only a minute's duration often suffice (Johnson, 1965). (The isolation of a crowding effect in laboratory experiments does not exhaust the complexity of nature. The tactile response in crowding may be a primary stimulus, but dwindling food can provoke contact between aphids in many ways not directly related to overwhelming density [for example, by increasing the "restlessness" of aphids—Way, 1973, p. 80].) A few experiments have also indicated that food has an effect independent of density (Mittler, 1973), though Way (1973, p. 80) deems it "of minor ecological importance." (As in most surveys of natural history, there are also a few annoying counter-cases—aphids that seem to shift to wingless forms with dwindling food, for example.) As Kennedy and Stroyan write: "Because the exceptionally favorable food situation for sap feeders is short-lived in any plant, the success of aphids depends on the timely production of forms less degenerate than the apterous vivipara" (1959, p. 146).

Lees (1966) has noted that juvenile hormone suppresses the growth of wings—a further indication that the wingless morph represents a juvenile condition. If juvenile hormone is applied to winged nymphs of *Myzus persicae,* the adult wings are small and other wingless characters are expressed (Johnson, 1974). The corpora allata of *Brevicoryne brassicae* are smaller during development in winged than wingless forms.

Additional Cases of Progenesis with a Similar Ecological Basis

In my cursory examination of an immense and unfamiliar literature, I have been impressed with the common association of facultative progenesis and *r* selection for the exploitation of ephemeral and superabundant resources. The progenetic phase can be switched off by an automatic response to crowding or dwindling resources. Mobile forms are then produced for the discovery of new resources, although the development of complex sensorimotor systems requires a longer generation time, often with an increased number of molts. This pattern arises again and again in completely unrelated groups, including the following:

1. The mite *Siteroptes graminum* is also a fungal feeder (Rack, 1972). When food is superabundant, it becomes sexually mature at an early stage (Fig. 53). Moreover, the progenetics are "born" sexually mature since larval stages develop in the mother's body. As resources

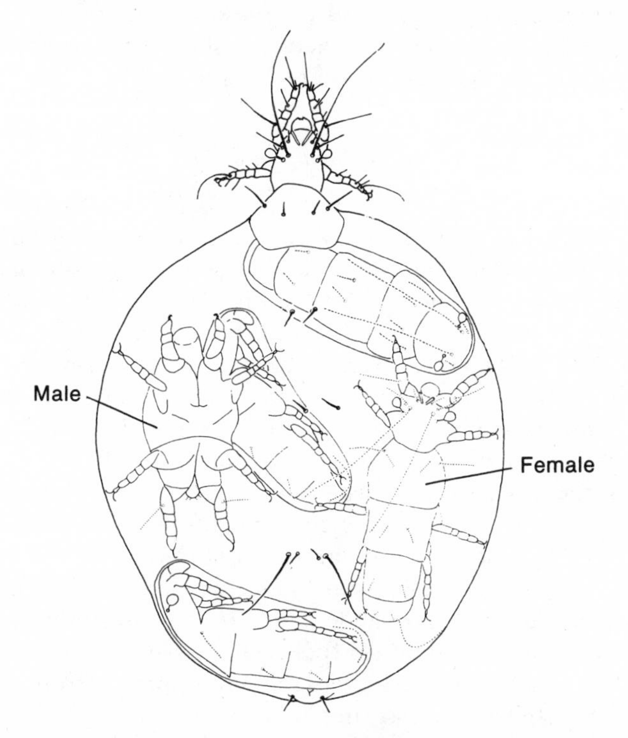

Fig. 53. Progenesis in the mite *Siteroptes graminum.* A sexually mature female with one male and four females growing within her body. (From Rack, 1972.)

dwindle, progenetic reproduction is abandoned and the regular sexual cycle is resumed, with larvae molting to normal adults. Although this normal cycle omits the nymphal stages, Rack (p. 164) asserts that the progenetic cycle is much shorter and that progenetic forms are more fecund than normal adults.

2. Cousin (1938) studied the entomophagous chalcid wasp *Melittobia chalybii.* The typical form takes 90 days to metamorphose into a normal, winged adult with a long life span. But a "second form" develops in 14 days and has a short imaginal life. It is progenetic, with small wings and hypertrophied genitals functional right at eclosion. Eggs deposited on a nonparasitized host develop into second-form progenetics. Eggs from the same source deposited on a host already infected with *Melittobia* develop into the typical form.

3. When uncrowded on abundant rice plants, the plant hopper *Nilaparvata lugens* develops into a brachypterous (paedomorphic) feeding form. If the plant wilts or becomes too crowded, a normal-winged imago develops. The brachypterous imago is flightless; the normal-winged form flies to find new resources. The brachypterous imago is progenetic rather than neotenic since it has a shorter larval life than the normal-winged form (Kisimoto, 1956).

Neotenic Solitary Locusts: Are They an Exception to the Rule?

Many species of locusts exhibit a continuous polymorphism reflecting their alternating life styles as solitary grasshoppers and migratory swarmers. The end forms, with obvious reference to their habits, have been named the solitaria and gregaria phases (Key, 1950; Kennedy, 1956; Uvarov, 1961; Dempster, 1963, for reviews). Of the transition to phase gregaria, Dempster writes: "Phase changes are complex and affect many characters such as size, shape, color, rate of development, behavior and fecundity; but basically they reflect a change in the locust's metabolism to favor greater mobility" (1963, p. 513). Gregaria locusts migrate in daylight hours as nymphs and adults; in comparison with solitaria locusts (Kennedy, 1956; Dempster, 1963), they are smaller and longer winged, have higher metabolic rates, eat more but retain less for growth and storage, and contain less water but more fat (to fuel their flight).

Kennedy (1956) argued, with much support from previous workers, that solitaria locusts should be viewed as a juvenilized phase, gregaria locusts as accentuated adults. He based his arguments on a large suite of characters, not just the larger wings of the gregaria phase; his conclusions have been generally accepted.

The paedomorphosis of solitaria locusts has been affirmed by all available information on the role of juvenile hormone in locust phases. Doane (1973) reviewed a series of experiments illustrating the role of enhanced juvenile hormone in the retention of larval characters by solitaria locusts. Joly and Joly (1953) implanted extra corpora allata into gregaria nymphs in simulated gregaria environments and obtained adult phenotypes nearer to solitaria in both pigmentation and biometrical indices; controls remained in typical gregaria phase. These results have been affirmed in many later studies (Joly, 1958; Novak, 1966, p. 93; Doane, 1973). Carlisle and Ellis (1959) noted that the ventral cephalic glands (presumed homologs of the lepidopteran prothoracic glands) disappear in gregaria adults but not in solitaria adults of the desert (*Schistocerca gregaria*) and migratory (*Locusta migratoria migratoroides*) locusts. These glands generally atrophy in holometabolous insects after the disappearance of juvenile hormone determines the adult molt. Their maintenance in solitaria adults implies a continued role for juvenile hormone in these paedomorphs.

The mechanism for transition from solitaria to gregaria has been discussed extensively, and a consensus finally seems to be emerging. Crowding is the primary determinant; locusts will remain in phase solitaria until their density reaches a critical value. The transition extends over several generations, but its direction is well marked from the beginning. The transitional generations are often termed "phase transiens" (Albrecht, 1962). The chemical determinant seems to be a gregarization pheremone secreted in the feces of both solitaria and gregaria nymphs. Nolte et al. write: "If the pheremone is secreted in the feces it must be in liquid form in the gut and could perform its function efficiently only for a congregated mass of hoppers where an accumulation of the pheremone would result in raising it above the threshold value; with solitaries the value would not be reached" (1970, p. 470). The pheremone induces transformation, but it does not cause initial crowding. The impetus for initial crowding has not been ascertained with satisfaction, nor has it been widely discussed of late. To understand the relationship of locust phases to general theories of life-history strategy, we will have to know whether initial crowding arises from high intrinsic rates of increase in solitaria locusts or from restriction and deterioration of solitaria habitats. In any case, once the phenotypic transformation has occurred, the change in habit is set as well; for solitaria locusts avoid conspecifics, while gregaria seek them. Gregaria locusts reared in crowded conditions produce more of an attracting pheremone than those reared solitary (Norris, 1970).

So far, we seem to have a familiar pattern. Solitaria locusts are paedomorphic; they represent an efficient feeding phase; their transformation to a mobile morph is induced by crowding, a sure sign of dwindling resources. By all rights, the solitaria locusts should be *r* selected and progenetic. In fact, they are neotenic. Kennedy (1961) has summarized the differences between phases in habitats and life histories. The relevant distinctions include:

1. Gregaria eggs of *Schistocerca* hatch sooner.
2. In *Locustana,* egg diapause is much less frequent for gregaria parents.
3. Gregaria larvae are up to several times heavier than solitaria locusts at hatching; they survive longer when starved.
4. Gregaria larvae complete their postembryonic development more rapidly than solitaria locusts; they hatch in a more differentiated state and their life cycle usually includes one fewer molt (Albrecht and Blackith, 1957).
5. The maturation of gregaria adults is more rapid. Mature males of gregaria locusts secrete a pheremone that accelerates the maturation of newly ecdysed adults in both sexes (Loher, 1960; Norris, 1970). Flight accelerates maturation as well—another impetus to more rapid maturation of gregaria adults (Highnam and Haskell, 1964). (*Locusta migratoria migratoroides* is an exception because crowding seems to inhibit maturation; but Highnam and Haskell [1964] believe that the accelerating effect of flight outweighs this inhibition.)
6. An imaginal diapause induced by changing day-lengths is prevented by crowding in *Schistocerca* and induced by isolation in *Nomadacris.*
7. Solitaria locusts live longer.
8. Gregaria locusts are less fecund with fewer ovarioles in each ovary, a smaller proportion of functioning ovarioles in each ovarian cycle, and fewer ovarian cycles. Solitaria locusts may lay two to four times as many eggs as gregaria (though combined hatchling weight may be the same, because gregaria embryos are larger).

Solitaria locusts are more fecund, as an hypothesis of *r* selection might predict. But they are slower developing in all respects—eggs take longer to hatch; larval development is extended in time and frequency of molting; maturation of adults is delayed; length of life is increased; tendency to diapause is enhanced. In short, solitaria locusts clearly achieve their paedomorphosis by neoteny, not by progenesis. If the solitaria phase is *r* selected, as its more sedentary role in feeding might suggest, then our general hypothesis is in trouble. And a true exception it may well be, though two plausible rescues seem to be available with a minimum of special pleading:

1. Are the fecundity and development of gregaria locusts so

strongly determined by their highly peculiar and specialized mode of life that the framework of *r-K* theory does not apply to them? For paedogenetic midges and parthenogenetic aphids, there is reason to believe that the winged, mobile form is a primitive phyletic state, and that progenetic morphs have evolved for rapid maturation in *r*-selected parts of the habitat. But the situation, according to most entomologists, is reversed for locusts: the solitaria grasshopper seems to be the ancestral morph. We might better explain the longer life cycle of solitaria not by asking why its development has been retarded, but by accepting it as the primitive condition and asking why gregaria locusts are accelerated.

The reduced fecundity of gregaria locusts probably represents a necessary trade-off for the construction of such an efficient flying machine. Albrecht et al. favor such an explanation and speak of the "price for mobility" (1958, p. 410). In thinking that solitaria locusts might be *r* selected for increased fecundity, we may be approaching the problem in the wrong way. I suggest that gregaria locusts are structurally constrained to reduce their fecundity as an adaptation for sustained flight.

As for accelerated development, it may arise in part as a by-product of the need for synchronized reproduction in swarms. The maturation pheremone not only accelerates reproduction in *Schistocerca gregaria,* it also (or perhaps primarily) serves to synchronize maturation in swarming populations (Strong, 1970; Nolte et al., 1970). (I do not know if the most efficient device for synchronization also involves a general acceleration; this would be a good subject for model building.) In addition, accelerated development may result from a need to produce larvae already rather advanced at hatching for "greater stamina and foraging capacity in their often more difficult environment" (Kennedy, 1961, pp. 82–83). Moreover, accelerated maturation may be, in part, an adaptation to maintain sufficient capacity for population increase, given a structurally constrained reduction in fecundity. All in all, gregaria locusts are so derived and specialized for a highly peculiar mode of life that their accelerated maturation may be, in itself, a specialization quite unrelated to the generality of *r* and *K* selection. Finally, this capacity for more rapid development is not always realized in swarming gregaria locusts. Unfavorable environments may impose delays. Waloff (1962) states that *Schistocerca gregaria* roams over great areas with poor, erratic, and seasonal rainfall, and a variety of temperature regimes. Breeding is linked with rainfall and new adults may have to migrate thousands of miles before they find favorable sites for maturation.

2. Are solitaria locusts like, or unlike, the progenetic, feeding

morphs of midges and aphids in general ecology? Do they, in fact, have a higher *r* than gregaria locusts?

Almost all recent literature has assumed (without documentation) that solitaria locusts must have superior capacity for rapid increase in population size, because of their higher fecundity. No one has considered the effect of precocious maturation in gregaria locusts. Since speed of maturation is usually more important than fecundity in determining *r*, it is possible that gregaria locusts usually have higher *r*, or that solitaria fecundity matches gregaria precocity to yield no general differences in *r* between the phases. I have not found data that would permit a quantitative assessment. The increased fecundity of solitaria locusts provides no argument for assuming that they represent an *r*-selected phase.

In an early version of phase theory, an *r*-selected status for solitaria could not be avoided—for solitaria was seen as a growth phase for a favorable environment; populations expanded in their stable environment until they became so crowded that they transformed to gregaria and swarmed out of it. But this charmingly simple vision of self-regulating cycles has been abandoned.* It now appears that crowding is induced not by the high *r* of an expanding population (as in progenetic midges and aphids) but by a deterioration of the solitaria habitat (Key, 1950, pp. 398–399; Dempster, 1963, p. 514; Gunn, 1960; Kennedy, 1956). Key characterizes the solitaria habitat in outbreak areas as patchy for favorable vegetation and climatically unstable. Uvarov (1957) argues that locusts prefer a complex heterogeneous habitat with varied subenvironments to match the requirements of their different stages. Locusts oviposit on bare ground relatively free of plant cover, but nymphs and adults require plants for food and shelter. Moreover, oviposition sites are relatively dry (water-logged eggs will not hatch), while feeding areas are ideally more humid. Environmental fluctuations will tend to favor one or the other of these subenvironments and lead to the crowding of locusts in developmental stages that inhabit the deteriorating subenvironment. Crowding is not a result of general increase in population but of

* Lea (1969) has recently revived it to explain the transformation of *Locusta pardalina* in the South African veld. He believes that initial solitaria populations inhabit the most favorable parts of their potential habitat. As population growth begins to crowd the subhabitat, solitaria locusts are pushed (by their natural repulsion to one another) into less favorable but still acceptable regions of the potential habitat. When the entire potential habitat is occupied at fairly uniform and high density, living space is exhausted and further intrinsic increase of numbers forces a transformation to gregaria. If Lea is correct (and if the behavior of *L. pardalina* represents a general pattern for locusts and not just an unusual adaptation of a single species), then my entire argument under point 2 is completely invalid.

concentration into spaces restricted by unfavorable climates. Uvarov claims that most species (but not all) swarm during drought after the feeding nymphs and adults are "concentrated in the reduced areas of better vegetation which leads to the appearance of the gregarious phase" (Uvarov, 1957, p. 23).

Dempster argues (1963, p. 514) that locusts and grasshoppers generally move only when a habitat deteriorates. He cites Waloff's five-year study of British nonswarming grasshoppers; they moved only in the year that drought greatly restricted their habitat. Dempster also supports Uvarov's theme in his study of gregarization in *Dociostaurus maroccanus:* "The components of a patchy vegetation are anything but stable, and changes in their relative extent are to be expected. It is likely that these will sometimes cause the crowding of one or other of the stages in the locust's life cycle, and therefore the development of gregarious behavior" (1957, p. 58; see also Merton, 1959, p. 113).

Are solitaria populations generally expanding? They must expand at an early stage when a few gregaria locusts have colonized a new site and transformed to solitaria. But this phase of increase may occupy a very short part of the duration of most solitaria populations at a site. The solitaria populations may well be relatively stable in number during much of their existence. One could even take an extreme view and argue that they are often *K* selected to avoid overcrowding a favorable habitat; for if they overcrowd it, they automatically transform to gregaria and swarm out. Dempster (1963, p. 513) states that solitaria locusts generally avoid conspecifics and may be territorial; this may represent an adaptation to crowded environments with populations regulated by density-dependent effects. In another argument for regulation of solitaria populations by density-dependent mortality, Greathead has calculated that observed rates of parasitism and predation are sufficient to maintain population equilibrium for solitaria, but not for "large gregarious populations under optimum conditions at the height of a plague" (1966, p. 248; see also Stower and Greathead, 1969).

I would love to know (though I can find no data) whether solitaria locusts exhibit different rates of maturation during the initial colonizing and later stable phases at a site. I would also welcome information on whether the enormous populations of single swarms (often measured in tens of billions of individuals—Gunn, 1960, p. 287) represent a build-up before swarming or merely result from the aggregation of solitaries from a large deteriorating habitat. Magor (1970) has described how swarms of the Australian locust *Chortoicetes terminifera* can form by the drifting and concentration of locusts against timber barriers. Waloff records the formation of some swarms in *Schistocerca* "from direct concentrating effects of meso-scale convergent air-flow on mobile adults" (1966, pp. 8, 83). When a build-up is required, does it occur in solitaria phase or in gregaria phase? If it occurs in gregaria, the accelerated maturation of gregaria may be an *r* adaptation after all. Stortenbecker (1967) states that swarms of *Nomadacris* leave an outbreak area in the intermediate transiens phase and only reach gregaria after a few generations of swarming. But Stortenbecker does not record either the life history characters of the transiens locusts, or the period of greatest population increase. Are the transiens locusts already at maximal numbers, or does the main increase occur within the swarm after transformation to gregaria? Waloff (1966, p. 88) records a marked increase for populations already in gregaria phase during the 1949–1950 plague of *Schistocerca*. She believes that small swarms of some 10^7 to 10^8 locusts (presumably in phase gregaria) might build through a succession of favorable breeding periods to swarming populations of 10^{10} and perhaps even 10^{11} individuals!

In any case, I feel confident that the analogy between locusts and gall midges or aphids is superficial. Each has a more solitary and a more mobile phase; in each, the transition is induced by crowding and by dwindling resources. But there is a key difference: progenetic midges and aphids represent an *r*-selected adaptation to ephemeral, superabundant resources; populations expand rapidly by accelerated maturation and exhaust the habitat by their own growth. Solitaria locusts live in a good (Kennedy, 1956, calls it "soft")—but not a superabundant—habitat. They become crowded when external conditions restrict this habitat, not as a result of their own growth. Populations are probably not expanding during most of their history.

In another potential exception, Steffan (1973) has found an interesting polymorphism in the Hawaiian sciarid dipteran *Plastosciara perniciosa.* The micropterous, strongly paedomorphic imago oviposits in the pupal chamber. The larvae burrow and feed, remaining below the surface of the substrate. Fourth instar larvae construct a communal pupal chamber with at least one male and one female. Within this chamber, larvae pupate, mate after eclosion, oviposit, and die. Larvae of the next generation eat the dead adults and leave the chamber. The macropterous form of *P. perniciosa,* on the other hand, is a normal fly; it arises when the micropterous habitat deteriorates and the species must move. Dr. Steffan informs me (personal communication, 1975) that both forms have the same number of molts and apparently take the same amount of time to develop an imago. Thus, the paedomorph is not progenetic. Maturation time is not altered; somatic development is retarded. Is the micropterous habitat simply favorable, without providing superabundant and ephemeral resources? Are micropterous populations fairly stable over many generations? If so, the micropterous imago is like a solitaria grasshopper and poses no threat to my general hypothesis. A mushroom or a leaf sounds like an ephemeral, superabundant resource, but a well-protected piece of subterranean turf does not.

Amphibian Neoteny

The facultative paedomorphosis of several *Ambystoma* species has been known for more than a century—ever since Duméril, having reported (with some surprise) the sexual maturation of axolotls, compounded his astonishment by observing the transformation of other specimens (Chapter 6). In most species with facultative paedomorphosis, the ease and frequency of transformation can be correlated with geographic, climatic, and ecological gradients. These correlations provide a key for determining the adaptive significance of paedomorphosis and for ascertaining whether it represents a progenetic truncation by precocious maturation or a neotenic retardation of somatic development.

Most of the early interpretations were mechanistic and nonevolutionary. In *Ambystoma gracile,* for example, frequency of paedomorphosis is far greater in cold Rocky Mountain ponds than in lowland habitats, where most specimens transform (Snyder, 1956). Early interpretations included thyroid inhibition by low iodine in cold ponds and retardation of growth rate by low temperatures, causing a loss in sensitivity of tissues to thyroid hormone (Sprules, 1974a). These mechanisms may well apply, but they leave open the issue of whether any adaptive significance can be ascribed to consistent patterns in the frequency of paedomorphosis.

Sprules (1974a) has reviewed the evidence for adaptive significance and has confirmed and extended a widespread consensus in recent literature. Paedomorphosis with larval reproduction is most common in ponds where the surrounding terrestrial environment is harsh—severe fluctuations in temperature, lack of suitable cover or food, and low humidity, for example—and where water is permanent and predators (fish) are rare or absent. For *A. gracile,* Sprules argues that paedomorphosis is more common in Rocky Mountain ponds because they are more permanent and free from predaceous fish than lowland ponds. Wilbur (1971) records a short larval life with early transformation for *A. laterale* in transient, vernal ponds. He cites several reports of massive mortality in *Ambystoma* following the drying up of ponds. In temporary ponds of New Jersey, *A. tigrinum* metamorphoses in 75 days; larvae are born in the spring and transform in the summer before the ponds dry. In a pond that dried two weeks after the transformation of its larvae, metamorphosis had occurred more rapidly and at smaller sizes than usual (Hassinger et al., 1970). But in permanent ponds of the Western United States and Mexico, larvae overwinter for one or two seasons and often become sexually mature. The famous Xochimilco population of *A. mexicanum* is notoriously difficult to transform (it is the source of laboratory stocks for most experimental work on axolotls[1]). In nature, it inhabits deep, permanent, clear water. Nearby populations living in shallow, temporary, and turbid ponds are rarely paedomorphic and can be easily induced to transform (Smith, 1969). In a temporary pond near Crater Lake, *A. macrodactylum* metamorphosed in a single two-month season; in permanent ponds, larval life may extend through two seasons (Farner and Kezer, 1953). In a related taxon, *Notophthalmus viridescens,* Brandon and Bremer (1966) link paedomorphosis to life in permanent, fish-free ponds. The Japanese salamander *Hynobius lichenatus* metamorphoses during its first year in temporary pools but may never transform in cold lakes (Wilbur and Collins, 1973). The high frequency of paedomorphosis for *Triturus alpestris* in desert and high altitude ponds correlates with the hostility of the surrounding terrestrial environments (Wilbur and Collins, 1973).

Wilbur and Collins (1973, p. 1312) restate the generality in ecological terms: in salamanders with facultative paedomorphosis, transformation becomes rarer and more difficult as aquatic habitats become more predictable and terrestrial habitats more hostile and unproductive. Or, as Garstang wrote succinctly:

> But when a lake's attractive, nicely aired, and full of food
> They cling to youth perpetual, and rear a tadpole brood.
> (1951, p. 62)

All this clearly points to *K* selection for the more paedomorphic populations. The paedomorphs are not colonizers of newly available resources; they are not trying to "hold on," by rapid reproduction, in rigorous, fluctuating, environments with density-independent mass mortality. Rather, they seem to inhabit highly favorable ponds with stable, nonexpanding populations. In anything approaching an *r*-selective environment, they transform. Wilbur and Collins' recent demographic model (1973, p. 1313) for facultative paedomorphosis depends upon density-dependent competition and implies *K* selection. Animals with an initial growth advantage stay in the lake and continue to grow. With growth and aging, their tissues become less sensitive to thyroid hormone, and metamorphosis becomes unlikely. Slower-growing animals metamorphose to escape the density-dependent effects of competition with larger larvae destined to become paedomorphic.

If my general hypothesis is worth anything, these paedomorphic salamanders must be neotenic rather than progenetic. They must retain larval features by a delay in somatic development. Their sexual maturity must come at the same age, or later, than that of transforming individuals in the same population. The selective advantage of delayed somatic development is obvious in forms with such a profound metamorphosis. If they transform, they are forced to trade a favorable pond for harsh ground. To stay in the pond, they must delay somatic development to a point where sexuality intervenes before transformation occurs. The selection is not, as in *r* situations, directly upon generation time, but upon a definite morphology required to maintain life in a stable environment.

Unfortunately, data are extraordinarily hard to come by. There is no dearth of information confirming that metamorphosers transform long before their neotenic brood-mates reach sexual maturity (Snyder, 1956), but sexual maturity may occur long after transformation. I can find no explicit studies of comparative maturation time in transformers and neotenics of a single population. One must clutch at straws by selecting some ambiguous comments from descriptive studies in natural history. Such information as I have found all confirms the hypothesis.

W. G. Sprules has kindly considered the issue for me, and believes that neotenics and transformers probably reach sexual maturity at about the same time. He writes:

> To the best of my knowledge these larvae reach sexual maturity at the same age as normal or transformed salamanders do. In my own laboratory experiments I found that larval *Ambystoma gracile* reached sexual maturity at the same age as described by Neish (1971) for natural populations of transformed individuals of the same species. As my laboratory animals were similar in many other aspects to the natural populations (i.e. had similar growth curves, metamorphosed at the same age, etc.) I feel that this comparison is valid. (personal communication, 1975)

Of *A. gracile* in Marion Lake, British Columbia, Efford and Mathias write: "The few individuals that do normally metamorphose in the lake come back to breed in the spring about the same time as the neotenous adults breed" (1969, p. 734). Anderson and Worthington maintain that both neotenic and transformed individuals of *A. ordinarium* "reach sexual maturity at about the same size (presumably age)" (1971, p. 172). If neotenics and transformers maintain the same relationship between size and age in *Dicamptodon ensatus,* then Nussbaum's assertion supports my hypothesis: "Transformed *D. ensatus,* of a given population, normally mature at about the same sizes as do paedogenes [neotenics] of the same population" (1976, p. 51).

Tilley (1973) has provided excellent support, by analogy, in his study on maturation time in the nonneotenic salamander *Desmog-*

nathus ochrophaeus. In North Carolina populations of the Mt. Mitchell region, maturation is slower in high-elevation than in low-elevation populations; among low-elevation populations, maturation is slower in woodland than in rockface populations. Since the growth rates of all populations are similar, delayed maturation increases the age-specific fecundity of late-maturers (fecundity correlates positively with body size). Tilley relates late maturation to *K*-selective, stable environments where advantages of increased fecundity are not counterbalanced by high, random mortality before maturation. Delayed maturation would be a poor strategy in *r*-selected, low-elevation, rockface populations that suffer heavy mortality, apparently random with respect to age at maturation.

I believe that there is macroevolutionary information in this contrast of progenetic insects with neotenic salamanders. Wassersug (1975) has linked the maintenance of complex life cycles with short-term environmental perturbations (frogs with direct development live in relatively aseasonal environments). The progenetic stages of insect life cycles have not become permanent, for they are adapted to rapid expansion in ephemeral habitats. They would make no sense as permanent, simple life cycles; for they provide no exit from a habitat exhausted by their own rapid growth (except via transformation to the normal, nonprogenetic stage). Neotenic salamanders, on the other hand, have repeatedly abandoned their facultative status for an irreversibly determined, permanent larval life (Fig. 54). Moreover, this has occurred in many distantly related groups (see list in Dent, 1968, pp. 282–283). The *K*-selective regime of a favorable and relatively permanent pond might well invite the final suppression of an unused capacity for transformation. Wassersug (1975) argues that the transition from complex to simple life cycles should occur only in very stable, predictable environments.* Wilbur and Collins (1973, p. 1312) have

* I would rather argue that such transitions will only succeed in the long run if they occur in stable environments. Since evolution is so notoriously opportunistic, a speciation event for a simpler life cycle may occur for short-term benefit in any kind of environment (I would be surprised if no paedogenetic gall midge in history had ever abandoned its adult). In theories of speciation, we often forget that current diversity may only reflect differential success, not the actual number of speciation events. Permanent progenetic insects may arise as often as permanent neotenic salamanders. But if they rarely survive more than a few generations (and almost never radiate), we will find few (if any) of them at any one time. This argument assumes, as I believe (Eldredge and Gould, 1972), that speciation can be very rapid (as defined by its criterion of reproductive isolation), and that the infrequency of successful speciation in nature is not a result of its difficulty or its very slow occurrence, but only of the high improbability for successful incorporation into crowded ecospace once a species has formed (it can rarely exist for long as a small, peripheral isolate).

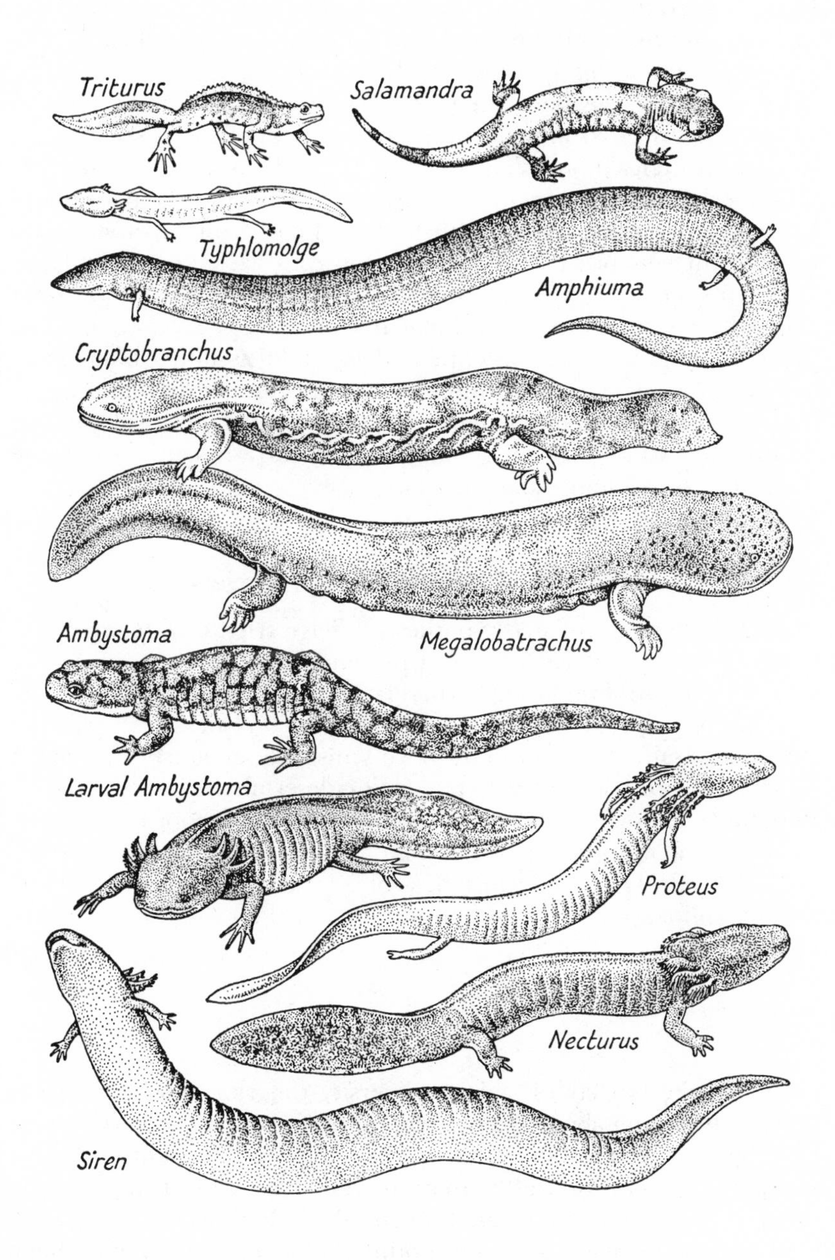

Fig. 54. Various urodele amphibians, most neotenic and including all major genera mentioned in text. Not all to same scale. (From Young, 1962.)

characterized the environments of perennibranchiate salamanders in just such terms: many perennibranchiates are cavernicolous,[2] while others inhabit permanent lakes and streams.

In summary, the permanent fish-free pond and the ephemeral mushroom are environments of a very different order. A salamander will not crowd itself out of a good pond when life is so difficult outside it—especially when the pond may persist as a favorable environment for several thousand years. A mushroom or a leaf, on the other hand, appears infrequently (relative to a tiny insect's size and mobility) and offers much in the very short run. Since it won't last long anyway (and since the next superabundant patch will not appear in the same place), it might as well be exploited to build as much biomass as quickly as possible, so that some fortunate bearer of parental genes may survive to find the next mushroom. In this metaphorical simplification, the *r* strategy of progenetic insects contrasts sensibly with the *K* strategy of neotenic salamanders.

The Ecological Determinants of Progenesis

As a further test of my hypothesis, I have tried to categorize the environments for all clear cases of progenesis known to me from the literature. I find a remarkable correlation between progenesis and *r*-selective habitats (except in cases where selection works primarily for small size itself). I do not know of any instance in which juvenilized morphology per se is clearly the major determinant of progenesis. Rapid maturation or small body size are the primary objects of selection; juvenile morphology is often an incidental (though not inadaptive) by-product. This conclusion is contrary to the conventional view that morphology is of prime importance in the evolution of heterochrony.

Unstable Environments

During the heyday of stability-diversity theory a few years back, much was written about the correlation of low organic diversity with unstable, unpredictable habitats that impose frequent, density-independent, mass mortality upon the few species able to survive. In these environments, fine-tuned morphological specializations would provide no benefit—for there is nothing stable enough to tune them to, and selection must operate primarily for sheer survival against frequent disruption. High *r* may yield enough offspring (and produce them with sufficient rapidity) to get a few through any crisis and to provide for rapid recolonization of resources following a mass mortality. Such harsh and unstable environments should favor *r* selection and progenesis for rapid maturation. Margalef (1949)[3] discusses several brackish-water progenetic subspecies of the marine amphipod *Gammarus locusta.* He regards the quicker generation time of these progenetic forms as the primary determinant of their ability to colonize such ephemeral waters (p. 46).

Snyder and Bretsky (1971) have reviewed the classical subject of "dwarfed faunas" in paleontology. Some may be artifacts of post-mortem transport and accumulation of small juvenile shells. Previously, faunas of genuinely dwarfed adults had always been interpreted as the direct effect of an unfavorable circumstance: deficient oxygen, abnormal salinity, or insufficient food (Tasch, 1953; Hallam, 1965). No one had proposed that dwarfing might be an active, adaptive strategy rather than a passive sign of exhaustion in deteriorating habitats. Snyder and Bretsky have reinterpreted the famous dwarfed fauna of the basal Maquoketa Formation (Ordovician) as a progenetic assemblage adapted by *r* selection so that generation times are shortened by the acceleration of sexual maturation. They regard the basal Maquoketa environment as one of high stress (fluidity of substrate, deficiency of oxygen, and abnormal salinity), but of frequently superabundant food (as inferred from phosphate deposits presumably derived from algal stands or high planktonic productivity). Such unexploited productivity implies a density-independent mortality and reduced intraspecific competition, thus favoring *r* selection among potential herbivores. Britton and Stanton (1973) eliminate post-mortem concentration of juveniles and physiological stunting as causes for molluscan dwarfism in the Lower Cretaceous Del Rio Formation of East Central Texas. The "seasonal catastrophes" of their inferred environment point to *r* selection for progenesis as a possible cause. Mancini (1974) has also spoken favorably of progenesis as an explanation for diminutive size in a soft-substrate Cretaceous fauna suffering very high (and probably random) mortality by foundering and fouling.

Baid (1964) has discussed progenesis with larval reproduction in the brine shrimp *Artemia salina* from Sambar Salt Lake, Rajasthan, India. This lake is exceedingly salty and reaches a maximum depth of only 0.61 m after heavy rainfall; it dries up completely during the summer. Progenesis is facultative and occurs during periods of high mortality when larval reproduction is favored. Baid (p. 175) presents a general argument linking progenesis with hostile environments.

Colonization

Under colonization, I link two phenomena often considered separately: the exploitation of renewed abundant resources by the survivors of an in situ population crash,[4] and the chance dispersal of a few colonizing immigrants to new areas with no competitors. Both possess the common feature of presenting to a few individuals a superabundant regime of resources; *r* strategists should prevail in such a circumstance, since rapid increase in numbers will be so strongly favored. Progenesis represents one of the easiest and most rapid pathways to high *r*; it should be very common among colonizers.

My previous examples among insects all fall into this category. The newly grown mushroom and the unexploited leaf are superabundant, uncolonized resources. Wyatt (1964) presents a striking example of how effective accelerated maturation can be among the paedogenetic gall midges. *Mycophila speyeri* and *M. barnesi* infest the same beds of cultivated mushrooms. They differ in little else but their generational cycles: newly hatched *M. speyeri* larvae reproduce in five to six days; *M. barnesi* takes seven to eight days. The first flush of affected mushrooms begins five weeks after spawning, and *M. speyeri* can destroy up to 80 percent of it; thereafter, the density of *M. speyeri* falls. *M. barnesi* does not reach damaging density during the first flush. It may attack the second flush a week later in a moderate way, but it often destroys the third and later flushes. Wyatt (1964) attributes this difference entirely to the more rapid generation time of *M. speyeri.*

J. G. Blower has published several works on the correlation of progenesis and colonizing ability in myriapods. Blower and Gabbutt (1964) contrasted the life histories of *Cylindroiulus punctatus* and *C. latestriatus.* Both species reach their seventh stadium after two years. Progenetic *C. latestriatus* breeds in this stadium, producing about one-half as many eggs as *C. punctatus,* which breeds in its third year at the eighth or ninth stadium. Blower and Gabbutt assessed the relative effect of laying half as many eggs, but laying them a year earlier, by constructing theoretical curves assuming a constant mortality of 80 percent from birth to maturation. Since *C. latestriatus,* by this calculation, outproduces *C. punctatus* until the sixth year, Blower and Gabbutt felt that its superior colonizing ability could be explained by progenesis (the effect may be even more pronounced; *C. punctatus* probably suffers greater mortality between birth and maturity since it spends an additional year in the immature state). Blower (1969) then contrasted three pairs of related species, each containing one small progenetic form and one normal form (Brolemann, 1932, has emphasized the importance of progenesis among myriapods). In each pair, the progenetic form is a better colonizer. Blower and Miller (1974) compared two common British iuline myriapods, *Iulus scandinavius* and *Ophyiulus pilosus. O. pilosus* matures a year earlier than *I. scandinavius,* in a progenetic state with juvenilized morphology. *O. pilosus,* unlike *I. scandinavius,* has had outstanding success as a colonizer in parts as distant as North America and New Zealand.

Turtonia minuta is a strongly progenetic veneracean bivalve 1–2 mm long; in size and form, it is remarkably similar to the spat of *Venerupis* (Fig. 55). Both outer demibranchs and a posterior inhalent siphon are lacking—a sure sign of early ontogeny in other veneraceans. It inhabits the tidal zone, attached to algae, and seems to be a classical *r* strategist with wide tolerance for environmental fluctuation (Ockelman, 1964, p. 141). Like many of the tiniest progenetic *r* strategists, it broods only a few eggs, in capsules attached to the progenetic byssal threads. It is also the best colonizer among veneraceans and the most widely distributed species of its group (Ockelman, 1964). In another example among bivalves, Soot-Ryen (1960) has described (without recognizing it as such) a remarkable fauna of progenetic species from Tristan da Cunha, an isolated group of islands on the Mid-Atlantic Ridge. It includes ten species; all are tiny (rarely longer than 5 mm) and probably progenetic. *Cyclopecten perplexus,* for example, has a maximum diameter of 1 mm and, like *Turtonia,* has developed only the inner demibranch of its gills. Seven or eight of the ten species brood a small number of young. Most have their closest affinities with species found in New Zealand—a long float. (I am not suggesting that they arrived as progenetic forms; I can imagine no impetus for this, and the initial float can be better accomplished by a larger, longer-lived animal with planktonic larvae. I suggest that progenesis occurred as a common response to superabundant resources in one of

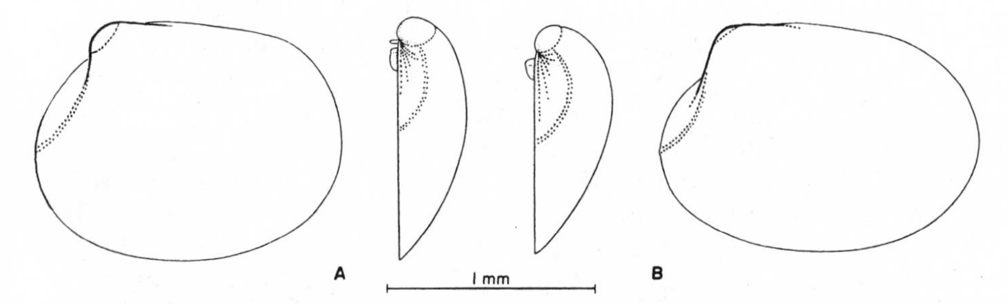

Fig. 55. Progenetic adult *Turtonia minuta* (*B*) compared with juvenile *Venerupis pullastra* (*A*). (From Ockelman, 1964.)

the world's emptiest ecospaces [shallow water environments around a maximally isolated group of islands].)

When the reproductive pair in a termite colony dies, secondary reproductives (also called, and spelled, "neoteinics") can be quickly recruited by inducing sexual maturity in larvae or nymphs. In some species, an entire caste (the pseudergates) is kept in a larval state, despite repeated molting, with potential for rapid development into either soldiers or secondary reproductives as the need arises (Lebrun, 1961). The "idling" pseudergates cannot be properly labeled as either progenetic or neotenic, but the speed with which they can be recruited as secondary reproductives suggests a progenetic strategy. Gay (1955) reported a high frequency of paedomorphic secondary reproductives in two species of *Coptotermes* recently established as successful colonists in New Zealand. Yet the same species in their native territory of Australia very rarely develop secondary reproductives. This latent capacity for rapid recruitment of reproductives has evidently been exploited by the expanding colonists.

Parasites

In defining progenesis, Giard linked its frequent occurrence with parasitism. He cited a general antagonism between generation and growth, arguing that parasites often need rapid and copious generation, but do not require many adult organs: "Thus we see that a very great number of parasitic animals are progenetic" (1887, p. 24). De Beer (1958, pp. 64, 90) maintained this association by dismissing progenesis as unimportant in evolution because it is commonly associated with such "degenerate" developments as the simplified morphology of parasites.

Progenesis among parasites may be recognized by two standards of comparison. Some species are progenetic with respect to normal, free-living relatives; they have "traded" adult morphology for one of the outstanding attributes of parasites—devotion of energy to reproduction. Other species are progenetic with respect to related parasites. These have usually developed a simpler life cycle by maturing in an intermediate host (Baer, 1971, p. 126).

The conditions of parasitic life often impose a regime of *r* selection. An uninfected host is an unexploited resource; if parasites are small relative to their host, a period of rapid increase in numbers may follow the first infestation. If the host cannot support many parasites (because of their large size or debilitating effect), high *r* might still be favored in order to produce progeny rapidly for dispersal to other hosts. Any host is an evanescent resource (for it will die a natural death even if the parasite regulates its effect in order to preserve it). Moreover, hosts are generally like the gall midge's mushroom or the aphid's leaf—patchy in distribution and hard to find. Juvenile mortality will usually be extremely high; the quicker and more often a parasite can reproduce, the better. (I am well aware that this generality is rife with exceptions. Parasites with complex life cycles, for example, may spend most of their time in *K*-selective regimes. The general link of parasitism with *r* selection has been noted by King and Anderson [1971].)

The fecundity of some parasites is as legendary as that of aphids. The tapeworm *Diphyllobothrium latum* can produce up to two billion eggs in ten years (Stunkard, 1962). The common roundworm may produce 65 million eggs a year with a combined weight 1,500 times that of the parasite itself (Cheng, 1964). The parasitological literature has stressed fecundity alone as the primary determinant of adaptation for easier dispersal, but the framework of *r-K* theory suggests that precocious maturation may play an equally important role—being first is often a better strategy than trying often.

Many parasitic copepods are progenetic in comparison with free-living species. Normal copepods pass through six nauplius and five copepodid stages before molting to an adult. The parasitic Lernaeopodidae shorten this sequence (Gurney, 1942). *Achtheres ambloblistis* exhibits "a marked concentration of development"; the nauplius and metanauplius stages are passed in the egg, and are "so fused as to be indistinguishable" (Wilson, 1911, p. 224). The animal hatches as a first copepodid, its only swimming stage. After a short swimming period of 24–48 hours, it attaches to the gill arches of a fish, and transforms to a second copepodid, which molts directly to an adult (corresponding to the third copepodid of normal development). *Salmincola salmonea* goes even further and becomes sexually mature as a first copepodid (de Beer, 1958, p. 64). A mite progenetic on millipedes has been described by Kethley (1974), and others progenetic upon moths by Treat (1975, p. 58). Kinzelbach (1971) and Ivanova-Kasas (1972) have discussed the progenesis of parasitic strepsipteran insects. Many progenetic erycinacean bivalves are parasitic.

Several groups of cestodes are progenetic with respect to other parasitic cestodes (Ginetsinskaya, 1944; Dubinina, 1960). Stunkard (1962) has characterized them as "merely precocious progenetic larvae that have dropped the strobilate stage." Progenesis, in all cases, has accompanied the deletion of terminal hosts from a complex life cycle; reproduction then occurs in an intermediate host of ancestors (Smyth, 1969). The genus *Archigetes* of the progenetic order Caryophyllidea (Cheng, 1964, p. 342) first infests an oligochaete and reaches its terminal host when a fish ingests the infested worm (Wisniewski, 1928; Mackiewicz, 1972). Some species maintain a capacity for facultative reproduction in the oligochaete (Calentine, 1964, on *A. iowensis,* which may mature in the seminal vesicles of *Limnodrilus hoffmeisteri*). In other species, maturation in the oligochaete has become obligatory; "appropriate" fish will not be infected even if they eat parasitized worms (Nybelin, 1963, on *A. sieboldi;* Kennedy, 1965, on *A. limnodrili*).

The cestodarian genus *Amphilina* is parasitic in the coelom of fishes and tortoises. Since these are common sites of some eucestodan larvae (and since *Amphilina* maintains a larval morphology), Cheng (1964, p. 302) regards the genus as progenetic. In a not implausible bit of paleopoetry, Janicki (1930) speculates that the ancestors of *A. foliacea,* now a parasite of sturgeons, became sexually mature in this ancient fish when the ichthyosaur that once supported its final stage became extinct. Other species of *Amphilina* can attain sexual maturity in their first crustacean host (Stunkard, 1962, p. 28). Finally, the pseudophyllidean genus *Bothrimonus* often reaches sexual maturity in its single intermediate host, usually a brackish-water gammarid.

I have already discussed the progenesis of *Polystomum,* a trematode. *Polystomum integerrimum* usually matures in the bladder of a frog after initial attachment to the gills of its tadpole; there it becomes sexually mature after three years (Caullery, 1952). But if a *Polystomum* attaches to a sufficiently young tadpole, it grows rapidly on the gills and reproduces there as a progenetic form. The normal parasites are far more fecund (Williams, 1961), but the progenetics can pass through several generations before the normals mature. Some species that parasitize fully aquatic amphibians have become permanently progenetic. The closely related *Sphyranura* lives on the gills of perennibranchiate *Necturus* (Williams, 1961). *Protopolystoma xenopi* infests the aquatic frog *Xenopus laevis* (Williams, 1961), where it matures in the bladder (*Xenopus* is a normal frog with no external gills). Infection may be direct (through the cloacal opening) and the mature trematode is strongly paedomorphic (organs last to differentiate in normal ontogeny are the first to disappear in these progenetics).

Male Dispersal

Ghiselin (1974) has identified and presented impressive documentation for a common reproductive strategy among males in many small, marine invertebrates. These males are progenetic dwarfs, often minute compared with their females (Fig. 56). Ghiselin identifies the following conditions as prerequisites for this variety of sexual selec-

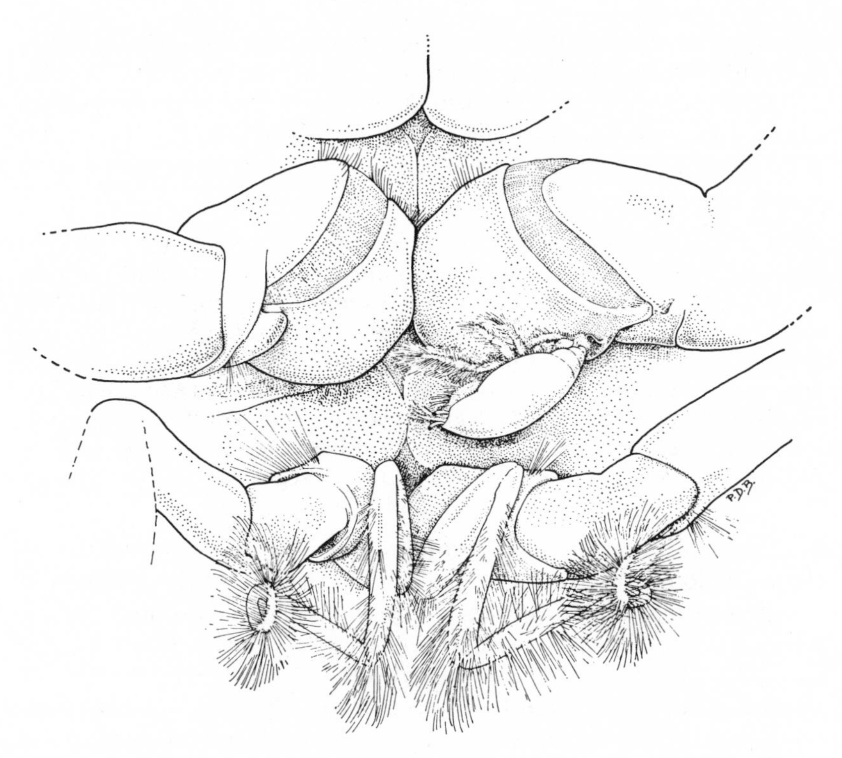

Fig. 56. Progenetic dwarf male (2.5 mm) of *Emerita rathbunae* attached to first segment of left third walking leg of large female (38 mm). (From Efford, 1967.)

tion by competition for dispersal among males: low population density (to preclude counter-pressure for combat to enlarge males), restricted mobility at larger sizes (especially when adult stages are sessile), and a premium on long female life with no pressure for long male life. This progenetic strategy is only important in the sea where current transport through a nutrient-rich medium permits the wide dispersal of tiny animals. Progenetic dwarf males are most common among deep-sea animals, certain parasites (Baer, 1971), and sedentary or sessile filter feeders.

Efford (1967) has studied dwarf males in several species of the sand crab *Emerita.* In *E. analoga,* with males of half the female length, females overwinter in large numbers and become sexually mature in their second year (some in their first year). Males mature during their first year and mate with females of their own year and the previous one. In other species (*E. rathbunae* and *E. emeritus*), males may be only one-tenth the length of females.

As in most cases of dwarfism (Gould, 1971), these males more closely resemble geometrically scaled-down adults than larvae.[5] But they all retain several important larval characters, left behind so to speak by their precocious maturation. Efford writes:

> They are certainly precociously mature although this maturity is achieved when they have a juvenile adult form rather than a larval form. They do, however, retain certain larval characters. Commonly they are very small and soft and in this resemble the megalopae rather than the large adults; they also show a general simplicity of the appendages associated with their small size. This is demonstrated particularly by the antennae which are simple and do not have the regularly arranged, closely packed setal net of the large animals. In both *E. talpoida* and *E. emeritus* the small males retain the stumps of the pleopods normally only present in the zoeae, megalopae and the females. (1967, p. 89)

I doubt that any primary selective significance can be attributed to these paedomorphic features. If Ghiselin is right, the determinant of progenesis must be precocious maturation (and perhaps small size for more efficient transport by currents)—"first come, first service," to cite Ghiselin's aphorism. The partial retention of larval features is a developmental consequence of selection for a life-history strategy involving precocious maturation. These features are not inadaptive; they are probably not even "neutral"; but I would be surprised if selective pressures for their evolution played any important role in the evolution of dwarf males.

Progenesis as an Adaptive Response to Pressures for Small Size

Although previous attention has been focused on larval morphology, progenesis involves two other events of potential evolutionary import: precocious maturation and small size. I have been arguing, in this section, for the primary significance of precocious maturation as an ecological strategy. Other examples indicate that selection is principally for small size, and again, that partially larval morphology may be a passive consequence of this primary need.

Surlyk (1974) has studied the progenesis of *Aemula inusitata,* a brachiopod from the very fine grained Cretaceous chalk deposits of Northwestern Europe. It is very small (maximum length of 7 mm), short lived (as deduced from growth lines), and paedomorphic (the lophophore remains in the schizolophous state so characteristic of early ontogeny in complex forms). Its progenesis cannot easily be attributed to the advantages of juvenilized morphology. Complexity of lophophores is a mechanical consequence of body size; greater coiling and bending is an allometric adaptation to provide enough feeding surface for the increased body volumes of large animals. The simple lophophore of *Aemula* is a secondary consequence of its small size, not an adaptation selected in its own right. Moreover, Surlyk argues that early maturation was not the primary determinant of progenesis, for *Aemula* is not an *r* strategist. If forms part of a stable ecosystem and persists throughout the Maaestrichtian Chalk in small but regular numbers. Small size itself must have been the primary object of selection. The fine-grained, soft-bottom chalk substrate provided no attachment sites for these pedunculate brachiopods; they depended upon sparsely distributed hard objects, organisms often not much larger than themselves. Large brachiopods would have had no place to attach (Fig. 57).

Thorson (1965) describes a dwarfed, paedomorphic form of the gastropod *Capulus ungaricus.* It lives as a commensal on the shell margin of *Turritella communis* in areas where *Turritella* is the only potential host. *T. communis* is a relatively small and very narrow gastropod—*Capulus* has little room for attachment on its favored site. Progenesis has truncated ontogeny at a size small enough for successful attachment throughout life.

Progenesis for small size is most clearly (and most often) found in

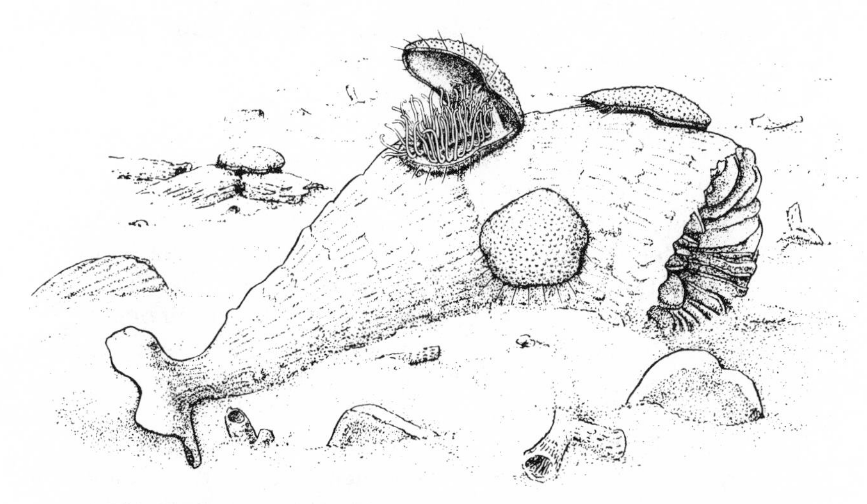

Fig. 57. Progenetic brachiopod *Aemula inusitata* attached to a scleractinian coral *Parasmilia excavata.* No large attachment sites were available on the chalk sea floor. (From Surlyk, 1974.)

the interstitial fauna—a remarkable assemblage of the world's tiniest metazoans, adapted to live between the grains of marine sands and muds (Swedmark, 1964). Many are as small as protozoans and have, perforce, so few cells that they cannot differentiate many complex organs.

The interstitial environment provides some basis for a claim that *r* selection might be favoring rapid reproduction as a major determinant of progenesis. Swedmark (1964, 1968) and McIntyre (1969) have emphasized the physical instability of interstitial habitats. McIntyre believes that, relative to the macrofauna, interstitial animals possess "greater ability . . . to flourish in areas of environmental stress" (1969, p. 282). Swedmark speaks of continuous rearrangement by winds, waves, and currents—"a very dynamic environment which is changing continuously" (1968, p. 139). He also notes that: "Such physical factors as temperature and salinity vary greatly in many interstitial biotopes, particularly in the tidal zone. This implies that the interstitial organisms are physiologically adapted to endure both the seasonal variation and the often rapid changes that occur in connection with ebb and flood. The littoral interstitial fauna is therefore eurythermal and euryhaline" (1964, pp. 33–34).

Still, one can't help assuming that a rigid requirement for small size must be the primary determinant of progenesis among interstitial organisms (although *r* selection for rapid maturation in unstable environments might provide a preadaptation for traversing incipient stages of intermediate size—in terms of size alone, being a hundred times bigger than a sand grain may represent very little "improvement" on being a thousand times bigger).

Many authors have commented upon the high frequency of progenesis among interstitial organisms (Swedmark, 1968; Serban, 1960; several articles in the symposium on meiofauna, *Smith. Contr. Zool.,* vol. 76, 1971). The comments would be even more numerous were it not for the residual tendency (particularly among continental biologists) to identify any simplification of structure as a mark of phyletic antiquity. It would be remarkable indeed if so many archaic creatures reduced their size so precipitously and ended up in one of the world's most peculiar biotopes. Rather, small size and simple structure are linked through a recent event of progenesis for adaptation to a very specialized environment.

Swedmark (1958; 1964, p. 607) has noted a clear case of progenesis by true truncation. The polychaete *Psammodrilus balanoglossoides* has a volume twenty-five times that of its close relative *Psammodriloides fauveli* (Fig. 58). In its morphology, the smaller species is little more than a sexually mature version of a juvenile stage of the

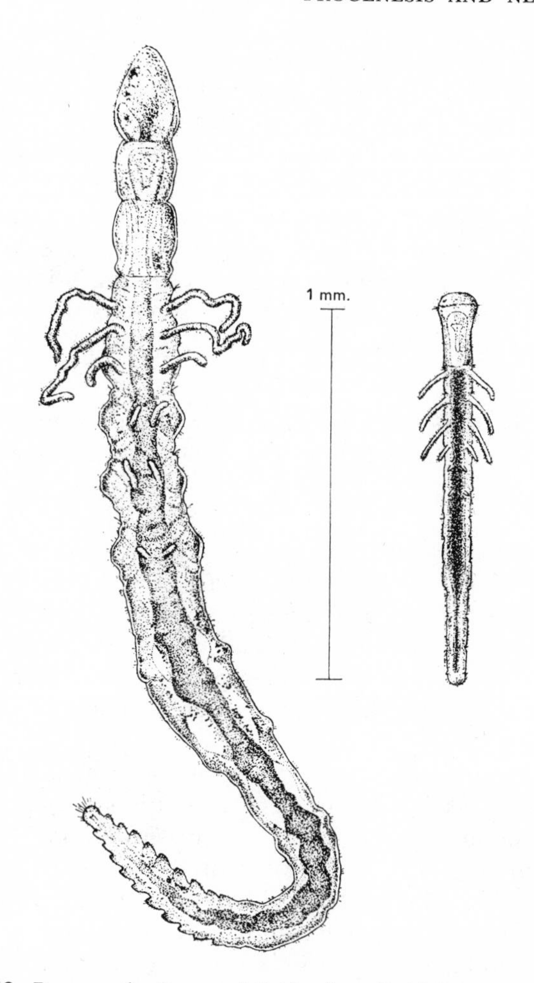

Fig. 58. Progenetic *Psammodriloides fauveli* (right) compared with its close relative *Psammodrilus balanoglossoides*. (From Swedmark, 1964.)

larger animal. (It differs only in the absence of some organs not needed at its size but already differentiated in *Psammodrilus* for use at larger sizes reached in ontogeny.) Other cases of striking progenesis include copepods with reduced larval molts (Serban, 1960) and mature ascidians easily confounded with juvenile stages of the Mogulidae (Monniot, 1971).

Among major groups confined to the interstitial realm, Swedmark (1964) has listed four taxa of ordinal rank or higher: the Gnathostomidula (generally ranked as a phylum), Actinulida (an order of hydrozoans), Mystacocarida (Crustacea), and Acochlidiacea (Mollusca). It is a testimony to the importance of heterochrony that all these groups have received a progenetic interpretation. The Actinulida are generally regarded as progenetic (Hadzi, 1963, p. 94). Hessler (1971, p. 87) has urged a progenetic interpretation of the Mystacocarida in opposition to previous claims for their "primitive" nature. Riedl (1969, p. 856) points to the paedomorphic features of gnathostomulids in suggesting their possible progenetic origin from much larger fossil conodonts. Odhner (1952) suggests that the family Microhedylidae of the Acochlidiacea may be progenetic.

The Role of Heterochrony in Macroevolution: Contrasting Flexibilities for Progenesis and Neoteny

Progenesis

Progenesis has usually been dismissed as an agent of degeneration with no evolutionary importance. I have tried to rescue it from this charge by demonstrating its immediate significance for precocious maturation in *r*-selected regimes (and, in some cases, for small body size). But de Beer's dismissal was more an assessment of retrospective significance in macroevolution than a denial of adaptive value in local environments.

The link of progenesis to *r* selection might seem to affirm the usual denial of retrospective significance. "Evolutionary importance" is an ambiguous notion (to put it mildly); in conventional writing about macroevolution, a process partakes of this recondite "importance" when it contributes to the classical material of macroevolutionary success: slow, continuous, and sustained trends toward more complex morphology and greater diversity. These trends are the material of *K* selection; *r* selection, with its emphasis on production of offspring at the expense of perfection in morphology, can only serve as a brake upon such "progress." As Dobzhansky wrote in his prophetic article of 1950, long before theoretical ecology codified the concept: "Physical factors, such as excessive cold or drought, often destroy great masses of living beings, the destruction being largely fortuitous with respect to the individual traits of the victims and the survivors . . . Indiscriminate destruction is countered chiefly by development of increased fertility and acceleration of development and reproduction, and does not lead to important evolutionary advances" (p. 220; see also Murphy, 1968, p. 402). Levin has noted the restriction of recombination in *r*-selected plants. *K*-selected genomes, he argues, do not congeal because pathogens and herbivores exert intense pressure and track any new defense evolved by the plant. Levin concludes: "In *r*-selected species the recombination system usually favors immediate fitness at the expense of flexibility, while in *K*-selected species the recombination system favors flexibility in lieu of immediate fitness" (1975, p. 437).

Yet, if we survey the speculative literature, we find that the most "fertile" paedomorphs—those that escape from an ancestral adult specialization to serve as progenitors of a new higher taxon—are not *K*-selected, neotenic animals. They are small, progenetic larvae with precocious sexual maturity—Müller's larva of turbellarians, the six-legged larva of myriapods, the tunicate tadpole. Some authors have seen the contradiction and have labored mightily to transform these larvae into neotenic animals—by arguing, for example, that their maturation was not precocious, but that they grew into giant plankters with persistent larval form and matured at the usual time. Thus, Bone dismisses the tunicate tadpole from our ancestry: "This type of larva is not a suitable one from which to derive the chordates by neoteny. It is the result of selection maintaining and specializing a short larval phase, and is physiologically quite unlike the type of larva which underwent neoteny, whose whole history must have been of longer and longer periods spent in the plankton while the neoteny gradually appeared" (1960, p. 256). Still, in most phyletic speculation, these paedomorphic progenitors are small larvae and, as such, must generally be progenetic.

Clark has faced this issue squarely in his discussion on the paedomorphic origin of pseudocoelomates. (But his devotion to morphological explanations based on retrospective significance bars him from appreciating the immediate advantages of progenesis. He argues that "poorly controlled" development in these primitive forms permits an easy dissociation of sexual maturation from somatic development, but it never occurs to him that this dissociation, with an acceleration of maturation by progenesis, might be adaptive in itself.)

> The pseudocoelomates are often small animals and appear originally to have been benthic, either sessile, or creeping at the surface of the substratum, or living as an interstitial fauna. It cannot be argued that they evolved in response to the selective advantages of prolonging the pelagic larval phase . . . The fact that paedogenesis [= progenesis] occurred several times in the history of the early Metazoa suggests, rather, that the timing of sexual maturation with respect to somatic development was not very precisely controlled in these animals. (1964, p. 221)

Theories of macroevolution have not faltered in explaining sustained trends within adaptive zones and common *Baupläne*. But they have floundered badly in trying to apply the Darwinian, continuationist perspective to transitions between fundamentally different designs in the origin of higher taxa—for how can these transitions be gradual and under continuous selective control? The names of Cuvier, Goldschmidt, and Simpson are rarely linked, but on one thing they all agreed—pure gradualism with conventional control by selection cannot extend across the gaps in basic design. Cuvier, of course, denied that the gaps could be bridged at all: the correlation of parts would not allow it. Goldschmidt postulated a macromutation that would bridge the gaps by sheer good fortune at a single leap: the hopeful monster. Simpson (1944) invoked genetic drift as a contributing agent to quantum evolution across a discontinuity between adaptive zones: the inadaptive phase. It is not likely that such a profound transition can occur by continuous gradation mediated only by direct selection upon morphology.

The major theme of this chapter has been an argument that progenesis is selected not primarily for morphology but by the need for precocious maturation as a life-history strategy. The morphology of progenetic forms is not inadaptive (the very fact that wingless aphids and paedogenetic gall-midges cannot fly keeps them in their superabundant resource and spares them the energy needed to produce "expensive" sensorimotor devices). But morphology is simply not the primary ingredient of many progenetic adaptations. The redirection of selection towards the timing of maturation might well release the rigid selection usually imposed upon morphology. Morphology would then no longer be fine tuned to a changing environment. In fact, it may matter little what a progenetic organism looks like (though it must still function adequately), so long as it reproduces as rapidly as possible. This "unbinding" of morphology from its usual selective control may set the macroevolutionary significance of progenesis. If selection must continually superintend morphology as adaptive zones are bridged, then the conventional dilemma applies. If morphology can be unbound by the redirection of selection elsewhere, then these difficult transitions may occur within a purely selectionist framework. It is the object of selection that changes—not the control by selection itself. I envisage progenesis as one of the few processes that can accomplish such a redirection of selection and lead to the unbinding of morphology.

Two other properties of progenesis reinforce the claim that it plays a role in the rapid origin of higher taxa:

1. If progenetic morphologies were nothing more than perfect juvenile stages, then they would have little potential in macroevolution. More is needed than a simple unbinding of morphology from

rigid selection. New morphologies are required to exploit the opportunity.

The potential of progenesis resides not merely in the unbinding of morphology, but also in the provision of new morphological combinations, allowing experimentation within this rare "freedom." No progenetic animal, so far as I know, is simply a sexually reproducing larva. Differentiation and maturation are dissociable, to be sure (Chapter 7), but their usual correlation is rarely disrupted completely. If maturation is accelerated, the new adult may resemble an ancestral larva in most features. But maturation itself carries a train of consequences for differentiation of somatic structures. Major changes in form and growth rates are correlated with maturation in most animals. Some adult features will accompany the precocious maturation of progenesis; others will be sloughed off the end of ontogeny. The progenetic animal is a dynamic mixture of adult, juvenile, and intermediate features. Thus, *Turtonia minuta* has an adult ligament with larval gills and siphons. And Efford's dwarf males of *Emerita* are largely adult, but have larval antennae and pleopods. The mechanics of development provide new morphological combinations in the very circumstance that promotes their persistence and permutation by unbinding selection from morphology. From a macroevolutionary point of view, progenesis is not an automatic degeneration. It may well establish a laboratory for morphological experimentation.

2. Gene duplication has been granted an important role in the origin of evolutionary novelties (Britten and Davidson, 1971; Markert et al., 1975) because it provides "extra" genetic material freed from the need to function in only one way and therefore available for experimental change. An analogous process must occur in progenesis when genes formerly expressed only in the ancestral adult stage become, in de Beer's term, "unemployed," by precocious maturation of a larva. Thus, into the laboratory mix of unbound morphology and novel combinations, we must throw a set of unusually transformable genes. Stir these three ingredients together often enough in the history of life and they may congeal every once in a great while to yield a new higher taxon.

Darwinian theory has been overly burdened by a rigid insistence upon very slow, continuous, adaptive transformations—an unwarranted extrapolation from directional selection upon single loci in local populations to the origin of new designs. Progenesis is a perfectly orthodox (though unfamiliar) mechanism that permits rapid transition for very little initial genetic input, and that frees morphology to experiment not by releasing selective control altogether (and abandoning Darwinism), but by directing it elsewhere.

Hardy made the point unwittingly in discussing vertebrate origins: "If a naturalist from some other planet had visited the earth and seen the little paedomorphic chordates in their very early days he might well have failed to see any future in them" (1954, p. 151). Quite right; for if these hypothetical animals existed at all, they were progenetic forms adapted for precocious maturation to some immediate environment. But they also had unusual genetic and morphological flexibility (to be appreciated, if utilized, only in retrospect).

I do not deny that this creative role in macroevolution is very rarely exploited by progenetic organisms. A simplified life cycle probably decreases flexibility far more often than unbound morphology enhances it. In fact, the usual genetic result of any paedomorphosis will not be increased opportunity from the freeing of adult genes, but reduced flexibility from their loss. I have demonstrated this myself in paedomorphic land snails from Bermuda (Gould, 1969). These paedomorphs have arisen several times but have always become extinct in a short time; the parental stock of nonpaedomorphs persists throughout the Pleistocene. In one case, the paedomorphs converge upon the form and habitat of nonpaedomorphs (following the Holocene extinction of nonpaedomorphs). Several details of this convergence suggest the loss of adult genes: the shell becomes thick again, but the callus is not redeveloped; the color bands return to their nonpaedomorphic strength, but a variant only expressed in the late stage of nonpaedomorphs is never displayed by the paedomorphic convergers. Similarly, a few perennibranchiate salamanders now live in unstable aquatic environments that favor transformation (Wilbur and Collins, 1973). These animals are committed to their juvenile forms; their tissues have lost sensitivity to thyroxin. They can only survive by developing such special adaptations as burrowing and aestivation in the face of drought. (These two examples involve neoteny rather than progenesis. But the same point holds in either case: loss of an adult stage [either by truncation or retardation] usually leads to the loss of genes transcribed only late in the ontogeny of ancestors.)

But extreme rarity thrives on the immensity of geological time. Twenty successes in a million opportunities could have established most of the major designs of organisms. Komai has objected to the invocation of progenesis in explaining the origin of large groups: "It is, therefore, hard to believe that so exceptional a phenomenon should have determined the course of evolution of such a large group as the Ctenophora" (1963, p. 184). But current size is irrelevant. The Ctenophora—if the taxon as defined is worth anything—had a monophyletic origin in a single species. Only one creative progenesis is required for the entire phylum.

I believe that evolutionary biologists shy away too readily from hypotheses that invoke improbable events and excuse them by citing the immensity of time. The criterion should be cogency in theory, not frequency of occurrence. Goldschmidt's hopeful monsters were rejected because they were untenable in theory, not because they needed to succeed but rarely. An extremely rare creative role for progenesis in the rapid origin of higher taxa is not only tenable in theory; it might even save Darwinism from an embarrassing situation usually swept under the rug of orthodoxy—the difficulty of explaining transitions between major groups if the transitions must be gradual and under the continual control of selection upon morphology.

Neoteny

Evolutionary trends toward greater size and complexity form the classical subject matter of "progessive" evolution as it is usually conceived—the slow and gradual fine tuning of morphology under the continuous control of natural selection. These trends display three common features marking them almost inevitably as primary products of *K*-selective regimes:[6]

1. A primary role for morphology in adaptation—usually leading to increased complexity, improvement in biomechanical design, or at least the continual exaggeration of specialized structures with clear functions.

2. A general tendency to increasing size—Cope's rule. Hairston et al. (1970, p. 685) have linked phyletic increase in size with the density-dependent regulation of *K* regimes.

3. In most cases, a delay in the absolute time of maturation. This property has rarely been mentioned, but I regard it as unavoidable. Larger animals with a generally increased level of morphological differentiation almost surely mature later than their much smaller and more generalized ancestors. Bonner (1974, p. 27) has emphasized the usual association of increased size and delayed reproduction. Sexual maturation usually marks the termination (or at least the pronounced slowdown) of both size increase and differentiation.

The usual heterochronic result of such trends is probably hypermorphosis—an extension or extrapolation of ancestral allometries. With a delay of maturation, differentiation can proceed beyond its ancestral level into the larger sizes of descendants. The irony of this

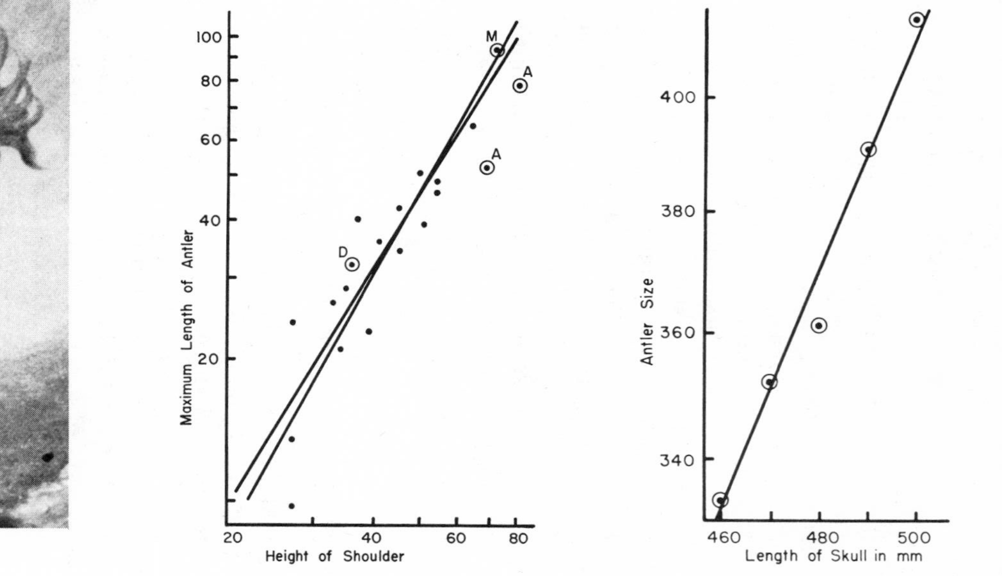

Fig. 59. Hypermorphosis in the Irish Elk. This giant deer was the largest known cervine and had the largest antlers (absolutely and relatively) of any deer. (*A*) Figure from Millais, 1906. (*B*) Inter- and intraspecific positive allometry of antlers in cervine deer and among adult stags of *Megaloceros* (the Irish Elk). Left: Positive allometry in all cervine deer. Line of higher slope is the reduced major axis, lower slope is least squares y on x. The smaller, presumably related fallow deer (*D*) has a higher positive deviation than the Irish Elk (*M*). Irish Elks have the "right" antler size for a deer of their body size—the antlers are hypermorphic. Right: Similar positive allometry among adult stags of Irish Elks. Data grouped at 10 mm intervals of skull length; logarithmic scales. (From Gould, 1974.)

conclusion is that hypermorphosis is a mechanism of recapitulation—previously adult levels of differentiation are overstepped by descendants and become their juvenile features. McGuire writes: "It is in these groups most commonly affected by phyletic growth that ontogenetic development tends to parallel phylogeny" (1966, p. 896). Recapitulation has usually been linked with acceleration (Chapter 7). Yet I suspect that the primary determinant of hypermorphosis is retardation—of sexual maturation rather than somatic differentiation. The usual tie of hypermorphosis with *K*-selective regimes (where we anticipate a delay in maturation) provides the justification for this assertion.

The problem with hypermorphosis as a common ingredient in phyletic trends is that it must generally restrict evolutionary potential through overspecialization—it leads evolution into its famous blind alleys far more often than it promotes an evolutionary novelty with potential for increased diversity. To be sure, Rensch (1959) has emphasized a potentially creative role for hypermorphosis. An extrapolated allometry need not provide "more of the same" in a restricted and specialized manner. Allometric exaggerations may lead directly to a change in function. Rensch's favorite example is the vertebrate brain. Large vertebrates have absolutely larger (though relatively smaller) brains than small vertebrates (Jerison, 1973; Gould, 1975). Larger brains imply more neurons and greater intelligence (Rensch, 1958). The extrapolation of a brain/body relationship to larger body sizes in phyletic sequences may produce a brain with qualitatively new capacities.

Nonetheless, most hypermorphoses are, ultimately, restrictive specializations. The immense antlers of the "Irish Elk" (actually a Pan-Eurasian giant deer) cannot have encouraged the flexibility that long-term evolutionary success requires. They are clearly hypermorphic (Fig. 59): *Megaloceros* is the largest cervine deer and positive allometry of antler size with increasing body size can be demonstrated both in the interspecific regression of all cervids and in the intraspecific relationship among males of a single *Megaloceros* population (Gould, 1974).[7]

We are in a dilemma quite similar to that for progenesis: the most common heterochrony associated with an ecological strategy provides clear immediate significance but seems unpromising as a source of new evolutionary directions and subsequent diversity (retrospective significance). The general result of *r*-selected progenesis is reduction of flexibility by simplification of structure and loss of adult genes. The usual heterochronic effect of *K*-selective regimes is hypermorphosis, accompanied by restricted flexibility. This restriction is a result of specialization in structure produced by extrapolated differentiation following delayed maturation. In fact, I suspect that the loss of flexibility associated with these common heterochronies is one of the most important reasons for the extreme rarity of sustained evolutionary trends accompanied by increasing diversity.[8]

I have suggested that some progeneses may gain retrospective significance, because development usually works to mix adult and youthful characters in an *r*-selected organism relatively free from rigorous selection upon its morphology. I cannot claim a similar unbinding of morphology from selection in *K* regimes, for fine control of morphology in crowded and competitive ecospaces is a primary attribute of *K* selection; but I do believe that development also furnishes a path for increased flexibility in *K*-selected heterochrony.

The delay in maturation that often accompanies *K* selection need not lead to hypermorphosis. Hypermorphosis assumes the dissociation of maturation and somatic differentiation: ancestral rates of differentiation are maintained and delayed maturation permits their extrapolation beyond ancestral conditions. But maturation and differentiation may be strongly tied in development; delay of one leads to retardation of the other if the morphology of that delay is favored in a *K* regime. If both are delayed at the same rate, we obtain a proportioned giant and no heterochrony in the strict sense. But slow development often leads to the retention of juvenile growth rates and proportions. A general "matrix of retardation" (Chapter 10) presents juvenile features to selection at ancestral adult sizes. If advantageous, they can be retained and delayed even further to yield neoteny. Through neoteny, the evolutionary plasticity of unspecialized juvenile structures can be preserved in *K*-selected regimes. A larger, late-maturing animal with such morphological flexibility may be a primary candidate for unrestricted evolutionary change with great potential for novelty in adaptation and increase in diversity. As Linnaeus wrote in his perfunctory, though charming, diagnosis of the genus *Homo*: *nosce te ipsum* ("know thyself").

In summary, both modes of paedomorphosis—progenesis and neoteny—may represent an "escape from specialization," but not in the sense usually reserved for that phrase. In progenesis, the "escape" is made by morphology itself.[9] In neoteny, *K*-selected trends are "rescued" from overspecialization by the linking of retarded somatic development with delayed maturation (Table 8). In both cases, morphology is freed from a constraint that would provide immediate benefit at the expense of long-run flexibility. Creative progenesis must be an exceedingly rare event; creative neoteny, though less common than hypermorphosis, occurs often—for it requires no more than a direct selection for juvenile features in *K* regimes. It is rarer than hy-

Table 8. The macroevolutionary significance of heterochrony.

	Usual heterochronic event with strong immediate advantages	Less common heterochronic event with great macro-evolutionary potential
r selection	Progenesis, with specialization and simplification accompanying precocious maturation	Progenesis, with an experimental set of juvenile and adult characters freed from rigid selection upon morphology (exceedingly rare)
K selection	Hypermorphosis, with extended differentiation and increase in specialized complexity accompanying delayed maturation	Neoteny, with delayed maturation linked to retarded differentiation and retention of flexible juvenile morphology (moderately common)

permorphosis only because evolutionary trends usually lead to increasing differentiation of ancestral structures (see, for example, M. Williams, 1976, on how, under regimes of sexual selection in vertebrates, the advantages to males of appearing older will yield greater differentiation in phylogeny along ancestral ontogenetic pathways).

I envision two distinct macroevolutionary roles for the two forms of paedomorphosis—linked in a complex way with their differing immediate significances as *r* strategies for precocious maturation and *K* strategies for juvenile morphology. Progenesis may lead to the origin of higher taxa in a rare and serendipitous fashion when morphology is released from rigorous control by selection. Neoteny provides evolutionary flexibility as a moderately common pathway to adaptation in *K* environments; its occurrence is promoted by the common developmental correlation of delayed maturation with retarded somatic development. I believe that the key to human evolution lies in such a correlation (Chapter 10).

The Social Correlates of Neoteny in Higher Vertebrates

Delayed maturation has been correlated with crowding and intense intraspecific competition in several groups of birds. Ashmole (1963) proposed an individual advantage for delayed maturation in tropical oceanic birds, to circumvent Wynne Edwards' (1962) claim that it represents an individual sacrifice for group advantage. Most tropical oceanic birds lay a clutch of only one egg; their numbers are regulated by intense competition for food at spatially restricted breeding sites. A pronounced delay in maturation is usually associated with this reduction in clutch size and its necessary correlate of low mortality. The royal albatross, for example (Ashmole, 1963, p. 469), has the lowest adult mortality recorded for any bird and the longest known period of immaturity (nine years). Ashmole recognizes that a tendency towards early breeding can only be neutralized if birds attempting to breed when young suffer for their efforts and raise fewer successful progeny. Food is so short in the breeding area, he argues, that only the most efficient birds can raise any young successfully. Efficiency is a matter of learning, and several years of experience at a site are required for success in strong competition with conspecifics.

In grackles of the genus *Quiscalus,* polygamous males are generally excluded from breeding in their first year (though females breed successfully); this exclusion is correlated with a failure to attain full adult plumage. However, males of monogamous *Quiscalus quiscula* develop full plumage and reproduce in their first year (Selander, 1965). Wiley (1974) has detected in grouse a similar association among polygamy, large body size, strong sexual dimorphism, and delay in male maturation. Assuming that the sex ratio is near 1:1, the difference in maturation for males of polygamous and monogamous species must be a result of "crowding" in an unconventional sense. Each monogamous male will be able to attract a female, and breeding can begin early. In polygamous species, many males must be excluded and a longer period of growth and experience may be a prerequisite for breeding success.

These cases do not involve neoteny: birds, when adult, do not retain juvenile features (so far as the literature records), and there is no delay of somatic development relative to reproduction. We do, however, note the basic features of delayed maturation associated with *K* selection: stable, crowded environments, populations near carrying capacity, and intense intraspecific competition. In mammals, this association is often accompanied by neoteny.

Geist (1971) has specifically invoked neoteny in his behavioral studies of mountain sheep. The case has particular interest because neoteny occurs in two modes, each adaptive in a different sense: somatic development is retarded with respect to sexual maturation, and somatic development, when completed, remains in a juvenile state relative to ancestral conditions. When they first reach maturity (at 1.5 to 2.5 years), males look very much like fully adult females. But, unlike

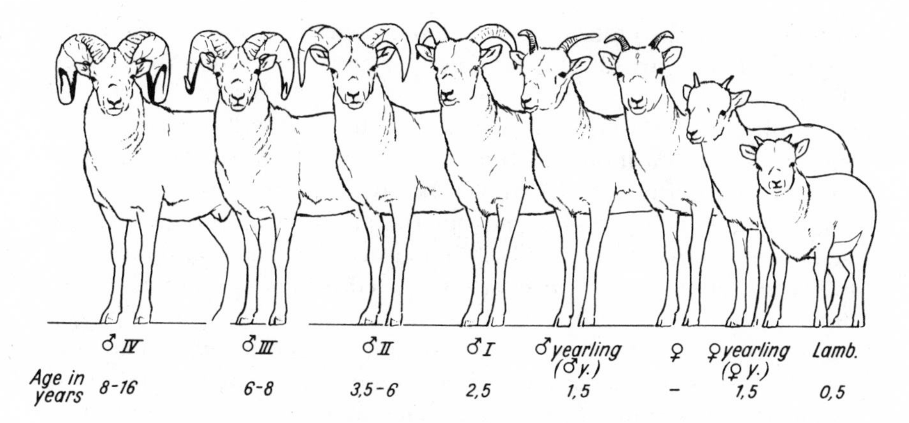

Fig. 60. Neoteny in bighorn sheep. Although sexually mature by class I (2.5 years) or earlier, somatic differentiation is retarded and males continue to develop through to class IV. These age-determined differences in form allow males to recognize rank by appearance (rather than by battle) and enhance stable and complex social structure. (From Geist, 1971.)

most mature mammals, they are nowhere near their final body size or level of somatic differentiation. They grow, continue their allometries (particularly horn growth), and mature behaviorally for a full 5 to 6 years beyond their sexual maturation. This long period of postreproductive change establishes an extensive series of distinguishable forms among mature males and allows them to maintain a dominance hierarchy on largely visual criteria: "In consequence one finds age-dependent horn-size differences among rams, which can be used by the animals to predict dominance rank of strangers. In other words, horns function as rank symbols" (Geist, 1971, p. 349).

Moreover, adult males at their final differentiation are still neotenic relative to ancestral sheep: they have lost both the neck ruff and the beard typical of old males in primitive sheep (p. 332). Geist also presents some behavioral tests of neoteny at final differentiation (p. 333). He compares Rocky Mountain bighorns (the most advanced race of an advanced species) with Stone's sheep (the most advanced race of a more primitive species). In both species, as individual rams mature they display their horns more, kick subordinates more often with their front legs, inflict fewer butts and mounts against subordinates, and display more often to equal-sized opponents. If bighorns are neotenic with respect to Stone's, fully adult males should display less, kick less, butt and mount more, and not display to equal-sized opponents as often—all of which they do.

This general juvenilization of behavior permits a more gregarious society; Geist regards his bighorn rams as more "cowardly" than other sheep and argues that they recognize a hierarchy better in declining more often to fight equals: "It appears that neotenization of sheep leads to a simplification and reduction of 'ritual' in mountain sheep. The mature individual resembles behaviorally the juvenile of the original parent population and not the adult. It appears then that neotenization produces a more 'immature' behavior, although the individuals are more highly evolved" (p. 345). Neotenization is clearly linked, as my general hypothesis requires, with intense intraspecific competition and relatively stable environments, both external and internal: "The social behavior of sheep appears to be an adaptation to create and maintain a predictable social environment . . . Thus rams do not fight for females, but for dominance or, conversely, for a reduction of uncertainty in social status" (p. 351).

Geist's pattern of delayed somatic development for structures of display and intraspecific combat is quite common among social mammals. Male giraffes use their heads as clubs in intraspecific fighting with other males. The various horns and exostoses of their upper skulls grow throughout life (Spinage, 1968). Elephant tusks also grow continuously; in males the rate actually increases after puberty, while in females the rate remains constant (Lawes, 1966).

Estes has recently generalized this conclusion in his comparison of territorial and nonterritorial African bovids. In territorial forms, growth virtually ceases at maturation. In nonterritorial species, where "reproductive success is based on absolute dominance rather than territorial dominance" (1974, p. 196), "male rank hierarchies are based on differential size achieved through prolonging growth in adults" (p. 195). By growth, Estes refers not only to increase in bulk, but also differentiation in structure; the hierarchies are "evidently based on continuing growth in the male, not only in body size but also in armament" (p. 196).

The African buffalo, for example, is adolescent at 2.5 to 3 years, but complete dentition is not achieved before 5 years, males are not fully armed before 6 or 7 years, they are not powerful enough to compete for breeding until 7 or 8 years, and they continue to increase in size beyond the 15th year: "Such an extended size range, apart from promoting an orderly and evolving rank hierarchy in closed herds, should facilitate the visual assessment by strangers of bulk and power . . . The bovine male rank hierarchy is based on body size as a function of age" (p. 196). "Bovid (and perhaps many other) male dominance hierarchies are based on a special physiological mechanism—continuing growth of males (only)—whereby differential size as a function of age helps determine rank" (p. 167). Estes regards nonterritorial neoteny as the most successful social strategy among African bovids: "The most socially advanced—i.e. the most gregarious—tend to achieve the highest population densities and to be ecologically dominant" (p. 166).

The most general argument linking neoteny with delayed maturation and *K* selection among mammals has been developed with great care, and over several decades, by Adolf Portmann and his school (for example, Müller, 1969, 1973; Wirz, 1950; Mangold-Wirz, 1966). (These works deserve to be far better known among English-speaking scientists. They have been neglected, in part because the articles are long, difficult, and invariably *auf Deutsch*; in part because Portmann's non-Darwinian perspective has evoked little sympathy. But the data can be easily extracted from their unorthodox context, and should be considered on their own merit.) Portmann and his students believe that large litters, rapid growth, and early birth of altricial (helpless and undeveloped) young represent the primitive condition among mammals. There has been a pervasive trend in advanced groups towards reduction in litter size, slower development, and lengthening gestation times, leading to the birth of precocial (advanced and capable) young. (The secondary altriciality of human babies is an interesting special case—see p. 369.) Retardation of development becomes the key adaptation in this change of "ontogenetic type." (J. B. S. Haldane was fond of arguing that reduction in litter size would decrease intrauterine competition for rapid development.) Lillegraven (1975, p. 715) has noted the general advantages of extended gestation in discussing the marsupial-placental transition. He emphasizes the increase in evolutionary flexibility provided by greater opportunities for heterochrony in the protected uterine environment.

The argument, so far, includes no neoteny among advanced adults. But Portmann then proposes that this complex of advanced characteristics, mediated by retardation in development, permits the evolution of higher cerebralization. For example, Eisenberg (1974, pp. 4–5) points out that litter size among Malagasy tenrecoids varies from over 30 in *Tenrec ecaudatus* to one or two in *Microgale talazaci*. *M. talazaci* has the most extended development, greatest recorded longevity, and relatively largest brain of any terrestrial tenrecoid. (Portmann insists on a "rule of precedence" [*Praezedenzregel*], whereby change in ontogenetic type must precede increase in cerebralization—but this seems poorly supported, even in work of his own students [for example, Müller, 1969, 1973].) This increase in brain size is a neotenic effect, intimately linked with a general retardation in developmental rate. The growth of the brain follows a characteristic pattern in mammalian ontogeny: strong prenatal positive allometry followed by equally strong postnatal negative allometry. The easiest developmental pathway to increasing brain size is a prolongation of embryonic growth rates and stages to later ages and sizes through a retardation in general development. Since Portmann's change in ontogenetic type involves just such a retardation, and since it brings increased cerebralization in its wake, I regard this increase in brain size as a neotenic effect (this argument is developed more fully in Chapter 10, in connection with human evolution).

Other neotenic characters are usually associated with increased brain size in slowly growing mammals with small litters. The margay, *Felis wiedii,* lives deep in the tropical forests of Central and South America (a good *K* environment). The adults have the relatively large eyes and short muzzles so characteristic of juvenile felines (R. Fagen, personal communication). It has a gestation period of 83 to 84 days and usually gives birth to a single young (ocelots have much higher twinning frequencies; litter size of domestic cats averages four). I would make a vulgar appeal to intuition and suggest that our association of "cute" features with adult mammals of high intelligence offers strong support for the correlation of neotenic characters with large brains. Our concept of "cute" is strongly determined by the common traits of babyhood: relatively large eyes, short face, smooth features, bulbous cranium. The preservation of this complex in advanced, adult mammals argues for neoteny.

Several American ecologists and ethologists have related Portmann's ontogenetic types to their immediate ecological situations of *r* and *K* selection rather than to the vague, general, and inherent progressive tendencies that Portmann favors. Eisenberg has studied ontogenetic type and cerebralization among edentates, noting that the small litters, slow development, and intense parental care of big-brained species strongly suggest *K* selection:

> A species, such as *Dasypus novemcinctus,* reproduces at an early age and typically has 4 to 8 young. Many members of the family Dasypodidae [armadillos] have similarly high reproductive rates. This suggests that many of the species are adapted to make use of transient resources at rapidly fluctuating densities. On the other hand, the anteaters and sloths typically have a single young; they show a prolonged gestation, and a long period of parental care. In their capacity to reproduce, they suggest a *K* strategy where a species is adapted to a stable niche and there is no special advantage to having a high reproductive rate, rather emphasis is placed upon the ability of an individual to retain a home range for its exclusive use. In short, reproduction is geared to maintain the population at a stable carrying capacity. (1974, p. 7)

Sloths and anteaters have much larger brains than armadillos. "This correlation of brain development, litter size, parental care, and longevity is a complex which shows itself again and again in other taxa having evolved convergently in response to several inter-related selective pressures" (p. 5).

Fagen has linked the occurrence of play in mammals to large brains, slow maturation, and *K* selection. He argues that *r* habitats with catastrophic environmental variability require increasingly rapid development and eliminate the protracted juvenile period "whose existence appears to be a necessary condition for play" (1974, p. 855). Among rodents and dasyurids, small, rapidly maturing species do not play, while larger, more slowly maturing species do. Eayrs (1964) has demonstrated that artificially accelerated neural maturation leads to impaired intelligence in adult rats. The extent of play in early ontogeny correlates with levels of sociality in canids: "A striking relationship emerges from comparative developmental studies on canids, namely that the more social canids fight less and play more very early in life than do the less social canids. The delay in the appearance of rank-related aggression may be responsible for the development of a coordinated social group" (Bekoff, in press; see also Bekoff, 1972, p. 424).

By now, this associated complex of characters—neoteny, large brains, *K* selection, slow development, small litters, intense parental care, large body size—must have suggested a look in the mirror. A neotenic hypothesis of human origins has been available for some time, but it has been widely ridiculed and ignored. Nonetheless, I believe that it is fundamentally correct and that the framework I have established may help to vindicate it.

—10—

Retardation and Neoteny in Human Evolution

The beast and bird their common charge attend
The mothers nurse it, and the sires defend;
The young dismissed, to wander earth or air,
There stops the instinct, and there ends the care.
A longer care man's helpless kind demands,
That longer care contracts more lasting bands.
Alexander Pope, *Essay on Man,* 1733

The Seeds of Neoteny

With the consummate arrogance that only an American millionaire could display, Jo Stoyte set out to purchase his immortality. Dr. Obispo, his hired scientist, discovered that the fifth earl of Gonister had, by daily ingestion of carp guts, prolonged his life into its third century. Stoyte and Obispo rushed to England, broke into the earl's quarters and discovered to Stoyte's horror and Obispo's profound amusement that man in his allotted three score years and ten is but an axolotl in its pond. We are neotenic apes and the fifth earl had grown up:

"A foetal ape that's had time to grow up," Dr. Obispo managed at last to say. "It's *too* good!" Laughter overtook him again. "Just look at his face!" he gasped . . . Mr. Stoyte seized him by the shoulder and violently shook him . . . "What's happened to them?" "Just time," said Dr. Obispo airily. Dr. Obispo went on talking. Slowing up of developmental rates . . . one of the mechanisms of evolution . . . the older an anthropoid, the stupider . . .

the foetal anthropoid was able to come to maturity . . . It was the finest joke he had ever known. Without moving from where he was sitting, the Fifth Earl urinated on the floor.

So wrote Aldous Huxley in his novel *After Many a Summer Dies the Swan*—a few years after his brother Julian's important work on delayed metamorphosis in amphibians and on the heels of Louis Bolk's fetalization theory of human origins.

If a good public press is the sign of a theory's impact, then the proponents of human neoteny should be satisfied with the suffusion of their theory into popular consciousness—though its influence can only be described as modest relative to that once wielded by the opposing concept of recapitulation (Chapter 5). At least, human neoteny has motivated a long defense of Rudolf Steiner's mysticism (Poppelbaum, 1960) and played an important role in several treatises in the current fad for "pop ethology" (Jonas and Klein, 1970; Shepard, 1973).

The notion of human neoteny has its roots in two obvious facts: the striking resemblances between juvenile pongids and adult humans and the obliteration of this similarity during pongid ontogeny by strong negative allometry of the brain and positive allometry of the jaws (Fig. 61). Louis Bolk did not discover these phenomena in his fetalization theory of the 1920s; they had been recognized as soon as juvenile pongids had reached the zoos and museums of Europe. Moreover, the subject was not merely treated as an incidental observation; for Etienne Geoffroy Saint-Hilaire (1836a, 1836b), in describing the form and behavior of a young orang-utan recently brought to Paris, placed his observations firmly in the context of recapitulatory theory.

Geoffroy began by noting the extent of ontogenetic allometry and retracting his earlier conclusion, based on museum skeletons, that young and old orangs represented two separate genera:

Could we have dared hope in 1798 that such different crania, one taken from a juvenile, the other from an adult, would reveal the fact of successive development in a single species? For we have a distance greater than that between the genera *Canis* and *Ursus*. (1836a, p. 94)

He then attributes these allometries to the early cessation of growth in the orang brain:

The skull of a young orang strongly resembles that of a human child. The cranial vault, which faithfully represents the form of an organ it protects, could be taken for that of a small human were it not for the more forward position of the maxillaries and the much larger cutting teeth. But it happens, as

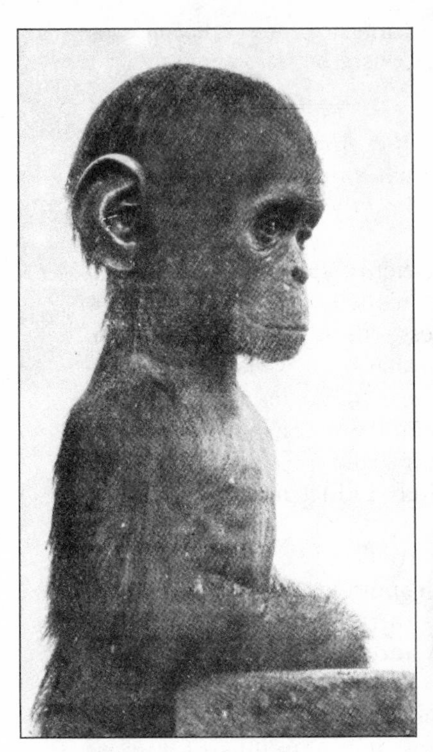

Fig. 61. Baby and adult chimpanzee from Naef, 1926b. Naef remarks: "Of all animal pictures known to me, this is the most manlike" (p. 448).

a result of the progress of age, that the contents practically cease to increase while the container grows strongly and continually. (1836b, pp. 6–7)

Geoffroy extended his observations to the habits and behavior of his young orang:

In the head of the young orang, we find the childlike and gracious features of man . . . We find the same correspondence of habits, the same gentleness and sympathetic affection, also some traits of sulkiness and rebellion in response to contradiction . . . on the contrary, if we consider the skull of the adult, we find truly frightening features of a revolting bestiality [*formes vraiment effroyable et d'une bestialité révoltante*]. (1836a, pp. 94–95)

Yet humans display an "inverse system of development" (1836b, p. 7), for our brain does not so quickly cease its growth: both form and behavior remain close to the juvenile state. Could human ontogeny represent an "arrest of development" with respect to the next lower link in the chain of being? Geoffroy preferred to save recapitulation by regarding the adult orang as anomalous—as a form developed "too far" in its own ontogeny and therefore not recapitulated during human development: "This new revelation of such a great departure from the rules [of recapitulation] that we have discovered is a teratological fact of the greatest importance" (1836a, p. 94).

The resemblance of adult humans to juvenile apes was treated as an anomaly throughout the heyday of recapitulation. But single ugly facts, despite Huxley's aphorism, do not destroy great theories. Recapitulationists simply followed Haeckel's strategy in arguing that the anomalies were few, recognizable, and unimportant (Chapter 6). But man posed a special problem: if the most "important" of all species had evolved by retardation, then exceptions to recapitulation could not all be relegated to insignificance. E. D. Cope considered the problem in great detail and admitted that many human features had evolved by retardation. But he quickly added that these retarded features were not involved in our superiority, and that the progressive features of our mental development displayed acceleration and recapitulation:

> I have pointed out that in the structure of his extremities and dentition, he agrees with the type of Mammalia prevalent during the Eocene period. Hence in these respects he resembles the immature stages of those mammals which have undergone special modifications of limbs and extremities . . . I have also shown that in the shape of his head man resembles the embryos of all Vertebrata, in the protuberant forehead, and vertical face and jaws. In this part of the structure most Vertebrata have grown further from the embryonic type than has man, so that the human face may be truly said to be the result of retardation. Nevertheless, in the structure of his nervous, circulatory, and for the most part, of his reproductive system, man stands at the summit of the Vertebrata. It is in those parts of his structure that are necessary to supremacy by force of body only, that man is retarded and embryonic. (1896, pp. 204–205)

Cope's defense of the mainstream prevailed, and recapitulatory thinking dominated concepts of human evolution until the collapse of Haeckel's doctrine itself (Chapter 6). In the best known of more recent usages, Wood Jones sought man's origin among the tarsioids, arguing that the early ontogenetic fusion of the maxillary-premaxillary suture precluded an origin from monkeys, apes, and even from australopithecines (for these forms retain the suture longer). Writing about the fusion of the two bones, Wood Jones deemed it "probable that its early ontogenetic accomplishment is a guarantee of its early phylogenetic acquirement" (1929, p. 319).

The Fetalization Theory of Louis Bolk

The collapse of Haeckel's biogenetic law and the rise of Garstang's neoteny virtually guaranteed that a century of observations would be gathered to yield a paedomorphic theory of human origins. J. Kollmann (1905), inventor of the term "neoteny," paved the way with a curious theory that humans had originated from pygmies who had simply retained their juvenile features during phyletic size increase. The pygmy progenitors, Kollmann added, probably arose from juvenile apes that had lost the ancestral tendency to regress (*zurücksinken*) during ontogeny to lower levels of cephalization: "The juvenile orang-utan is doubtlessly better qualified for human ancestry than the ape of Trinil" (1905, p. 19).[1]

Louis Bolk (1866–1930), professor of human anatomy at Amsterdam developed the inevitable idea in a long series of papers (1915, 1923, 1924, 1926a, 1926c, 1929, for example), culminating in his pamphlet of 1926, *Das Problem der Menschwerdung*. Bolk's theory of fetalization set the stage for all later discussion. The subsequent debate has been murky and confused because Bolk's arguments have been presented outside the context of his philosophical positions. His insight has been ridiculed in the light of modern doctrine and dismissed in toto because he linked valid and important data to evolutionary views now rejected. In presenting this detailed exposition of Bolk's views, I am trying to establish a ground for the acceptance of his basic notion by referring the old observations that lie at its core (and that date at least to Geoffroy) to a philosophical context of current orthodoxy. I shall try to support three statements: (1) It is irrelevant that Bolk's evolutionary theory seems outdated or even foolish today; his theory was reasonable in his time and he supported it cogently. (2) The data that he presented can survive the collapse of his explanatory structure. (3) We must try to identify the "philosophical baggage" that underlies all theories—both to understand why a man says what he does and to aid in rescue operations when new philosophies require a separation of baby from bath water.

I want to rescue Bolk's data—and his basic insight—from the evolutionary theory to which he tied it and from which it has not been adequately extracted.

Bolk's Data

To support the argument that we evolved by retaining juvenile features of our ancestors, Bolk provided lists of similarities between adult humans and juvenile apes: "Our essential somatic properties, i.e. those which distinguish the human body form from that of the other Primates, have all one feature in common, viz they are fetal conditions that have become permanent. What is a transitional stage in the ontogenesis of other Primates has become a terminal stage in man" (1926a, p. 468). In his most extensive work Bolk (1926c, p. 6) provided an abbreviated list in the following order:

1. Our "flat faced" orthognathy (a phenomenon of complex cause related both to facial reduction and to the retention of juvenile flexure, reflected, for example, in the failure of the spheno-ethmoidal angle to open out during ontogeny).
2. Reduction or lack of body hair.
3. Loss of pigmentation in skin, eyes, and hair (Bolk argues that black peoples are born with relatively light skin, while ancestral primates are as dark at birth as ever).
4. The form of the external ear.
5. The epicanthic (or Mongolian) eyefold.
6. The central position of the foramen magnum (it migrates backward during the ontogeny of primates).
7. High relative brain weight.
8. Persistence of the cranial sutures to an advanced age.
9. The labia majora of women.
10. The structure of the hand and foot.
11. The form of the pelvis.[2]
12. The ventrally directed position of the sexual canal in women.
13. Certain variations of the tooth row and cranial sutures.

To this basic list, Bolk added many additional features; other compendia are presented by Montagu (1962), de Beer (1948, 1958), and Keith (1949). The following items follow Montagu's order (pp. 326–327) with some deletions and additions:

14. Absence of brow ridges.
15. Absence of cranial crests.
16. Thinness of skull bones.
17. Position of orbits under cranial cavity.
18. Brachycephaly.
19. Small teeth.
20. Late eruption of teeth.
21. No rotation of the big toe.
22. Prolonged period of infantile dependency.
23. Prolonged period of growth.
24. Long life span.
25. Large body size (related by Bolk, 1926c, p. 39, to retardation of ossification and retention of fetal growth rates).[3]

These lists from Bolk and Montagu display the extreme variation in type and importance of the basic data presented by leading supporters of human neoteny. There are obvious difficulties. The features, for example, are not all independent: the position of the orbits (17) is not unrelated to the expansion of the brain (7); prolonged infantile dependency (22) and prolonged growth (23) are aspects of a more basic retardation in temporal development. Moreover, two very different phenomena are mixed together: the slowing down of developmental rates (8, 20, 22, 23, 24) and the retention of juvenile shapes (all others). I shall explore the general problems of this enumerative approach in the next section.

The cataloguing of such similarities between "average" adult humans and juvenile apes dominates the literature on human neoteny. Two other arguments have, however, played an important, if subordinate, role.

1. The intermediate status of hominid fossils. The structural sequence of *Australopithecus africanus, Homo erectus,* and *Homo sapiens* exhibits a progressive retention of juvenile proportions by adults as the brain increases and the jaw decreases. Moreover, the juveniles of Taung and Modjokerto prophesy, so to speak, the proportions later attained by descendant adults; no reference to the idealized juvenile form of a hypothetical ancestor is needed. Although not emphasized by Bolk, this argument has been urged by Schindewolf (1929), Buxton and de Beer (1932), Overhage (1959), and Montagu (1962), and strongly supported in implicit arguments by Boyce (1964).

2. The differentiation of human races. Bolk (1926c, 1929) tried to buttress his general argument by claiming that the major morphological differences among human races had their origin in differing intensities of neoteny: "If the entire genus 'Homo,' after having arisen from its pithecoid ancestral form, would have been equally retarded in all its ramifications, the terminal shape of all people living today would be identical, save for the differences due to nutrition and environment. But if the factor of retardation has operated less actively in one group than in another, this must have given rise gradually to differences in the terminal form of these groups—to differentiation of the genus" (1929, p. 6).

As a white European in the racist tradition, Bolk labored mightily to place Caucasians on the pinnacle, despite an embarrassing compendium of features demonstrating more retarded development in other peoples, particularly Mongoloids (see list of Montagu, 1962, p. 331, and the uncomfortable special pleading of Bolk, 1929, p. 28, to dismiss the more paedomorphic features of Mongoloids). "Qualitative differences in fetalization and retardation are the base of racial inequivalence. Looked at from this point of view, the division of mankind into higher and lower races is fully justified . . . The white race appears to be the most progressive, as being the most retarded" (1929, pp. 25–27). "In his fetal development the negro [*sic*] passes through a stage that has already become the final stage for the white man" (1926a, p. 473). I have analyzed the racist argument in Chapter 5, and shall explore the subject no further here.

A different approach to our intraspecific differentiation has been little pursued but offers great promise: the comparison of intra- with inter-populational variation. Abbie (1958) calculated a significantly negative correlation between cephalic index and stature among male Australian aboriginals; similar data for intrapopulational variation of other peoples has been available since Boas' early study of 1899. Variation in adult size within a population follows the ontogenetic trend, with skulls becoming relatively longer as they depart from juvenile brachycephaly. But the correlation of mean cephalic index with mean stature for 50 different human groups is insignificantly positive (.211 for males, .132 for females, from Abbie, 1958). Differences between large and small peoples do not follow the ontogenetic trend of departure from juvenile proportions; the degree of brachycephaly of smaller peoples is retained or even increased among larger peoples. Abbie concludes: "With improvement in environment headform tends to adhere more to the fetal type: neither much longer nor much shorter than the mean fetal cranial index at about the middle of the human scale. It seems then, that paedomorphism is still an active factor in determining at least one human character" (1958, p. 203).

Bolk's Interpretation

Bolk emphasized that his list of features reflected two different phenomena: the physiological retardation of development ("the retardation hypothesis of anthropogenesis"—1926a, p. 470) and the somatic retention of ancestral juvenile proportions ("the fetalization theory of anthropogeny"—1926a, p. 469). The two phenomena are joined in "the narrowest causal connection" ("in engstem kausalen Zusammenhang"—1926c, p. 12) because delayed development prolongs fetal growth rates and conserves fetal proportions: "The essential in his form is the result of a fetalization, that of his life's course is the result of retardation. These two facts are closely related, for, after all the fetalization is the necessary consequence of the retardation of morphogenesis" (1926a, pp. 470–471). If fetalized form is a consequence of retarded development, then the key to human evolution lies in the cause of this retardation: "There is no mammal that grows so slowly as man, and not one in which the full development is attained at such a long interval after birth . . . What is the essential in Man as an organism? The obvious answer is: The slow progress of his life's course . . . This slow tempo is the result of a retardation that has gradually come about in the course of ages" (1926a, p. 470). "Human life progresses like a retarded film" (1929, p. 1).

Bolk conceived this physiological retardation as pervasive and general. This claim provided the central focus for later debate and set the basis for most precipitate rejections of human neoteny: "The biological basis of the fetalization principle consists of a checking of human development *in its entirety*" (1924, p. 344). "The body as a whole was transformed because its development stopped at a younger stage, a process that was brought about gradually of course" (1926a, p. 469). We are, in essence, a transitional stage in the early ontogeny of a primate ancestor, brought forward to adulthood and stabilized at large size by a general retardation of development: "Our ancestor already possessed all primary specific characteristics of contemporary man, but only during a short phase of its individual development" (1926c, p. 8).

But how could such a position be maintained? Man, in toto, is surely no primate fetus, for he bears unmistakable characters of recent adaptation to features of adult life, such as upright posture. To circumvent this objection, Bolk followed the same strategy that Haeckel had pursued in dividing characters into the essential and the exceptional. Haeckel had separated the palingenetic markers of true ancestry from adaptations to larval life that falsified the repetition of phylogeny in ontogeny. Bolk divided his characters into "primary" and "consecutive" (1926c, p. 5). Primary characters are those attained by the essential process of human evolution: retarded development and its attendant fetalization of form. Consecutive characters reflect the minor modeling of this primary form by adaptive requirements of human habits and environment. They are subsidiary and confusing; they must be identified and separated to reveal the basic cause of human evolution.

If our essential characters arose all together by a gradual and coordinated retardation of development, what caused the retardation itself? Bolk argues that a simple alteration of the endocrine system must have been responsible: "The gradual retardation of the life's course of the ancestors of Man with all the consequent effects, both as regards his morphological features and his functional properties, must have had for its immediate cause a modification of the action of the endocrine system of the organism . . . In controlling the intensity of the metabolism these hormones can inhibit or promote the growth. In Anthropogenesis an inhibitive action appears to have gradually increased in significance, the rate of development became slower, the progress of development more retarded" (1926a, pp. 471–472). Therefore, Bolk argues in his most striking phrase:

> If I wished to express the basic principle of my ideas in a somewhat strongly worded sentence, I would say that man, in his bodily development, is a primate fetus that has become sexually mature [*einen zur Geschlechtsreife gelangten Primatenfetus*]. (1926c, p. 8)

What a tenuous position for the crown of creation! An ape arrested in its development; holding the spark of divinity only through a chemical brake placed upon its glandular development; retaining a corporeal reminder, so to speak, of original sin—the potential of sinking again into the Tertiary abyss, should that brake ever be released: "You will note that a number of what we might call pithecoid features dwell within us in latent condition, waiting only for the falling away of the retarding forces to become active again" (1926c, p. 15). Pete, Aldous Huxley's young enthusiast, put it more pungently: "There's a kind of glandular equilibrium . . . Then a mutation comes along and knocks it sideways. You get a new equilibrium that happens to retard the developmental rate. You grow up; but you do it so slowly that you're dead before you've stopped being like your great-great-grandfather's fetus" (1939, p. 85).

Bolk's Evolutionary Theory

Bolk's notion of fetalization contains a host of elements that must clearly be rejected in the light of current knowledge. These include: (1) the division of characters into primary results of retarded development and secondary features of "merely" adaptive significance; (2) the insistence that retardation affects all essential features to the same degree in a single coordinated event; (3) the search for a cause of retardation in a simple chemical alteration of the glandular system; (4) the complete absence of any consideration for the adaptive significance of such retardation. With so much to question, we might conclude that the whole edifice is rotten to the core. But it is important to recognize that all these rejected contentions flow naturally from Bolk's view of evolution—and that this view, however untenable today, was popular and reasonable in its time.

In short, Bolk was not a Darwinian. He believed that inner factors controlled the direction of evolution by transforming entire organisms along harmonious and definite paths of vitalistic determination. If evolution were directed by the external factors of environment, then harmonious development would be impossible because individual characters have different adaptive requirements. Hence, all Darwinian adaptation must be secondary and superficial—mere surface molding upon an underlying transformation directed by inner agencies. There is no room for mosaic evolution in Bolk's system.

> For our primary physical features carry a common stamp. This condition is incompatible with the idea that they developed independently from each other in the course of time, each as a consequence of its own causal factors. The unitary character indicates a common cause . . . The fetalization of form cannot be the consequence of external forces, of influences that work upon the organism from the outside. It was not the effect of an adaptation to changing external conditions; it was not determined by a "struggle for life"[4]; it was not the result of a natural or sexual selection . . . It was an internal, functional cause . . . human evolution as the result of a unitary, organic principle of development. (1926c, p. 9)

In a later article, Bolk generalized his anti-Darwinian beliefs: "It becomes more and more apparent that the evolutionary factors of Darwin's and Lamarck's theories do not suffice to explain the origin of new species. This problem appears to us in a somewhat different light than to the preceding generation. There is a general tendency today to trace back the origin of new species to the action of internal factors" (1929, p. 3).

If internal forces direct evolution, then adaptive significance plays no essential role in the assessment of evolutionary change. Upright posture, for example, might be adaptive, but it arose as a secondary consequence of such fetalized features as retention of cranial flexure and the forward position of the foramen magnum (1926c, p. 6). The problem of human evolution reduces to the search for a single, internal cause: "Through this union of retardation as a result with internal secretion as a cause, the problem of human evolution becomes purely physiological" (1926c, p. 13).

I am continually struck by the power and persistence of ontogenetic metaphors in phyletic thinking. As yet another example, I note that Bolk defended his vitalistic belief in internally directed and coordinated trends by comparing evolution to ontogeny (1926c, p. 12): "Evolution is for organized nature what growth is for the individual; and [for the former] as for the latter, outer factors have only a secondary influence. They can never play a creative role, only one of modelling what is already there . . . What we understand as evolution is the manifestation of differentiation in the cosmic macroorganism."

A Tradition of Argument

When I survey the literature on paedomorphosis in human evolution, I cannot help but recall such catch phrases as "the dead hand of the past"—for Bolk's explanation and the rejected evolutionary theory hidden within it established a strong tradition of argument. This tradition, in subtle ways, still hinders the application of new and more fruitful approaches. Since Bolk insisted on a *harmonious* retardation of *all* "essential" features, the *experimentum crucis* has been cast in the following way: enumerate as many morphological features as you can and judge which are paedomorphic and which are not. If you regard a vast majority, or perhaps merely a "basic" list, as paedomorphic, you accept the hypothesis; if you disagree with this claim, you reject the hypothesis.

Thus, Weidenreich (1932) based his rejection on the fact that several adult features are closer to their fetal condition in Peking Man than in modern *Homo sapiens*. Since "Sinanthropus" is our direct ancestor, Weidenreich argues, the last stages of human evolution cannot have evolved by a general retardation in development. Ewer's rejection is squarely in the enumerative tradition: "The paedomorphic characters although undoubtedly a necessary element in the 'ascent of man' are secondary consequences of the nonpaedomorphic key characters" (1960, p. 180). And Delsol and Tintant (in press) write: "The several paedomorphic characters which incontestably accompany the appearance of man represent only a quite secondary element of the traits that mark this evolution. The principal elements of the phenomenon correspond to divergences."

Supporters of paedomorphosis scrutinize the same lists of features but apply different weights. Schindewolf classed as paedomorphic "the typical characters of the human skull—i.e. the characters that make up the essence of the human skull and that distinguish it so fundamentally from that of the lower mammals" (1929, p. 762). Konrad Lorenz argues: "The number of persistent juvenile characters in human beings is so large, and they are so decisive for his overall habitus, that I can see no cogent reason for regarding the general juvenescence of man as anything other than a special case of true neoteny" (1971, p. 180).

I believe that this enumerative tradition is the worst way to carry on the discussion. Modern evolutionary theory has driven Bolk's notion of harmonious development into obsolescence: to invoke it now as a criterion for judgment is to set up a straw man of surpassing weakness. No Darwinian supporter of retardation as a major element in human evolution can deny that many distinctive features are not paedomorphic; the concept of mosaic evolution practically requires such a belief. Thus Abbie, the staunchest advocate of paedomorphosis in recent years, writes: "It is the balance between these two [paedomorphosis and gerontomorphosis] in different parts of the body that produces the distinctively human form among primates and, to a large extent, the distinctions of physically different ethnic groups among humans" (1958, p. 202).

Other reasons for rejecting the enumerative tradition reinforce some of the basic themes of this book. In discussing the fall of Haeckelian recapitulation (Chapter 6), I advanced the general argument that theories in natural history are not made or broken by the enumeration of cases. The biogenetic law did not collapse under the weight of contradictory evidence; it fell because it became unfashionable in practice and untenable in theory. Haeckel knew all the empirical objections that his opponents advanced: he merely rejected them as insignificant or incorporated them as an expected and permissible category of exceptions. Bolk did much the same in distinguishing primary from consecutive features. Theories are tested by examining the processes that yield the results and by judging whether these processes are consistent with more general theories.[5] They are not proved or disproved by amassing a compendium of empirical results, listing those items that fit the theory and those that do not, and comparing the lists.

The cataloguing of results is especially inappropriate in this case. As I emphasized in Chapters 7–9, the results of heterochrony are particularly confusing because recapitulation and paedomorphosis arise from different processes with distinct evolutionary significances. Moreover, the central failure of de Beer's approach to heterochrony lay in his attempt to classify the results (in a large number of complex and inconsistent categories) rather than to assess the processes.

In short, Bolk's belief in a harmoniously correlated paedomorphosis of all essential features is untenable in theory. The evolutionary direction of each feature is controlled by natural selection; the capacity for independent variation of characters is very great. An absence of paedomorphosis in some characters is inevitable and poses no threat to the notion that paedomorphosis played a central role in human evolution. If a large suite of human features are paedomorphic, we must seek the adaptive significance of retarded development in our evolution. We cannot hope to understand this significance by compiling lists of morphological features. Such lists are tabulations of results. We cannot simply enumerate them and hope to work backwards by induction to a correct view of heterochrony in human evolution. We must focus on the processes themselves. The processes of heterochrony are acceleration and retardation. Our paedomorphic features are linked to retardation in development; it is to this retardation that we must fix our attention.

Retardation in Human Evolution

Although Bolk's theory has been rejected, his central insight may be reinstated as the foundation of a modern analysis: retardation in development must be distinguished from fetalization of form—the first is a process that may or may not yield the second as a result. Curiously, this point has generally been lost in the enumerative tradition that followed Bolk's pronouncements. The evidences of retarded development have simply been included within lists of paedomorphic characters, and Bolk's distinction has been blurred.

I believe that human beings are "essentially" neotenous, not because I can enumerate a list of important paedomorphic features, but because a *general, temporal retardation of development has clearly characterized human evolution. This retardation established a matrix within which all trends in the evolution of human morphology must be assessed.* This matrix does not in itself guarantee a central role for paedomorphosis,[6] but it certainly provides a mechanism for such a result, if this result be of selective value. This mechanism was utilized again and again in human evolution because retarded development carried a set of potential consequences with it: prolongation of fetal growth rates leading to larger sizes* and the retention of juvenile proportions. Is not such a system the proximate cause for evolutionary increase of the human brain?

Slijper (1936) based a famous critique of human neoteny upon the

* Or, if the feature is one that develops late in ontogeny, the retention of fetal growth tendencies usually leads to a reduction in size and the retention of juvenile proportions; an example is the human face. Several authors have fallen into the trap of requiring a correspondence in size and internal arrangement between juvenile ancestor and adult descendant before making an assessment of paedomorphosis. Thus, since the human brain has added so many billions of neurons to the feature of similar shape in ancestral juveniles, paedomorphosis is denied. "In the evolution of man from his ancestors, neoteny has taken place in respect of several features which show fetalization. At the same time, of course, in other directions, the evolution of man has involved progressive change of vast importance, some of which, however, might not have been possible (e.g. the development of the brain) had it not been for certain features of neoteny (e.g. the delay in closing the sutures of the skull)" (De Beer, 1958, p. 76). But this represents another pitfall in the use of static morphology as a criterion; rates and processes of growth are the appropriate standard—and the human brain is paedomorphic because it has increased by prolonging to later times and larger body size (even past birth) the characteristic positive allometry generally confined to fetal stages in primates and other mammals.

fact that retardation does not automatically entail a retention of juvenile features. I have supported this claim in Chapter 9 by arguing that retardation confined to the gonads produces hypermorphosis with recapitulation. Mosaic evolution affirms the possibility of complete dissociation between sexual maturation and somatic development. Nonetheless, many aspects of somatic development are usually linked with maturation. Hubbs (1926) tried to generalize the rule that slow growth in fishes implies a retardation in somatic differentiation. Wigglesworth's (1952) studies on *Rhodnius* confirm this correlation: in experimental regimes of lowered temperature, increased duration of the nymphal stage was accompanied by a mild degree of paedomorphosis. In humans, retardation has affected many systems in concert, somatic as well as germinal.

The postulate that human form is paedomorphic has not been placidly accepted; the status of nearly every feature has been doubted and debated (see next section of this chapter). On the other hand, the statement that a general retardation characterizes human evolution can scarcely be denied. As Bolk remarked: "The life course of man runs slowly; this is a fact that can be easily ascertained through direct comparison" (1926c, p. 19).[7]

Our retardation relative to apes and other primates has been extensively documented, but another aspect of retardation has been widely ignored in discussions of human heterochrony: the general retardation of development in primates versus other mammals—a phenomenon that is historically prior to later differentiation within the primates and that establishes a tendency that is merely extended by the pronounced retardation of humans.

Standard allometric plots of gestation period, age at sexual maturity, and lifespan versus body size (Fig. 62, for example) express this retardation. Primates live longer and mature more slowly than other mammals of comparable body size (Sacher, 1959, p. 128). We reach puberty at about 60 percent of our final body weight, chimpanzees at slightly less than 60 percent. Most laboratory and farm animals reach puberty at about 30 percent of final weight (Bryden, 1968).

Some recent studies indicate that retardation begins early in human development and increases continually throughout embryogenesis, at least by comparison with nonprimate mammals. Otis and Brent (1954) compared the appearance of 147 stage marks in the prenatal development of mouse and human. The sequential order is essentially the same in both species, but early stages take two to four times as long to develop in humans while later stages take five to fifteen times as long. Adolph (1970) studied the appearance of 16 physiological features in the development of 12 species of mammals. Again the order of development is about the same, but while a mouse day equals

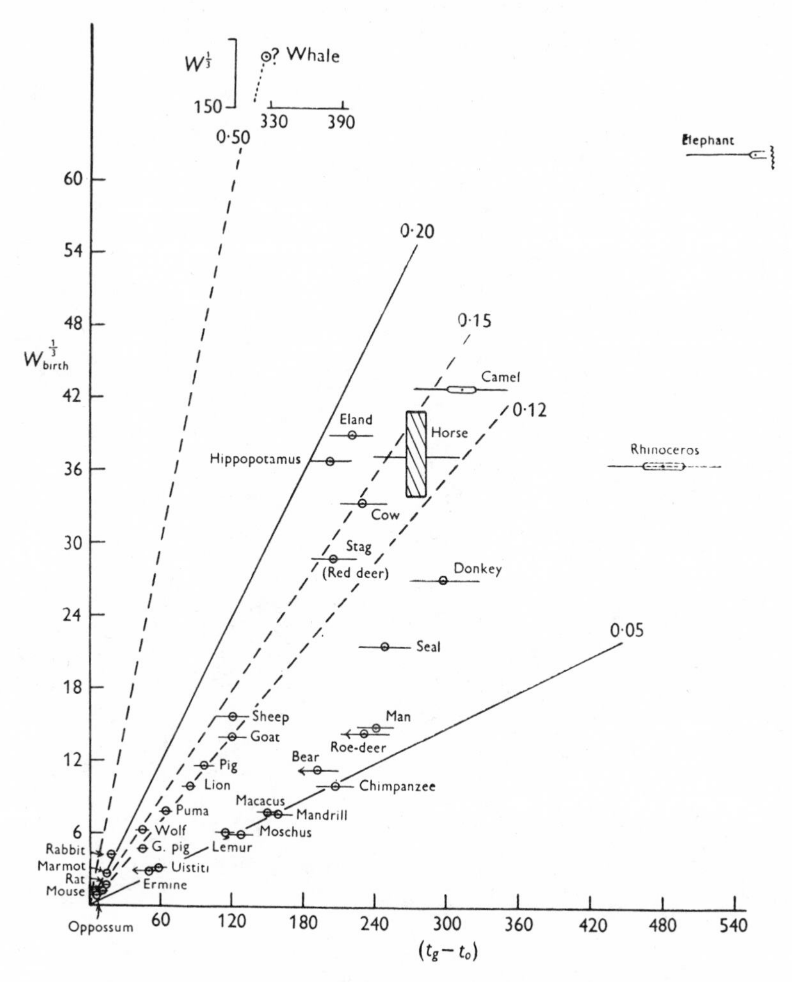

Fig. 62. Retardation in primates expressed as lengthening of gestation relative to birth weight. For any given birth weight, the average primate has a much longer gestation period than the average nonprimate mammal. (From Huggett and Widdas, 1951.)

about 4 human days early in embryogenesis, this eventually increases to about 14 human days as human developmental rates slow down.

This pattern of retardation continues during primate evolution: apes are generally larger, mature more slowly, and live longer than monkeys and prosimians. "The total slowing up of human development" (Abbie, 1958, p. 209) relative to all other primates is even more marked. Table 9 (from Abbie, 1958, with addition of one column from Reynolds, 1967) summarizes the data on temporal retardation of development in humans. (Figures of other authors differ slightly, but even the extremes of variation do not alter the general pattern.)

Table 9. Retardation in humans. (From Abbie, 1958; last column from Reynolds, 1967.)

Primate	Gestation (weeks)	Complete hair covering	Carpal ossification centers at birth	1st dentition (months)	2nd dentition (years)	Growing period (years)	Life span (years)	Sexual maturity, female (years)
Macaque	24	During gestation	All centers	0.6–5.9	1.6–6.8	7	25	—
Gibbon	30	Onset during gestation completed after birth (Gibbon, Orangutan, Chimpanzee, Gorilla)	2–3	1.2–?	?–8.5	9	33	—
Orangutan	39		2–3	4.0–13.0	3.5–9.8	11	30	—
Chimpanzee	34		2	2.7–12.3	2.9–10.2	11	35	9
Gorilla	37		—	3.0–13.0	3.0–10.5	11	35	6–7
Homo sapiens	40	Never completed	0	6.0–24.0	6.0–20.0	20	70	13

The length of human gestation provides a single important exception to the general retardation of our development with respect to other higher primates. Human gestation is slightly longer than that of great apes, but the differences are small and do not match the retardation in other phases of human development. Are our babies, then, born before their time?

Portmann (1941, 1945), in a series of widely neglected but fascinating essays, argues that the general trend in mammalian evolution leads from large litters of rapidly growing, highly undeveloped (altricial) young to small, slowly developing litters of advanced (precocial) young. Human neonates are a striking exception. We are the most highly evolved mammals, in Portmann's sense; our litter sizes could scarcely be smaller; and our general rates of development are unmatched for their slowness. Yet human babies are helpless and undeveloped at birth—we are secondarily altricial.

In assessing our general retardation relative to pongids, Portmann determined that we "should" have a gestation period of 21 months. In fact, he argues that we actually do! Our growth rates during the first postnatal year follow fetal tendencies of other primates, but our birth is accelerated, and we spend our first year as extrauterine embryos: "Human growth follows the mammalian norm, but birth occurs much earlier than this norm would imply" (1941, p. 516). In Portmann's psychogenic view of evolution, this precocious birth must be a function of mental requirements. He argues that humans, as learning animals, need to leave the dark, unchallenging womb to gain access, as flexible embryos, to the rich extrauterine environment of sights, smells, sounds, and touches. He ridicules the argument (1945, p. 51) that something so coarsely mechanical as difficulty in parturition might have anything to do with precocious birth. But when one contemplates the radical redesign that human females would have to undergo in order to give birth to year-old babies, the link of "early" birth to difficult parturition seems quite reasonable.

In a literal sense, I do not regard Portmann's theory as more than an insightful metaphor. But I believe that a fundamental truth lies hidden within it. Our time of birth has been accelerated relative to most other systems during the course of our evolution; but this acceleration has been obscured because general retardation is so strong that the *absolute* time of our gestation is still longer than that of any other primate. I suspect that Portmann is approximately right in arguing that we would spend 21 months in utero if our retardation in gestation matched the slowdown in our other systems.

Why, then, has our gestation been accelerated? The major reason must be another aspect of retarded development—our anomalously large birth weight produced by an extension of rapid early embryonic growth well beyond the time of its cessation in other primates (Oliver and Pineau, 1958). At birth, our brains are still growing at fetal rates (see pp. 371–374); if this increase continued in utero, heads would soon become too big for successful parturition. Leutenegger (1972) plotted fetal weight versus maternal weight for 15 anthropoid species (Fig. 63). Human neonates display the highest positive deviation from the curve (predicted weight, 2200 g, actual weight, 3300 g—this cannot be explained by our slightly longer gestation relative to large pongids since the positive deviation is equally strong in a plot of fetal weight versus days to term, as shown in Fig. 62).

If the large size of our neonates reflects, as I believe, the failure of fetal growth rates to slow down as term approaches, then we must note, as support for paedomorphosis, the large suite of retarded morphologies associated with this retention of fetal growth tendencies. A pronounced delay in skeletal maturation is the most outstanding example. Schultz comments: "Though man grows *in utero* to larger sizes than any other primate, his skeletal maturation has progressed less at birth than in any monkey or ape for which relevant information has

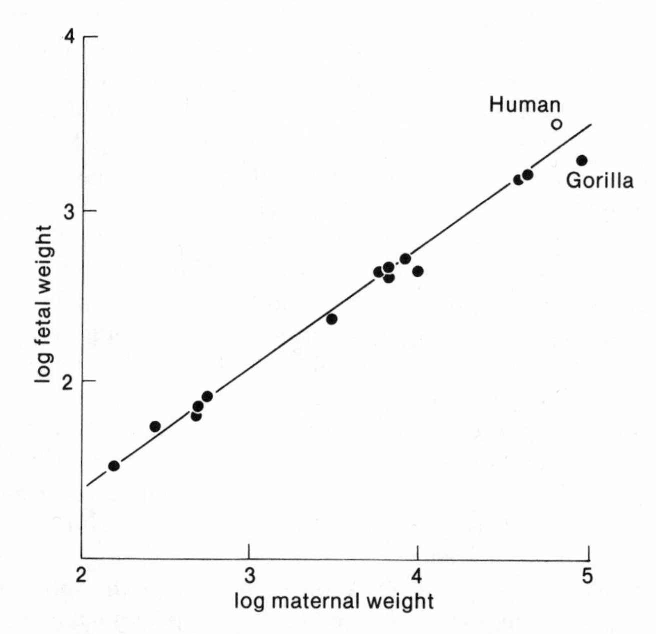

Fig. 63. Fetal weight versus maternal weight (grams) for 15 anthropoid species. Human neonates show the largest positive deviation and are born "too heavy" due to longer maintenance of rapid fetal growth rates. (Redrawn from Leutenegger, 1972.)

become available" (1949, p. 200). Only in humans are the epiphyses of long bones and digits still entirely cartilaginous at birth; carpal ossification centers are usually entirely lacking in newborn humans. "Such a state of ossification exists in macaque fetuses of the 18th week and at term (24 weeks) the limb bones of macaque fetuses have already become ossified to an extent which is recorded in man not until years after birth" (p. 201).

In practically all human systems, postnatal growth either continues long past the age of cessation in other primates, or the onset of characteristic forms and phenomena is delayed to later times. The brain of a human baby continues to grow along the fetal curve; the eruption of teeth is delayed; maturation is postponed; body growth continues longer than in any other primate; even senility and death occur much later. "It is evident," Schultz writes, "that human ontogeny is not unique in regard to the duration of life *in utero,* but that it has become highly specialized in the striking postponement of the completion of growth and of the onset of senility" (p. 198). A distinguished expert on the growth of children states: "Man has absolutely the most protracted period of infancy, childhood and juvenility of all forms of life, i.e., he is a neotenous or long-growing animal. Nearly thirty percent of his entire life-span is devoted to growing" (Krogman, 1972, p. 2). McKinley has compared survivorship curves, estimated for *Australopithecus africanus* and for *A. robustus,* with data for a preurban population of *Homo sapiens;* he finds them to be "much the same" (1971, p. 417). The pattern of human retardation may be quite ancient in its origin.

The retention of juvenile proportions in adults can often be linked in a simple way to retarded development in time. The brains of most mammals are essentially fully formed at term (see Hubbert et al., 1972, on cattle). Primates prolong brain growth into early stages of postnatal ontogeny.* *Macaca mulatta* achieves 65 percent of final cra-

* I am speaking of crude increase in mass, not the differentiation of brain components. Patterns of differentiation are very complex. Harry Jerison (personal communication) writes that all neurons may be represented by tightly packed nuclei in the early neonatal brain. He suspects that evolutionary increase in neuronal number occurs by the prolongation of normal rates for cell division to later times—a potential effect of general retardation. The marked postnatal increase in size of the human brain involves no further genesis of neurons, but a proliferation of glial cells, the growth of dendrites and axons, and the myelinization of axons. G. Cahill (personal communication) argues that the time needed to myelinate neurons and complete their growth requires a retardation in reproductive development. In a series of fascinating papers, G. Sacher also argues that our general retardation arises from the requirements of an enlarged brain (Sacher and Staffeldt, 1974, p. 606). I am consciously bypassing the issue of causal direction and historical primacy in discussing our large brain and retarded development. The neotenic hypothesis applies whether brain enlargement precedes more general retardation or vice versa.

nial capacity by birth, chimpanzees 40.5 percent, and humans only 23 percent (Leutenegger, 1972, estimates the range of *Australopithecus africanus* at 24.9 to 36.9 percent, again finding the intermediate degree of retardation so common in hominid fossils). Chimps and gorillas reach 70 percent of final capacity early in the first year, while we do not attain this value until early in our third year. Chest circumference surpasses head circumference at about 35 days in chimps and not until the second year in humans (Catel, 1953).

Count (1947) studied the ontogenetic allometry of brain size in humans and other mammals. The two-part curve he plotted for brain weight versus body weight has a high fetal slope (usually greater than 1) and a very flat postnatal slope (about 0.1); these are generally connected by a curved intermediate portion. This intermediate portion corresponds roughly to birth. We have preserved this basic pattern with one crucial quantitative modification: we have prolonged the high fetal slope well into postnatal life and achieved thereby our remarkable cephalization. Holt et al. (1975) studied four primate species and reached the remarkable conclusion that all have the same prenatal brain-body curve (same slope and same position—Fig. 64). Humans, as Count concluded, simply prolong the high prenatal slope of this "universal" primate curve well into postnatal ontogeny. Holt et al. (1975, p. 28) conclude that the departure from high slope occurs just before birth in *Semnopithecus,* at 150 days of gestation in laboratory macaques, just after birth in chimpanzees, and not until two years after birth in humans. Figure 65 shows that the "prenatal" curves are identical for humans and macaques, but the curve is extended in humans.

Even Franz Weidenreich, a general opponent of fetalization, admitted that our brain reached its impressive size through retention of fetal growth rates (his emphasis upon rates rather than static shapes should be noted):

> The essential difference which characterizes the ontogenetic development of man, on one hand, and anthropoids, on the other, is not the retention of fetal features in the first case and their abandonment in the latter but the preservation of the fetal *growth proportions* for a longer period in man and their early reversal in the great apes. In man as also in anthropoids, there primarily exists in ontogenetic evolution[8] an allometry between cranium (brain) and face in that the growth of the cranium prevails and that of the face lags behind (positive allometry). Later the conditions change. While in man the growth proportions remain almost the same until the end of the growth, in anthropoids the brain lags behind and the growth of the face becomes dominant (negative allometry). (1941, pp. 415–416)

The prolongation of fetal growth rates for the brain is but one aspect of general retardation in skull features. Delayed closure of the

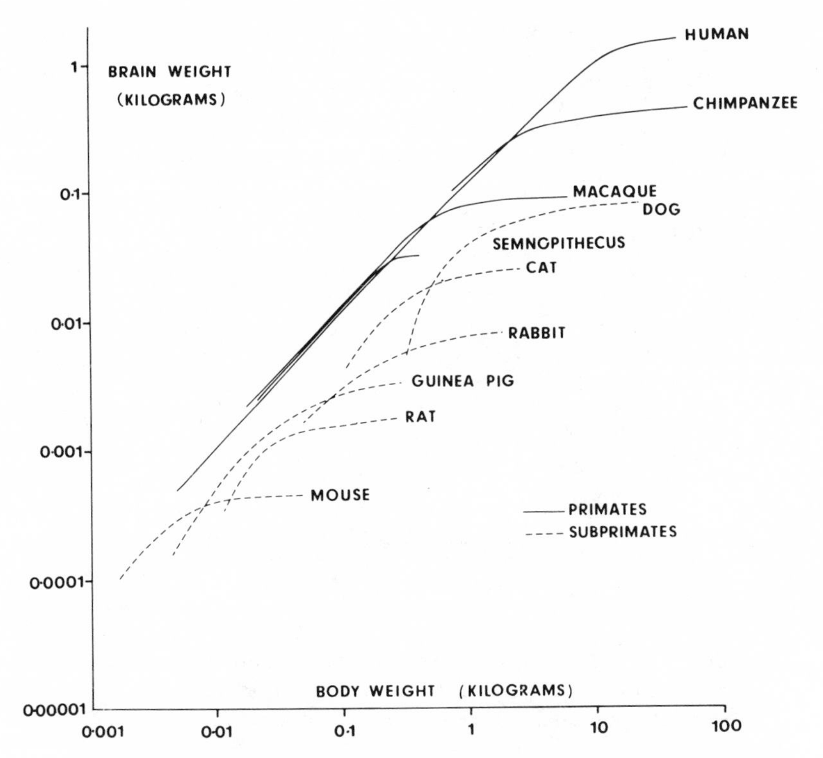

Fig. 64. Ontogenetic brain-body curves for several mammals. Four species of primates follow the same curve, but humans extend the period of high prenatal slope well into postnatal ontogeny, achieving thereby a markedly higher encephalization. (From Holt et al., 1975.)

cranial sutures is equally marked. The most active period of closure occurs between 25 and 30 years of age, while the circummeatal sutures, in some skulls, are still undergoing active closure during the ninth decade of life (Todd and Lyon, 1924, 1925a, 1925b). There is no difference in times of closure between Negro and white skulls.[9] Since the sutures close long after the brain has completed its growth, the retention of open sutures cannot be viewed as the mere mechanical correlate of an expanding brain; the phenomena are at least partly independent.

Laird (1967, 1969) has recently extended the argument for retardation from its familiar territory of shapes and rates to the form of growth curves themselves. She believes that weight versus age data for the ontogeny of all birds and nonprimate mammals can be fit by a "Gompertz-linear" model (essentially Gompertzian early in life with

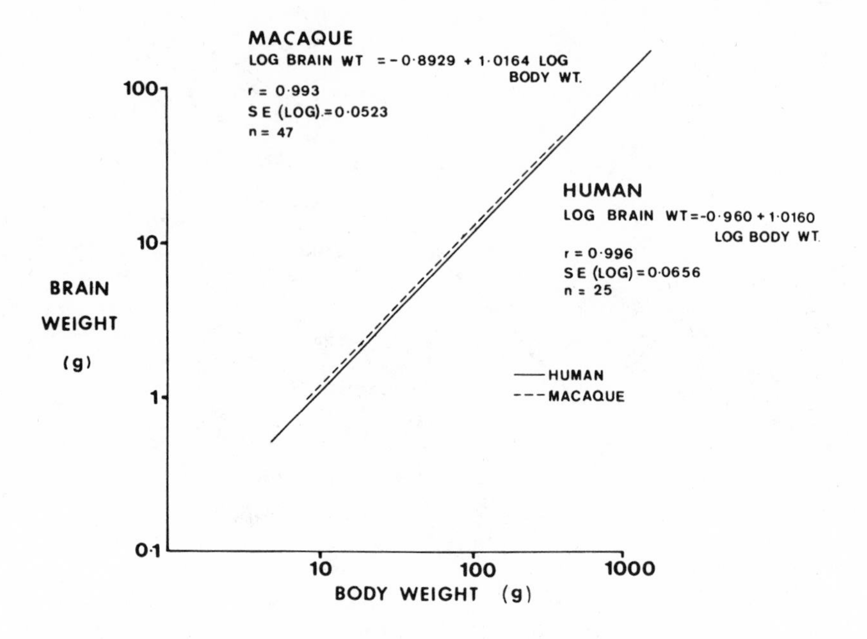

Fig. 65. Prenatal brain-body curves for humans and macaques are identical in slope and position. Humans extend the curve well into postnatal ontogeny. (From Holt et al., 1975.)

extension to a linear component, usually of low slope, later on); sexual maturation characteristically occurs during curvilinear growth. In humans, a linear component of higher slope has been added to an original Gompertz-linear phase, while a new Gompertzian phase (the adolescent growth spurt) follows these two previous stages. Laird would homologize the Gompertzian adolescent growth spurt with the entire postnatal curve of birds and nonprimate mammals, thus tying retardation in human development to the interpolation of a linear phase and the subsequent delay of the curvilinear phase during which maturation occurs.

The idea is intriguing, but I doubt that it can stand. It is difficult enough to recognize homology in complex morphology; confidence in the identity of growth curves with such simple Gompertzian shapes cannot be very high. Moreover, Laird did not use truly longitudinal data. The longitudinal study of Bock et al. (1973) fits human growth data with two logistic components (prepubertal and pubertal); they reproduce Laird's intermediate linear period by diminishing the first logistic function while the second increases.

Without belaboring the point any further, I think it fair to state that general retardation in human development (relative to other primates) is a fact. This matrix of retardation assumes special importance in relation to an observation that Schultz and Starck, the most careful and diligent students of comparative primate morphology, have made independently: the differences in form between humans and apes are almost entirely quantitative; they are produced by alterations in rates and durations of ontogenetic development. Schultz writes: "Adult man possesses numerous, detailed morphological characteristics which are entirely due to minor phylogenetic changes in the *rate* of growth and development in the corresponding bodily parts, and are not caused by any deviation from the general *direction* of developmental differentiation common to at least all higher primates" (1950, pp. 440–441). Starck notes that "nearly all observed morphological differences between fetal skulls of *Homo* and *Pan* are of a quantitative nature" (1960, pp. 633–634).

If the morphological distinctions that we have accumulated during the course of our evolution are all attributable to quantitative alterations in developmental rates, then heterochrony must characterize our emergence. And if these quantitative alterations have occurred within a matrix of general retardation in ontogenetic development, then we have an "enabling criterion" for paedomorphosis on a large scale. The early stages of ontogeny are a storehouse of potential adaptation, for they contain countless shapes and structures that are lost through later allometries. When development is retarded, a mechanism is provided (via retention of fetal growth rates and proportions) for bringing these features forward to later ontogenetic stages. Of our big toe, and its role in upright posture, for example, Schultz writes: "Early in development, when the digits have just separated on the embryonic, plate-like hands and feet, the thumb and the great toe show as yet no sign of rotation in any of the primates. The well-known rotation of the first digit, necessary for effective opposability, develops gradually during ontogeny and reaches widely different degrees of perfection in the adult hand and feet of the various groups of primates" (1949, p. 209). Plantigrade locomotion in a large, upright primate clearly required a strong, unrotated big toe. Such a stage existed already as a transient, and almost necessary, phase in the differentiation of digits. Within a general matrix of retarded development, this preadaptation could easily be retained in adulthood.

The key to a proper understanding of human paedomorphosis is not the simple enumeration of putatively juvenile shapes in human adults, but rather the documentation of a general matrix of retarded development within which adaptive paedomorphosis can readily occur.* Given the fact of this matrix, it would be surprising indeed if paedomorphosis had not played a dominant role in human evolution (Table 9).

Morphology in the Matrix of Retardation

Although I have tried to establish general retardation as an enabling criterion for massive human paedomorphosis, an assessment of our neoteny still requires a discussion of morphological features. I shall not pursue this subject in the enumerative tradition. Rather, I shall treat morphology in the context of three objections that have unjustifiably weakened the theory of our paedomorphic origins and relegated one of the major factors of human evolution to short and ambiguous paragraphs in text books. I believe that these objections are both ill-founded in theory and untenable in fact.

Of Enumeration

The most popular argument for denying our paedomorphic nature rests upon the enumeration of specific features that do not fit the hypothesis. This contention takes two forms: (1) An external similarity in form between juvenile primates and adult humans need not reflect a "passive" paedomorphic retention; it may signify the active acquisition of a new feature yielding the same result. (2) Many important human features do not resemble the juvenile stages of primates. I shall deal with the first contention by discussing the specific issue—cranial flexures—that led to its assertion; I shall treat the second by attempting to show that many nonpaedomorphic features are, nonetheless, direct consequences of retardation in development.

CRANIAL FLEXURES. The literature on orientation of the parts of the skull with respect to each other and of the entire skull to the rest of the body is among the most confusing that I have ever encountered. Conclusions depend strongly on which points or planes are taken as references (Frankfort horizontal versus basicranial axis, for example), and a bewildering array of angles have been proposed and inconsistently used. Yet the subject is of great importance, since the relative

* As I have emphasized continually, not all types of retardation imply paedomorphosis; hypermorphosis (with recapitulation) is also an aspect of retardation. But hypermorphosis involves a delay in reproductive maturation alone, dissociated from all other aspects of somatic development. Human retardation is a pervasive phenomenon of almost all systems, somatic and germinal. General retardation of this sort entails extensive paedomorphosis as an almost ineluctable consequence.

orientation of parts changes so strongly during primate ontogeny, and since we seem clearly to retain a set of orientations characteristic of fetal or juvenile stages of most lower primates.

Most authors use orientations in the sagittal plane relative to the basicranial axis (basion to prosphenion of Fig. 66—from Zuckerman, 1954, whose definitions I follow in this section). Three angles have been commonly used to express the flexure of the basicranial axis with respect to other parts of the skull: (1) The spheno-ethmoidal angle (basicranial axis with prosphenion-nasion) measures the orientation of the facial skeleton with respect to the cranium. The angle increases during the embryology of most mammals and approaches a straight angle as the snout projects far forward of the cranium. In humans, the angle remains relatively small as the reduced face retains a more ventral position. (2) The spheno-maxillary angle (basicranial axis with prosphenion-prosthion) measures the position of the alveolar portion of the face. This character can display a high degree of independence from the spheno-ethmoidal angle. The secondary alveolar prognathism of great apes, for example, reflects the evolution of a massive dentition and is largely independent of the rest of the facial skeleton (Scott, 1958). (3) The foramino-basal angle (basicranial axis with plane of foramen magnum) records the position of the skull with respect to the body. The angle is nearly straight in mammalian

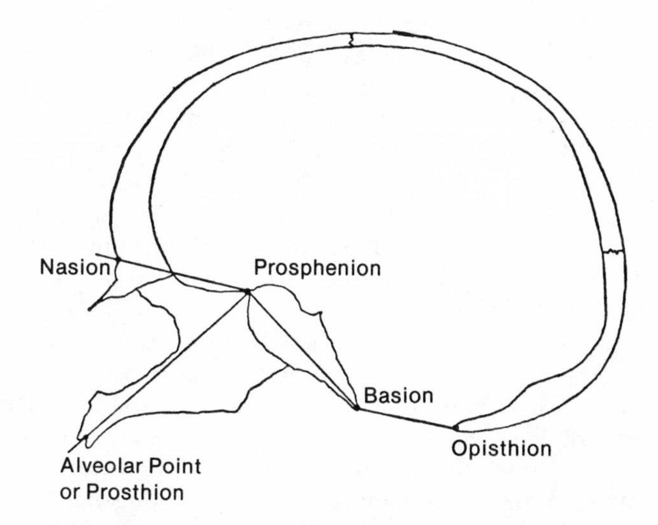

Fig. 66. Human skull in sagittal section showing standard points for anthropometric measurements. (From Zuckerman, 1954.)

fetuses, as the long axis of the head remains approximately perpendicular to the vertebral column (the vertebral column is at a right angle to the plane of the foramen magnum—thus, a straight angle between basicranial axis and foramen magnum implies a perpendicular orientation of skull to body). During ontogeny, the foramen magnum moves posteriorly while its plane acquires a progressively sharper angulation with the basicranial axis; thus, the head and the vertebral axis come to lie in the same plane. In humans, however, the fetal condition is retained throughout life.

Thus, we seem to retain characteristically juvenile values for all major flexures between skull parts and the basicranial axis. The case of the foramen magnum is particularly instructive: Zuckerman (1954) showed that when the human skull is oriented on the Frankfort horizontal (eye-ear plane), the fetal plane of the foramen magnum points slightly forward; this orientation becomes horizontal as maturity is reached. In gorillas and chimpanzees, the plane always faces slightly backwards, and its angle increases from 5 degrees to 20–30 degrees during growth. We follow the same ontogenetic trend as pongids, but we begin in a "more fetal" condition and change our original orientation less during growth. The same is true for the position of the foramen magnum. Schultz (1955) computed the ratio of nasion-condylion (nasion to line connecting center of the occipital condyles) to nasion-opisthocranion (nasion to most posterior point of skull with nasion-basion axis horizontal) projected to the nasion-basion axis; the higher this ratio, the more posterior the position of the foramen magnum. The value of the index (ratio × 100) is low and similar in newborn hominids and pongids (66.1 in chimpanzees, about 64 in humans). It increases during ontogeny in all higher primates, markedly in pongids (reaching values of 77–101 at adulthood), but only very slightly in humans. Again, the trend is the same, but we stray very little from juvenile proportions: "The later aboral migration of the condyles during juvenile growth, though only slight, follows in its direction the very marked postnatal changes typical of apes and monkeys" (Schultz, 1955, p. 114).

This traditional assessment of paedomorphosis has been challenged recently in a long series of papers by German and Swiss anatomists on the ontogeny of flexure in man, particularly as it affects the orientation of face and cranium (Biegert, 1957; Hofer, 1957; Kummer, 1952, 1960; Starck, 1954, 1960; Starck and Kummer, 1962; Vogel, 1964, 1968). To summarize their findings: (1) All fetal mammals have a prebasal kyphosis[10] at the junction of presphenoid and ethmoid bones (that is, at the anterior end of the basicranial axis). (2) During ontogeny this kyphosis decreases, the spheno-ethmoidal angle opens out, and the face comes to lie in front of the cranium. (3) While the prebasal kyphosis is decreasing in human ontogeny, another kyphosis develops *within* the basicranial axis between the basisphenoid and presphenoid bones at the level of the dorsum sellae. This second kyphosis produces a secondary decrease in the spheno-ethmoidal angle following the earlier increase conditioned by straightening of the prebasal kyphosis. (4) The "fetal" value of the spheno-ethmoidal angle in human adults does not reflect the retention of a fetal condition (strong prebasal kyphosis), but arises from development of the new, sellar kyphosis. It is a new feature, not a paedomorphic retention: "In the skull base of the human fetus, the kyphosis lies in a very different position than in the adult . . . We have a new feature [*Neubildung*], and not the persistence of a characteristic fetal trait" (Kummer, 1952, p. 122). These authors have used this single contention as the basis for a campaign against the hypothesis of fetalization: "Man's course of individual development proceeds in an entirely different direction than in *Pan* and other apes. The form of the human skull-base cannot be derived by a process of fetalization from that of a pongid. In man, propulsive processes doubtlessly play a decisive role in the development of the skull-base" (Starck and Kummer, 1962, p. 206).

I do not doubt that the new, sellar kyphosis produces a postnatal decrease in the spheno-ethmoidal angle. But this may not be the whole story (the German tradition of excellence in descriptive morphology is combined with a general avoidance of quantification, and this may have hindered a full assessment). Does the original prebasal kyphosis disappear as fully in humans as in other primates? If it does not, then we may still have a fundamental fetal retention to which the postnatal decrease in spheno-ethmoidal angle (produced by the sellar kyphosis) merely adds emphasis. Quantitative studies of Zuckerman (1955) and Ashton (1957) seem to confirm this. The postnatal decrease in spheno-ethmoidal angle[11] measures about 10 degrees in humans (Zuckerman, 1955). Ashton (1957, p. 72) refers to this as a "slight postnatal closure" and documents the considerably lower value of the angle in newborn human babies compared with other primates. Since the angle has been *increasing* throughout prenatal ontogeny (by 20 degrees between the 10th and 40th week in humans), the effect of the sellar kyphosis has yet to be noted[12] and the relatively low angle of newborn humans is a paedomorphic retention (though it increases in prenatal development, it increases less than in other primates).

And yet, even if Starck et al., were completely correct in their claim, would they have produced so damaging a blow against the hypothesis that humans are fundamentally neotenous? Their attack was directed only against Bolk's extreme view that *all* primary human features are equally retarded in development: "If fetalization were a fundamental process in human evolution, this would imply that the specific characteristics of man, *in their entirety* [*in ihrer Gesamtheit*], would be results of a process of retardation. This is, as we have said, not at all the case" (Starck and Kummer, 1962, p. 208). No modern supporter of human neoteny would have anticipated anything else. No Darwinian could ever expect a complete and harmonious retardation during a slow evolutionary transformation directed by selection. The issue is not whether exceptions to a general paedomorphosis exist (they must); it is, rather, their extent. If they should overwhelm the paedomorphic features in frequency and importance, then we could maintain no general hypothesis of neoteny. I have suggested that such an overwhelming is unlikely within the matrix of general retardation that has characterized human evolution.

Abbie, the staunchest supporter of human neoteny in recent years, has strongly affirmed the concept of dissociability:

> The skull is composed of a mosaic of features which, within wide limits, can vary independently of one another . . . What applies to the skull applies equally to the rest of the body. That, too, appears to comprise a mosaic of independently variable features held together only in loose harmony. An apparently incongruous assemblage of physical characters proves a stumbling block only to those obsessed by too rigid a preconception of what the line of human evolution should have been. (1952, p. 81)

"EXCEPTIONS" AS RESPONSES TO RETARDATION. I now turn to the other side of the argument by enumeration: features of adult humans that do not resemble the juvenile condition of other primates. These include features that seem to represent an extension of general ontogenetic trends in primates (the chin, as a presumptive case of acceleration, is most often emphasized) and traits that seem to be deviations from standard ontogenetic patterns (order of tooth eruption is the example most commonly advanced). When these features are merely listed, divorced from their adaptive significance or the mechanics of their origin, they seem exceptional because they do not represent the retention of juvenile proportions. Nonetheless, I believe that many of them are direct consequences of retarded development and, as such, are among the complex of features that reflect the fundamental heterochrony of human evolution. I shall consider five cases:

1. Order of dental eruption. In rapidly growing primates, all three molars appear before the first deciduous incisor is shed (Clements and Zuckerman, 1953). Nursing ends early, and grinding teeth are

soon needed for the mastication of food (Schultz, 1949, p. 202). This pattern is not retained in humans; molars are strongly delayed in appearance and all deciduous teeth are replaced before the molar row is complete. Schultz interprets this alteration as an adaptive requirement of delayed development:

It is tempting to speculate that this human distinction is the result of some natural selection, directly connected with the extreme prolongation of the period of growth in man. The deciduous teeth of man are not more durable than those of other primates, yet they have to serve in the former for much longer periods than in the latter. Hence this newly acquired precedence for the replacement of milk teeth over the addition of molars is undoubtedly beneficial, if not necessary, for man. (1950, p. 440)

2. The human chin. Monkeys and apes have no projecting chin; we develop one gradually. Since we pass beyond an initial chinless state, our adult chin seems to be a direct, recapitulatory exception to paedomorphosis (as strong supporters of human neoteny admit: Bolk, 1926c, p. 329; Abbie, 1958, p. 204).

The jaw grows with positive allometry during the postnatal ontogeny of all primates, including *Homo sapiens.* The allometric increase is very large in chimps and gorillas, and very small in humans. Thus, our jaw remains small and retains the relative proportions surpassed by other primates. But all parts of the jaw are not equally retarded in development. The lower or basal part is retarded far less than the upper, alveolar part (Weidenreich, 1904; Bolk, 1924, p. 342; Abbie, 1958, p. 204). In fact, Enlow (1966) has shown that the alveolar portion of our jaw (forward of the mental foramina) is an area of bone resorption, while in *Macaca* the entire jaw undergoes deposition. The sharply reduced growth of the alveolar region is presumably related to the extreme reduction of our anterior dentition. Similar effects can be produced by experimental removal of teeth in other mammals. Riesenfeld extracted the incisors and excised the temporalis and masseter muscles in rats. This operation produced mandibular shortening and resorption, and led to the formation of a chin (admittedly a weak one). Riesenfeld attributes the chin to "hypofunction and mandibular shortening" (1969, p. 248).[13] In toothless rats, this effect is phenotypic and "genuine" (that is, the rats use their jaws less than normal individuals of their species). In man, it is genetic and relative (humans chew as much as they "should," but we may speak of "hypofunction" relative to the jaw's operation in other primates). The morphological effect might well be the same nonetheless, and I regard this as an ingenious explanation for the efficient cause of chin formation.

I do not see how we can rightly regard the human chin as an example of acceleration and recapitulation. The chinless state is not a specific feature that appears earlier and earlier in development. We carry no gene for a pointed chin. Man does not "grow beyond" an ancestral state to create a chin by adding new tendencies of growth to an older pattern. Rather, the chin arises because one part of the jaw is more strongly retarded in its development than another. The chin is not paedomorphic—it does not represent the retention of a juvenile feature. But it is a consequence of retarded development, not a recapitulatory exception to it. This classic exception fits comfortably within a neotenic hypothesis of human evolution.

3. Early obliteration of the premaxillary-maxillary suture. Strong retardation of the jaws has many other consequences. The early obliteration in humans, and later closure in other primates, of the premaxillary-maxillary suture has been widely cited as an example of recapitulation (it formed, in this context, the basis for Wood Jones' claim that humans arose from tarsioids). But Ashley Montagu (1935) has correlated this closure to the simple reduction of the premaxillary; this, in turn, is a consequence of retardation in ontogenetic development of the jaws—by remaining small, anterior parts of the jaw may be overgrown by posterior structures.

4. Prominence of the external nose. The nose holds second place to the chin for citations as an exception to paedomorphosis. But Glanville has proposed, following an early suggestion of Schultz, that the prominent nose is yet another example of retardation in development of the jaws. As the lips are drawn back by decrease in size of the dental arch, "the nose retains its original position and thus, in a merely passive manner, projects relatively more" (1969, p. 36). (Glanville also notes an empirical correlation between strong prognathism and a broad and short nose; he provides a functional explanation linking our prominent nose to our paedomorphic orthognathy.)

5. Possible cases of hypermorphosis. The attempt to enumerate exceptions to human neoteny must focus upon recapitulated structures produced by true acceleration—since only these are contrary to retardation. I have emphasized that other recapitulatory characters are related to retardation: they arise by hypermorphosis as extensions to ancestral ontogeny (shifting the end stages of previous ontogenies into juvenile phases of descendants), and they depend upon the *retardation* of maturation that makes such an extension of growth possible (see pp. 341–345). Of course, such characters are exceptions to lists of paedomorphic features, but they arise from the same heterochronic phenomenon and, in a classification by process rather than morphology, would be in the same category. The matrix of retardation is pri-

mary: mosaic evolution operates within it to bring forward by neoteny the adaptive traits of ancestral juveniles and to develop by hypermorphosis the advantageous traits of an extended ontogeny.

Abbie (1948, 1958) has classed as hypermorphic a recapitulatory feature of great importance: our long and strong legs which he attributes to a postnatal reversal in axial gradient that has sufficient time to operate because our period of growth has been so greatly extended by retardation:

> Bodily growth and differentiation are governed by an axial gradient whose maximum intensity is at the head end of the body. As differentiation proceeds, the intensity at the head end gradually drops and that at the hind end rises, and if the process lasts long enough the hind end catches up. Monkeys and apes never grow for as long as man and their inferior extremities are relatively stunted in comparison. Man, therefore, ends up with longer lower limbs and generally greater stature. (1948, p. 41)

Schultz affirms this statement by pointing to the common condition of humans and great apes at birth: "The human peculiarity of proportionately long lower extremities develops only late in ontogeny, being entirely lacking at the time of birth when man is still a comparatively short-legged primate" (1949, p. 206). Most pathologies that accelerate maturation or retard growth tend to produce relatively short limbs—hypergonadism, malnutrition, achondroplasia, and hypothyroidism, for example (Abbie, 1958).

Schultz has presented another possible case of hypermorphosis: In quadrupedal monkeys, the cavity of the thorax is suspended underneath the vertebral column. In humans, the vertebral column shifts towards the center of the chest cavity, "a much more advantageous position for the efficient support of the weight of the upper parts of the erect body" (1950, p. 442). This internal position is attained gradually in ontogeny; at birth, we are "in this respect still more monkey-like than typically human" (p. 442). Newborn chimps and humans differ very little in this feature, but the vertebral column of adult chimps is more external than ours. We may attain our more internal position by prolonging a common ontogenetic trend into an extended period of growth.

Of course, we also possess several recapitulatory traits produced by true acceleration in development; moreover, some of these are important components of upright posture, a fundamental human adaptation. Examples include the early fusion of the sternebrae to produce a sternum; the pronounced bending of the spinal column at the lumbo-sacral border; the fusion of the centrale with the naviculare; and several aspects of pelvic shape (Schultz, 1949, 1950). In presenting this discussion, I have no intention of falling into the Bolkian trap of all or nothing; I merely want to argue that most of the classic "exceptions" to human paedomorphosis are really consequences of retarded development, the central phenomenon of our heterochronic evolution.

Of Prototypes

A hidden assumption of much writing on human paedomorphosis is that the great apes are appropriate surrogates for human ancestors. Human traits are judged as paedomorphic if they correspond to juvenile states of chimps and gorillas. The use of pongids as prototypes has inspired two very different, almost contradictory, objections: first, that they are inappropriate as prototypes; and second, that when they are employed as prototypes, they confute the hypothesis. I shall uphold the first objection, use it to deny the second, and then argue that the best prototype among living forms is the human fetus or juvenile itself.

We did not evolve from any primate bearing much resemblance to the adult forms of chimps and gorillas. The great apes, in their ontogeny, develop many peculiar features confined uniquely to them and having little to do with simian phases of human ancestry. If we are paedomorphic with respect to these specializations, this fact is irrelevant to events in our own phylogeny.

Many of the "simian" features that we have supposedly "avoided" by paedomorphosis are actually the consequences of a specialized adaptation in pongids: strong alveolar prognathism, with its correlates, including brow ridges and the sagittal crest (Naef, 1926a; Bolk, 1926c, p. 31; Scott, 1963). "This alveolar growth," Scott writes, "is of course related to the massive development of the dentition and appears to be a late specialization among the anthropoid apes" (1963, p. 131). Scott argues that the correlates of alveolar growth are necessary adaptations to support a massive dentition in animals that had already undergone a paedomorphic reduction of the face:

> Massive alveolar processes are associated with large cheek teeth, prominent canines, powerful muscles of mastication, a well-developed simian shelf or mandibular torus, and well-developed brow ridges especially in animals with low retreating foreheads. Other features of the facial skeleton such as the nasal, lacrimal and ethmoid bones and turbinate processes are not involved in the development of this skeletal masticatory mask. The superimposed masticatory skeleton which reaches its greatest development in the male gorilla is a functional necessity absent in animals such as the dog and the pig in which the strong tubular snout, extending back between the orbital cavities, provides an adequate skeletal anvil to resist the moving mandible. (1963, p. 132)

We may accept this argument while continuing to affirm the paedomorphosis of the human face, on two accounts. First, our direct ancestors possessed the same combination of massive dentition with alveolar prognathism and its functional correlates. *Australopithecus africanus,* at less than half our adult body weight, had absolutely larger teeth (Pilbeam and Gould, 1974). (The situation is further exaggerated in robust australopithecines, off the line of our ancestry; *A. boisei* had teeth much larger still, strong brow ridges, and a sagittal crest.) Second, our primary facial reduction has occurred by retardation relative to a common primate condition representing a general ancestral pattern. In primates lacking secondary alveolar prognathism and its correlates, the face still begins relatively small and increases in ontogeny with positive allometry to project further and further in front of the cranium. We follow the same trend, but it is much less pronounced and adult humans retain the juvenile proportions of generalized primates. Schultz has shown that only very small monkeys of the genera *Saimiri* and *Cebus* have faces as relatively small and orthognathous as ours; in all these forms, "the fetal conditions are retained with very little change" (1955, p. 117).

German and Swiss anatomists have continued their attack on human neoteny by pointing to differences in ontogenetic trends between humans and great apes. Starck and Kummer (1962), for example, have compared cranial ontogeny of humans and chimpanzees by transformed coordinate analysis of skulls oriented on the Frankfort horizontal. They emphasize the differences (Fig. 67): the relative height of the cranial vault decreases throughout human ontogeny, as the original squares of the coordinate network are transformed to rectangles. In chimps, the frontal squares are converted to rectangles of opposite shape as relative shortening exceeds relative lowering. The sellar kyphosis of humans produces a basally concave bending of coordinate lines around the ear region; this tendency is absent in chimps. (Starck and Kummer might also have mentioned such features as development of the human chin and nose). They conclude: "Despite similar points of origin in ontogeny, development in both cases follows different directions. The skull of an adult human resembles no developmental stage of the pongid skull, but is rather the result of progressive processes of morphogenesis" (p. 641; see also Kalin, 1965).

I regard these objections as irrelevant to the hypothesis of human neoteny. First of all, the living great apes are not our ancestors and their ontogenetic allometries are not those of our forebears. Second, and more important, these objections are based on a false criterion for the recognition of paedomorphosis. Consider a juvenile condition and a set of developmental allometries leading away from it: the ques-

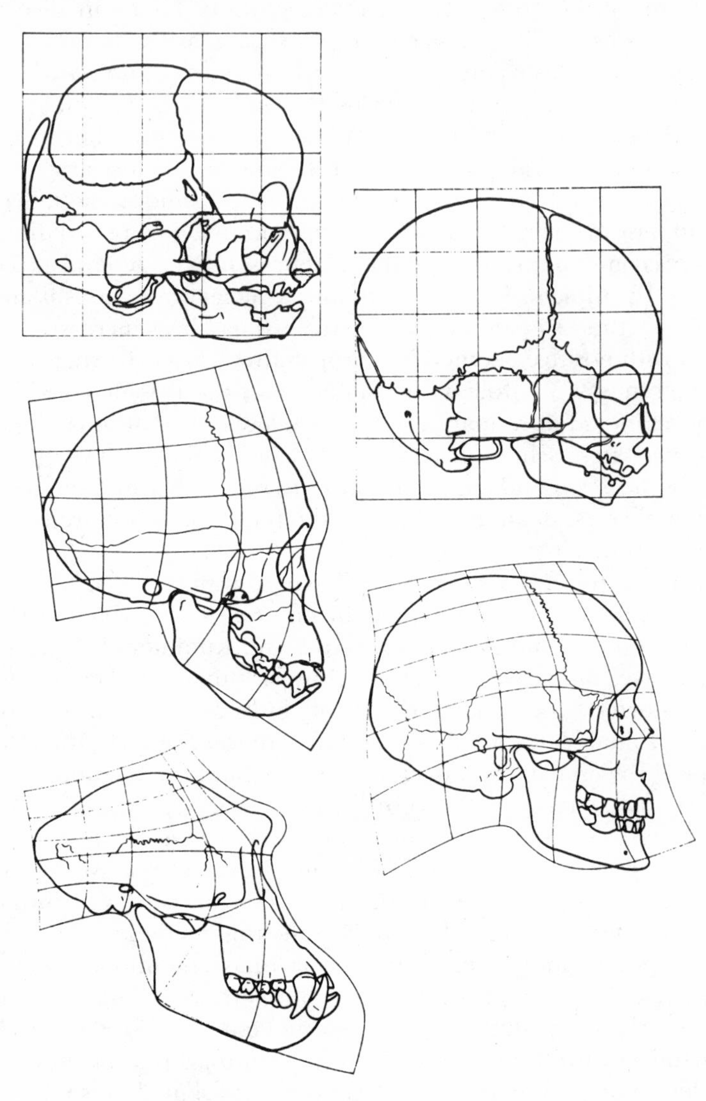

Fig. 67. Human neoteny displayed on transformed coordinates. (*A*) Growth of a chimpanzee. (*B*) Growth of a human skull. The beginning fetal skulls are very similar. The direction of transformation is the same (negative allometry of cranium, positive allometry of face and jaws). But the adult human skull departs far less from the common juvenile form than does the adult chimpanzee. (From Starck and Kummer, 1962.)

tion we must pose is not "in which direction does allometry proceed?" but rather, "how far does it go?" The best measure of paedomorphosis is the *extent* to which an adult descendant resembles an ancestral juvenile.

What juvenile among living primates is most similar in form to the young stages of our forebears? The answer must be: our own juvenile form itself.* This statement may seem surprising, but it actually reflects a non-issue. Von Baer's laws predict, and observation richly confirms, that fetuses of different primates are very similar to each other, while corresponding adults display considerable divergence. Early stages of complex ontogenies are the most conservative of evolutionary phenomena. If we choose a sufficiently early stage, the fetus of practically any higher primate (human and chimp included) can

* This would be false only if our distinguishing characters were fully formed in the embryo—either because they arose as juvenile adaptations later brought forward or because they had been transferred back after arising late in development. But all students of primate anatomy agree that the major differences between humans and other higher primates arise gradually during ontogeny. In this context, it is interesting that the great German paleontologist Otto Schindewolf developed an entire theory of evolutionary mode around the first possibility. As a macromutationist, he believed that major evolutionary novelties arose rapidly in juvenile stages and moved slowly forward to characterize adults. He named the process "proterogenesis"—the law of the early ontogenetic origin of types (*Gesetz der frühontogenetischen Typentstehung*). L. S. Berg (1926) proposed a similar scheme in his principle of "the precession of phylogeny by ontogeny." Although Schindewolf did apply this law to human evolution, he did so in a way that affirmed the early ontogenetic similarity of higher primates. He held that all primates had undergone the primary ontogenetic introduction—that is, all possessed as a common fetal stage the set of progressive characters that could lead to a new stage of evolution. This stage had arisen "bei Urvertretern des Primatenstammes auf frühontogenetischen Stadium spontan and sprunghaft als Neubildung" (1929).

But all nonhuman primates lose these progressive traits during ontogeny because they retain the ancestral set of allometries that lead away from them and back towards animality. We alone have retained these features by neoteny and realized the exalted promise so long unfulfilled in the early ontogenies of all primates. Moreover, since we maintain the same set of allometries leading away from juvenile excellence—though in incomparably weaker degree than any other primate—we may look to our fetus for signs of future progress. Schindewolf does not even shrink from identifying this fetal form with Nietzsche's Übermensch: "Drastically expressed, juvenile man, in the form of his skull, is not only a perfected man [*Vollmensch*], but already a superman [*Übermensch*] . . . This is the thought that Nietzsche has already expressed in his philosophical doctrine of the superman" (1929, pp. 727, 755). But, Schindewolf argues in a eugenic diatribe, we may never proceed further along the path to hominization because society preserves the defective germ plasm that selection would otherwise weed out. In what must be considered, in retrospect, a rather grim statement for Germany in 1929 (though no different from what many American eugenecists were preaching at the same time), Schindewolf attributed the decline in our racial stock to "the zeal for reproduction among spiritually defective men" ("gestig defektor Menschen in deren Zulassung zur Fortpflanzung").

serve as a reasonable prototype. Starck (1960, pp. 633–634), for example, has noted the "entirely astounding way" ("geradezu erstaunlicher Weise") in which the fetuses of humans and chimps resemble one another.

We might as well use the juvenile stage of each species as its own prototype and judge our relative paedomorphosis by the following criterion: do we as adults depart less from our own early form than other higher primates do from theirs (Fig. 68).[14]

The hypothesis that we have diverged least from our own embryo is an ideal subject for multivariate analysis of total morphologic pattern. Brodie followed the ontogeny of 21 males from 3 months to 8 years of age via x-rays of the skull; the primary conclusion of this longitudinal study affirmed the relative constancy of skull form during this period (1941, p. 251). Baer considered total pattern in a subjective way and concluded: "The form of the human skull, in contrast to that of the anthropoid, appears to undergo relatively little change during ontogenetic development" (1954, p. 120).

Before electronic computers permitted the widespread application of multivariate biometry, the pictorial technique of transformed coordinates (Thompson, 1942) was the favored approach to analysis of total pattern. Various studies comparing the extent of ontogenetic transformation in humans and apes confirm the hypothesis of restricted human allometry in a striking way. This seems particularly evident in the analysis of Starck and Kummer (1962), despite the decision of these authors to focus upon some minor differences in direction (Fig. 67). The coordinate lines are clearly more altered, in toto, between chimp fetus and chimp adult than between human fetus and human adult. Moreover, the basic direction of allometry in both ontogenies is quite similar; we have diverged less upon the same general path (as Havelock Ellis realized in 1894, p. 518). In both species, the brain grows with negative allometry and the face with positive allometry. The extent of change is much greater in chimpanzees: coordinate lines are pinched together over the cranium and vastly expanded over the jaws. We display the same compression and expansion, but their extent is extremely limited and form never departs very far from the fetal condition.

Boyce (1964) performed a multivariate analysis on 99 skull characters for male, female, and juvenile pongids and fossil and recent hominids. The first axis of a principal components analysis separated adult apes from adult humans. The second axis separated juveniles from adults by criteria of ontogenetic allometry. Juvenile apes, juvenile humans, and the juvenile australopithecine from Taungs occupy one extreme position on this axis; adult apes and adult australopithe-

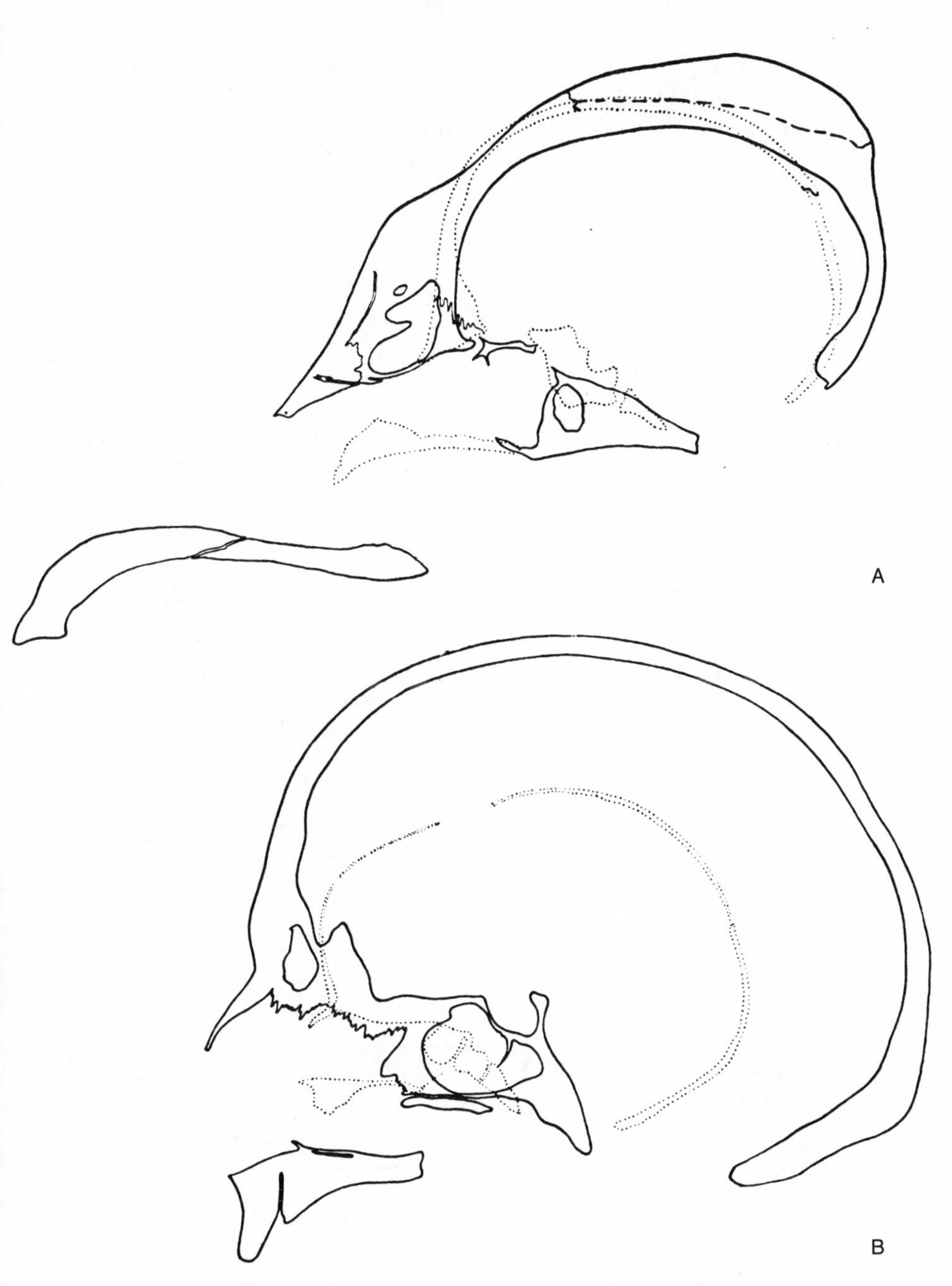

Fig. 68. (*A*) Median sections through skulls of juvenile (dotted) and adult orang-utans. (*B*) Same for humans. Human adult skulls depart far less from their own juvenile forms than do those of any other higher primate. (From Weidenreich, 1941.)

cines group at the other extreme. The skulls of adult human females and males, despite their large size, fall far closer to the juveniles. The unusual position of adult humans on an axis expressing ontogeny demonstrates our retention, in a multivariate sense, of a morphology associated with juvenile stages of other higher primates. A cluster analysis, emphasizing differences in shape for the calculation of distances, places pongid juveniles closer to adult humans than to their own adult forms (Fig. 69).

Of Correlation

The morphological argument seeks to overwhelm by sheer quantity: it operates under the hidden assumption that the enumerated

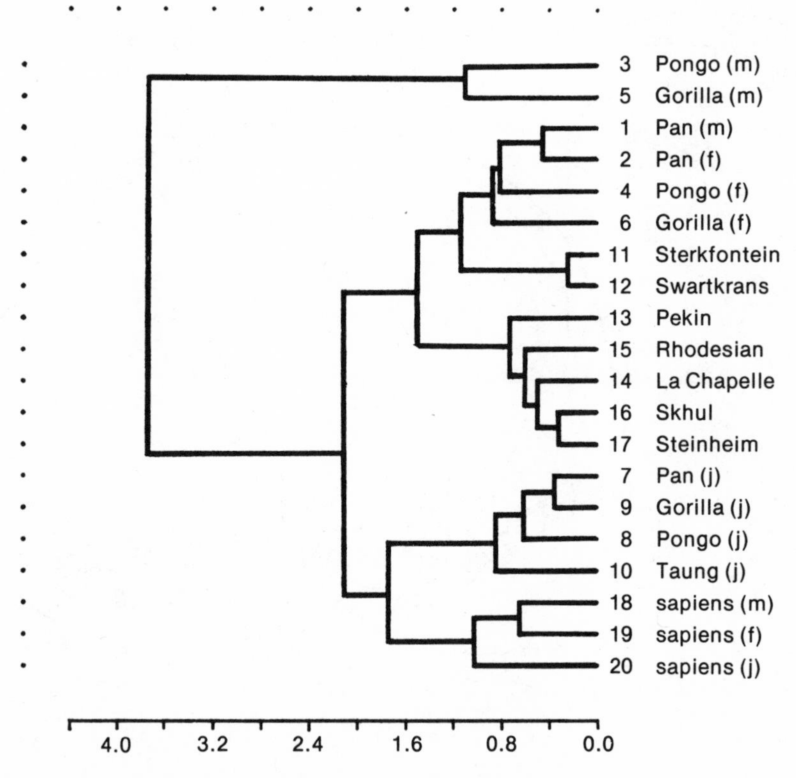

Fig. 69. Cluster analysis of pongid and hominid skulls emphasizing differences in shape over variation in size. Adult humans are grouped with pongid and australopithecine juveniles. (From Boyce, 1964.)

features are independent, so that each item reinforces the general claim. This is Ko Ko's strategy:

> I've got a little list
> And they'll none of them be missed.

It would be very damaging to this argument if many or all the favored items were mechanical correlates of a single feature.

In a famous article, Weidenreich (1941) made just such a claim, explicitly to counter Bolk's theory. He regarded all the paedomorphic features of our skull as directly imposed by mechanical pressures of a large brain. The brain, he noted, grows with strong negative allometry during the postnatal ontogeny of all primates (indeed of all vertebrates—see Gould, 1975); adults in a static series of related races or species (breeds of dogs, small and large primates) exhibit the same relationship. Small animals (juveniles or adults of small species) have relatively large brains (Fig. 70); we are the only large primate with a brain relatively as big as that of small monkeys or juvenile stages of large monkeys. Thus, in both juvenile monkeys and adult humans, the brain fills the cranial cavity and exerts pressure upon the entire skull. The inflated skull is molded like a balloon by its expanding contents: the foramen magnum assumes an anterior position as the occipitally expanding brain flows around it (Delattre and Fenart, 1963, p. 533; Biegert, 1957, pp. 183–184). The anterior cranial flexure is retained as the expanding frontal lobes inhibit facial rotation and keep the spheno-ethmoidal angle at its relatively low value (Moss, 1958). In the generation of enough bone to cover an expanding cranial vault, any propensity for brow ridges or a sagittal crest is suppressed, and the bones of the vault remain thin and unfused. Even jaws and teeth are inhibited by the "pressing" needs of an expanding brain: relatively large-brained animals have small faces.

Weidenreich then argues that the suite of resemblances in skull form between juvenile apes or monkeys and adult humans only reflects the fact that both have relatively large brains for different reasons: young apes because they are small animals on the lower end of a negatively allometric trend, adult humans because the brain has increased in the course of evolution. There is no need to postulate either delayed development or selection for juvenile proportions; all that needs explaining is the evolutionary increase of the brain itself:

> Fetal features to which Bolk refers as, for instance, orthognathism, central position of the foramen magnum, persistence of the cranial sutures are conditioned by a relatively large brain and brain case and persist in their original character during the postnatal life only because brain and brain case retain their predominance in growth or, in other words, they remain positively allo-

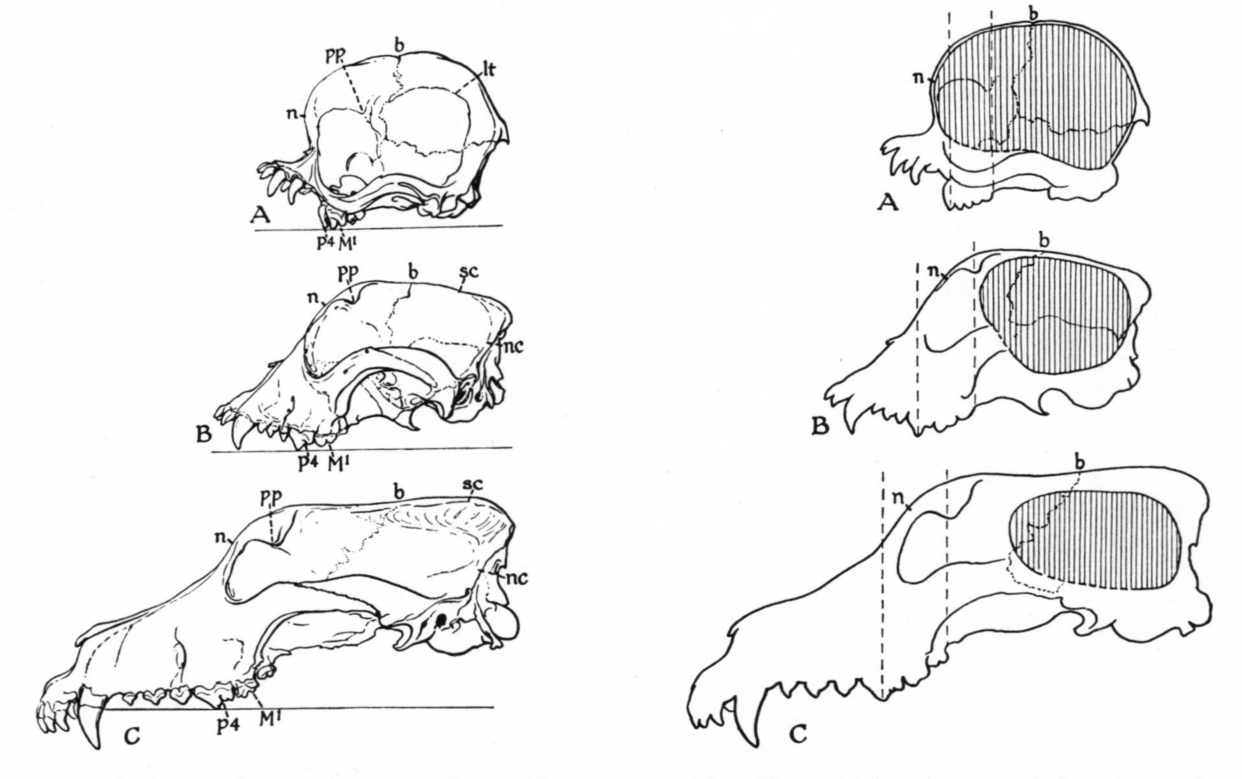

Fig. 70. Skulls of King Charles spaniel (*A*), English bulldog (*B*), and Irish wolfhound (*C*) to show correlation of short face with vaulted cranium, and same skulls in sagittal section showing comparative dimensions of cranial cavities. From Weidenreich, 1941, who argues that the large brain, by it own mechanical pressures, causes the correlated features usually seen as independent signs of human neoteny (short face, vaulted cranium, low foramen magnum).

metric; however, they are of transitory nature in the great apes because their prevalence becomes lost or, in other words, brain and brain case change to negative allometry. (Weidenreich, 1941, pp. 415–416)

Weidenreich did not utterly deny the theme of delayed development or prolonged juvenility. He was quite happy to attribute the brain's expansion to retention of rapid fetal growth rates: "The changes may be brought about by a gradual alteration of the growth rate, that is to say, by retaining the original fetal rate over an ever more extended period" (p. 427). Nonetheless, his argument would reduce the enumerative value of all paedomorphic skull features to that of a single item—the enlarging brain itself.

There is much of value in Weidenreich's views, particularly as an antidote to an overly atomistic view of anatomical parts. Moss (1958), for example, confirmed the brain's role in inhibiting the rotation of the anterior cranial flexure by excising the anterior part of the brain in rats and noting the increased rotation.[15] Nonetheless, as a general postulate, Weidenreich's notion has been extensively and conclusively refuted. The delayed closure of the skull sutures cannot be entirely ascribed to an expanding brain, since they may remain open for decades after the brain has ceased its growth. But most critical attention has been devoted to Weidenreich's central claim that a large brain inhibits dental and facial development—for a strong hypothesis of neoteny demands at least the independence of these two fundamental systems. Weidenreich's major illustration involved a sequence of dog breeds of different sizes. He chose the small Pekingese with its reduced face and jaws for one extreme of his series and implied that it could serve as a model for all small dogs. But other dwarfed dogs—dachshunds, miniature schnauzers, and Scottish terriers, for example—have short heads with very long jaws (Abbie, 1947), while shortened jaws of some small breeds "are probably related to achondroplasia and the reduction of the facial skeleton is a consequence of this rather than an increase in brain size" (Scott, 1958, p. 342; see also Starck, 1954).

Developmental studies have consistently detected independence of brain and face. Baer (1954) stained growing rat bone with alizarin red and detected two separate systems of skull growth: an early rapid expansion of the brain case in conjunction with brain growth, and a slow growth of longer duration resulting in elongation of the cranial base and face. Baer and Harris extended this conclusion to humans: "Two discrete systems are evident in the growth of the skull: a rapidly growing neural system essentially completed by adolescence, and a facial system of slower growth and longer duration" (1969, p. 39). Zuckerman associated facial growth with dental development and denied that the brain could strongly influence the form of the permanent dentition: "Since the growth of the face is associated with the eruption of the permanent teeth, changes in the shape and size of the brain, which for all practical purposes may be regarded as having ceased growing before the jaws assume their mature proportions, cannot be regarded as responsible for the major developmental transformations of the skull" (1954, p. 366). Van der Klaauw (1945) extended the notion of facial and cranial independence to the lower vertebrates. Starck (1954, pp. 170–171) reviewed the German literature on this independence. Baer concluded, in terms very favorable for the hypothesis of neoteny (since it affirms independent causes for paedomorphosis of the brain and face):

> The interpretation here advanced is in conflict with Weidenreich's theory, namely, that the brain is the sole determinant of form in the mammalian skull and that the enlargement of the brain has a depressing influence on the growth of the face . . . The form of the skull is primarily the result of the interaction of two fundamental systems of growth . . . The brachycephalization of the human skull would appear to be the product of an adaptive trend in which the size of the brain and the size of the face have differential selective value. (1954, pp. 121–122)

Another tradition of argument correlates many paedomorphic features of the skull with the attainment of upright posture. (This serves both as another kind of claim involving correlation and as a further counter to Weidenreich's emphasis upon the brain.)

DuBrul noted a series of "man-like" characters in some rodents and rabbits that have small brains but share the human habit of keeping the head erect: pronounced vaulting of the dorsum, strong angulation between facial and cranial axis, and relatively ventral position of the foramen magnum and occipital condyles.[16] He concluded:

> Parallel changes have occurred in three reputedly unrelated orders, Lagomorpha, Rodentia, and Primates. Only among the primates can the development of the brain be adduced as an important factor. The single common factor among the three orders is the change in bodily orientation. It appears as though the Lagomorpha had made the phylogenetic experiment to separate the influence of progressive encephalization upon the shape of the skull from that of posture and locomotion. (1950, p. 293)

The classical explanation relating size and orientation of the face and foramen magnum to upright posture invokes the adaptive significance of a well-balanced skull: a small face reduces the weight that an erect creature must bear upon its spinal column, while a central foramen magnum with an expanded occiput and a shortened face equalizes the load on either side (Adams and Moore, 1975). Weidenreich himself invoked this general argument before developing his theory on the mechanical correlates of brain size (1924). Schultz (1955) denies that upright posture demands a central foramen magnum, because great apes stand erect with relatively more weight in their unbalanced heads than we bear. Giraffes, camels, llamas, some antelopes, and some dogs hold their spine erect but carry their heads entirely in front of it; powerful muscles are sufficient to hold the "unbalanced" skull. Biegert (1957, p. 193) has argued that upright posture is conditioned by post-cranial adaptations of the pelvis and limbs, not by modification of the skull. But erect posture is not the characteristic stance of apes, and the fact that other mammals follow different adaptive strategies to support an erect head does not deny the potential significance of a well-balanced skull. Brues (1966) has suggested that the significance of balance should be sought not in classical arguments of static load, but in the advantages it confers for rapid rotation. The greater importance of pelvic features does not preclude a role for cranial modifications.

I confess that my preference for a connection between paedomorphic skull features and upright posture has to do with the very different status of this claim and Weidenreich's argument about the brain. Weidenreich's correlation is causal: the expanding brain mechanically molds the skull—general neoteny is precluded as an explanation. The correlation to upright posture is only a statement about adaptive requirements.[17] It prescribes no mechanism for the attainment of features so required (in fact, it does not exclude per se the invocation of mechanical effects imposed by an increasing brain—though DuBrul's more specific argument does attempt such an exclusion).

We are therefore free to speculate about the mechanisms that have brought, in a coordinated way, so many juvenile features into the adult form of our skull. The extreme atomism of "bean-bag" genetics might seek an independent efficient cause for each, tying their coordinated appearance only to adaptive requirements. But I share D'Arcy Thompson's conviction that complex organic pattern can usually be reduced to fewer and simpler generating factors (Thompson, 1942; Gould and Katz, 1975; Raup, 1966, on generating the range of form of coiled shells with only four parameters; Vermeij, 1973, on the evolutionary significance of "parameterization").

As an example of reduction to simpler generating factors, several lines of evidence have recently converged to indicate the influence of a shortened skull base upon other aspects of skull form. DuBrul and Laskin (1961) excised the spheno-occipital synchondrosis (between the basisphenoid and the basioccipital in the posterior part of the skull base) in rats and obtained an impressive series of "man-like" changes: increased shortness and roundness of the skull, curvature of the cranial roof, forward displacement of the occipital condyles, ventral and forward rotation of the plane of the foramen magnum, and, as a putative mechanical cause of all these features, shortening of the cranial base. Vogel (1964) reported a similar series of changes in a pathological specimen of *Cercocebus* which lacked the spheno-occipital synchondrosis. Riesenfeld (1969) noted that spheno-occipital chondrodystrophy causes cranial shortening in chick embryos, and that short-headed dwarfs of goats, sheep, and cattle exhibit premature closure of the same synchondrosis. The link to a similar suite of features in humans does not involve a malfunction of the synchondrosis, but relies upon the common feature of cranial base shortening: this is a normal trait in humans, and a pathological effect produced by malfunction of the synchondrosis in other cases. In humans, it may merely reflect the relative expansion of the brain around an unmodified base (Vogel, 1964), thus recalling Weidenreich's argument in a more limited context.

The relative shortening of the skull base may well be responsible for several paedomorphic features of the cranial vault, but it provides no explanation for the other major paedomorphic complex of the skull: the reduced face and dentition. Moreover, shortening of the cranial base itself may be conditioned by retarded development—either directly by differential delay in growth of the base or indirectly by the relatively greater growth of the brain due to prolongation of fetal growth rates.

The entire attempt to use correlation as an argument against neoteny suffers from a most curious omission—it has neglected everything below the neck, except as it provides a pedestal for mounting the head. Even Weidenreich's extreme position only attempted to debunk the neotenic explanation of *skull* form by tying all supposed retardations to mechanical effects of the brain. Our propensity for focusing on the brain is an unhappy legacy from the history of Western philosophy, as Frederick Engels (1876, in 1954) showed so well in asserting both the primacy of upright posture and the dignity of labor. Many of the strongest evidences for human neoteny are postcranial: reduction of hair and pigment, position of the big toe and the female genital tract. Yet these have been ignored utterly in all discussions of correlated development as an argument against human neoteny. The correlation of these postcranial characters with paedomorphic features of the skull cannot possibly be attributed to mechanics of growth in the brain or cranial base. The common factor of all these adaptive features is retarded development.

The Adaptive Significance of Retarded Development

Our paedomorphic morphology is a consequence of retarded development; in this sense, we are neotenous. I have tried to avoid the tradition of discussing human heterochrony by enumerating paedomorphic features, for I believe that retardation itself is the primary phenomenon—both because is serves as the ground for paedomorphic morphology and because it constitutes a central feature of human adaptation in its own right.

Retardation gains much of its adaptive significance in human evolution as the ground of paedomorphic morphology. To repeat an argument advanced throughout this chapter: our paedomorphic features are a set of adaptations coordinated by their common efficient cause of retarded development. We are not neotenous only because we possess an impressive set of paedomorphic characters; we are neotenous because these characters develop within a matrix of retarded development that coordinates their common appearance in human adults.

The early stages of ontogeny are a reservoir of potential adaptation. Juvenile features may be adaptations in their own right or simple topological consequences of morphogenetic development from simplicity to complexity. DuBrul and Laskin, for example, tie many fetal features to requirements of "close packing" in egg or womb: "The fetal head of the rat is rounded as in most mammals; evidently it fits the largest mass—the brain—into the smallest space—a sphere—within the constricting confines of the uterus. The face is diminutive, tucked in below and in front of the brain case. The skull need but 'unbend' during growth to make the long rectangular shape of the adult" (1961, p. 122). The same features are adaptive in adult humans for different reasons: the cranial flexure for upright posture, the small ventral face for balance and as a response to reduced dentition, the spherical shape of the cranial vault for economy in ossification (Abbie, 1947, p. 25). The availability of these features as transient stages in juvenile ancestors and the existence of a mechanism (retarded development) for their transfer to adult descendants establishes their preadaptive value.[18] As Bolk (1915) noted in an early paper: "Bipedal walk found in the primitive, fetal, central position of the foramen magnum a lucky condition, sympathetic to its trend."

In some cases, we can make a stronger claim: as size increased in human evolution, retardation provided the only "escape" from an ancestral allometry and the only path to a favored adaptation. In an ingenious argument, marred by poor statistical definition and a confusing mixture of ontogenetic and static data, Hemmer (1969) plotted braincase length versus facial length for *Australopithecus* and *Homo sa-*

piens (Fig. 71). He calculated a similar slope of strong negative allometry for each group. As body size increased in evolution, a larger braincase could not be achieved simply by extrapolating the australopithecine line—for the face would increase even more rapidly, "with compression of the braincase and formation of strong crests by the enormously developed jaw musculature, making impossible any increase in braincase capacity" (p. 180). The ancestral correlation had to be *dissociated*: points for descendants with relatively larger braincases had to lie above the australopithecine line and had, therefore, to display a paedomorphic morphology (on a criterion of standardization by common size). In situations of negative allometry, any point above the ancestral line is paedomorphic because an isometric line passed through it intersects the ancestral curve at a smaller size (Gould, 1971)—that is, an ancestor of the same shape is smaller in size (and generally more juvenile). The obvious mechanism for an upward transposition of the allometric line in *Homo sapiens* is a prolongation to larger sizes of rapid fetal growth rates for the brain. This permits the subsequent negative allometry to start at a larger braincase size in

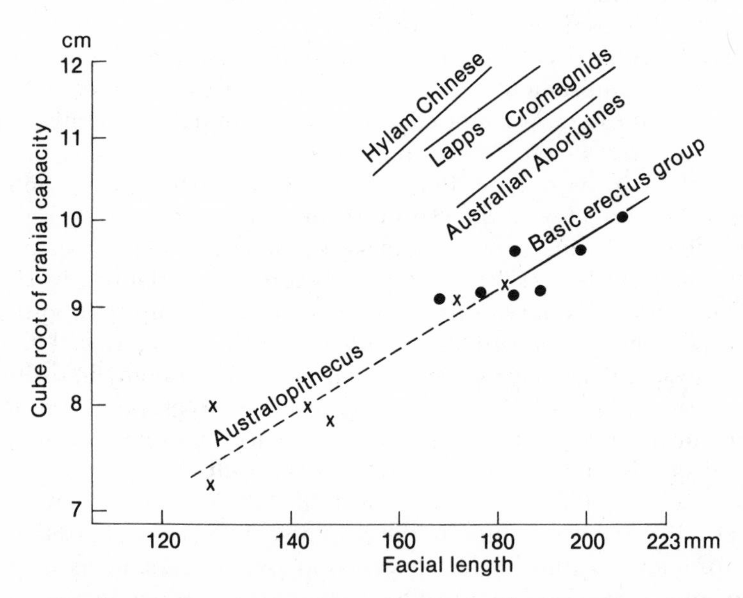

Fig. 71. Cube root of cranial capacity versus facial length for hominids. Size and shape are dissociated by retardation since line of constant shape (slope = 1) intersects ancestral curve at smaller size than descendant curve (see Fig. 30). (Redrawn from Hemmer, 1969.)

descendants; again, a shift in timing underlies a major change in form. "It seems probable that the prolongation of embryonic growth beyond birth into the first year of life found in *H. sapiens* was not present to the same extent in *Australopithecus* and that this factor is responsible for the allometric level of *H. sapiens*" (Hemmer, 1969, p. 180).

But we should not focus on morphology in discussing the adaptive significance of retarded development. This retardation, of itself and apart from any morphological correlates or consequences, has been a factor of paramount importance in human evolution. In fact, though I have prefaced this section with long discussions of morphology, I believe that the temporal delays themselves are the most significant feature of human heterochrony.

In asserting the importance of delayed development, I have no desire to enter the largely meaningless debate on historical primacy. I assume that major human adaptations acted synergistically throughout their gradual development (Bielecki, 1969). The interacting system of delayed development–upright posture–large brain is such a complex: delayed development has produced a large brain by prolonging fetal growth rates and has supplied a set of cranial proportions adapted to upright posture. Upright posture freed the hand for tool use and set selection pressures for an expanded brain. A large brain may, itself, entail a longer life span.[19] Two elements of the complex have long been recognized among the australopithecines—an essentially complete development of erect posture, and brains far larger than those of comparably sized pongids. Mann (1975) has now presented evidence for the third element from his studies of dental eruption and skeletal maturation in South African australopithecines—these primitive hominids had already evolved an extended childhood.

As a new theme for assessing the evolutionary significance of heterochrony, I have been trying to deemphasize the traditional arguments of morphology while asserting the importance of life-history strategies. In particular, I have linked accelerated development to *r*-selective regimes and identified retarded development as a common trait of *K* strategists (Chapters 8 and 9). I have also tried to link *K* selection to what we generally regard as "progressive" in evolution, while suggesting that *r* selection generally serves as a brake upon such evolutionary change. I regard human evolution as a strong confirmation of these views.

To begin with, we belong to a class of animals in which *K* selection dominates (Pianka, 1970). We evolved in the generally *K*-selective tropical regime. We belong to an order of mammals distinguished by their propensity for repeated single births,[20] intense parental care, long life spans, late maturation, and a high degree of socialization—a point-for-point agreement with Pianka's listing of traits common to *K* strategists (1970).

Human evolution has emphasized one feature of this common primate heritage—delayed development, particularly as expressed in late maturation and extended childhood. This retardation has reacted synergistically with other hallmarks of hominization—with intelligence (by enlarging the brain through prolongation of fetal growth tendencies and by providing a longer period of childhood learning) and with socialization (by cementing family units through increased parental care of slowly developing offspring). It is hard to imagine how the distinctive suite of human characters could have emerged outside the context of delayed development. This is what Morris Cohen, the distinguished philosopher and historian, had in mind when he wrote that prolonged infancy was "more important, perhaps, than any of the anatomical facts which distinguish *homo sapiens* [*sic*] from the rest of the animal kingdom" (1947, p. 174).

Delayed development has generally been portrayed in a particularistic context as a special feature of human evolution. I regard it rather as the quintessence of a common ecological tendency—and as a confirmation that *K* selection favors retardation. In this most extreme case of retardation among animals, we also note, in our inevitable arrogance, the correlation of retardation with "progressive" tendencies in evolution—whether we measure progress by the martial concept of dominion or by the more harmless observation that no other species has ever been able to bore its members with such tedious disquisitions on their evolutionary estate.

The arguments for the significance of delayed maturation are old and familiar. I shall present no detailed justification, for this has been done often and has never, to my knowledge, been ably challenged (supporters have included Keith, Washburn, Montagu, de Beer, Lorenz, and many others). I shall, rather, emphasize just how old the arguments are, for this has never been recorded in the scientific literature, and it may be of more than antiquarian interest to readers who believe that great ideas are identified by their persistence through series of conceptual revolutions. The arguments can be reduced to two general claims:

1. From the child's standpoint: the newborn human child is about as dependent a creature as we find among placental mammalian infants. This dependency is then extraordinarily prolonged, and the child requires intense parental care for many years. The flexibility of childhood persists during more than a decade of necessarily close

contact with adults. The adaptive premium thus placed on learning (as opposed to innate response) is unmatched among organisms. Krogman, at the close of a career spent in studying the growth of children, stated: "This long-drawn-out growth period is distinctively human; it makes of man a learning, rather than a purely instinctive, animal. Man is programmed to *learn* to behave, rather than to react via an imprinted determinative instinctual code" (1972, p. 2) Holloway would turn the argument around: "An inversion of this statement is suggested: the function of prolonged youth or growth was to provide a better brain to accomplish the learning" (1970, p. 308). Again, I regard these debates about historical primary as fruitless and agree with de Beer's assessment that both factors acted in concert: "Delay in development enabled him to develop a larger and more complex brain, and the prolongation of childhood under conditions of parental care and instruction consequent upon memory-stored and speech-communicated experience, allowed him to benefit from a more efficient apprenticeship for his conditions of life" (1959, p. 930).

Jacobson (1969) has identified a neurological basis for the importance of delayed development. During their ontogeny, individual neurons develop highly specific and determined synaptic connections. In an earlier stage of ontogeny, these connections are modifiable; this flexible period occurs at varying times for different types of neurons. Human development is so strongly retarded that even mature adults retain sufficient flexibility for our adaptive status as a learning animal.

> During ontogeny, there is evidence of a progressive reduction of the capacity to form new neuronal connections and to modify existing ones. This reduction occurs at different times in different classes of neurons, so that those which are generated late in ontogeny and those which mature slowly have the greatest degree of modifiability in the mature animal. According to this theory, the modifiability of neuronal connections in the adult is regarded as a continuation of developmental processes that are much more pronounced in embryos. (Jacobson, 1969, p. 547)

The correlation of maturation with loss of plasticity (mental as well as physical) has long been recognized. Schelling, the philosophical mentor of German romantic biology, argued in the late eighteenth century that higher organisms strive to delay as long as possible the onset of sexual maturation and its attendant differentiation and rigidification. Lower organisms fail in this quest and find themselves locked into low positions on the *scala naturae*. Higher organisms postpone this inevitable fate and reach higher levels of organization. Haldane once suggested facetiously that Jesus be viewed as the prophet of human neoteny for his statement (Matthew 18:3): "Except ye be converted and become as little children, ye shall not enter the kingdom of heaven."

Konrad Lorenz, in particular, has repeatedly emphasized the persistently "juvenile" character of our behavioral flexibility. He asks: "What is the source of this remarkable persistent juvenile characteristic of investigative curiosity in human beings, which is so fundamental to the essence of humanity?" (1971, p. 279) Lorenz answers his own rhetorical question by arguing that "behavioral neoteny" is but another consequence of the developmental retardation that permitted our morphological neoteny as well: "The constitutive character of man—the maintenance of active, creative interaction with the environment—is a neotenous phenomenon" (p. 180). "Human exploratory inquisitive behavior—restricted in animals to a brief developmental phase—is extended to persist until the onset of senility" (p. 239).

My favorite illustration of flexibility in human neoteny comes not from a scientific treatise, but from the lowly badger who tells a parable of creation in T. H. White's novel *The Once and Future King.* At first, the badger relates, God created only a set of embryos, looking perfectly alike in the best tradition of von Baer. He called them before his throne and offered them any specialization they desired for an adult form. One by one, they chose their weapons, their defenses, and their insulation. Finally, the human embryo approached the throne:

"Please God," said the embryo, "I think that you made me in the shape which I now have for reasons best known to Yourselves and that it would be rude to change. If I am to have my choice, I will stay as I am. I will not alter any of the parts which you gave me . . . I will stay a defenceless embryo all my life, doing my best to make myself a few feeble implements out of the wood, iron, and the other materials which You have seen fit to put before me . . ." "Well done," exclaimed the Creator in delighted tone. "Here, all you embryos, come here with your beaks and whatnots to look upon Our first Man. He is the only one who has guessed Our riddle . . . As for you, Man . . . You will look like an embryo till they bury you, but all the others will be embryos before your might. Eternally undeveloped, you will always remain potential in Our image, able to see some of Our sorrows and to feel some of Our joys. We are partly sorry for you, Man, but partly hopeful. Run along then, and do your best."

2. Delayed maturation is also significant from the parents' standpoint (though still, of course, to benefit children and perpetuate parental genes). Young, dependent children require an organization of adults to support them and guide their growth. Moreover, the period of dependence is so protracted that subsequent children gen-

erally appear before earlier offspring acquire much independence. Pair-bonding must have been enhanced by this continual resupply of dependent offspring, and many authors have seen in delayed development a primary impetus for the origin of the human family. A high degree of socialization is characteristic of primates in general; this trend must have been accentuated by the evolution of offspring that require a protracted period of adult attention and education to ensure their survival.

Bolk, who shunned the subject because it smacked of adaptation, nonetheless remarked in a footnote: "In the very long duration of the period during which the human child must be nourished and protected by its parents, do we not have the natural cause for the origin of the human family and, therefore, the basic element of all human society?" (1926c, p. 20) Versluys and Lorenz wrote: "Protracted development must have greatly strengthened the bond between parents and children . . . Retardation is the biological basis for societal life [*Die Retardation ist die biologische Grundlage des gesellschaftlichen Lebens*]" (1939, in Overhage, 1959, p. 27).

This theme is also an old one. Lovejoy (1959, p. 215) has traced it to John Locke's *Second Treatise of Government* (1689). Locke emphasized both lengthening infancy and the consequent reinforcement of family ties by new children born while older siblings are still dependent upon parents—"wherein one cannot but admire the wisdom of the great Creator who . . . hath made it necessary that society of man and wife should be more lasting than that of male and female among other creatures, that so their industry might be encouraged, and their interest better united, to make provision and lay up goods for their common issue." The historian Herder, another godfather of German romantic biology, strongly supported this argument in the late eighteenth century:

> To destroy the savagery of men and to habituate them to domestic intercourse, it was necessary that infancy in our species should continue for some years. Nature held them together by tender bonds, so that they might not separate and forget one another, like the beasts that soon reach maturity. The father becomes the teacher of his son, as the mother has been his nurse, and thus a new tie of humanity is formed. Herein lay the ground of the necessity of human society, without which it would have been impossible for a human being to grow up, and for the species to multiply. Man is thus born for society; this the affection of his parents tells him, this the years of his longer infancy show. (in Lovejoy, 1959, p. 213)

Heroic couplets, whatever their literary merit, are at least an excellent format for summary statement. I can imagine no more succinct an epitome of the significance of prolonged infancy than Alexander Pope's lines of 1733:

> A longer care man's helpless kind demands,
> That longer care contracts more lasting bands.

I am led, in conclusion, simply to reassert Bolk's claim: "What is the essential in man as an organism? The obvious answer is: the slow progress of his life's course."

—11—

Epilogue

Although the differences between humans and chimps may be quantitative only, the two species as adults do not look much alike and their adaptive differences are, to say the least, profound (no monkey, despite the common metaphor, will ever type—much less write—the *Iliad*). Yet King and Wilson (1975), reviewing evidence for the astoundingly small difference in structural genes between the two species, have found that the average human polypeptide is more than 99 percent identical with its counterpart in chimps. Moreover, much of the difference can be attributed to redundancies in the genetic code or to variation in nontranscribed regions (pp. 114–115). For 44 structural loci, the average genetic distance between chimps and humans is *less* than the average distance between sibling species barely, if at all, distinguishable in morphology—and far less than the distance between *any* measured pair of congeneric species.

What, then, is at the root of our profound separation? King and Wilson argue convincingly that the decisive differences must involve the evolution of regulation; small changes in the timing of development can have manifold effects upon a final product: "Small differences in the timing of activation or in the level of activity of a single gene could in principle influence considerably the systems controlling embryonic development. The organismal differences between chimpanzees and humans would then result chiefly from genetic changes in a few regulatory systems, while amino acid substitutions in general would rarely be a key factor in major adaptive shifts" (p. 114). Differences in regulation may evolve by point mutations of regulatory genes or by a rearrangement of gene order caused by such familiar chromosomal events as inversion, translocation, fusion, and fission. Studies of banding indicate that at least one fusion and ten large inversions and translocations separate chimps and humans.

Of the nature of our regulatory differences, King and Wilson profess ignorance: "Most important for the future study of human evolution would be the demonstration of differences between apes and humans in the timing of gene expression during development of adaptively crucial organ systems such as the brain" (p. 114). We cannot, of course, trace our differences in regulation to any specific part of the genome. But I submit that we do know the nature of the regulatory changes that separate us from chimps: they operate to slow down our general development (thereby setting our major adaptive differences from other higher primates). Human development, in general, is retarded (Chapter 10), while maintenance of fetal growth rates leads to hypertrophy of organs developing with early positive allometry in uterine ontogeny (the brain), and to reduction of parts which grow with positive allometry only in the postnatal stages of other primates (the face).

The most important event in evolutionary biology during the past decade has been the development of electrophoretic techniques for the routine measurement of genetic variation in natural populations. Yet this imposing edifice of new data and interpretation rests upon the shaky foundation of its concentration on structural genes alone (*faute de mieux,* to be sure; it is notoriously difficult to measure differences in genes that vary only in the timing and amount of their products in ontogeny, while genes that code for stable proteins are easily assessed). Structural genes may make up less than 3 percent of sea urchin genomes (E. Davidson, personal communication). If most evolutionary events are controlled by changes in regulation, then how are we to measure and discover their basis? The literature already contains two kinds of arguments for the importance of regulation in evolutionary change:

1. Arguments based on the relative frequency of a common, microevolutionary event. In the large literature on bacterial mutations permitting growth on novel substrates, the most common change is not an alteration in structural genes for the evolution of new enzymes, but a marked increase in rate of production for enzymes already present but repressed in ancestral clones (Wu, Lin, and Tanaka, 1968; Hegeman and Rosenberg, 1970; Hall and Hartl, 1974). A change in regulation leads to the derepression of pre-existing genes; an enzyme formerly synthesized only when induced by an external medium can now be generated intrinsically. New substrates that do not induce the enzyme are now available for colonization. "Mutations

that confer constitutive synthesis of an originally inducible enzyme or group of enzymes appear to be the most commonly-observed mechanism by which bacteria acquire the ability to grow on a novel substrate" (Hegeman and Rosenberg, 1970, p. 432).

2. Inferences about macroevolution from the molecular biology of modern organisms, or from models of molecular processes. Wilson and his colleagues have compared rates of change in structural genes and chromosomes for slowly evolving frogs and rapidly evolving mammals (Wilson, Sarich, and Maxson, 1974; Wilson, Maxson, and Sarich, 1974). The two groups do not differ in rates of protein evolution for structural genes, but change in chromosome number has proceeded 20 times more rapidly in mammals than in frogs. Hybrid inviability also develops far more rapidly in mammals. Gene arrangement (a regulatory phenomenon) may be more important than point mutation as a source for defining differences in morphology, physiology, and behavior.

Britten and Davidson (1971) present a model for ascribing evolutionary novelties, not just changes in rates, to alterations in regulation. They speculate, for example, that at some remote protochordate level, an integrator gene set *A* regulated the production of circulating blood cells (in Britten and Davidson's scheme, integrator genes are regulators that make activator RNA; activator RNA complexes with a receptor gene to cause transcription of an adjacent producer [structural] gene). A distant integrator element *B* might have regulated the ontogeny of cuticle (synthesis of heme-controlling globulin for trapping oxygen is an important function of cuticular cells). If *B* were translocated into integrator gene set *A*, the heme compound would be synthesized with the blood cells and might be incorporated within them. Britten and Davidson tie these effects to the dispersal of repetitive DNA throughout the genome. Most speculation on the evolutionary significance of repetitive DNA has centered on the potential for mutational change without sacrifice of original function (for the unchanged copies still synthesize the essential enzyme). But Britten and Davidson show that the entrance of identical copies into different regulatory arrangements might play an even more important role.

Stebbins (1973) has tied the evolutionary success of eukaryotes to the evolution of their probable mode of regulation—general repression with specific activation (in prokaryotes, groups of genes are activated and deactivated by the products of specific inhibitor loci; inhibitors are often affected by small molecules entering cells from the external medium):

The analogy to man-made power systems is obvious. If a small cabin is to be lighted either by candles or a few hand lanterns, efficiency demands that each lamp is lighted and extinguished separately by turning on and off the power or fuel that supplies it. In a large office building, on the other hand, the only efficient system is one in which the power is always there, but is uniformly repressed or "off" unless it is turned on by means of specific switches for each light or group of lights. (pp. 228–229)

(See Davies, 1973; Kenney et al., 1973; and Ruddle, 1973 for summaries of recent information about the nature of regulation.)

In a particularly forceful statement, Zuckerkandl (in press) advances the case for regulatory change as the essential ingredient of evolution:

> If it were possible to take judiciously chosen structural genes and put them together in the right relationship with regulatory elements, it should be possible to make any primate, with some small variations, out of human genes . . . Likewise it should be possible to make any crustacean out of the genes of higher Crustacea. This fairy tale may be conservative. It is told only to emphasize that structural genes are building stones which can be used over again for achieving different styles of architecture, and that evolution is mostly the reutilization of essentially constituted genomes. We may not generally have been tempted to take such a view, impressed as we were by the constant structural divergence of genes throughout evolution. Yet, by and large, this divergence resembles Brownian motion. With notable, all important exceptions, the essential properties of the protein are not changed.

The classical data of heterochrony have been widely ignored and regarded as old-fashioned. But I believe that they have a central role to play in the growing discussion on the evolutionary significance of changes in gene regulation. I predict that this debate will define the major issue in evolutionary biology for the 1980s. I also believe that an understanding of regulation must lie at the center of any rapprochement between molecular and evolutionary biology; for a synthesis of the two biologies will surely take place, if it occurs at all, on the common field of development.

The data of heterochrony are data about regulation; they define changes in the timing of development for features held in common by ancestors and descendants. We do not know what percentage of all regulatory events are heterochronic—for we have not learned how to assess the regulatory component of evolutionary novelties. But the data of heterochrony represent the only confident estimate that classical macroevolutionary morphology can supply for the importance of changes in regulation. (The estimate, moreover, is a favorable one for testing any hypothesis that affirms the importance of regulation, for it is a conservative, minimal measure.) These data can provide evidence in both categories cited above—assessments of relative frequency *and* inferences from macroevolution.

Throughout this book, I have tried to demonstrate that heterochrony is extremely important in evolution—both in frequency of occurrence and as the basis of significant evolutionary change. I hope that I have added thereby some support for the belief that alterations in regulation form the major stuff of evolutionary change.

The reconciliation of our gradualistic bias with the appearance of discontinuity is a classical problem of intellectual history. We have sought to reduce the external phenomena of saltation to an underlying continuity of process—to reduce the qualitative to the quantitative. Philosophies of change and progress have wrestled with this dilemma and have tried to resolve it by formulating such laws as the "transformation of quantity to quality" of the Hegelian dialectic: the addition of quantitative steps will lead eventually to a qualitative leap—reduce the food of Aesop's ass and it will eventually die; oppress the workers more and more, and revolution will ensue (in Engels' material reformulation of Hegel's law).

External discontinuity may well be inherent in underlying continuity, provided that a system displays enough complexity. The evolution of consciousness can scarcely be matched as a momentous event in the history of life; yet I doubt that its efficient cause required much more than a heterochronic extension of fetal growth rates and patterns of cell proliferation. There may be nothing new under the sun, but permutation of the old within complex systems can do wonders. As biologists, we deal directly with the kind of material complexity that confers an unbounded potential upon simple, continuous changes in underlying processes. This is the chief joy of our science.

NOTES
BIBLIOGRAPHY
GLOSSARY
INDEX

Notes

2. The Analogistic Tradition from Anaximander to Bonnet

1. Events in the second half of the cycle have won a place for Empedocles among the forerunners of evolution. As love begins to amalgamate the elements, scattered parts of bodies form and these join in a variety of combinations. A sort of "natural selection" then acts to preserve the favorable combinations. I regard any comparison between this scheme and later evolutionary thinking as purely gratuitous; Empedocles' system, needless to say, was for another time and another purpose.

2. Meyer (1935, p. 382) also states that Hunter's illustrations for the development of the chick have never been fully published; yet they are included in Owen (1841).

3. Adelmann (1966) has argued that Malpighi's position lay neither in the camp of evolution or epigenesis (his late seventeenth-century works predated the explicit formulations of a predominantly eighteenth-century debate). Yet he was generally cited, by Haller, for instance, as a supporter of preformationism.

4. I do not wish to imply, of course, that preformationism was without its difficulties, or that its debate with epigenesis revolved only around conflicting philosophies. Preformationism faced a series of empirical dilemmas: If the embryo is preformed in the egg, what is the role of the sperm? If the sperm merely triggers the growth of a homunculus, how can the intermediate form of a mule be explained? Bonnet regarded this as a telling point and argued weakly: "This production already existed in miniature, but in the form of a horse in the ovaries of the mare. How was this horse metamorphosed? In particular, where did its long ears come from? . . . I say that elements of the seminal fluid acted on those of the germ; the semen of the ass contains more

particles capable of enhancing the development of ears than does that of the horse" (1762, p. 23). How can anomalies of development be explained in any but an epigenetic way? Of Siamese twins, Bonnet argued: "The germs, at first almost fluid and quite gelatinous for a longtime thereafter, are very penetrable. If they happen to touch each other, they are mixed up at least in part" (1764, p. 178).

5. Bonnet envisioned his illusory transmutation as a series of large, discrete alterations (in analogy to metamorphosis)—not as a succession of small, continuous "changes."

6. Whitman (1894), in his brilliant analysis of Bonnet's system, interprets this aspect of Bonnet's remarks on the history of life differently. He believes that Bonnet viewed each *individual* as a series of three encapsulations, with the "germ of restitution" as the innermost layer. Each individual may therefore live in each of three worlds (former state, present condition, and future state). In my reading, each individual of any world bears a germ of restitution that survives the death of the original body and waits for resurrection till the end of time—that is, each individual lives but twice. Change of form from one world to the next occurs because God arranged the sequence of encapsulations in the ovary of each original creature to yield a new form at each physical revolution of the globe.

Bonnet is more zealous in developing the spiritual consequences of his general view than in presenting clear explanations for the details of his system. Moreover, his text (1769) contains inconsistencies and bears signs of composition at different times. I confess that both interpretations (Whitman's and my own) seem consistent with various parts of the manuscript. Whitman clearly based his reading on part 6 of *La palingénésie philosophique* (1769, pp. 236–262). Here Bonnet (p. 254) presents a "model" for an earth history of but three stages (while admitting that many more may have occurred). He seems to advocate the "resurrection" of each individual from each world: "Je dis que les Êtres Organisés du premier Monde, ne furent détruits qu'en apparence: ils se conserverent dans ces Germes impérissables, destinés des l'Origine des Choses à peupler le second Monde" (p. 256).

Yet I regard several points emphasized elsewhere in the *Palingénésie* and in Bonnet's other works as inconsistent with Whitman's interpretation and more in accord with my reading. First, against Whitman's "germ trinity" (1894c, p. 263) for three worlds, and against Bonnet's simplified model (1769, p. 254) stands Bonnet's continued insistence upon several past worlds (pp. 174, 179, 247, 252, 253, and 257). Second, Bonnet (1769, p. 210) draws a distinction between germs of restitution residing in bodies that lived in a former world and those freed from bodies that had never tasted life (they will enjoy the future world in different ways: the former by comparison with the previous state it had witnessed, the latter with all the joys of a tabula rasa exposed to perfection). Yet if each *individual* has a germ for existence in each world, every germ of restitution (the last term in the series for each individual) will have experienced some taste of a former life. Finally, the human germ of restitution, unlike the homunculi preformed in ovaries, resides in the corpus callosum of the brain (1764, pp. 88–90). It is a different kind of germ from all others, not the last term of a series. We are never told where the germ for a resurrected individual of our own world resided in the body of a creature in the former world. I prefer to believe that the individuals of our world were simply members of the normal sequence of ovarian encapsulations—but endowed with different forms and arranged to appear at a physical revolution of the globe.

3. Transcendental Origins, 1793–1860

1. Cohen (1947, p. 13) and Hampson (1968) emphasize the influence of advances in embryology upon a dynamic view of human history, while Crocker (1959) and Temkin (1950) record the impact of new social and political thinking upon embryological concepts. Collingwood claims that the primary causal link ran from history to science: "Modern cosmology could only have arisen from a widespread familiarity with historical studies, and in particular with historical studies of the kind which placed the conception of process, change, development in the center of their picture and recognized it as the fundamental category of historical thought" (1945, p. 10).

2. Kielmeyer is a very little known figure. Besides the 1793 address, he published only one other work, an address to a Stuttgart scientific society in 1834; moreover, he later recanted much of his Naturphilosophie in a letter to his pupil Cuvier (Meyer, 1939). Yet, he was evidently regarded as a dominant influence in his time. Schelling, the guiding spirit of Naturphilosophie, called the 1793 address "eine Rede, von welcher das künftige Zeitalter ohne Zweifel die Epoche einer ganz neuen Naturgeschichte rechnen wird" (Gode von Aesch, 1941). A gold medal presented to him was inscribed: *Graiis Aristoteles, Harveus Brittanis Teutonum populis Kielmeyerus erit.* ("Kielmeyer shall be to the German people what Aristotle was to the Greeks and Harvey to the British"—Meyer, 1939.) For an analysis of Kielmeyer's version of recapitulation, see Coleman (1973).

3. Although Russell (1916, p. 215) includes a few Naturphilosophen (notably Meckel) among pre-Darwinian evolutionists, most members of this school did not believe in physical descent. Neither, of course, did they accept a static order, created once and for all time; generally, they postulated a vital spirit or force that would generate ever higher forms, each as a separate and fresh attempt to reach the goal of man. This was the position of the philosophical mentors Schelling and Herder (Temkin, 1950, pp. 240–241). Oken draws an analogy to ontogeny: just as offspring arise from parents only after reduction to the formlessness of initial embryonic stages (note Oken's extreme epigenesis, 1847, pp. 190–191), so must any higher species originate from "the initial chaos" (p. 190), not from a slightly lower ancestor. "The history of generation is a retrogression into the Absolute of the Organic, or the organic chaos—mucus, and a new evocation from the same" (pp. 190–191). Serres, the strongest supporter of recapitulation among the French transcendentalists, was a strong opponent of evolution; he even classified man in a separate kingdom because he felt that our capacity for evolutionary change (physical perfectibility of "lower" races) distinguished us sharply from all

other organisms. From teratology, Serres argued that the deformities produced by a lack of vital force resembled adults of lower organisms, while monstrosities arising from an excess of such force never produced a higher stage, but only a duplication of adult structures; the type cannot be transcended.

4. Goethe wrote that Shakespeare and Spinoza were his greatest influences. After them he cited Linnaeus, but "durch den Widerstreit, zu welchem er mich aufforderte" (in Oppenheimer, 1967, p. 136). He continued: "Denn indem ich sein scharfes, geistreiches Absondern, seine treffenden, zweckmässigen, oft aber willkürlichen Gesetze in mich aufzunehmen versuchte, ging in meinem Innern ein Zwiespalt vor: Das, was er mit Gewalt auseinanderzuhalten suchte, musste, nach dem innersten Bedürfnis meines Wesens, zur Vereinigung anstreben."

5. Many claims made by Naturphilosophen and often ridiculed from today's perspective followed naturally and sensibly from romantic assumptions. Oken's long glorification of zero (1809, pp. 5–16) sounds absurd out of context ("das höchste Princip, auf das sich alles Einzelne, alles Endliche, alle Zahl der Mathematik reducieren lässt, und von dem alles begründet ist"—p. 4); it has often been used to discredit him completely as a scientist. In fact, since all development must start from the original chaos that zero represents, this figure is the ground of all being and the historical antecedent of all existence. Huschke's homology of plants and inverted animals inspires easy ridicule: "The root might correspond to the mouth of animals . . . Leaves and blossoms would be compared to the branchiae and genitals with their nearness to the anus . . . The height of organic development in man reaches an attitude directly opposed to that of plants and lower animals" (in Gode von Aesch, 1941, p. 231). But it arises from a belief in the structural unity of all life and a desire to arrange organisms in a single chain of ascent (plant, quadruped, man, corresponding to positions of organs: inverted, horizontal, erect).

6. Physio-philosophy was the nineteenth century's usual English translation of Naturphilosophie. Tulk's translation of Oken is entitled: Principles of Physiophilosophy.

7. For example, Nordenskiold: "His speculations were as grotesque as they were irrational," (1929, p. 287). Haeckel had the following uncomplimentary words to say: "Oken verlor sich, bei allen seinen Verdiensten, doch nur allzuleicht and allzutief in unbestimmten und mystischen naturphilosophischen Träumereien, und brachte noch dazu diese phantastischen Einbildungen in einer so dunkeln orakelhaften Weise vor, oft so leichtfertig die empirische Basis verlassend, dass die bald emporkommende exact-empirische Schule Cuvier's sich gar nicht mehr um ihn bekümmerte" (1866, 2:161).

8. Even so, since "each class takes its starting point from below" (1847, p. x) as it begins to replay the sequence of senses after adding the characteristic organ of its class, the highest animal of a lower class may be more advanced in general organization than the lowest animal of the next higher class. Yet it remains on a lower rung of the sequence because it lacks the key organ that marks the next stage of organic advance. Oken maintains that "the classes stand one above the other, but yet that each recommences from below, so that

the lower animals of a higher class are more stunted or rudimental than the upper ones of a lower class. Thus the salamanders are more rudimental, that is, they have organs more imperfect than the sharks . . . Nevertheless these stunted animals stand higher than those of the lower classes, because they are characterized by a higher organ . . . The lowest man is still higher than the uppermost ape." Oken's arguments for equating classes with organs or senses are often exceedingly weak, even within his own context. Thus, the class Mammalia represents sight (even though animals in many lower classes have eyes) because: "For the first time, in the Mammalia, the eyes are movable and covered with two perfect lids, without the other organs of sense having suffered degradation through this completion of the eyes" (1847, p. 496).

9. Oken must have had second thoughts about these words, for the third edition ends with this expanded version (in Newspeak) of an earlier statement: "The art of war is the highest, most exalted art; the art of peace and of justice, of the spiritual condition of man and mankind—the principle of peace."

10. I do not mean to suggest that these categories are unrelated, for ideas often encourage or even prescribe the styles of their pronouncement. There was, for example, a major argument among the Naturphilosophen regarding the extent to which conclusions could justifiably be drawn from general premises without empirical support. Schelling's most dedicated disciples, Oken and C. G. Carus, believed strongly in the prior imposition of ideas upon the world of observations; their writing is speculative, brilliant, and sweeping. Meckel shared their beliefs about the nature of the universe (single perfecting tendency, unity of law and structure), but felt compelled to display these beliefs as consequences of (or, at least, as overwhelmingly supported by) observation; his style is systematic, dry, and often heavy.

11. Not the mechanist's argument that all phenomena should be reduced to laws of physics and chemistry, of course, but rather the opposite: since man is the measure of all things, the laws of his development shall extend down to everything else.

12. Life itself seems less integrated in lower organisms, for these display a much stronger reproductive force (desire to separate) than higher creatures and often reproduce by complete division of the body (1811b, p. 68).

13. Herein, of course, lies the basic difference between two biologies that embraced recapitulation for completely different reasons. Recapitulation, to Naturphilosophen, was a simple consequence of a priori beliefs that demanded no justification in mechanical terms and prompted few questions about efficient causes. To later evolutionists, recapitulation became an observation that demanded a causal explanation.

14. I must reemphasize a previous point. The deformity of Figure 3 is clearly human; it is not, by any stretch of the imagination, a mollusk. It is quite inviable. But the external form of a clam is not at issue: the clam merely represents a stage of the unilinear sequence of organic advance. A human fetus sharing the same key features occupies the same position, however radically it differs in other ways. This example also illustrates Serres' "law of centripital development"—that ontogeny proceeds, so to speak, from the outside

in. Cutaneous respiration is prior to and more primitive than gills and lungs because it is external.

15. As Oken so correctly noted in his review of the *Entwickelungsgeschichte:* "Der Haupteinwurf kommt übrigens nicht von der Masse einzelner Thatsachen, die er aufführt and deren es wohl noch viele geben mag, sondern von der Idee . . . die Fortbildung der Thierklassen sei einreihig" (1829, p. 209). Nonetheless, Severtsov wrote: "Das von Baerische Gesetz ist dennoch kein aprioristischer Satz im Sinne L. Okens und seiner Anhänger, sondern eine ausserordentlich scharfsinnige Verallgemeinerung der von ihm beobachteten embryologischen und taxonomischen Tatsachen" (1927, p. 150).

16. Haeckel placed both these examples in his category of cenogenesis (falsifications of ancestral development): the placenta is a transient adaptation to embryonic life; the early eruption of incisors is a heterochrony (a displacement in the time of appearance of a specific organ).

17. This would scarcely have bothered Oken, who saw recapitulation only as the repetition of essential features in the abstract, unilinear procession of added organs.

18. Both these cases are often cited today as possible examples of paedomorphosis: an evolutionary phenomenon opposite to recapitulation (descendants retain the juvenile features of ancestors). The large human brain may represent a prolongation of embryonic growth rates, while insects may have evolved from paedomorphosis of juvenile myriapods (de Beer, 1958).

19. The dependence of this version of recapitulation upon an ascending unilinear classification was the favored target of most of its critics. Milne-Edwards (1844) applied Cuvier's classification to refute recapitulation, and Lereboullet, supporting Milne-Edwards against Serres in a prize essay on differences between vertebrates and invertebrates, wrote: "The results which I have obtained are diametrically opposed to the theory of the zoological series constituted by stages of increasing perfection, a theory which tries to demonstrate in the embryonic phases of the higher animals a repetition of the forms which characterize the lower animals, and which has led to the assertion that the latter are permanent embryos of the former" (in Russell, 1916, p. 207). In the next sentence, he uses von Baer's first argment: "The embryo of a vertebrate shows the vertebrate type from the very beginning, and retains this type throughout the whole course of its development."

20. Although vertebrates would be ranked as the highest type in any scheme proposed by humans, von Baer regards members of the highest grades of differentiation within other types as more advanced (under his criterion of greater heterogeneity) than some vertebrates. "In fact, I believe that bees are more highly organized than fishes, though according to another type" (p. 208).

21. It is only this early nineteenth-century version of recapitulation, "the Meckel-Serres law," that requires a unilinear ascent of organisms as its mechanism. Although Haeckel equated the fertilized ovum with our amoeboid ancestor, evolutionists usually invoked recapitulation for specific lineages only. Evolutionists sought a mechanism for recapitulation in heredity's laws (see

Chapter 4); they did not require any particular arrangement of organisms to guarantee its operation.

22. Meyer (1956, p. vi) speaks of the "evil influence" of Naturphilosophie, but holds that von Baer escaped its spell because he was an "outstanding, objective scientist" (p. 65). Oppenheimer and Lovejoy present more sophisticated views in intriguing contrast:

> Von Baer's own embryology, for the first time, for all of its emphasis on the relationship of the special to the general, was an embryology in which the metaphysical became subordinate to the biological in the sense of modern embryology, and became an embryology which proceeded from embryological facts and phenomena towards embryological concepts, rather than in the reverse direction. (Oppenheimer, 1967, pp. 145–146)

> For those conceptions were not derived simply from his observations of the successive shapes assumed by fertilized ova; they were interwoven with, and conceived by him to be necessary deductions from, certain metempirical and essentially metaphysical theorems, from which, *inter alia,* it followed (as he believed) that *any* recapitulation of ancestral forms is, not merely unverified, but impossible *a priori.* Being impossible, recapitulation was, of course, also unverified by empirical evidence. (Lovejoy, 1959, p. 443)

See also Cohen (1947, pp. 208–209) and Severtsov (1927, p. 150).

23. In an appraisal of Tiedemann's life and work, Agassiz rated him as a more convincing exponent of recapitulation than Meckel (Lurie, 1960, p. 412, note 63).

24. The same man who had influenced von Baer so strongly in Würzburg. In 1826, the Bavarian government established a university at Munich and brought to it several outstanding scientists, including Döllinger from Würzburg and Oken from Jena.

25. Agassiz's last essay, "Evolution and Permanence of Type" (1874) is a refutation of Darwinism based primarily on the immutability of Cuvier's four *Baupläne.*

26. A revision of this sort could not be reconciled with the explanation for recapitulation given by most Naturphilosophen; for they required a single direction of organic development to propel an embryo's repetition of lower adult stages. Freed from this philosophical baggage, such a revision entails only a limitation in recapitulation's scope, not a sacrifice of any theoretical point.

27. In heterocercal tails, the upper lobe is larger than the lower and the vertebral column extends into the upper lobe. By contrast, the homocercal tail of teleosts (higher bony fishes) has equal lobes and the vertebral axis terminates at the base of the tail.

28. *Pentacrinus* is a living form, permanently attached by a stalk as an adult; the Comatulae are stalked as juveniles but become free-floating as adults.

29. Russell states (1916, p. 230) that J. V. Carus inferred the probability of evolution from Agassiz's threefold parallelism. In the *Origin,* Darwin does not

accept recapitulation; he devotes all his embryological passages to an evolutionary reinterpretation of von Baer's views. He argues, for example, that related animals are more similar as embryos because evolutionary changes are usually introduced late in ontogeny and are inherited by progeny at the same stage and not, as von Baer had claimed, because a general law of development prescribes that the general must precede the special. Thus, similarity of embryonic structure reveals community of descent. (See Oppenheimer's excellent work [1959] on Darwin's use of embryology.) Yet Darwin did accept Agassiz's parallel between geology and embryology as an argument for evolution (see pp. 70–74 for the reconciliation of Darwin's support of Agassiz with von Baer's views): "Agassiz insists that ancient animals resemble to a certain extent the embryos of recent animals of the same classes; or that the geological succession of extinct forms is in some degree parallel to the embryological development of recent forms. I must follow Pictet and Huxley in thinking that the truth of this doctrine is far from proved. Yet I fully expect to see it hereafter confirmed . . . For this doctrine of Agassiz accords well with the theory of natural selection" (1859, p. 338). No evolutionist could resist the threefold parallelism. Lurie has misrepresented Darwin's position in writing: "Had Darwin followed von Baer and not Agassiz, modern embryology would not have had to rescue von Baer's interpretations from the obscurity in which they were placed by the triumph of Darwinism and by the ideas of such subsequent advocates of the Agassiz position as Ernst Haeckel" (1960, p. 288). If we must blame, we shall have to cite Haeckel, not Darwin. Darwin's position *is* von Baer's refurbished in evolutionary dress.

In later years, Agassiz saw that his argument was being snatched by evolutionists for an alien purpose. Rather than fight for his interpretation, he simply stopped teaching it. Though he despised Haeckel and all he stood for (Lurie, 1960, p. 412), Agassiz never defended his concept of recapitulation against its new interpretation. Of this, his student Alpheus Hyatt wrote: "In his personal talks with his students or in his lectures, I cannot remember that [recapitulation] was ever treated directly by anything more than incidental references . . . Nevertheless, I must have got directly from him, subsequently to 1858, the principles of this branch of research . . . I soon began to find that the correlations of the epembryonic stages and their use in studying the natural affinities of animals were practically an infinite field for work and discovery." (1897, p. 216)

4. Evolutionary Triumph, 1859–1900

1. The major change in taxonomic practice occurred in the 1930s when "population thinking," with its emphasis on variation, replaced previous typologies that involved a search for essential form. Huxley referred to this change as the "new systematics"; species were redefined as morphologically variable populations and the practice of coining specific names for all differences in form was thankfully curtailed (though not eliminated). One can argue that Darwin's words furnish sufficient justification for such an alteration in practice; nonetheless, it did not occur in his day.

2. De Beer has succinctly summarized Darwin's views in relation both to von Baer and the recapitulationists in commenting upon this and other passages in the essay of 1842: "Here, Darwin is following von Baer in the latter's law of embryonic resemblance . . . Darwin's adoption of this view shows that he rejected the transcendental theories of Serres and of Meckel on which Haeckel later based his theory of recapitulation . . . Species do not pass through the adult stages of their ancestors during their development, and there is no pressing back into earlier stages of development of characters which first appeared at later stages. In other words, Darwin's argument is free from the objections which beset Haeckel's theory of recapitulation" (1958b, pp. 16–17).

3. That is, most evolutionary changes. No recapitulationist took the extreme view that this rule had no exceptions. One had, rather, to prove that the exceptions were few and unimportant. Haeckel, for example, specifically classified all evolutionary changes into additions that preserved ancestral ontogenies intact (palingenetic) and modifications that interrupted ancestral ontogenies (cenogenetic—that is, an adaptation acquired by a larval form that influences the subsequent course of development).

4. This is too good to resist in the original German.

> In diesem Geistes-Kampf, der jetzt die ganze denkende Menschheit bewegt . . . stehen auf der einen Seite unter dem lichten Banner der Wissenschaft: Geistesfreiheit und Wahrheit, Vernunft und Cultur, Entwicklung und Fortschritt; auf der anderen Seite unter der schwarzen Fahne der Hierarchie: Geistesknechtschaft und Lüge, Unvernunft und Rohheit, Aberglauben und Rückschritt . . . Denn die Entwickelungsgeschichte ist das schwere Geschütz im "Kampf um die Wahrheit"! Ganze Reihen von dualistischen Trugschlüssen stürzen unter den Kettenschüssen dieser monistischen Artillerie haltlos zusammen und der stolze Pracht-Bau der römischen Hierarchie, die gewaltige Zwingburg der "unfehlbaren" Dogmatik, fällt wie ein Kartenhaus ein.

5. Thus, Remane dismisses Haeckel's statement that phylogeny is the mechanical cause of ontogeny. "Dieser Satz," he proclaims, "ist Unsinn" ("This statement is nonsense"—1960, p. 309). Huxley and De Beer speak of Haeckel's view that "phylogeny was the mechanical cause of ontogeny, whatever Haeckel may have meant by such a statement" (1934, p. 8), while Shumway writes of Haeckel's belief "that an adult might in some occult sense determine the course of development in the offspring" (1932, p. 96). Even the brilliant interpreter Russell wrote (for once, I think, incorrectly): "It was, for instance, mere rhetoric on Haeckel's part to proclaim that phylogeny was the mechanical cause of ontogeny" (1916, p. 314).

6. Although a mechanist and reductionist, Haeckel vehemently denied the third part of a common trilogy: materialism. Monism, he claimed, taught that matter and spirit are one: "Monism knows neither the matter without spirit of which materialism speaks, nor the spirit without matter that spiritualism upholds. Monism accepts neither spirit nor matter in the usual sense, but only One; the two are the same" (1866, 2:451).

7. Both the lineage and the individual are simply termed an "organism" (*Organismus*).

8. In the 1860s, as Mayr (1959) and Oppenheimer (1959) have noted, German used the same word—*Entwicklungsgeschichte*—to designate both ontogeny and phylogeny. The use of a single word for two processes may well provoke a tendency to seek relations between them. "Until now, we have by developmental history [*Entwicklungsgeschichte*] understood only that of the individual organic form . . . But this ontogeny is only one major branch of biogeny or the all-encompassing 'Entwicklungsgeschichte der Organismen.' . . . Paleontological developmental history of species and lineages stands as a second major branch . . . Germ-history [*Keimesgeschichte*] and stem-history [*Stammesgeschichte*], ontogeny and phylogeny, are in my opinion two sciences that stand in the narrowest and most immediate causal union" (Haeckel, 1876, p. 9).

9. Haeckel conferred upon these claims the status of one of his hereditary "laws"—though it must be said that these so-called laws are little but a systematization of common folklore and experience. Among his "laws of progressive transmission," the first (*lex hereditas adaptatae seu accommodatae*) affirms the inheritance of acquired characters itself; the second (*lex hereditas constitutae*—the law of established transmission) holds that the heritability of an acquired character is in proportion to the force with which (and length of time during which) the character is impressed (1866, 2:186–191).

10. Haeckel is extrapolating from a mammalian model. However, organisms that grow throughout life generally display a marked reduction in rate at adulthood (both of increase in size and change of form).

11. Haeckel was keenly aware of the philosophical differences between "hard" physical sciences and the complex, holistic, synthetic sciences of natural history. See the account in Chapter 6 of his debate with Wilhelm His.

12. This point is vital because Haeckel's critics often sought unfairly to undermine his biogenetic law by citing a small number of exceptions (and attacking the straw man of the law's absolute universality). A compendium of exceptions exceeding the number of valid examples would, of course, have been another matter; but this was never provided.

13. "The struggle for existence has had just as profound an influence on the freely moving and still immature young forms as on the adult forms. Hence . . . palingenesis is much restricted by cenogenesis" (1905, p. 498).

14. This is not willful obfuscation. All science was basically ignorant of heredity's operation before the Mendelian rediscovery of 1900. It is legitimate, in this context, to postulate that regularities in the results of inheritance reflect unknown laws of its operation.

15. To which Haeckel's "heterotopy" and "heterochrony," as categories of cenogenesis, are exceptions.

16. This formulation is clearly taken from Fritz Müller's famous conclusion (1864, in 1915, p. 250) since the ideas and vocabulary are almost identical: Development is "verwischt, indem die Entwicklung einen immer geraderen Weg vom Ei zum fertigen Thiere einschlägt, und sie wird häufig gefälscht durch den Kampf ums Dasein, den die freilebenden Larven zu bestehen haben." Haeckel acknowledges his debt to Müller for this formulation elsewhere (1874, p. 293).

17. The British embryologist F. M. Balfour writes: "When the life history of a form is fully known, the most difficult part of the task is still before the scientific embryologist. Like the scholar with his manuscript, the embryologist has by a process of careful and critical examination to determine where the gaps are present, to detect the later insertions, and to place in order what has been misplaced" (1880, p. 4).

18. Cope (1887, p. 7) listed it among his four laws of animal structure. The others are: homology, successional relation (evolution) and teleology (adaptation).

19. I do not mean to underestimate his contribution, staggering in volume and usually in quality as well, towards gathering the data of vertebrate paleontology and working out the phyletic lineages of American Mammalia. Cope published hundreds of short papers on fossil vertebrates, and his *Tertiary Vertebrata* (1883c)—known as "Cope's Bible" to all modern practitioners of the field—is more than twice as thick as the Manhattan telephone directory!

20. It is often said that Neo-Lamarckism is a term poorly chosen because its advocates followed Lamarck only in the less important notions of use and disuse and inheritance of acquired characters—for these produce the side branches, not the steps of the main stem—and not in the more basic concepts of perfecting principles and responses to felt needs. This is incorrect. In their early work, both Cope and Hyatt accepted Lamarck's primary distinction between progressive adaptations and specific deflections.

21. This distinction of specific and generic characters, so foreign to our modern definitions, must be understood in order to make sense of many of Cope's statements. For example, "The very frequent absence of the posterior molars (wisdom teeth) has been recently found to characterize a race in India. Should this peculiarity prove constant, this race would with propriety be referred to as a new genus of Hominidae [since it has lost a character and is retrogressing down the main trunk of the lineage], as we have many cases of very similar species being referred to different genera" (1871, in 1887, p. 180).

22. In later writings, as his belief in the mechanical origin of acquired characters grew and his faith in Darwinian processes diminished yet further, Cope largely dropped his early distinction between the causes of specific and of generic characters. He came to believe that all new characters arose by effort, either as a conscious (or protoconscious) response to a felt need; or by "kinetogenesis," the mechanical acquisition of a trait by use (fixation of calluses on feet as a permanent, inherited character, for example).

23. I find no reference to the third, and least important—heterotopism.

24. It is purely coincidental that the idea of a secular increase in the earth's oxygen has reappeared and gained acceptance in the last 30 years. Cope's increase occurs during the Phanerozoic (last 600 million years); he postulates it to justify the increasing frequency of acceleration over retardation; and he bases it on an incorrect argument (removal of carbon dioxide to produce coal does not reduce its abundance in the atmosphere; oceans are the main reservoir of CO_2; they liberate enough to redress the balance if terrestrial proc-

esses remove it). Today we believe that the increase is a Precambrian phenomenon (levels have been fairly stable for the past 600 million years); it arises because the primitive atmosphere contained little or no oxygen and only acquired it progressively after the evolution of photosynthesis.

25. Hyatt wrote to Darwin in November, 1872: "My relations with Prof. Cope are of the most friendly character; and although fortunate in publishing [the law of acceleration] a few months ahead, I consider that this gives me no right to claim anything beyond such an amount of participation in the discovery, if it may be so called, as the thoroughness and worth of my work entitles me to" (F. Darwin, 1903, pp. 339–340).

26. He did not, of course, deny that true retardation sometimes occurs. See letter to C. Darwin, quoted in F. Darwin (1903, p. 341).

27. Hyatt's writings on this subject are not easy to follow. Darwin became quite intrigued and engaged Hyatt in an extensive correspondence about acceleration and senescence (F. Darwin, 1903, pp. 339–348). In his first letter, Darwin wrote: "I confess that I have never been able to grasp fully what you wish to show, and I presume that this must be owing to some dulness on my part" (p. 339).

28. Fanciful (and forced) as Hyatt's notion may seem to us, J. B. S. Haldane took it very seriously and proposed a genetic explanation for Hyatt's claim that apparently youthful (though actually senile) features often follow a long history of acceleration: "I suggest that the gerontic straightness of degenerate forms was due to the pushing back of DZ [delayed zygotic] genes governing embryonic development into the Z^4 [adult] and Z^3 [juvenile] stages. The adult *Bactrites* formed an uncoiled shell for the same reason that its ancestors formed a straight protoconch. The same argument could be applied to other cases where, after a long evolutionary history of acceleration, embryonic characters appear in the later stage of life-history" (1932, p. 17). This phenomenon, Haldane suggests, "might lead to racial 'second childhood'" (p. 19).

29. Lest we view this statement as an antidote to the rampant scientific racism of his age, Hyatt adds later that the backwards step is, indeed, only morphological. The larger cranium that reduced prognathism entails is a spiritual advance of a rather great order.

30. I now include Haeckel with the American Neo-Lamarckists because his views on the causes of recapitulation lie wholly within the Lamarckian part of his complex and contradictory beliefs.

31. Butler was Britain's chief defender of this theory. He championed it in three books (*Unconscious Memory,* 1880; *Life and Habit,* 1877; and *Luck or Cunning?,* 1887); these, along with *Evolution Old and New,* form the basis of his anti-Darwinian crusade.

32. There are, as always, hints of the argument in many earlier works (Butler, 1880, p. 62), but all major adherents credited Hering with primary authorship. Hering delivered his views in a lecture to the Imperial Academy of Sciences at Vienna on May 30, 1870. It was published by Karl Gerold's Sohn Vienna under the title: *Das Gedächtniss als allgemeine Funktion der organisierten Materie* (Memory as a Universal Function of Organised Matter). Samuel Butler translated it as a chapter of his work *Unconscious Memory.*

33. This work, *Die Perigenesis der Plastidule,* has always been translated as *The Perigenesis of the Plastidule.* But Haeckel used *Plastidul* as the singular for his designation of life's atom, and his title therefore uses the plural.

34. Hartog (1920) assumes that this naturalist is F. Müller, but I do not see why it should not have been Haeckel. Haeckel's *Generelle Morphologie* was published in 1866, and was surely better known than Müller's work.

35. All evolutionists granted at least an executioner's role to natural selection—the removal of the unfit. Darwinians believed that natural selection, by accumulating favorable variants, could also create the fit.

36. Total ontogeny is generally not shortened because new features are added to the end of ancestral ontogenies that are being condensed.

37. A "lean animal specter"—Kleinenberg's characterization of gastraea (1886, p. 2).

38. Weismann and all serious students of recapitulation—Oppel, 1891; Mehnert, 1891, 1897; and Keibel, 1895, 1898, for example—accepted Cope's redefinition of recapitulation or rediscovered it for themselves: that is, it should be applied to individual organs, not to entire animals, and that Haeckel's important category of exceptions—his "heterochronies"—could therefore be taken as examples of recapitulation involving unequal rates of acceleration among organs. Weismann wrote: "In the course of the phylogeny, numerous time-displacements of the parts and organs in ontogeny must result, so that ultimately it is impossible to compare a particular stage in the embryogenesis of a species with a particular ancestral form. Only the stages of individual organs can be thus compared and parallelized" (1904, p. 174).

39. In many cases, these are not the taxonomic designations used for these insects today. To avoid any confusion, I am using Weismann's names.

40. There is a discrepancy in years between this account and his statement in the autobiography.

41. Itself not uninfluenced by von Baer's embryology. Milne-Edwards was a leading supporter of von Baer's principle of differentiation against the unilinear advance from simplicity to complexity preached by the transcendental recapitulationists. See Milne-Edwards (1844).

5. Pervasive Influence

1. The typical heads-I-win-tails-you-lose argument of an incontrovertible racism. The very behavior that would be regarded as heroic for a white man—think of how many great Western heroes died with courage in excruciating pain—demands a different interpretation when the victim is an Indian. In this case, he is no hero because he does not feel the pain of his martyrdom.

2. Lombroso argues that children, despite a common impression to the contrary, have a natural penchant for alcohol. That scientists fail to note it merely reflects a class bias; for middle- and upper-class parents never give their children the opportunity to indulge this natural vice. "One who lives among the upper classes has no idea of the passion babies have for alcoholic

liquor, but among the lower classes it is only too common a thing to see even suckling babies drink wine and liquors with wonderful delight" (1895, p. 56).

3. This remarkable passage is the only attack I can find leveled by a recapitulationist against the pseudoscientific justification of political positions based on racial rank. It is, of course, an attack on the polygenist theory, not on the concept or use of recapitulation.

4. Note the full threefold parallelism of recapitulation: paleontology ("fossil" record of historical ancestors), comparative anatomy (modern "savages"), and ontogeny.

5. "The progress of our race has been a progress in youthfulness" (1894, p. 519).

6. Surely incorrect as a generality since human jaws become progressively prognathous with increasing age—see Chapter 10.

7. Hrdlicka argues that this and other animal traits of babies are "temporary reminiscences of and connections with man's ancestral past. They do not appear to prejudice in the least the further normal development of the child both physically and mentally . . . It seems just to conclude that just as the human child before birth recapitulates, more or less, various phases of its physical ancestry, so the child after birth recapitulates and uses for a time various phases of its prehuman ancestral behavior" (1931, p. 92).

8. President Teddy Roosevelt, strongly impressed by Hall's views on the education of pre-adolescents, stated in a letter: "I must write you to thank you for your sound common sense, decency and manliness in what you advocate for the education of children. Over-sentimentality, over-softness, in fact, washiness and mushiness are the great dangers of this age of this people. Unless we keep the barbarian virtues, gaining the civilized ones will be of little avail" (in Ross, 1972, p. 318).

9. "J'ai peu travaillé en psychologie les rapports entre l'ontogenèse et la phylogenèse car, psychologiquement, l'enfant explique l'adulte davantage que l'inverse."

10. Almost all of Piaget's work applies only to the type of knowledge that he calls "logico-mathematical." Piaget has had less success in attempting to establish an ontogeny for ethical learning or biological knowledge, for example.

11. Note also, Goethe's comment to Eckermann in 1827: "Wenn auch die Welt im ganzen vorschreitet, die Jugend muss doch immer wieder von vorn anfangen und als Individuum die Epochen der Weltkultur durchlaufen" (in Schmidt, 1909, p. 156).

12. Herbart was a philosophical empiricist of the *tabula rasa* school. It is hard to imagine how he could have supported the biogenetic law with its necessary consequence of inherited racial memory. Herbart died in 1841.

13. This becomes a major theme in Freud's last work, *Moses and Monotheism.* The belief in a single, omnipotent God represents the oedipal stage of attachment to the father. The child's oedipal desire to kill his father and possess his mother reflects an historical act of parricide—in this case, the Jewish people killed their leader Moses. Both the oedipal attachment and the crushing guilt for parricide emerge in the elevation of the father to absolute status as an omnipotent God:

The psychoanalyses of individuals have taught us that their earliest impressions, received at a time when they were hardly able to talk, manifest themselves later in an obsessive fashion, although those impressions themselves are not consciously remembered. We feel that the same must hold good for the earliest experiences of mankind. One result of this is the emergence of the conception of one great God. It must be recognized as a memory—a distorted one, it is true, but nevertheless a memory. It has an obsessive quality; it simply must be believed. As far as the distortion goes, it may be called a delusion; in so far as it brings to light something from the past, it must be called truth. The psychiatric delusion also contains a particle of truth; the patient's conviction issues from this and extends to the whole delusional fabric surrounding it. (1939, p. 167)

14. Freud scholars have generally identified a role for recapitulation in Freud's thought. But, as Frank Sulloway maintains in his forthcoming book on *Freud as Psychobiologist,* this traditional historiography has continually insisted that the biogenetic law was little more than a late addition to Freud's psychoanalytic interests. Sulloway's thorough study of Freud's intellectual development during the crucial years of psychoanalytic "discovery" proves that psychoanalytic theory emerged in full cooperation with Freud's a priori belief in recapitulation. Furthermore, Sulloway has shown that recapitulation was a prevalent belief in the literature on sexual pathology, child psychology, and neurology that Freud read and annotated during the 1880s and 1890s.

6. Decline, Fall, and Generalization

1. For example, Morgan: "Of its own weight of contradictions the method fell into disrepute, and towards the end of the century was replaced by experimental work" (1932, p. 174).

2. Marshall, one of recapitulation's champions, spoke of these exceptions with the same literary metaphor that Darwin had used in discussing the inadequacy of the fossil record: "It [embryology] is indeed a history, but a history of which entire chapters are lost, while in those that remain many pages are misplaced and others are so blurred as to be illegible; words, sentences, or entire paragraphs are omitted, and worse still, alterations or spurious additions have been freely introduced by later hands, and at times so cunningly as to defy detection" (1891, p. 832). Such a catalogue of expected exceptions makes a theory rather refractory to inductive disproof.

3. Ironically, Kleinenberg (1886, p. 2), the author of this critical phrase, was a recapitulationist. In coining this epithet, he was simply criticizing Haeckel's decision to invoke a purely hypothetical ancestor in place of an actual coelenterate that might better serve as the ancestral metazoan (since its adult structure recalled the germ-layers of early development in higher organisms). Huxley had first linked the structure of adult coelenterates to the germ-layers of developing embryos in higher groups (in arguing for homology, not for descent): "It is curious to remark that throughout the outer and the inner membranes appear to bear the same physiological relation to one

another as do the serous and mucous layers of the germ; the outer becoming developed into the muscular system and giving rise to the organs of offense and defense; the inner, on the other hand, appearing to be more closely subservient to the purpose of nutrition and generation" (1849, p. 425). See excellent discussion of this issue in Oppenheimer, 1967, pp. 261–271.

4. Haeckel's name has been resurrected by many authors (for example, Whittaker, 1969) to designate the kingdom of procaryotic organisms (bacteria and blue-green algae) lacking cell organelles. Many biologists view the break between procaryotes and nucleated eucaryotes (true protists and all metazoa and metaphyta) as the most profound in life's history. One popular theory postulates the origin of eucaryotes from the symbiotic association of procaryotes. In this view, the nucleus and mitochondrion are homologous with entire procaryotic organisms (Margulis, 1970).

5. For a previous attempt to codify the laws of cenogenesis and link them, as a unitary process, with palingenesis, see Goette (1884).

6. Haeckel is only mentioned once, and fleetingly, in this article. Since Haeckel was still alive and strongly supporting his own nation in the midst of World War I, his French colleague probably regarded the very mention of his name as anathema.

7. A. C. Hardy, for example, wrote of Garstang's paedomorphosis: "He added to zoological thought an idea which, I believe, will come to be classed among the more original and profound of those put forward in the last half century" (in Garstang, 1951, p. 1). But we cannot blame a man very strongly for lavishing too much praise upon his father-in-law.

8. The generic name of the axolotl, properly spelled *Ambystoma.* See note 8, Chapter 8.

9. Several genera of salamanders are permanently larval in morphology; no metamorphosis has ever been observed. Because they retain external gills throughout life, they are called perennibranchiate.

10. The date of this reference is often cited differently. The article was written in 1883; the volume number of its journal is for 1884, but the article was published in 1885.

11. Lancelet is the common name for *Amphioxus.*

12. In this article, Bolk argues that recapitulated stages of no apparent use to embryos (notochord and gill slits of mammalian fetuses) secrete hormones necessary for the next and definitive stages of development. In this way he sought to counter the argument that all "nonfunctional" palingenetic stages should be eliminated, thus hopelessly truncating the phyletic record preserved in embryonic development.

13. He named this phenomenon *Dissogonie* because he believed that the sexual organs degenerated at metamorphosis to be reformed anew in the adult. Thus, the animal reproduced at two distinct phases of its life. Gary Freeman (personal communication) informs me that this is generally not true. Sexual maturity does first occur in larvae, but the gonads often persist through metamorphosis and throughout life.

14. Thus engendering a heroic confusion that I shall attempt to resolve in

the next chapter by distinguishing paedomorphosis, neoteny, paedogenesis, and several related terms.

15. To be sure, of the earlier and rather different version of the Naturphilosophen. But von Baer's critique was directed primarily towards the repetition of *adult* ancestral stages and this had been carried forward *in toto* by Haeckel and his evolutionary school.

16. Or, more accurately for this case, forgotten. Von Baer's work had a great influence in curbing the speculative excesses of Naturphilosophie. When recapitulation arose again in evolutionary garb, von Baer's challenge, though applicable as ever, was rarely urged against it. Morgan (1903, p. 75) stated that only he and Hurst (1893) had attacked the biogenetic law from this standpoint.

17. Against such an attitude, Haeckel directed his most magnificent, and untranslatable, prose:

> Die grosse Mehrzahl der Naturforscher . . . begnügt sich mit der blossen Kenntniss derselben; sie sucht die unendlich mannigfaltigen Formen, die äusseren und inneren Gestaltungsverhältnisse der thierischen und pflanzlichen Körper auf und ergötzt sich an ihrer Schönheit, bewundert ihre Mannigfaltigkeit und erstaunt über ihre Zweckmässigkeit; sie beschreibt und unterscheidet alle einzelnen Formen, belegt jede mit einem besonderen Namen und findet in deren systmatischer Anordnung ihr höchstes Ziel . . . So gleicht denn leider die wissenschaftliche Morphologie der Organismen heutzutage mehr einem grossen wüsten Steinhaufen, als einem bewohnbaren Gebäude. Und dieser Steinhaufen wird niemals dadurch ein Gebäude, das man alle einzelnen Steine inwendig und auswendig untersucht und mikroskopiert, beschreibt und abbildet, benennt und dann wieder hinwirft . . . Sie begnügen sich damit, die organischen Formen (gleichgültig ob die äussere Gestalt oder den inneren Bau) ohne sich bestimmte Fragen vorzulegen, oberflächlich zu untersuchen und in dicken papierreichen und gedankenleeren Büchern weitläufig zu beschreiben und abzubilden. Wenn dieser ganz unnütze Ballast in den Jahrbüchern der Morphologie aufgeführt und bewundert wird, haben sie ihr Ziel erreicht . . . Erst wenn die Betrachtung der Gestalten sich zur Erklärung erheben wird, erst wenn aus dem bunten Chaos der Gestalten sich die Gesetze ihrer Bildung entwickeln werden, erst dann wird die niedere Kunst der Morphographie sich in die erhabene Wissenschaft der Morphologie verwandeln können. (1866, pp. 3–7)

18. G. E. Allen (1975) has epitomised the development of twentieth century biology as "the history of the successive introduction of experimental methods from the physical sciences through physiology into previously descriptive areas such as embryology, heredity and the origin of species." In several fields, embryology in particular, this movement had its roots in the late 1880s and 1890s.

19. F. M. Balfour (1851–1882) was the Pergolesi of nineteenth-century English biology. He is not well known today only because his early death in an

alpine climbing accident prevented an attainment of the eminence that all had foreseen. Darwin, in receiving Balfour's great treatise on comparative embryology, replied: "I am proud to receive a book from you, who, I know, will some day be the chief of the English Biologists" (in Foster and Sedgwick, 1885, p. 23).

20. His' treatise is presented as a series of 17 letters to a friend—hence its demonstrative style.

21. "Mechanik der Entwickelung"—His does not use Roux's term *Entwicklungsmechanik.*

22. His was by no means immune to invective in return. He accused Haeckel of shocking dishonesty in repeating the same picture several times to show the similarity among vertebrates at early embryonic stages in several plates of the *Natürliche Schöpfungsgeschichte:* "I myself have grown up in the belief that, among all qualifications of a scientist, reliability and unconditional respect for factual truth is the only one that can never be lacking . . . In my judgment, he [Haeckel], by his way of waging battle, has himself renounced the right to be counted as a peer in the company of earnest researchers" (1874, p. 171).

23. Haeckel's aversion to most mathematical studies in natural history is well displayed in his sarcastic comment upon the work of Victor Hensen (who had earned Haeckel's wrath by attacking the biogenetic law within Haeckel's own domain of planktonic organisms): He "tries to give an 'exact' explanation of the phenomena of marine life by counting how many million individuals of each species can live in a cubic mile of sea water" (1905, p. 862).

24. As did many other recapitulationists. Lankester, for example, wrote of "the ultimate goal of biology which is the accounting for the phenomena of living matter or protoplasm by reference to the laws of chemistry and physics" (1877, p. 432).

25. Driesch wrote his dissertation under Haeckel. Roux studied at Jena and attended Haeckel's lectures. Roux and Driesch were not, of course, the first scientists to manipulate an embryo. Even Haeckel had once done some experimental work. Oppenheimer writes: "It is no secret, though it is not commonly bruited about, that in 1869 Haeckel himself . . . published the results of experimental division of siphonophore larvae, demonstrating that half-larvae were able to form whole organisms" (1967, p. 6). Experimental methods did not become dominant in embryology until the 1890's.

26. In this section, I shall only discuss the impact of Roux's methodology upon the status of recapitulation. I have neither the space nor competence to chronicle the achievements of experimental embryology and the controversies that developed within it. This topic has been treated superbly by Oppenheimer (1967).

27. In using this phrase, I am distinguishing between the intellectual attitude of publications and the social relations of men. The tactic of experimental embryology was simply to ignore recapitulation as irrelevant in their writings. Recapitulationists were another matter. Since they dominated embryology, they had to be displaced before experimental methods could triumph.

28. Conklin uses the term "evolution" in its older sense to specify the embryological theory of predetermination. He is not, of course, speaking of Darwin or phylogeny.

29. Oscar Hertwig, more epigenetic than preformationist (1906), devoted the final chapter of his massive *Handbuch* to phylogeny and the issue of recapitulation. He, at least, had not chosen simply to ignore the subject. Hertwig's opposition to recapitulation reflects another mechanistic and experimental attitude. He argued that the earliest stages of vertebrate development are similar not because they repeat a common ancestry, but simply because there is no other physical way to develop a many-layered structure from an initial cell: "The main reason why certain conditions in the development of animals recur with such great constancy and always in essentially the same way is that they provide, under all circumstances, the necessary preconditions through which alone the following, higher stages of ontogeny can be built up. The single-celled organism can, by its very nature, only transform itself to a many-celled organism by the process of cell division. Thus, all metazoa must begin their ontogeny with a process of segmentation, and similar statements could be made about each following stage" (1901, p. 57; see also 1906).

30. Eugenio Rignano (1911, pp. 14–18), an extreme recapitulationist, defended the biogenetic law on this account against attacks based on the egg's organization.

31. Lankester developed this argument to avoid a thorny problem connected with his belief that the initial segregation of germ layers occurred by delamination of the blastula. In living organisms, it occurs by invagination of the gastrula. This invagination formed the basis of Haeckel's famous "gastraea theory," and Lankester introduced precocious segregation in order to maintain his "planaea theory" against this powerful argument. Invagination of the gastrula, Lankester claimed, is a secondary consequence of the early distinction now made in first cleavage between precursors of the germ layers. When a modern embryo becomes a blastula, the fate of its cells are already sealed; the precursors of endoderm can aggregate and invaginate. But this determination is only the result of a secondary precocious segregation. It does not reflect the situation of ancestral Blastaea; the cells of Blastaea must, at first, have been equipotential. Lankester then argues, in the best speculative tradition, that delamination is much more likely than invagination as a process for the phyletic separation of germ layers.

32. The empirical (and unsatisfactory) status of these laws is well illustrated by Perrier and Gravier's (1902, p. 151) comments on "tachygenesis" (their name for the law of condensation). Their monograph is a 250-page list of supposed cases, but they can provide no other justification for their principle beyond: "Tachygenesis is a constant mode of heredity's action. Whatever its nature may be, tachygenesis must be considered as one of the qualities of heredity or, to put it another way, one of the conditions according to which this more general ensemble of poorly known (but certainly knowable) causes that we call heredity must operate, at least in part."

33. This point acquired even greater force under the DeVriesian macromutational theory for the origin of new species—a theory that remained pop-

ular among Mendelians until the modern synthesis of the 1930s. Mutational substitution is not only a mechanism for the evolution of new characters; it is also *the* device for the production of new species. Thus, the primary evolutionary event itself does not conform to recapitulatory expectations. Morgan wrote: "Such a conception does not fall easily into line with the statement of the biogenetic 'law'; for actual experience with discontinuous variation has taught us that new characters that arise do not add themselves to the end of the line of already existing characters but if they affect the adult characters they change them without, as it were, passing through and beyond them. I venture to think that these new ideas and this new evidence have played havoc with the biogenetic 'law'" (1916, pp. 18–19).

7. Heterochrony and the Parallel of Ontogeny and Phylogeny

1. The choice of points for standardization depends upon the problem under discussion. If a lineage's evolution is characterized only by an alteration of its larvae, we would not want to depict its phylogeny as the sequence of (unchanged) adults, but would choose the appropriate juvenile stage as our standardized point. This notion of a standardized point also permits us to view the problem of recapitulation and paedomorphosis in an appropriately widened light. If phylogeny is rigidly construed as a sequence of adults, then these phenomena are studied only with respect to adult characters. But characters arise at all stages of ontogeny, and we may speak of recapitulation and paedomorphosis for any character with respect to the stage of ontogeny at which it appeared in ancestors. We do this by choosing the stage of ancestral appearance as our point of standardization, and seeing if our character appears earlier (recapitulation) or later (paedomorphosis) in descendants.

2. Cope (1870, in 1887, p. 154) clearly recognized this distinction when he wrote: "'acceleration' means a gradual increase of the rate of assumption of successive characters in the same period of time. A fixed rate of assumption of characters, with gradual increase in the length of the period of growth, would produce the same result—*viz.* a longer developmental scale and the attainment of an advanced position."

3. I have been distressed to find that the penchant for coining new terms is almost as prevalent here as among those chief sinners, the students of fossil hominids. The number of terms available for the various phenomena included under paedomorphosis is simply staggering—neoteny, paedogenesis, progenesis, proterogenesis, fetalization, and epistasy to mention just a few. Since I am not writing as an antiquarian, I trust I will be forgiven if I forgo the compilation of a synonymy.

4. *Portunion* is a parasitic crustacean with normal appendages in its youthful stages; these appendages degenerate as it assumes a parasitic adult existence.

5. Louis Dollo placed it among his laws of evolution (Gould, 1970).

6. Nothing in this section should be construed as an attack upon the idea of correlation. Dissociation and correlation are different issues; I view dissocia-

tion as the change or disruption of correlations that exist for all the reasons commonly adduced (genetic linkage and mechanical pressures, for example).

7. There are, of course, some very prevalent correlations among these fundamental processes. Even these can often be dissociated experimentally. The attainment of sexual maturity often marks the great reduction or cessation of growth. A high degree of differentiation often seems to preclude mitosis (neurons, muscle cells). But Goss writes: "The majority of cell types are capable of division despite the broad spectrum of differentiated states that are represented. Only in extreme cases do the differentiated features of a cell make mitosis impossible" (1964, p. 60).

8. The designation of a proper measure for body size is not an easy matter. Mosimann (1970) and Sprent (1972) discuss some pitfalls and provide suggestions. Jolicoeur (1963) suggested that the first principal component of the covariation matrix of logarithmically transformed data might provide the best general assessment of size in multivariate situations. Some authors have plotted individual organ sizes against this overall measure of size (Matsuda and Rohlf, 1961).

9. I shall limit this discussion to a simple bivariate case (a single ratio for shape). Clarity and aesthetics notwithstanding, a clock of many more hands can be easily envisioned.

10. I am not unaware of the fact of intraspecific variation; nor do I wish to construct a typology for ontogeny. The idealized bivariate curve is simply the best statistic for estimating the values of any subsequent specimen.

11. The "outset" depends upon the problem under consideration. It may be conception, birth, or the end of major differentiation in fetal development. I do not, however, wish to imply that the choice of outset is purely "relative." There is an extensive literature on the consequences of shifting an origin in allometric studies (Angleton and Pettus, 1966, for example).

12. This value should be close to 1 since the attached portion of the coiled valve is flat. It is less than 1 here because Burnaby's equation (the basis of Table 2) was calculated for larger *Gryphaea* and fails in distant, backward extrapolation.

13. This probably happens rather infrequently. Acceleration of sexual development will rarely leave other characters unaltered since patterns of growth and form are usually linked to maturation.

8. The Ecological and Evolutionary Significance of Heterochrony

1. The name given to the ontogeny of an entire colony as opposed to that of its individual members.

2. De Beer gives a curiously reversed and incorrect interpretation because he treats the entire colony as an individual: "These evolutionary novelties appear first in early stages of colony-formation and become prolonged into later stages in subsequent phylogeny . . . This is a case of 'colonial neoteny'" (1958, pp. 86–87). Proximal thecae are indeed the first-formed parts of the

colony, but they are the *oldest individuals* of a *group,* not the juvenile stages of a single entity! The spread of originally adult characters to younger stages is recapitulation.

3. The same point is emphasized in the one common macroevolutionary concept based on the difference between immediate and retrospective significance—preadaptation (though the term is unfortunate because it implies a foreknowledge of retrospective significance). Preadaptation is a standard (and quite legitimate) way out of the classical dilemma of incipient usefulness: How can an intricate structure arise if it cannot work properly until fully formed? The answer given is that a similar structure functioned in quite a different way for ancestors. Thus, the reptilian quadrate and articular carried immediate significance for their function in articulation of the jaw. Their small size and position near the hyomandibula preadapted them for a transformation into the malleus and incus of the mammalian ear. This "accidental" capacity for such a transformation constitutes their retrospective significance. The acanthopterygian pectoral fin could never have made it as a tetrapod limb; but the crossopterygian fin, with its rotatable central axis, was admirably suited for such a transformation (retrospective significance), even though it was working quite well for a very different purpose in ancestors (immediate significance). Operationally, the concept of preadaptation simply holds that marked functional change can often occur with minor structural change.

4. The importance of juvenile hormone in heterochrony may be enhanced by another property. The hormone not only functions in the control of somatic differentiation, but also plays a role in sexual maturation through its gonadotropic effect. After disappearing in pupae to permit the adult transformation, juvenile hormone is again secreted by adult females to promote the synthesis of yolk proteins and their accumulation in developing oocytes. It may also affect the activity of accessory sex glands in some male insects (Schneiderman, 1972).

5. Although most manipulations reported in the literature are done in laboratories, several natural experiments may be cited as well. Parasites provide several examples (their tampering with development is well known in such common phenomena as parasitic castration). Johnson (1959) studied parasitism of the aphid *Aphis craccivora* by a tiny braconid wasp. When unhatched parasitic eggs developed within the aphid's body, nymphal characters were retained in the adult instar. Johnson suggests that juvenile hormone from the developing embryos might be diffusing into the blood of the host.) An opposite effect is exerted by hatched parasites. The aphid is killed in its fourth nymphal instar, but this instar has already developed some accelerated features of the unrealized adult molt.

6. I do not mean to imply that this generation is by any means automatic or universal. Willis (1974) has emphasized the repeated failure of experiments designed to produce supernumerary larvae by massive overdoses of juvenile hormone in several species.

7. The tadpole stage of many frogs can be lengthened markedly in time

and extended to giant sizes, but none have ever become sexually mature. Wassersug (1975) considers the reasons for lack of complete neoteny in frogs.

8. In older literature, the generic name is often spelled *Amblystoma* and some modern authors have retained the "l" despite a 1963 I.C.Z.N. decision validating *Ambystoma*. Tschudi coined *Ambystoma* in 1838, but Agassiz assumed that he had made a grammatical error in intending *Amblystoma* (from amblys = wide, and stoma = mouth). This seems reasonable to me, but Stejneger in 1907 managed to invent a justification to rescue Tschudi from a charge of classical ignorance (*Ambystoma* as a contraction of Anabystoma—to cram into the mouth). Contrived as this attempt at rescue may be, the Commission was willing to seize upon any straw for retaining both priority and established usage. This and other more momentous issues in the biology of *Ambystoma* are reviewed by Smith (1969).

9. Thyroxin (and its stimulators) are not the only hormones involved in amphibian metamorphosis and heterochrony. Grant (1961) has implicated prolactin in the "second metamorphosis" of the eastern spotted newt *Notophthalmus viridescens* from a terrestrial eft to a permanently aquatic adult.

10. Frisch (1972, and Frisch and Revelle, 1971) has argued convincingly for a control of menarche by critical weight in human females. Since the timing of maturation is a primary input to heterochrony, its immediate control by such environmental factors as nutrition supports the ideas developed in this chapter. (Long-term evolutionary pattern in menarche may be another matter, of course.)

9. Progenesis and Neoteny

1. This work rarely has anything to do with paedomorphosis per se. The axolotl has become a "standard preparation" for much research in general physiology.

2. Wake strongly implies that many of the cavernicolous plethodontines are progenetic. He argues (1966, p. 79) that selection for paedomorphosis included an acceleration of sexual maturation. But he presents no evidence that these species mature more rapidly than either their presumed ancestors or any modern relatives. His claim seems to be an inference based on the idea that complete paedomorphosis is likely to be degenerative while "differential metamorphosis" (paedomorphosis of selected organs) may be progressive—and that the degenerative mode of paedomorphosis is progenesis (rejected on pp. 324–341 of this book). Wake regards these cavernicolous forms as evolutionary dead ends (and I have no quarrel with this statement).

3. This is an appropriate place to remind readers of a regrettable confusion in terminology that pervades the study of heterochrony. Margalef entitled his paper: "Importancia de la neotenia en la evolucion de los crustaceos de agua dulce"—using neoteny in the neutral meaning of any juvenilization in morphology. Although de Beer recognized the difference between progenesis and neoteny, he recommended neoteny as the term for any juvenilization because he regarded progenesis as degenerative and therefore unim-

portant in evolution. I (with many others) have used paedomorphosis for the neutral descriptor of juvenilized morphology—while I reserve progenesis and neoteny for the different processes that lead to juvenilization. Margalef uses neoteny as a synonym for my paedomorphosis; the mechanism of paedomorphosis for his fresh- and brackish-water crustaceans is clearly progenesis.

4. Krebs and Myers (1974), for example, argue that microtine rodents maintain high reproductive rates by reducing the age of sexual maturation (and not by increasing litter size) during the expansion phase following a population crash. They present no evidence of a juvenilization in morphology. Accelerated maturation is not produced by progenesis, but the effect and adaptive significance are similar.

5. By analogy, I am reminded of the remarkable and recurrent dwarfism of large mammals on small Pleistocene islands. Dwarf elephants evolved independently on the Mediterranean islands of Malta, Sicily, Crete, and Cyprus (Leonardi, 1954; Accordi and Colacicchi, 1962) and on the Pacific islands of the Celebes, Flores, and Timor (Hooijer, 1967). *Elephas falconeri* was only 0.9 m high! Dwarfed hippos, each independently derived from the *Hippopotamus amphibius* stock, evolved on Sicily, Malta, Cyprus, and Crete (Boekschoten and Sondaar, 1966, 1972). The dwarf giant-deer of Crete and Sardinia also evolved independently from large ancestors (Boekschoten and Sondaar, 1966, p. 36). These animals are not primarily progenetic. They display a few minor juvenile features (long retention of the deciduous fourth premolar with no eruption of its final counterpart in some of the hippos; reduced tusks in some of the elephants). They are, for the most part, remarkably well-proportioned miniatures of large parental forms (Westoll, 1950, p. 500; Gould, 1971). Proportioned dwarfism is not heterochrony in the strict sense, but it shares with heterochrony the common mechanism of dissociation between growth and development. Dwarfism must have been extremely rapid as an immediate adaptation to the radically different ecology of an uncrowded island. Boekschoten and Sondaar (1966, p. 39) suggest "the order of a millenium" for some of the hippos. I would not be surprised if the reduction in size (though not some of the accompanying specializations for changed diet and locomotion) occurred much more quickly. Proportioned dwarfing can arise as a single Mendelian mutation (Lambert and Sciuchetti, 1935, for rats; Johnson et al., 1950, for cattle; Bennett, 1961, for mice; McKusick and Rimoin, 1967, on the pedigree of General Tom Thumb).

Proposals for the significance of such dwarfism range from degeneration (Leonardi, 1954) to restriction of food (Boekschoten and Sondaar, 1966). I wonder, in analogy with the small size of progenetic organisms, if early maturation in a context of *r* selection might not be involved. The islands were empty of predators and large competitors when the elephants and hippos arrived. Early reproduction for rapid *r* may have been the best strategy for any individual. Proportioned dwarfing may have been the most rapid and most readily available genetic path to early maturation.

The major problem with this explanation (and most others) is that it does not explain an apparent counter-case—the tendency for small mammals to increase in size on the same islands (Hooijer, 1967; Freudenthal, 1972, on a remarkable, geometrically scaled giant insectivore). Further size reduction might not be favored for small forms, especially if there is an optimum mammalian size, from which many mammals in crowded ecosystems are pushed in both directions. On initially uncrowded islands, both large and small animals converge towards the optimum. For the large animals, this tendency carries the additional benefit of earlier maturation.

6. My own preference (Eldredge and Gould, 1972) is to view these trends not as gradual sequences within a single phylum, but as a selection from an essentially random set of discontinuous speciation events. The link to *K* selection will hold in either case. In the classic view (phyletic gradualism), the entire trend is governed by selection upon *K* strategists. In the alternative presented by Eldredge and myself (punctuated equilibria), the set of speciations providing a random input to the trend may present as many *r* strategists as *K* strategists. But the trend—since it leads to increased size, delayed maturation and morphological specialization—must incorporate *K* strategists differentially.

7. The allometric parameters of these relationships differ, of course; only the direction of change is the same. A judgment of hypermorphosis does not require (and cannot expect) a constancy in allometric exponents. The parameters of power functions are as variable in evolution as any morphological structure (Gould, 1966). Rates will be modified in hypermorphosis, but developmental directions must be similar. As Rensch has argued: "The important thing in allomorphosis is not the absolute constancy of allometric exponents, but the much more frequent constancy of a general allometric tendency" (1971, p. 17).

8. This claim for rarity may seem surprising, but it is the common (though often suppressed) knowledge of all paleontologists. Since the rare trends sustain our interest, we emphasize them in our writing; but nearly 100 percent of all species become extinct without issue and play no part in such trends.

9. The prestige of behavioral biology has led to a reassertion of Lamarck's insight in its congenial Darwinian context: a shift in behavior must generally precede an alteration in morphology. I have no quarrel with this statement in general: control of morphology by selection virtually demands it. Against this functional view stands the structuralist contention (of Saint-Hilaire among others) that shifts in behavior must *follow* an altered morphology. Creative progenesis may provide one of the few cases for primacy of morphological change. This primacy may be crucial for any theory of rapid transition across adaptive zones.

10. Retardation and Neoteny in Human Evolution

1. Thus, Dubois's *Pithecanthropus* was debarred from human ancestry as a disharmonious type with perfected bipedal locomotion but too small a brain. This insistence on correlated modification of the entire body within true phyletic lineages also underlay Bolk's concept of fetalization and, from our perspective, produced most of its problems.

2. These last two points are, I confess, vague and confusing. But I have translated Bolk's words literally.

3. This is a fairly complete list of the "basic" features usually presented to support the contention that humans are paedomorphic. Other characters are usually the concomitants or the causes of features presented above (much, for example, has been written about the role of various cranial flexures in setting the position of the foramen magnum and the disposition of the face and jaws).

4. These words are in English within the German original—an obvious reference to Darwin and an attack upon his beliefs.

5. To cite the collapse of the biogenetic law again: the universal acceleration that it required proved to be inconsistent with the precepts of Mendelian genetics.

6. Retardation in maturation can lead to hypermorphosis and recapitulation in other circumstances.

7. The original, for those who like extended German adjectives: "Das ist eine durch direkte Vergleichung leicht festzustellende Tatsache."

8. Used in the old sense of ontogeny rather than phylogeny.

9. Agassiz, in slavery days, had once argued that sutures of Negroes close during mid-childhood years. He believed that education for blacks beyond this age would be not only useless, but also potentially dangerous, since an overstimulated brain might put undue pressure on a rigid and inflexible vault.

10. A bending with concave side towards the rest of the body, as opposed to a lordosis; usually used to describe flexures in the spinal column.

11. Measured either from the pituitary point (and therefore including the sellar kyphosis in the pituitary point-nasion limb of the angle) or in the traditional way from the prosphenion (and including the sellar kyphosis within the basicranial axis).

12. Since this kyphosis reverses the prenatal trand and produces a decrease in spheno-ethmoidal angle.

13. Riesenfield thinks that he has thereby refuted various claims for the adaptive significance of the chin advanced, for example, by DuBrul and Sicher, 1954 and Scott, 1963. Yet Riesenfeld seems unaware that his ingenious identification of "hypofunction" as an efficient cause does not challenge any proposal for the adaptive significance (or final cause) of the structure thus produced.

14. I do not mean to deny utterly the issue of similar directions in ontogenetic allometry. If all higher primates develop in the same direction and humans develop least, then the case for paedomorphosis is more striking. But mosaic evolution practically guarantees that paths of ontogeny will diverge. If we depart least from a common juvenile form, but depart along a different direction, we are still the most paedomorphic of higher primates.

15. Primarily the olfactory lobes in rats. The olfactory lobes are relatively much reduced in humans and the expanded frontal lobes assume this position.

16. DuBrul's representatives included the jack rabbit and the large South

American hystricomorph *Dolichotus*. Of *Dolichotus,* he remarked (1950, p. 290): "Its alert, motionless posture is a tense, squatting position very similar to that of the jack rabbit, with the head held high on an erect cervical vertebral column."

17. It is interesting, in this regard, that Bolk regarded adaptive ties of paedomorphic features to upright posture as contrary to his ideas. Bolk was committed to the view that neoteny arose from internal factors working in a coordinated way upon the entire body in the absence of selection.

18. A structure is defined as preadaptive only if it serves an ancestor in different ways than a descendant; the concept of adaptive shift integrates the apparently finalistic notion of preadaptation into Darwinian theory.

19. Herbert Spencer advanced this argument, though with no real evidence to support it: "Other things equal, the less-evolved types or organisms take shorter times to reach their complete forms than do the more evolved . . . There is reason for associating this difference with the difference in cerebral development. The great costliness of the brain, which so long delays human maturity, as compared with mammalian maturity . . ." (1886, 1:52). Sacher (1966) has reasserted this claim with extensive documentation of a partial correlation (with body size removed) of brain size and life span within groups in which neoteny is not a common cause of both features.

20. Schultz (1948) recorded the prevalence of single births among primate species. He related this tendency to the difficulty of carrying several large offspring about in trees. Haldane (1932) argued that single births strongly favored (or at least permitted) delayed development since uterine competition among several growing offspring would favor accelerated development.

Bibliography

Abbie, A. A. 1947. Headform and human evolution, *J. Anat.* 81: 233–258.
——— 1948. No! No! a thousand times no! *Aust. J. Sci.* 11: 39–42.
——— 1952. A new approach to the problem of human evolution, *Trans. Roy. Soc. S. Austr.* 75: 70–88.
——— 1958. Timing in human evolution, *Proc. Linn. Soc. New South Wales* 83: 197–213.
Accordi, B., and R. Colacicchi. 1962. Excavations in the pygmy elephants cave of Spinagallo (Siracusa), *Geol. Romana* 1: 217–229.
Acher, R. A. 1910. Spontaneous constructions and primitive activities of children analogous to those of primitive man, *Am. J. Psychol.* 21: 114–150.
Adams, L. M., and W. J. Moore. 1975. Biomechanical appraisal of some skeletal features associated with head balance and posture in the Hominoidea, *Acta Anat.* 92: 580–594.
Adelmann, H. B. 1966. *Marcello Malpighi and the evolution of embryology,* 5 vols. (Cornell Univ. Press, Ithaca, N.Y.).
Adolph, E. F. 1970. Physiological stages in the development of mammals, *Growth* 34: 113–124.
Agassiz, E. C. 1885. *Louis Agassiz: his life and correspondence* (Houghton Mifflin, Boston), 794 pp.
Agassiz, L. 1849. *Twelve lectures on comparative embryology* (Henry Flanders, Boston), 104 pp.
——— 1857. *Essay on classification* (from *Contributions to the natural history of the United States,* vol. 1), ed. E. Lurie (Harvard Univ. Press, Cambridge, 1962).
——— 1874. Evolution and permanence of type, *Atlantic Monthly* 33: 92–101.
Albrecht, F. O. 1962. Some physiological and ecological aspects of locust phases, *Trans. Roy. Entomol. Soc. London* 114: 335–375.

Albrecht, F. O., and R. E. Blackith. 1957. Phase and molting polymorphism in locusts, *Evolution* 11: 165–177.

Albrecht, F. O., M. Verdier, and R. E. Blackith. 1958. Determination de la fertilité par l'effet de groupe chez le criquet migrateur (*Locusta migratoria migratoroides* R. et F.), *Bull. Biol. France Belg.* 92: 349–427.

Allen, G. E. 1975. *Life sciences in the twentieth century* (John Wiley, New York).

Amlinskii, I. E. 1955. *Zhoffrua Sent-Iler i ego bor'ba protiv Kyuv'e* (Izdat, Akad. Nauk, Moscow), 424 pp.

Anderson, J. D., and R. D. Worthington. 1971. The life history of the Mexican salamander *Ambystoma ordinarium* Taylor, *Herpetologica* 27: 165–176.

Angleton, G. M., and D. Pettus. 1966. Relative-growth law with a threshold, *Perspect. Biol. Med.* 9: 421–424.

Arber, A. 1948. Analogy in the history of science, in M. F. Ashley-Montagu, ed., *Studies and essays in the history of science and learning offered in homage to George Sarton* (Henry Schuman, New York), pp. 221–233.

Aristotle. 1910. *Historia animalium,* tr. D'Arcy W. Thompson (Clarendon Press, Oxford).

——— 1965. *De generatione animalium,* tr. Arthur Platt (Clarendon Press, Oxford).

Ashmole, N. P. 1963. The regulation of numbers of tropical oceanic birds, *Ibis* 103: 458–473.

Ashton, E. H. 1957. Age changes in the basicranial axes of the Anthropoidea, *Proc. Zool. Soc. London* 129: 61–74.

Auden, W. H. 1971. Craftsman, artist, genius, *The Observer* 9377 (April 11)): 9.

Autenrieth, H. F. 1797. *Observationum ad historiam embryonis facientium, pars prima* (Tübingen).

Avebury, Lord (John Lubbock). 1870. *The origin of civilization and the primitive condition of man* (Longmans, London), 380 pp.

Baer, J. G. 1971. *Animal parasites* (World Univ. Library, London), 256 pp.

Baer, K. E. von. 1828. *Entwicklungsgeschichte der Thiere: Beobachtung und Reflexion* (Bornträger, Königsberg), 264 pp.

——— 1864. Das allgemeine Gesetz der Natur in aller Entwickelung, in *Reden gehalten in wissenschaftlichen Versammlungen,* vol. 1 (Karl Röttger, St. Petersburg), pp. 37–74.

——— 1866. Über Prof. Nic. Wagner's Entdeckung von Larven, die sich fortpflanzen, Herrn Garren's verwandte und ergänzende Beobachtung und über die Pädogenesis überhaupt, *Bull. Acad. Imp. des Sciences St. Petersbourg* 9: 63–137.

——— 1876. Ueber Darwin's Lehre, in *Reden gehalten in wissenschaftlichen Versammlungen,* vol. 2, *Studien aus dem Gebiete der Naturwissenschaften* (Karl Röttger, St. Petersburg), pp. 235–480.

Baer, M. J., and J. E. Harris. 1969. A commentary on the growth of the human brain and skull, *Am. J. Phys. Anthrop.* 30: 39–44.

Baid, I. 1964. Neoteny in the genus *Artemia, Acta Zool.* 45: 167–177.

Baldwin, J. M. 1906. *Mental development in the child and the race,* 3rd ed. (Macmillan, New York), 477 pp.

Balfour, F. M. 1880. Address to the Department of Anatomy and Physiology of the British Association for the Advancement of Science, in M. Foster and A. Sedgwick, eds., *The works of F. M. Balfour,* vol. 1 (Macmillan, London, 1885), pp. 698–713.

——— 1880–1881. *A treatise on comparative embryology,* 2 vols. (Macmillan, London), 591 pp., 792 pp.

Balinsky, B. I. 1970. *An introduction to embryology* (W. B. Saunders, Philadelphia), 725 pp.

Ballard, J. G. 1965. *The drowned world* (Penguin, London), 171 pp.

Baluk, W., and A. Radwanski. 1967. Miocene cirripeds domiciled in corals, *Acta Palaeont. Polonica* 12: 457–509.

Barton, A. D., and A. K. Laird. 1969. Analysis of allometric and non-allometric differential growth, *Growth* 33: 1–16.

Bateson, W. 1886. The ancestry of the Chordata, *Quart. J. Microscop. Sci.,* pp. 535–571.

——— 1922. Evolutionary faith and modern doubts, *Science* 55: 55–61.

Bather, F. A. 1893. The recapitulation theory in palaeontology, *Natural Science* 2: 275–281.

Beecher, C. E. 1893. Some correlations of ontogeny and phylogeny in the Brachiopoda, *Am. Nat.* 27: 599–604.

Bekoff, M. 1972. The development of social interaction, play, and metacommunication in mammals: an ethological perspective, *Quart. Rev. Biol.* 47: 412–434

——— In press. Mammalian dispersal and the ontogeny of individual behavioral phenotypes, *Am. Nat.*

Bennett, D. 1961. Miniature, a new gene for small size in the mouse, *J. Hered.* 52: 95–98.

Berg, L. S. 1926. *Nomogenesis, or evolution determined by law* (MIT Press, Cambridge, 1969), 477 pp.

Berrill, N. J. 1955. *The origin of the vertebrates* (Clarendon Press, Oxford), 257 pp.

Biegert, J. 1957. Der Formenwandel des Primatenschädels, *Morph. Jb.* 98: 77–199.

Bielicki, T. 1969. Deviation-amplifying cybernetic systems and hominid evolution, *Mater. i Pr. Anthrop.* 77: 57–60.

Blower, J. G. 1969. Age structures of millipede populations in relation to activity and dispersion, *Syst. Assoc. Publ.* 8: 209–216.

Blower, J. G., and P. D. Gabbutt. 1964. Studies on the millipedes of a Devon oak wood, *Proc. Zool. Soc. London* 143: 143–176.

Blower, J. G., and P. F. Miller. 1974. The life-cycle and ecology of *Ophyiulus pilosus* (Newport) in Britain, *Symp. Zool. Soc. London* 32: 503–525.

Boas, J. E. V. 1896. Ueber Neotenie, in *Festschrift zum siebzigsten Geburtstage von Carl Gegenbaur,* vol. 2 (W. Engelmann, Leipzig), pp. 1–20.

Boas, F. 1899. The cephalic index, *Am. Anthrop.* 1: 448–461.

Bock, R. D., H. Wainer, A. Petersen, D. Thissen, J. Murray, and A. Roche. 1973. A parameterization for individual human growth curves, *Human Biol.* 45: 63–80.

Bodenstein, D. 1953. Studies on the humoral mechanisms in growth and metamorphosis of the cockroach *Periplaneta americana*: I, transplantation of integumental structures and experimental parabioses, *J. Exp. Zool.* 123: 189–232.

Boekschoten, G. J., and P. Y. Sondaar. 1966. The Pleistocene of the Katharo Basin (Crete) and its hippopotamus, *Bijdragen tot de Dierkunde* 36: 17–44

——— 1972. On the fossil Mammalia of Cyprus. *Proc. Kon. Nederl. Akad. Weten. Amsterdam B* 75: 1–20.

Bolk, L. 1915. Über Lagerung, Verschiebung und Neigung des Foramen magnum am Schädel der Primaten, *Z. Morph. Anthrop.* 7: 611–692.

——— 1923. The problem of orthognathism, *Proc. Section Sciences Kon. Akad. Wetens. Amsterdam* 25: 371–380.

——— 1924. The chin problem, *Proc. Section Sciences Kon. Akad. Wetens. Amsterdam* 27: 329–344.

——— 1926a. On the problem of anthropogenesis, *Proc. Section Sciences Kon. Akad. Wetens. Amsterdam* 29: 465–475.

——— 1926b. La récapitulation ontogenetique comme phénomène harmonique, *Arch. Anat. Hist. Embryol.* 5: 85–98.

——— 1926c. *Das Problem der Menschwerdung* (Gustav Fischer, Jena), 44 pp.

——— 1929. Origin of racial characteristics in man, *Am. J. Phys. Anthrop.* 13: 1–28.

Bolton, F. E. 1899. Hydropsychoses, *Am. J. Psychol.* 10: 169–227.

Bone, Q. 1960. The origin of the chordates, *J. Linn. Soc. London* 44: 252–265.

Bonner, J. T. 1965. *Size and cycle* (Princeton Univ. Press, Princeton), 219 pp.

——— 1974. *On development* (Harvard Univ. Press, Cambridge), 282 pp.

Bonnet, C. 1762. *Considérations sur les corps organisés* (Marc-Michel Rey, Amsterdam), 274 pp.

——— 1764. *Contemplation de la nature,* 2 vols. (Marc-Michel Rey, Amsterdam), 298 pp., 260 pp.

——— 1769. *La palingénésie philosophique,* 2 vols. (C. Philibert and B. Chirol, Geneva), 427 pp., 448 pp.

Bounhiol, J. 1938. Recherches experimentales sur la déterminisme de la metamorphose chez les lepidoptères, *Bull. Biol. France Belg.* 24: 1–199.

Bourdier, F. 1969. Geoffroy Saint-Hilaire versus Cuvier: the campaign for paleontological evolution, in C. Schneer, ed., *Towards a history of geology* (MIT Press, Cambridge), pp. 36–61.

Bovet, P. 1923. *The fighting instinct* (George Allen & Unwin, London), 252 pp.

Bowler, P. J. 1975. The changing meaning of "evolution," *J. History Ideas* 36: 95–114.

Boyce, A. J. 1964. The value of some methods of numerical taxonomy with reference to hominoid classification, in V. H. Heywood and J. McNeill, eds., *Phenetic and phylogenetic classification* (Systematics Assoc., London), pp. 47–65.

Brandon, R. A., and D. J. Bremer. 1966. Neotenic newts, *Notophthalmus viridescens louisianensis,* in southern Illinois, *Herpetologica* 22: 213–217.

Brinton, D. G. 1890. *Races and peoples* (N. D. C. Hodges, New York), 313 pp.

Britten, R. J., and E. H. Davidson. 1971. Repetitive and non-repetitive DNA sequences and a speculation on the origins of evolutionary novelty, *Quart. Rev. Biology* 46: 111–133.

Britton, E. R., and R. J. Stanton, Jr. 1973. Origin of "dwarfed" fauna in the Del Rio formation, Lower Cretaceous, east central Texas, *Geol. Soc. Am. Abstracts with Programs* 5: 248–249.

Brodie, A. G. 1941. On the growth pattern of the human head from the third month to the eighth year of life, *Am. J. Anat.* 68: 209–262.

Brolemann, H. W. 1932. La contraction tachygénétique des polydesmiens (myriapodes) et leurs affinites naturelles, *Bull. Soc. Zool. France* 57: 387–396.

Browne, T. 1642. *Religio medici* (Andrew Crooke, London), 159 pp.

Brues, A. M. 1966. "Probable mutation effect" and the evolution of hominid teeth and jaws, *Am. J. Phys. Anthrop.* 25: 169–170.

Brunst, V. V. 1955. The axolotl (*Siredon mexicanum*), *Lab. Invest.* 4: 45–64, 429–449.

Bryden, M. M. 1968. Control of growth in two populations of elephant seals, *Nature* 217: 1106–1108.

Buckman, S. S. 1894. Babies and monkeys, *Pop. Sci.* 46: 371–388.

——— 1899. Human babies: some of their characters, *Proc. Cotteswold Nat. Field Club* 13: 89–118.

Burnaby, T. P. 1965. Reversed coiling trend in *Gryphaea arcuata, Geol. Jour.* 4: 257–278.

Bury, J. B. 1920. *The idea of progress* (Macmillan, London), 377 pp.

Butler, N. M. 1900. Status of education at the close of the century, *Educ. Rev.* 19: 313–324.

Butler, S. 1877. *Life and habit* (A. C. Fifield, London, 1910), 310 pp.

——— 1880. *Unconscious memory* (A. C. Fifield, London, 1920), 186 pp.

——— 1887. *Luck or cunning?* (A. C. Fifield, London, 1920).

Buxton, L. H. D., and G. R. De Beer. 1932. Neanderthal and modern man, *Nature* 129: 940–941.

Calentine, R. L. 1964. The life cycle of *Archigetes iowensis* (Cestoda: Caryophyllaeidae), *J. Parasitol.* 50: 454–458.

Carlisle, D. B., and P. E. Ellis. 1959. La persistence des glandes ventral céphaliques chez les criquets solitaires, *C. R. Acad. Sci. Paris* 249: 1059–1060.

Carlquist, S. 1962. A theory of paedomorphosis in dicotyledonous woods, *Phytomorphology* 12: 30–45.

Carmichael, L. 1967. The relationship of gestation-duration and birth weight in primates, in D. Starck, R. Schneider, and H-J. Kuhn, eds., *Neue Ergebnisse der Primatologie* (Gustav Fischer, Stuttgart), pp. 55–58.

Carneiro, R. L. 1972. The devolution of evolution, *Social Biol.* 19: 248–258.

Carpenter, W. B. 1839. *Principles of general and comparative physiology* (John Churchill, London), 478 pp.

Carter, G. S. 1960. Comments on Mr. Bone's paper, *J. Linn. Soc. London Zool.* 44: 265–267.

Carus, C. G. 1835. *Traité élémentaire d'anatomie comparée suivi de recherches d'anatomie philosophique ou transcendante,* tr. from 2nd German ed. by A. J. L. Jourdan, 3 vols. (J. B. Bailliere, Paris), 519 pp., 508 pp., 639 pp.

Castle, W. E. 1896. The early embryology of *Ciona intestinalis, Bull. Mus. Comp. Zool.* 27: 203–280.

Catel, J. 1953. Ein Beitrag zur Frage von Hirnentwicklung und Menschwerdung, *Klin. Wschr.* 31: 473–475.

Caullery, M. 1952. *Parasitism and symbiosis* (Sidgwick & Jackson, London), 340 pp.

Chamberlain, A. F. 1900. *The child: a study in the evolution of man* (Walter Scott, London), 498 pp.

——— undated. *The contact of "higher" and "lower" races* (Clark Univ., Worcester, Mass.), 15 pp.

Chambers, R. 1844. *Vestiges of the natural history of creation* (J. Churchill, London), 390 pp.

Chancellor, W. E. 1907. *A theory of motives, ideals, and values in education* (Houghton Mifflin, Boston), 534 pp.

Charters, W. W. 1913. *Teaching the common branches: a textbook for teachers of rural and graded schools* (Houghton Mifflin, Boston), 355 pp.

Cheng, T. C. 1964. *The biology of animal parasites* (W. B. Saunders, Philadelphia), 727 pp.

Chun, C. 1880. *Die Ctenophoren des Golfes von Neapel und der angrenzenden Meeres-Abschnitte,* Fauna Flora Zool. Stat. Neapel, vol. 1, 313 pp.

——— 1892. Die Dissogonie, eine neue Form der geschlechtlichen Zeugung, in A. Bogdanow et al., eds., *Festschrift zum siebzigsten Geburtstage Rudolf Leuckarts* (W. Engelmann, Leipzig), pp. 77–108.

Clark, D. L. 1962. Paedomorphosis, acceleration, and caenogenesis in the evolution of Texas Cretaceous ammonoids, *Evolution* 16: 300–305.

Clark, G. R. 1968. Mollusk shell: daily growth lines, *Science* 161: 800–802.

Clark, R. B. 1964. *Dynamics in metazoan evolution: the origin of the coelom and segments* (Clarendon Press, Oxford), 313 pp.

Clements, E. M. B., and S. Zuckerman. 1953. The order of eruption of the permanent teeth in the Hominoidea, *Am. J. Phys. Anthrop.* 11: 313–332.

Cloud, P. E. 1948. Some problems and patterns of evolution exemplified by fossil invertebrates, *Evolution* 2: 322–350.

Cock, A. G. 1966. Genetical aspects of metrical growth and form in animals, *Quart. Rev. Biol.* 41: 131–190.

Cody, M. L. 1966. A general theory of clutch size, *Evolution* 20: 174–184.

Cohen, Morris. 1947. *The meaning of human history* (Open Court Publishing Co., La Salle, Ill.), 304 pp.

Cole, F. J. 1930. *Early theories of sexual generation* (Oxford Univ. Press, London), 230 pp.

Cole, L. C. 1954. The population consequences of life history phenomena, *Quart. Rev. Biol.* 29: 103–137.

Coleman, W., ed. 1967. *The interpretation of animal form* (Johnson Reprint Co., New York), 191 pp.

——— 1973. Limits of the recapitulation theory: Carl Friedrich Kielmeyer's critique of the presumed parallelism of earth history, ontogeny, and the present order of organisms, *Isis* 64: 341–350.

Collingwood, R. G. 1945. *The idea of nature* (Clarendon Press, Oxford), 183 pp.

Condorcet, M. de. 1793. *Esquisse d'un tableau historique des progrès de l'esprit humain* (Agasse, Paris), 389 pp.

Conklin, E. G. 1903. The cause of inverse symmetry, *Anat. Anz.* 23: 577–588.

——— 1905. The organization of cell-lineage of the ascidian egg, *J. Acad. Nat. Sci. Philadelphia* 13: 1–119.

——— 1928. Embryology and evolution, in F. Mason, ed., *Creation by evolution* (Macmillan, New York), pp. 62–80.

Cope, E. D. 1866. On the Cyprinidae of Pennsylvania, *Trans. Am. Phil. Soc.* 13: 351–399.

——— 1869. The origin of genera, reprinted in Cope, 1887, pp. 41–123.

——— 1870. On the hypothesis of evolution: physical and metaphysical, reprinted from *Lippincott's Magazine* in Cope, 1887, pp. 128–172.

——— 1871. The method of creation of organic types, reprinted from *Proc. Am. Phil. Soc.* in Cope, 1887, pp. 173–214.

——— 1872. Evolution and its consequences, reprinted from *Penn. Monthly Mag.* in Cope, 1887, pp. 1–40.

——— 1876. The theory of evolution, reprinted from *Proc. Acad. Nat. Sci. Philadelphia* in Cope, 1887, pp. 124–127.

——— 1878. The relation of animal motion to animal evolution, reprinted from *Am. Nat.* in Cope, 1887, pp. 350–358.

——— 1880. A review of the modern doctrine of evolution, reprinted from *Am. Nat.* in Cope, 1887, pp. 215–240.

——— 1883a. The developmental significance of human physiognomy, reprinted from *Am. Nat.* in Cope, 1887, pp. 281–293.

——— 1883b. The evolutionary significance of human character, reprinted from, *Am. Nat.* in Cope, 1887, pp. 378–389.

——— 1883c. *The Vertebrata of the Tertiary formations of the West,* Rept. U.S. Geol. Surv. The Territories, vol. 3, 1009 pp., 134 plates.

——— 1887. *The origin of the fittest* (Macmillan, New York), 467 pp.

——— 1889. On inheritance in evolution, *Am. Nat.* 23: 1058–1071.

——— 1896. *The primary factors of organic evolution* (Open Court Publishing Co., Chicago), 547 pp.

Costlow, J. D., Jr. 1966. The effect of eyestalk extirpation on larval development of the mud crab, *Rhithropanopeus harrisii* (Gould), *Gen. Comp. Endocrinol.* 7: 255–274.

——— 1968. Metamorphosis in crustaceans, in W. Etkin and L. I. Gilbert, eds., *Metamorphosis* (Appleton-Century-Crofts, New York), pp. 3–41.

Count, E. W. 1947. Brain and body weight in man: their antecedents in growth and evolution, *Ann. N.Y. Acad. Sci.* 46: 993–1122.

Cousin, G. 1938. La néotenie chez *Gryllus campestris* et ses hybrides, *Bull. Biol. France Belg.* 72: 79–118.

Crocker, L. G. 1959. Diderot and eighteenth-century French transformism, in B. Glass, O. Temkin, and W. L. Strauss, Jr., eds., *Forerunners of Darwin: 1745–1859* (Johns Hopkins Press, Baltimore), pp. 114–143.

Crofton, A. F. B. 1897. The language of crime, *Pop. Sci.* 50: 831–835.

Cuvier, G. 1828. *Le règne animal distribué d'après son organisation,* 2nd ed. (Fortin, Paris).

d'Agoty, de Gautier. 1752. *Observations sur l'histoire naturelle, sur la physique et sur la peinture* (Delaquette, Paris).

Darwin, C. 1859. *The origin of species* (John Murray, London), 490 pp.

——— 1909. *The foundations of the origin of species: two essays written in 1842 and 1844,* ed. F. Darwin (Cambridge Univ. Press, Cambridge), 263 pp.

Darwin, F. 1903. *More letters of Charles Darwin,* vol. 1 (John Murray, London), 494 pp.

——— 1908. President's address, *Rept. 78th Meeting Brit. Assoc. Adv. Sci. Dublin,* pp. 1–27.

Davenport, C. B. 1890. *Cristatella:* the origin and development of the individual in the colony, *Bull. Mus. Comp. Zool.* 20: 101–151.

Davidson, P. E. 1914. *The recapitulation theory and human infancy* (Columbia Univ. Teacher's College, New York), 105 pp.

Davies, D. D., ed. 1973. *Rate control of biological processes,* Symp. Soc. Exp. Biol. no. 27 (Cambridge Univ. Press, Cambridge), 583 pp.

Davis, W. P., and R. S. Birdsong. 1973. Coral reef fishes which forage in the water column, *Helgoländer Wiss. Meeresunters.* 24: 292–306.

de Beer, G. R. 1930. *Embryology and evolution* (Clarendon Press, Oxford), 116 pp.

——— 1940. *Embryos and ancestors* (Clarendon Press, Oxford), 108 pp.

——— 1948. Embryology and the evolution of man, in A. du Toit, ed., *Robert Broom Commemorative Volume,* Sp. Pub. Roy. Soc. S. Af., pp. 181–190.

——— 1954. The evolution of Metazoa, in J. Huxley, A. C. Hardy, and E. B. Ford, eds., *Evolution as a process* (George Allen & Unwin, London), pp. 34–45.

——— 1958a. *Embryos and ancestors* (Clarendon Press, Oxford), 197 pp.

——— 1958b. Darwin's views on the relations between embryology and evolution, *J. Linn. Soc. London* 44: 15–23.

——— 1959. Paedomorphosis, *Proc. XV Int. Cong. Zool.,* pp. 927–930.

——— 1960. Darwin's notebooks on transmutation of species, *Bull. Brit. Mus. Nat. Hist. Historical Series* 2: 25–73.

——— 1969. *Streams of culture* (J. B. Lippincott, Philadelphia), 237 pp.

de Beer, G. R., and W. E. Swinton. 1958. Prophetic fossils, in T. S. Westoll, ed., *Studies on fossil vertebrates* (Athlone Press, London), pp. 1–15.

De Garmo, C. 1895. *Herbart and the Herbartians* (Charles Scribner's Sons, New York), 256 pp.

Delattre, A., and R. Fenart. 1963. Etudes des projections horizontale et vertico-frontales du crâne au cours de l'hominisation, *L'Anthropologie* 67: 525–561.

Delsol, M., and H. Tintant. 1971. Discussions autour d'un vieux problème: les relations entre embryologie et évolution, *Rev. Quest. Sci.* 142: 85–101.

——— In press. Les relations entre embryologie et évolution.

Demetrius, L. 1975. Reproductive strategies and natural selection, *Am. Nat.* 109: 243–249.

Dempster, J. P. 1957. *The population dynamics of the Moroccan Locust* (Dociostaurus maroccanus *Thunberg) in Cyprus, Anti-Locust Bull.* 27, 60 pp.

——— 1963. The population dynamics of grasshoppers and locusts, *Biol. Rev.* 38: 490–529.

Dent, J. N. 1968. Survey of amphibian metamorphosis, in W. Etkin and L. I. Gilbert, eds., *Metamorphosis* (Appleton-Century-Crofts, New York), pp. 271–311.

Dent, J. N., J. S. Kirby-Smith, and D. L. Craig. 1955. Induction of metamorphosis in *Gyrinophilus palleucus, Anat. Rec.* 121: 429.

Dewey, J. 1911. Culture epoch theory, in P. Monroe, ed., *A cyclopedia of education,* vol. 2 (Macmillan, New York), pp. 240–242.

——— 1916. *Democracy and education* (Macmillan, New York), 434 pp.

Dixon, A. F. G. 1972. Crowding and nutrition in the induction of macropterous alatae in *Drepanosiphum dixoni, J. Insect Physiol.* 18: 459–464.

Doane, W. W. 1973. Role of hormones in insect development, in S. J. Counce and C. H. Waddington, eds., *Developmental systems: insects* (Academic Press, New York), pp. 291–497.

Dobzhansky, T. 1950. Evolution in the tropics, *Am. Sci.* 38: 209–221.

Donovan, D. T. 1973. The influence of theoretical ideas on ammonite classification from Hyatt to Trueman, *Univ. Kansas Paleontological Contr.* 62: 1–16.

Dubinina, M. N. 1960. O vozmozhnosti progeneza u plerotserkoidov remnetsov (Cestoda, Ligulidae), *Zool. Zhur.* 39: 1467–1477.

Dubois, E. 1896. On *Pithecanthropus erectus*: a transitional form between man and the apes, *Sci. Trans. Royal Dublin Soc.* 6: 1–18.

DuBrul, E. L. 1950. Posture, locomotion and the skull in Lagomorpha, *Am. J. Anat.* 87: 277–313.

——— 1971. On the phylogeny and ontogeny of the human larynx: a morphological and functional study, *Evolution* 25: 739–740.

DuBrul, E. L., and D. M. Laskin. 1961. Preadaptative potentialities of the mammalian skull: an experiment in growth and form, *Am. J. Anat.* 109: 117–132.

DuBrul, E. L., and H. Sicher. 1954. *The adaptive chin* (Thomas, Springfield, Ill.), 97 pp.

Duméril, A. 1865a. Reproduction, dans la Ménagerie des Reptiles au Muséum d' Histoire naturelle, des axolotls, batraciens urodèles à branchies persistantes, de Mexico (Siredon mexicanus vel Humboldtii), qui n'avaient encore jamais été vus vivants en Europe, *C. R. Acad. Sci.* 60: 765–767.

——— 1865b. Nouvelles observations sur les axolotls, batraciens urodèles de Mexico (Siredon mexicanus vel Humboldtii) nés dans la Ménagerie des Reptiles au Muséum d'Histoire Naturelle, et qui y subissent des métamorphoses, *C. R. Acad. Sci.* 61: 775–778.

——— 1867. Métamorphoses des batraciens urodèles à branchies extérieures du Mexique dits axolotls, observées à la Ménagerie des Reptiles du Muséum d'Histoire Naturelle, *Ann. Sci. Nat. Zool.* 7: 229–254.

——— 1870. Création d'une race blanche d'axolotls a la Ménagerie des Reptiles du Muséum d'Histoire Naturelle, et remarques sur la transformation des ces batraciens, *C. R. Acad. Sci.* 70: 782–785.

Easton, W. H. 1960. *Invertebrate paleontology* (Harper & Row, New York), 701 pp.

Eayrs, J. T. 1964. Effect of neonatal hyperthyroidism on maturation and learning in the rat, *Animal Behaviour* 12: 195–199.

Efford, I. E. 1967. Neoteny in sand crabs of the genus *Emerita* (Anomura, Hippidae), *Crustaceana* 13: 81–93.

Efford, I. E., and J. A. Mathias. 1969. A comparison of two salamander populations in Marion Lake, British Columbia, *Copeia,* pp. 723–736.

Eimer, G. H. T. 1890. *Organic evolution as the result of the inheritance of acquired characters according to the laws of organic growth* (Macmillan, London), 435 pp.

Eisenberg, J. F. 1974. *Phylogeny, behavior, and ecology in the Mammalia,* preprint for Burg Wartenstein Symposium no. 61 (Wenner-Gren Foundation, New York), 30 pp.

Eldredge, N., and S. J. Gould. 1972. Punctuated equilibria: an alternative to phyletic gradualism, in T. J. M. Schopf, ed., *Models in Paleobiology* (Freeman, Cooper & Co., San Francisco), pp. 82–115.

Elles, G. L. 1922. The graptolite faunas of the British Isles, *Proc. Geologists' Assoc.* 33: 168–200.

——— 1923. Evolutional palaeontology in relation to the Lower Paleozoic rocks, *Rept. Brit. Assoc. Adv. Sci.* 91: 83–107.

Ellis, H. 1894. *Man and woman* (Charles Scribner's Sons, New York), 561 pp.

——— 1910. *The criminal* (Charles Scribner's Sons, New York), 440 pp.

Engels, F. 1954. *Dialectics of nature* (Foreign Languages Publishing House, Moscow), 496 pp.

Enlow, D. H. 1966. A comparative study of facial growth in *Homo* and *Macaca, Am. J. Phys. Anthrop.* 24: 293–308.

Estes, R. D. 1974. Social organization of the African Bovidae, in V. Geist and F. Walther, eds., *The behavior of ungulates and its relation to management* (I.V.C.N., Morges, Switzerland), pp. 166–205.

Etkin, W. 1968. Hormonal control of amphibian metamorphosis, in W. Etkin and L. I. Gilbert, eds., *Metamorphosis* (Appleton-Century-Crofts, New York), pp. 313–348.

Evans, M. A., and H. E. Evans. 1970. *William Morton Wheeler, biologist* (Harvard Univ. Press, Cambridge), 363 pp.

Ewer, R. F. 1960. Natural selection and neoteny, *Acta biotheoretica* 13: 161–184.

Eysenck, H. J. 1971. *The IQ argument: race, intelligence and education* (Library Press, New York), 155 pp.

Fagan, R. 1974. Selective and evolutionary aspects of animal play, *Am. Nat.* 108: 850–858.

Falkner, F. 1966. *Human development* (W. B. Saunders, Philadelphia), 644 pp.

Farner, D. S., and J. Kezer. 1953. Notes on the amphibians and reptiles of Crater Lake National Park, *Copeia,* pp. 448–462.

Ferenczi, S. 1924. *Thalassa: a theory of genitality* (W. W. Norton, New York, 1968), 110 pp.

Ferri, E. No date. *Criminal sociology* (D. Appleton, New York), 284 pp.

Fischer, P., and D.-P. Oehlert, 1892. Sur l'évolution de l'appareil brachial de quelques brachiopodes, *C. R. Acad. Sci.* 115: 749–751.

Fiske, E. B. 1975. Study ties child learning to the history of science, *New York Times,* 15 June 1975, p. 43.

Ford, E. B., and J. S. Huxley. 1927. Mendelian genes and rates of development in *Gammarus chevreuxi, Brit. J. Exp. Biol.* 5: 112–134.

Foster, M., and A. Sedgwick, eds. 1885. *The works of Francis Maitland Balfour*, 4 vols., memorial edition (Macmillan, London).

Franz, V. 1927. *Ontogenie und Phylogenie: das sogenannte biogenetische Grundgesetz und die biometabolischen Modi*, Abh. Theorie Org. Ent., no. 3 (Julius Springer, Berlin), 51 pp.

Freud, S. 1905. *Three essays on the theory of sexuality* (Avon Books, New York, 1962), 174 pp.

——— 1910. The origin and development of psychoanalysis, *Am. J. Psych.* 21: 181–218.

——— 1913. *Totem and taboo* tr. James Strachey (W. W. Norton, New York, 1950), 172 pp.

——— 1916. *Introductory lectures on psychoanalysis* (George Allen & Unwin, London, 1961), 395 pp.

——— 1930. *Civilization and its discontents* tr. James Strachey (W. W. Norton, New York, 1961), 109 pp.

——— 1939. *Moses and monotheism* (Random House, New York), 178 pp.

——— 1963. *Three case histories* (Collier Books, New York), 319 pp.

——— 1964. Findings, ideas, problems (notes written in 1938, first published in 1941), in *Standard edition*, vol. 23 (Hogarth Press, London), pp. 299–300.

Freudenthal, M. 1972. *Deinogalerix koenigswaldi* nov. gen., nov. spec., a giant insectivore from the Neogene of Italy, *Scripta Geol.* 14: 1–19.

Frisch, R. E. 1972. Weight at menarche: similarity for well-nourished and undernourished girls at differing ages, and evidence for historical constancy, *Pediatrics* 50: 445–450.

——— and R. Revelle. 1971. Height and weight at menarche and a hypothesis of menarche, *Archives of Diseases in Childhood* 46: 695–701.

Froebel, F. 1887. *The education of man* (D. Appleton, New York), 332 pp.

Fryer, G. 1959. Some aspects of evolution in Lake Nyassa, *Evolution* 13: 440–451.

Gadgil, M., and W. H. Bossert. 1970. Life historical consequences of natural selection, *Am. Nat.* 104: 1–24.

Gadgil, M., and O. T. Solbrig. 1972. The concepts of *r* and *K* selection: evidence from wild flowers and some theoretical considerations, *Am. Nat.* 106: 14–31.

Galloway, J. J. 1957. Structure and classification of the Stromatoporoidea, *Bull. Am. Paleontol.* 37: 345–486.

Garstang, W. 1898. On some modifications of structure subservient to respiration in decapod Crustacea which burrow in sand; with some remarks on the utility of specific characters in the genus *Calappa*, and the description of a new species of *Albunea*, *Quart. J. Microscop. Sci.* 40: 211–232.

——— 1922. The theory of recapitulation: a critical restatement of the biogenetic law, *J. Linn. Soc. Zool.* 35: 81–101.

——— 1928. The morphology of the Tunicata, and its bearings on the phylogeny of the Chordata, *Quart. J. Microscop. Sci.* 72: 51–187.

——— 1946. The morphology and relations of the Siphonophora, *Quart. J. Microscop. Sci.* 87: 103–193.

——— 1951. *Larval forms with other zoological verses* (Basil Blackwell, Oxford), 76 pp.

Gasking, E. 1967. *Investigations into generation: 1651–1828* (Hutchinson, London), 192 pp.

Gasman, D. 1971. *The scientific origins of national socialism: social Darwinism in Ernst Haeckel and the German Monist League* (MacDonald, London), 208 pp.

Gay, F. J. 1955. The occurrence of functional neoteinics [sic] in *Coptotermes lacteus, Austr. J. Sci.* 18: 58–59.

Gegenbaur, C. 1888. Cänogenese, *Verh. Anat. Ges.* 2: 3–9.

——— 1889. Ontogenie und Anatomie in ihren Wechselbeziehungen betrachtet, *Gegenbaurs Morph. Jahrb.* 15: 1–9.

Geist, V. 1971. *Mountain sheep: a study in behavior and evolution* (Univ. Chicago Press, Chicago), 383 pp.

Geoffroy Saint-Hilaire, E. 1833. Divers mémoires sur de grands sauriens . . . Quatrième mémoire sur le degré d'influence du monde ambiant pour modifier les formes animales; question intéressant de l'origine des espèces téléosauriennes et successivement celle des animaux de l'époque actuelle, *Mem. Acad. Roy. Sci. Inst. France* 12: 63–92.

——— 1836a. Considerations sur les singes les plus voisins de l'homme, *C. R. Acad. Sci.* 2: 92–95.

——— 1836b. Etudes sur l'orang-outang de la ménagerie, *C. R. Acad. Sci.* 3: 1–8.

Geoffroy Saint-Hilaire, I. 1833. Rapport fait a l'Académie royale des Sciences, sur un mémoire de M. Milne-Edwards, entitulé: observations sur les changements de forme que les crustacés éprouvent dans le jeune age, *Ann. Sci. Nat.* 30: 360–372.

——— 1836. Des rapports de la teratologie avec les sciences anatomiques et zoologiques, *C. R. Acad. Sci.* 3: 708–714.

George, T. N. 1933. Palingenesis and palaeontology, *Biol. Rev.* 8: 107–135.

Ghiselin, M. 1974. *The economy of nature and the evolution of sex* (Univ. California Press, Berkeley), 346 pp.

Giard, A. 1887. La castration parasitaire et son influence sur les caractères extérieurs du sexe male chez les crustacés décapodes, *Bull. Sci. Departement du Nord* (later *Bull. Sci. France et Belg.*, later *Bull. Biol. France Belg.*) 18: 1–28.

——— 1905. La poecilogonie, *Bull. Sci. France et Belg.* 39: 153–187.

Giard, A., and J. Bonnier, 1887. Contribution a l'étude des bopyriens, *Trav. Inst. Zool. Lille et Stat. Mar. Wimereux* 5: 1–272.

Ginetsinskaya, T. A. 1944. Yavlenie neotenii u cestodes, *Zool. Zhur.* 23: 35–42.

Glanville, E. V. 1969. Nasal shape, prognathism and adaptation in man, *Am. J. Phys. Anthrop.* 30: 29–37.

Gode von Aesch, A. 1941. *Natural science in German romanticism* (Columbia Univ. Press, New York), 302 pp.

Goette, A. 1884. *Abhandlungen zur Entwicklungsgeschichte der Tiere,* vol. 2 (Leopold Voss, Hamburg), 214 pp.

Goldschmidt, R. 1918. A preliminary report on some genetic experiments concerning evolution, *Am. Nat.* 52: 28–50.

——— 1923. Einige Materialen zur Theorie der abgestimmten Reaktionsgeschwindigkeiten, *Arch. Entwicklungsmech.* 98: 292–313.
——— 1927. *Physiologische Theorie der Vererbung* (Julius Springer, Berlin), 247 pp.
——— 1938. *Physiological genetics* (McGraw-Hill, New York), 375 pp.
Goodrich, E. S. 1924. *Living organisms* (Clarendon Press, Oxford), 200 pp.
Goss, R. J. 1964. *Adaptive growth* (Logos Press, London), 360 pp.
Gould, S. J. 1966. Allometry and size in ontogeny and phylogeny, *Biol. Rev.* 41: 587–640.
——— 1968. Ontogeny and the explanation of form: an allometric analysis, in D. B. Macurda, ed., Paleobiological aspects of growth and development, a symposium, *Paleont. Soc. Mem.* 2: 81–98.
——— 1969. An evolutionary microcosm: Pleistocene and Recent history of the land snail *P.* (*Poecilozonites*) in Bermuda, *Bull. Mus. Comp. Zool.* 138: 407–532.
——— 1970. Land snail communities and Pleistocene climates in Bermuda: a multivariate analysis of microgastropod diversity, *Proc. N. A. Paleontol. Convention,* part E: 486–521.
——— 1971. Geometric scaling in allometric growth: a contribution to the problem of scaling in the evolution of size, *Am. Nat.* 105: 113–136.
——— 1972a. Allometric fallacies and the evolution of *Gryphaea:* a new interpretation based on White's criterion of geometric similarity, in T. Dobzhansky, et al., eds., *Evolutionary Biology* 6: 91–118.
——— 1972b. Zealous advocates, *Science* 176: 623–625.
——— 1974. The evolutionary significance of "bizarre" structures: antler size and skull size in the "Irish Elk," *Megaloceros giganteus, Evolution* 28: 191–220.
——— 1975. Allometry in primates, with emphasis on scaling and the evolution of the brain, *Contrib. Primatol.* 5: 244–292.
Gould, S. J., and M. Katz. 1975. Disruption of ideal geometry in the growth of receptaculitids: a natural experiment in theoretical morphology, *Paleobiology* 1: 1–20.
Grant, W. C., Jr. 1961. Special aspects of the metamorphic process: second metamorphosis, *Am. Zoologist* 1: 163–171.
Greathead, D. J. 1966. A brief survey of the effects of biotic factors in populations of the desert locust, *J. Appl. Ecol.* 3: 239–250.
Gudernatsch, J. F. 1912. Feeding experiments on tadpoles: I, The influence of specific organs given as food on growth and differentiation: a contribution to the knowledge of organs with internal secretion, *Arch. Entwicklungsmech.* 35: 457–483.
Gunn, D. L. 1960. The biological basis of locust control, *Ann. Rev. Entomol.* 5: 279–300.
Gurney, R. 1942. *Larvae of decapod crustacea,* Ray Soc. Publ. no. 129, 306 pp.
Haber, F. C. 1959. *The age of the world: Moses to Darwin* (Johns Hopkins Press, Baltimore), 303 pp.
Hadzi, J. 1952. Application of the principles of phylembryogenesis to the Protista, *Nature* 169: 1019.
——— 1963. *The evolution of the Metazoa* (Macmillan, New York), 499 pp.

Haeckel, E. 1862—1868. *Die Radiolarien (Rhizopoda radiaria): Eine Monographie,* 3 vols. (Georg Reimer, Berlin).
——— 1866. *Generelle Morphologie der Organismen: Allgemeine Grundzüge der organischen Formen-Wissenschaft, mechanisch begründet durch die von Charles Darwin reformirte Descendenz-Theorie,* 2 vols. (Georg Reimer, Berlin), 574 pp., 462 pp.
——— 1868. *Natürliche Schöpfungsgeschichte* (Georg Reimer, Berlin), 568 pp.
——— 1872. *Die Kalkschwämme: Eine Monographie* (Berlin).
——— 1874a. *Anthropogenie: Keimes- und Stammes-Geschichte des Menschen* (W. Engelmann, Leipzig), 732 pp.
——— 1874b. Die Gastraea-Theorie, die phylogenetische Klassification des Tierreiches und Homologie der Keimblätter, *Jena. Z. Naturwiss.* 8: 1–55.
——— 1875. Die Gastrula und die Eifurchung der Thiere, *Jena. Z. Naturwiss.* 9: 402–508.
——— 1876. *Die Perigenesis der Plastidule oder die Wellenzeugung der Lebenstheilchen* (Georg Reimer, Berlin), 79 pp.
——— 1879a. *Anthropogenie,* 3rd ed. (W. Engelmann, Leipzig), 770 pp.
——— 1879b. *Das System der Medusen* (Gustav Fischer, Jena), 672 pp.
——— 1892. *The history of creation,* 2 vols., tr. E. R. Lankester from 8th ed. of *Natürliche Schöpfungsgeschichte* (Kegan Paul, Trench, Trubner & Co., London), 422 pp., 544 pp.
——— 1899. *Die Welträtsel* (E. Strauss, Bonn), 473 pp.
——— 1905. *The evolution of man,* 2 vols., tr. J. McCabe from 5th ed. of *Anthropogenie* (Watts & Co., London), 905 pp.
Hairston, N. G., D. W. Tinkle, and N. M. Wilbur. 1970. Natural selection and the parameters of population growth, *J. Wildlife Management* 34: 681–690.
Haldane, J. B. S. 1932. The time of action of genes, and its bearing on some evolutionary problems, *Am. Nat.* 66: 5–24.
Hall, B. G., and D. L. Hartl. 1974. Regulation of newly evolved enzymes. I. Selection of a novel lactase regulated by lactose in *Escherichia coli, Genetics* 76: 391–400.
Hall, G. S. 1897. A study of fears, *Am. J. Psychol.* 8: 147–249.
——— 1904. *Adolescence: its psychology and its relations to physiology, anthropology, sociology, sex, crime, religion, and education,* 2 vols. (D. Appleton, New York), 589 pp., 784 pp.
Hallam, A. 1959. On the supposed evolution of *Gryphaea* in the Lias, *Geol. Mag.* 96: 99–108.
——— 1965. Environmental causes of stunting in living and fossil marine benthonic invertebrates, *Palaeontology* 8: 132–155.
——— 1968. Morphology, palaeoecology, and evolution of the genus *Gryphaea* in the British Lias, *Phil. Trans. Roy. Soc. London* 254: 91–128.
Haller, A. von. 1744. *Hermanni Boerhaave praelectiones academicae,* vol. 5, part 2, passages tr. by H. B. Adelmann, *Marcello Malpighi and the evolution of embryology* (Cornell Univ. Press, Ithaca, N.Y., 1966), pp. 893–900.
Hamann, O. 1891. Monographie der Acanthocephalen (Echinorhynchen), *Jena. Z. Naturwiss.* 25 (n.s.18): 113–231.

Hampson, N. 1968. *The enlightenment* (Penguin, London), 304 pp.

Hanson, N. R. 1969. *Perception and discovery* (Freeman, Cooper & Co., San Francisco), 435 pp.

——— 1970. Hypotheses fingo, in R. E. Butts and J. W. Davis, eds., *The methodological heritage of Newton* (Univ. Toronto Press, Toronto), pp. 14–33.

Hardy, A. C. 1951. Introduction to W. Garstang, *Larval forms with other zoological verses* (Basil Blackwell, Oxford, 1951), pp. 1–21.

——— 1954. Escape from specialization, in J. Huxley, A. C. Hardy, and E. B. Ford, eds., *Evolution as a process* (George Allen & Unwin, London), pp. 146–171.

Harper, J. L. 1967. A Darwinian approach to plant ecology, *J. Ecol.* 55: 242–270.

Harris, R. G. 1925. Further data on the control of the appearance of pupalarvae in paedogenetic Cecidomyidae (*Oligarces* sp.), *Trav. Stat. Zool. Wimereux* 9: 89–97.

Hartog, Marcus. 1920. Introduction to Samuel Butler, *Unconscious Memory* (A. C. Fifield, London), pp. xi–xxxviii.

Harvey, W. 1653. *Two anatomical exercitations concerning the circulation of the blood* (Francis Leach, London), 86 pp.

Hassinger, D. D., J. D. Anderson, and G. H. Dalrymple. 1970. The early life history and ecology of *Ambystoma tigrinum* and *Ambystoma opacum* in New Jersey, *Am. Midland Nat.* 84: 474–495.

Heath, H. 1928. A sexually mature turbellarian resembling Müller's larva, *J. Morph.* 45: 187–207.

Hegeman, G. D., and S. L. Rosenberg. 1970. The evolution of bacterial enzyme systems, *Ann. Rev. Microbiol.* 24: 429–462.

Hemmer, H. 1969. A new view of the evolution of man, *Current Anthrop.* 10: 179–180.

Herbart, J. F. 1895. *The science of education,* tr. H. M. and E. Felkin (D. C. Heath, Boston), 268 pp.

Herder, J. G. 1794. *Ideen zur Philosophie der Geschichte der Menschheit,* 4 vols. (C. G. Schmiedler, Carlsruhe)

Hering, E. 1870. Memory as a universal function of organized matter [Das Gedächtniss als allgemeine Funktion der organisirten Materie], tr. Samuel Butler, 1880, in *Unconscious Memory* (A. C. Fifield, London, 1920), pp. 63–86.

Hertwig, O. 1894. *The biological problem of today: preformation or epigenesis? The basis of a theory of organic development* (Macmillan, New York), 148 pp.

——— 1901. Einleitung und allgemeine Litteraturübersicht, in O. Hertwig, ed, *Handbuch der vergleichenden und experimentellen Entwickelungslehre der Wirbeltiere,* vol. 1, part 1 (Gustav Fischer, Jena), pp. 1–85.

——— 1906. Ueber die Stellung der vergleichenden Entwickelungslehre zur vergleichenden anatomie, zur Systematik und Descendenztheorie (Das biogenetische Grundgesetz, Palingenese und Cenogenese), in O. Hertwig. ed., *Handbuch der vergleichenden und experimentellen Entwickelungslehre der Wirbeltiere,* vol. 3, part 3 (Gustav Fischer, Jena), pp. 149–180.

Hessler, R. R. 1971. Biology of the Mystacocarida: a prospectus, *Smith. Contr. Zoology* 76: 87–90.

Highnam, K. C., and P. T. Haskell, 1964. The endocrine system of isolated and crowded *Locusta* and *Schistocerca* in relation to oöcyte growth, and the effects of flying upon maturation, *J. Insect Physiol.* 10: 849–864.

Hille Ris Lambers, D. 1966. Polymorphism in Aphidae, *Ann. Rev. Entomol.* 11: 47–78.

Hilzheimer, M. 1926. Historisches und Kritisches zu Bolks Problem der Menschwerdung, *Anat. Anz.* 62: 110–121.

His, W. 1874. *Unsere Körperform und das physiologische Problem ihrer Entstehung* (F. C. W. Vogel, Leipzig), 224 pp.

——— 1888. On the principles of animal morphology, *Proc. Roy. Soc. Edinburgh* 15: 287–298.

——— 1894. Ueber mechanische Grundvorgänge thierischer Formenbildung, *Arch. Anat. Physiol. Anat. Abt.*, pp. 1–80.

Hoagland, K. E. 1975. Reproductive strategies and evolution in the genus *Crepidula* (Gastropoda: Prosobranchia) (Ph.D. dissertation, Dept. of Biology, Harvard Univ.).

Hofer, H. 1957. Zur Kenntnis der Kyphosen des Primatenschädels, *Verh. Anat. Ges.* 54: 54–76.

Holloway, R. L., Jr. 1970. Neural parameters, hunting, and the evolution of the human brain, in C. R. Noback and W. Montagna, eds., *The primate brain,* Advances in primatology, vol. 1 (Appleton-Century-Crofts, New York), pp. 299–310.

Holmes, S. J. 1944. Recapitulation and its supposed causes, *Quart. Rev. Biol.* 19: 319–331.

Holt, A. B., D. B. Cheek, E. D. Mellits, and D. E. Hill. 1975. Brain size and the relation of the primate to the nonprimate, in D. B. Cheek, eds., *Fetal and postnatal cellular growth: hormones and nutrition* (John Wiley, New York), pp. 23–44.

Hooijer, D. A. 1957. Three new giant rats from Flores, Lesser Sunda Islands, *Zool. Mededelingen* 35: 300–313.

——— 1967. Indo-Australian insular elephants, *Genetica* 38: 143–162.

Hopkins, J. W. 1966. Some considerations in multivariate allometry, *Biometrics* 22: 747–760.

Hoskins, E. R., and M. M. Hoskins. 1919. Growth and development of Amphibia as affected by thyroidectomy, *J. Exp. Zool.* 29: 1–69.

Hrdlicka, A. 1931. *Children who run on all fours* (McGraw-Hill, New York), 418 pp.

Hubbert, W. T., O. H. V. Stalheim, and G. D. Booth. 1972. Changes in organ weight and fluid volumes during growth of the bovine fetus, *Growth* 36: 217–233.

Hubbs, C. L. 1926. The structural consequences of modifications of the developmental rate in fishes, considered in reference to certain problems of evolution, *Am. Nat.* 60: 57–81.

Huber, I. 1974. Taxonomic and ontogenetic studies of cockroaches (Blattaria), *Univ. Kansas Sci. Bull.* 50: 233–332.

Huggett, A. St. G., and W. F. Widdas. 1951. The relationship between mammalian foetal weight and conception age, *J. Physiol.* 114: 306–317.

Hurst, C. H. 1893. The recapitulation theory, *Natural Science* 2: 195–200, 364–369.

Huxley, A. 1939. *After many a summer dies the swan* (Penguin, London, 1955), 251 pp.

Huxley, J. 1923. Time relations in amphibian metamorphosis, with some general considerations, *Sci. Prog.* 17: 606–618.

——— 1932. *Problems of relative growth* (MacVeagh, London), 276 pp.

Huxley, J., and G. R. de Beer. 1934. *The elements of experimental embryology* (Cambridge Univ. Press, Cambridge), 514 pp.

Huxley, T. H. 1849. On the anatomy and affinities of the family of the Medusae, *Phil. Trans. Roy. Soc. London* 139: 413–434.

Hyatt, A. 1866. On the parallelism between the different stages of life in the individual and those in the entire group of the molluscous order Tetrabranchiata, *Mem. Boston Soc. Nat. Hist.* 1: 193–209.

——— 1870. On reversions among the ammonites, *Proc. Boston Soc. Nat. Hist.* 14: 22–43.

——— 1880. The genesis of the Tertiary species of *Planorbis* at Steinheim, in *Anniversary Mem. Boston Soc. Nat. Hist. (1830–1880)*, pp. 1–114.

——— 1889. *Genesis of the Arietidae, Bull. Mus. Comp. Zool.* 16(3), 238 pp.

——— 1893. Phylogeny of an acquired characteristic, *Proc. Am. Phil. Soc.* 32: 349–647.

——— 1897. Cycle in the life of the individual (ontogeny) and in the evolution of its own group (phylogeny), *Proc. Am Acad. Arts Sci.* 32: 209–224.

Ilan, J., J. Ilan, and N. G. Patel. 1972. Regulation of messenger RNA translation mediated by juvenile hormone, in J. J. Menn and M. Beroza, eds., *Insect juvenile hormones* (Academic Press, New York), pp. 43–68.

Ivanova-Kasas, O. M. 1972. Polyembryony in insects, in S. J. Counce and C. H. Waddington, eds., *Developmental systems: insects* (Academic Press, New York), pp. 243–271.

Jacobson, M. 1969. Development of specific neuronal connections, *Science* 163: 543–547.

Jaekel, O. 1909. Über die Agnostiden, *Z. Deutsch. Geol. Ges.* 61: 380–401.

Jägersten, G. 1972. *Evolution of the metazoan life cycle* (Academic Press, London), 282 pp.

Janicki, C. 1930. Über die jüngsten Zustände von *Amphilina foliacea* in der Fischleibeshöhle, sowie Generelles sur Auffassung des Genus *Amphilina* G. Wagen, *Zool. Anz.* 90: 190–205.

Jastrow, J. 1892. The natural history of analogy, *Proc. Am. Assoc. Adv. Sci.* 4: 333–353.

Jenkin, P. M. 1970. *Control of growth and metamorphosis* (Pergamon Press, Oxford), 383 pp.

Jenkinson, J. W. 1909. *Experimental embryology* (Clarendon Press, Oxford), 341 pp.

Jerison, H. J. 1973. *Evolution of the brain and intelligence.* (Academic Press, New York), 482 pp.

Johnson, B. 1959. Effect of parasitization by *Aphidus platensis* Brethes on the developmental physiology of its host, *Aphis craccivora* Koch, *Entomologia Exp. Appl.* 2: 82–99.

——— 1965. Wing polymorphism in aphids: II, interaction between aphids, *Entomologia Exp. Appl.* 8: 49–64.

Johnson, B., and P. R. Birks. 1960. Studies on wing polymorphism in aphids: I, the developmental process involved in the production of the different forms, *Entomologia Exp. Appl.* 3: 327–339.

Johnson, C. G. 1974. Insect migration: aspects of its physiology, in M. Rockstein, ed., *The physiology of Insecta,* 2nd ed., vol. 3 (Academic Press, New York), pp. 279–334.

Johnson, L. E., G. S. Harshfield, and W. McCane. 1950. Dwarfism: an hereditary defect in beef cattle, *J. Heredity* 41: 177–181.

Johnson, W. H., L. E. Delanney, T. A. Cole, and A. E. Brooks. 1972. *Biology* (Holt, Rinehart and Winston, New York), 909 pp.

Jolicoeur, P. 1963. The multivariate generalization of the allometry equation, *Biometrics* 19: 497–499.

Joly, L. 1958. Comparaison des divers types d'adultoides chez *Locusta migratoria* L., *Insectes Sociaux* 5: 373–378.

Joly, P., and L. Joly. 1953. Resultats de griffes de corpora allata chez *Locusta migratoria* L., *Ann. Sci. Nat. Zool.* 15: 331–345.

Jonas, D., and D. Klein. 1970. *Man-child: a study of the infantilization of man* (McGraw-Hill, New York), 362 pp.

Jordan, Z. A. 1967. *The evolution of dialectical materialism* (St. Martin Press, New York), 490 pp.

Judd, C. H. 1903. *Genetic psychology for teachers* (D. Appleton, New York), 329 pp.

Jung, C. G. 1916. *Psychology of the unconscious* (Kegan Paul, Trench, Trubner & Co., London), 566 pp.

——— 1954. *Psychology and education* (Princeton Univ. Press, Princeton), 151 pp.

Kaiser, P. 1969. Welche Bedingungen steuern den generationswechsel der Gallmücke *Heteropeza* (Diptera: Itonidida)? *Zool. Jb. Physiol.* 75: 17–40.

——— 1972. Über die hormonale Regelung der Pädogenese bei den viviparen Gallmücken (Diptera: Cecidomyiidae), *Entomol. Mitt. Zool. Mus. Hamburg* 4: 259–262.

Kälin, J. 1965. Zur Ontogenese und Phylogenese des Schädels bei den höheren Primaten, *Rev. Suisse Zool.* 72: 594–603.

Kammerer, P. 1924. *The inheritance of acquired characteristics* (Boni & Liveright, New York), 414 pp.

Kautsky, F. 1939. Die Erycinen des niederösterreichischen Miocaen, *Ann. Naturhist. Mus. Wien* B 50: 584–671.

Keibel, F. 1895. Normentafeln zur Entwickelungsgeschichte der Wirbeltiere, *Anat. Anz.* 11: 225–234.

——— 1898. Das biogenetische Grundgesetz und die Cenogenese, *Ergeb. Anat. Entwick.* 7: 722–792.

Keith, A. 1949. Foetalization as a factor in human evolution, in *A new theory of human evolution* (Watts, London), pp. 192–201.

Kennedy, C. R. 1965. The life-history of *Archigetes limnodrili* (Yamaguti) (Cestoda: Caryophyllaeidae) and its development in the invertebrate host, *Parasitology* 55: 427–437.

Kennedy, J. S. 1956. Phase transformation in locust biology, *Biol. Rev.* 31: 349–370.

——— 1961. Continuous polymorphism in locusts, *Symp. Roy. Entomol. Soc. London* 1: 80–90.

Kennedy, J. S., and H. L. G. Stroyan. 1959. Biology of aphids, *Ann. Rev. Entomol.* 4: 139–160.

Kenney, F. T., B. A. Hamkalo, G. Favelukes, and J. T. August, eds. 1973. *Gene expression and its regulation* (Plenum Press, New York), 557 pp.

Kethley, J. B. 1974. Developmental chaetotaxy of a paedomorphic celaenopsoid, *Neotenogynium malkini* n.g., n. sp. (Acari: Parasitiformes: Neotenogyniidae, n. fam.) associated with millipedes, *Ann. Entomol. Soc. Am.* 67: 571–579.

Key, K. H. L. 1950. A critique of the phase theory of locusts, *Quart. Rev. Biol.* 25: 363–407.

Kezer, J. 1952. Thyroxin-induced metamophosis of the neotenic salamanders *Eurycea tynerensis* and *Eurycea neotenes, Copeia,* pp. 234–237.

Kidd, B. 1898. *The control of the tropics* (Macmillan, New York), 101 pp.

Kielmeyer, C. F. 1793. Ueber die Verhältnisse der organischen Kräfte untereinander in der Reihe der verschiedenen Organisationen, die Gesetze und Folgen dieser Verhältnisse (reprinted in 1930 with notes by H. Balss), *Sudhoff's Archiv* 23: 247–267.

King, C. E., and W. W. Anderson. 1971. Age specific selection: II, the interaction between *r* and *K* during population growth, *Am. Nat.* 105: 137–156.

King, M. C., and A. C. Wilson. 1975. Evolution at two levels in humans and chimpanzees, *Science* 188: 107–116.

Kinzelbach, R. K. 1971. Morphologische Befunde an Fächerflüglern und ihre phylogenetische Bedeutung (Insecta: Strepsiptera), *Zoologica* 41: 1–128.

Kisimoto, R. 1956. Effect of crowding during the larval period on the determination of the wing-form of an adult plant-hopper, *Nature* 178: 641–642.

Khamsi, F., and J. T. Eayrs. 1966. A study of the effects of thyroid hormones in growth and development, *Growth* 30: 143–156.

Klaauw, C. J. van der. 1945. Cerebral skull and facial skull, *Arch. Neer. Zool.* 7: 16–37.

Klapper, Paul. 1912. *Principles of educational practice* (D. Appleton, New York), 485 pp.

Kleinenberg, N. 1886. Die Entstehung des Annelids aus der Larve von *Lopadorhyncus.* Nebst Bemerkungen über die Entwicklung anderer Polychaeten, *Z. wiss. Zool.* 44: 1–227.

Kleinsorge, J. A. 1900. *Beiträge zur Geschichte der Lehre vom Parallelismus der Individual- und der Gesamtentwicklung* (B. Engau, Jena), 42 pp.

Koestler, A. 1972. *The case of the midwife toad* (Random House, New York), 188 pp.

Koffka, K. 1928. *The growth of the mind,* 2nd ed. (Kegan Paul, Trench, Trubner & Co., London), 427 pp.

Kohlbrugge, J. H. F. 1911. Das biogenetische Grundgesetz: Eine historische Studie, *Zool. Anz.* 38: 447–453.

Kollman, J. 1885. Das Ueberwintern von europäischen Frosch- und Tritonlarven und die Umwandlung des mexikanischen Axolotl, *Verh. Naturf. Ges. Basel* 7: 387–398.

——— 1905. Neue Gedanken über das alter Problem von der Abstammung des Menschen, *Corresp.-Bl. Deutsch. Ges. Anthrop. Ethnol. Urges.* 36: 9–20.

Kollros, J. J. 1961. Mechanisms of amphibian metamorphosis: hormones, *Am. Zoologist* 1: 107–114.

Komai, T. 1963. A note on the phylogeny of the Ctenophora, in E. C. Dougherty, ed., *The lower Metazoa* (Univ. California Press, Berkeley), pp. 181–188.

Krebs, C. J., and J. H. Myers. 1974. Population cycles in small mammals, *Adv. Ecol. Res.* 8: 267–399.

Krogman, W. M. 1972. *Child growth* (Univ. Michigan Press, Ann Arbor), 231 pp.

Kryzanowsky, S. G. 1939. Das Recapitulationsprinzip und die Bedingungen der historischen Auffassung der Ontogenese, *Acta Zool. Stockholm* 20: 1–87.

Kummer, B. 1952. Untersuchungen über die ontogenetische Entwicklung des menschlichen Schädelbasiswinkels, *Z. Morph. Anthrop.* 43: 331–360.

——— 1960. Zum Problem der Fetalisation, *Zool. Anz.* 164: 391–394.

Laird, A. K. 1967. Evolution of the human growth curve, *Growth* 31: 345–355.

——— 1969. The dynamics of growth, *Research/Development* 20: 28.

Laird, A. K., A. D. Barton, and S. A. Tyler. 1968. Growth and time: an interpretation of allometry, *Growth* 32: 347–354.

Lambert, W. V., and A. Sciuchetti. 1935. A dwarf mutation in the rat, *Science* 8: 278.

Lankester, E. R. 1877. Notes on the embryology and classification of the animal kingdom: comprising a revision of speculations relative to the origin and significance of the germ-layers, *Quart. J. Microscop. Sci.* 17: 399–454.

——— 1880. *Degeneration: a chapter in Darwinism* (Macmillan, London), reprinted in W. Coleman, ed., *The interpretation of animal form* (Johnson Reprint Co., New York, 1967), pp. 57–132.

Laws, R. M. 1966. Age criteria for the African elephant *Loxodonta a. africana, E. Afr. Wildlife J.* 4: 1–37.

Lea, A. 1969. The population ecology of brown locusts, *Locustana pardalina* (Walker), on fixed observation areas, *Phytophylactica* 1: 93–102.

Lebedkin, S. 1937. The recapitulation problem, *Biol. Gen.* 13: 391–417, 561–594.

Lebrun, D. 1961. Evolution de l'appareil genital dans les diverses castes de *Calotermes flavicollis, Bull. Soc. Zool. France* 86: 235–242.

——— 1970. Intercastes expérimentaux de *Calotermes flavicollis* Fabr., *Insectes Sociaux* 17: 159–176.

Lees, A. D. 1966. The control of polymorphism in aphids, *Adv. Insect Physiol.* 3: 207–277.

——— 1967. The production of the apterous and alate forms in the aphid *Megoura viciae* Buckton, with special reference to the role of crowding, *J. Insect Physiol.* 13: 289–318.

Leonardi, P. 1954. Les mammifères nains du Pleistocène Méditérranéen, *Ann. Paléont.* 40: 189–201.

Letourneau, C. 1892. *Property: its origin and development* (Walter Scott, London), 401 pp.

Leuckart, S. 1821. Einiges über die fischartigen Amphibien, *Isis* 8 (1): 249–265.

Leutenegger, W. 1972. Newborn size and pelvic dimensions of *Australopithecus, Nature* 240: 568–569.

Levin, D. A. 1975. Pest pressure and recombination systems in plants, *Am. Nat.* 109: 437–451.

Levin, H. 1969. *The myth of the Golden Age in the Renaissance* (Indiana Univ. Press, Bloomington), 231 pp.

Lewontin, R. C. 1965. Selection for colonizing ability, in H. G. Baker and G. L. Stebbins, eds., *The genetics of colonizing species* (Academic Press, New York), pp. 77–91.

Liem, K. F. 1973. Evolutionary strategies and morphological innovations: Cichlid pharyngeal jaws, in S. J. Gould, ed., Evolutionary development of form and symmetry, *Syst. Zool.* 22: 425–441.

Lillegraven, J. A. 1975. Biological considerations of the marsupial-placental dichotomy, *Evolution* 29: 707–722.

Lillie, F. R. 1895. The embryology of the Unionidae, *J. Morph.* 10: 1–100.

Loeb, J. 1964. *The mechanistic conception of life,* ed. D. Fleming (Harvard Univ. Press, Cambridge), 216 pp.

Loher, W. 1960. The chemical acceleration of the maturation process and its hormonal control in the male of the desert locust, *Proc. Roy. Soc. B* 153: 380–397.

Lomax, A., and N. Berkowitz. 1972. The evolutionary taxonomy of culture, *Science* 177: 228–239.

Lombroso, C. 1887. *L'homme criminel. Criminel-né—fou moral—épileptique. Etude anthropologique et médico-legale* (Germer Bailliere, Paris), 682 pp.

——— 1895. Criminal anthropology applied to pedagogy, *The Monist* 6: 50–59.

——— 1911. *Crime: its causes and remedies* (Little, Brown and Co., Boston), 471 pp.

Lorenz, K. 1971. *Studies in human and animal behavior,* vol. 2 (Harvard Univ. Press, Cambridge, Mass.), 366 pp.

Lovejoy, A. O. 1936. *The great chain of being* (Harvard Univ. Press, Cambridge).

——— 1959. Herder: progressionism without transformism, in B. Glass, O.

Temkin, and W. L. Straus, Jr., eds., *Forerunners of Darwin: 1745–1859* (Johns Hopkins Univ. Press, Baltimore), pp. 207–221.

——— 1959. Schopenhauer as an evolutionist, in *Forerunners of Darwin: 1745–1859,* pp. 415–437.

——— 1959. Recent criticism of the Darwinian theory of recapitulation: its grounds and its initiator, in *Forerunners of Darwin: 1745–1859,* pp. 438–458.

Lurie, E. 1960. *Louis Agassiz: a life in science* (Univ. Chicago Press, Chicago), 449 pp.

Lyell, C. 1832. *Principles of geology,* vol. 2 (John Murray, London), 330 pp.

Lynn, W. G. 1961. Types of amphibian metamorphosis, *Am. Zoologist* 1: 151–161.

Lynn, W. G., and H. E. Wachowski. 1951. The thyroid gland and its functions in cold-blooded vertebrates, *Quart. Rev. Biol.* 26: 123–168.

MacArthur, R. W., and E. O. Wilson. 1967. *The theory of island biogeography* (Princeton Univ. Press, Princeton), 203 pp.

MacBride, E. W. 1914. *Text-book of embryology* (Macmillan, London), 692 pp.

——— 1917. Recapitulation as a proof of the inheritance of acquired characters, *Scientia* 22: 425–434.

Mackiewicz, J. S. 1972. Caryophyllidae (Cestoidea): a review, *Exptl. Parasitol.* 31: 417–512.

Madsen, F. J. 1961. On the zoogeography and origin of the abyssal fauna in view of the knowledge of the Porcellanasteridae, *Galathea Rep.* 4: 177–218.

Magor, J. I. 1970. *Outbreaks of the Australian plague locust* (Chortoicetes terminifera *Walk.*) *in New South Wales during the period 1937–1962, particularly in relation to rainfall, Anti-Locust Mem.* 11, 39 pp.

Mancini, E. A. 1974. Origin of micromorph faunas—Grayson Formation (Upper Cretaceous, Texas) (Ph. D. dissertation, Texas A. & M.), 328 pp.

Mangold-Wirz, K. 1966. Cerebralisation und Ontogenesemodus bei Eutheria, *Acta anat.* 63: 449–508.

Mann, A. E. 1975. Paleodemographic aspects of the South African australopithecines, *Univ. Penn. Publ. Anthrop.* 1: 171.

Margalef, R. 1949. Importancia de la neotenia en la evolucion de los crustaceos de agua dulce, *Publ. Inst. Biol. Aplicada Barcelona* 6: 41–51.

——— 1959. Mode of evolution of species in relation to their places in ecological succession, *Proc. XV Int. Congr. Zool.,* pp. 787–789.

Margulis, L. 1970. *Origin of eukaryotic cells* (Yale Univ. Press, New Haven).

Markert, C. L., J. B. Shaklee, and G. S. Whitt. 1975. Evolution of a gene, *Science* 189: 102–114.

Marshall, A. 1891. Development of animals, *Rept. Sixtieth Meeting Brit. Assoc. Adv. Sci.,* pp. 826–852.

Massart, J. 1894. Le récapitulation et l'innovation en embryologie végétale, *Bull. Soc. Roy. Bot. Belgique* 33: 150–247.

Matsuda, R. In press. The insect abdomen.

——— and F. J. Rohlf. 1961. Studies of relative growth in Gerridae (5): comparison of two populations (Heteroptera: Insecta), *Growth* 25: 211–217.

Matveiev, B. S. 1932. Zur Theorie der Rekapitulation. Über die Evolution der

Schuppen, Federn, und Haare auf dem Wege embryonaler Veränderungen, *Zool. Jb. Abt. Anat.* 55: 555–580.
Mayr, E. 1959. Agassiz, Darwin, and evolution, *Harvard Library Bull.* 13: 165–194.
——— 1965. Comments (to paper by E. Mendelsohn), *Boston Studies Phil. Sci.* 1: 151–155.
McCormick, R. 1973. Recapitulation: Freud and Jung (unpublished).
McGuire, O. S. 1966. Population studies of the ostracode genus *Polytylites* from the Chester Series, *J. Paleontology* 40: 883–910.
McIntyre, A. D. 1969. Ecology of marine meiobenthos, *Biol. Rev.* 44:245–290.
McKinley, K. R. 1971. Survivorship in gracile and robust australopithecines: a demographic comparison and a proposed birth model, *Am. J. Phys. Anthrop.* 34: 417–426.
McKusick, V. A., and D. L. Rimoin. 1967. General Tom Thumb and other midgets, *Sci. Am.* 217: 102–110.
McMurray, C. A. 1892. *The elements of general method based on the principles of Herbart* (Public School Publ. Co., Bloomington, Ill.), 200 pp.
McNaughton, S. J. 1975. *r*- and *K*-selection in *Typha, Am. Nat.* 109: 251–263.
Meckel, J. F. 1808. *Beyträge zur Geschichte des menschlichen Fötus. Beyträge zur vergleichenden Anatomie,* vol. 1 (Carl Heinrich Reclam, Leipzig), pp. 57–124.
——— 1811a. *Entwurf einer Darstellung der zwischen dem Embryozustande der höheren Tiere und dem permanenten der niederen stattfindenen Parallele: Beyträge zur vergleichenden Anatomie,* vol. 2 (Carl Heinrich Reclam, Leipzig), pp. 1–60.
——— 1811b. *Ueber den Charakter der allmähligen Vervollkommung der Organisation, oder den Unterschied zwischen den höheren und niederen Bildungen: Beyträge zur vergleichenden Anatomie* (Carl Heinrich Reclam, Leipzig), pp. 61–123.
——— 1821. *System der vergleichenden Anatomie,* 7 vols. (Rengerschen Buchhandlung, Halle).
Medawar, P. B. 1945. Size, shape, and age, in W. E. LeGros Clark and P. B. Medawar, eds., *Essays on growth and form presented to D'Arcy W. Thompson* (Oxford Univ. Press, Oxford), pp. 157–187.
Mehnert, E. 1891. Gastrulation und Keimblätterbildung der *Emys lutaria taurica, Morph. Arb.* 1: 365–495.
——— 1895. Die individuelle Variation des Wirbelthierembryo, *Morph. Arb.* 5: 386–444.
——— 1897. Kainogenese, *Morph. Arb.* 7: 1–156.
——— 1898. *Biomechanik* (Gustav Fischer, Jena), 177 pp.
Merton, L. F. H. 1959. *Studies in the ecology of the Moroccan locust* (Dociostaurus maroccanus *Thunberg) in Cyprus, Anti-Locust Bull.* 34, 123 pp.
Meyer, A. W. 1935. Some historical aspects of the recapitulation idea, *Quart. Rev. Biol.* 10: 379–396.
——— 1939. *The rise of embryology* (Stanford Univ. Press, Stanford), 367 pp.
——— 1956. *Human generation: conclusions of Burdach, Döllinger and von Baer* (Stanford Univ. Press, Stanford), 143 pp.

Millais, J. G. 1906. *Mammals of Great Britain and Ireland* (Longmans, Green and Co., London).

Milne-Edwards, H. 1844. Considérations sur quelques principes relatifs à la classification naturelle des animaux, *Ann. Sci. Nat. Zool.* (Ser. 3) 1: 65–99.

Mittler, T. E. 1973. Aphid polymorphism as affected by diet, *Bull. Entomol. Soc. New Zealand* 2: 65–75.

Monniot, F. 1971. Les ascidiens littorales et profondes des sediments meubles, *Smith. Contr. Zool.* 76: 119–126.

Monroe, W. S., J. C. DeVoss, and G. W. Reagan. 1930. *Educational Psychology* (Doubleday, Doran & Co., Garden City, N.Y.), 607 pp.

Montagu, M. F. A. 1935. The premaxilla in the primates, *Quart. Rev. Biol.* 10: 32–59, 181–208.

——— 1962. Time, morphology, and neoteny in the evolution of man, in M. F. A. Montagu, ed., *Culture and the evolution of man* (Oxford Univ. Press, New York), pp. 324–342.

Montgomery, T. H. 1906. *The analysis of racial descent in animals* (Henry Holt, New York), 311 pp.

Morgan, T. H. 1891. The growth and metamorphosis of Tornaria, *J. Morph.* 5: 407–458.

——— 1899. Some problems of regeneration, *Biol. Lectures, Mar. Biol. Lab. Wood's Holl* [1898], pp. 193–207.

——— 1903. *Evolution and adaptation* (Macmillan, New York), 470 pp.

——— 1909. Recent experiments on the inheritance of coat colors in mice, *Am. Nat.* 43: 494–510.

——— 1916. *A critique of the theory of evolution* (Princeton Univ. Press, Princeton, 197 pp.

——— 1932. *The scientific basis of evolution* (W. W. Norton, New York), 286 pp.

——— 1934. *Embryology and genetics* (Columbia Univ. Press, New York), 258 pp.

Morse, E. S. 1892. Natural selection and crime, *Pop. Sci.* 41: 433–446.

Mosier, H. D., Jr. 1972. Decreased energy efficiency after cortisone induced growth arrest, *Growth* 36: 123–131.

Mosier, H. D., Jr., and R. A. Jansons. 1969. Allometry of body weight and tail length in studies of catch-up growth in rats, *Growth* 33: 319–330.

——— 1971. Allometry of body weight and tail length after head x-irradiation in rats, *Growth* 35: 23–31.

Mosimann, J. E. 1970. Size allometry: size and shape variables with characterizations of the log-normal and generalized gamma distributions, *J. Am. Statistical Assoc.* 65: 930–945.

Moss, M. L. 1958. Rotations of the cranial components in the growing rat and their experimental alteration, *Acta anat.* 32: 65–86.

Müller, F. 1864. *Für Darwin,* in A. Möller, ed., *Fritz Müller. Werke, Briefe und Leben* (Gustav Fischer, Jena, 1915), pp. 200–263.

Müller, Fa. 1969. Verhältnis von Körperentwicklung und Cerebralisation in Ontogenese und Phylogenese der Säuger, *Verh. Naturf. Ges. Basel* 80: 1–31.

——— 1973. Zur stammesgeschichtlichen Veränderung der Eutheria-Ontogenesen, *Rev. Suisse Zool.* 79: 1599–1685.

Murphy, G. I. 1968. Patterns in life history and the environment, *Am. Nat.* 102: 390–404.

Murphy, J. 1927. *Primitive man* (Oxford Univ. Press, London), 341 pp.

Naef, A. 1926a. Zur Morphologie und Stammesgeschichte des Affenschädels, *Naturwiss.* 14: 89–97.

——— 1926b. Über die Urformen der Anthropomorphen und die Stammesgeschichte des Menschenschädels, *Naturwiss.* 14: 445–452.

Needham, J. 1933. On the dissociability of the fundamental processes in ontogenesis, *Biol. Rev.* 8: 180–223.

——— 1936. *Order and life* (Yale Univ. Press, New Haven), reprint ed. (MIT Press, Cambridge, 1968), 175 pp.

——— 1959. *A history of embryology,* 2nd ed., with A. Hughes (Cambridge Univ. Press, Cambridge), 304 pp.

Neish, I. C. 1971. Comparison of size, structure, and distributional patterns of two salamander populations in Marion Lake, British Columbia, *J. Fish. Res. Bd. Canada* 28: 49–58.

Nevesskaya, L. A. 1967. Problems of species differentiation in light of paleontological data, *J. Paleont.* 4: 1–17.

Newell, N. D. 1949. Phyletic size increase, an important trend illustrated by fossil invertebrates, *Evolution* 3: 103–124.

Newport, G. 1841. On the organs of reproduction, and the development of the Myriapoda, *Phil. Trans.* 131: 99–130.

Nikolei, E. 1961. Vergleichende Untersuchungen zur Fortpflanzung heterogener Gallmücken unter experimentellen Bedingungen, *Z. Morph. Ökol. Tiere* 50: 281–329.

Nolte, N. D., I. R. May, and B. M. Thomas. 1970. The gregarisation pheremone of locusts, *Chromosoma* 29: 462–473.

Noodt, W. 1971. Ecology of the Copepoda, *Smith. Contr. Zool.* 76: 97–102.

Nordenskiöld, E. 1929. *The history of biology* (Kegan Paul, Trench, Trubner & Co., London), 629 pp.

Norris, D. O., R. E. Jones, and B. B. Criley. 1973. Pituitary prolectin levels in larval, neotenic and metamorphosed salamanders (*Ambystoma tigrinum*), *Gen. Comp. Endocrinol.* 20: 437–442.

Norris, M. J. 1970. Aggregation response in ovipositing females in the desert locust, with special reference to the chemical factor, *J. Insect Physiology* 16: 1493–1515.

Norsworthy, N., and M. T. Whitley. 1918. *The psychology of childhood* (Macmillan, New York), 375 pp.

Novak, J. J. A. 1966. *Insect hormones* (Methuen, London), 478 pp.

Nybelin, O. 1963. Zur *Archigetes*-Frage, *Zool. Bidrag Uppsala* 35: 293–306.

Ockelman, K. W. 1964. *Turtonia minuta* (*Fabricius*), a neotenous vereracean bivalve, *Ophelia* 1: 121–146.

Odhner, N. 1952. Petits opisthobranches peu connus de la côte méditéranéenne de France, *Vie et Milieu* 3: 136–147.

Oken, L. 1806. *Beiträge zur vergleichenden Zoologie: Anatomie und Physiologie* (J. A. Göbhardt, Bamberg), 122 pp.

——— 1809–1811. *Lehrbuch der Naturphilosophie,* 3 vols. (F. Frommand, Jena), 228 pp., 180 pp., 374 pp.

——— 1829. Ueber "Entwickelungsgeschichte der Thiere," *Isis,* pp. 205–212.

——— 1847. *Elements of physiophilosophy,* tr. Alfred Tulk (Ray Society, London), 665 pp.

Olivier, G., and H. Pineau. 1958. Croissance prénatale comparée du macaque et de l'homme, *C. R. Acad. Sci. Paris* 246: 1292–1293.

Oppel, A. 1891. *Vergleichung des Entwicklungsgrades der Organe zu verschiedenen Entwicklungszeiten bei Wirbeltieren* (Gustav Fischer, Jena), 181 pp.

Oppenheimer, J. 1959. An embryological enigma in the *Origin of Species,* in B. Glass, O. Temkin, and W. L. Strauss, Jr., eds., *Forerunners of Darwin: 1745–1859* (Johns Hopkins Univ. Press, Baltimore), pp. 292–322.

——— 1967. *Essays in the history of embryology and biology* (MIT Press, Cambridge), 374 pp.

——— 1973. Recapitulation, in P. P. Weiner, ed., *Dictionary of the History of Ideas,* vol. 4 (Charles Scribner's Sons, New York), pp. 56–59.

——— 1975. When sense and life begin: background for a remark in Aristotle's *Politics* (1335b24), *Arethusa* 8: 331–343.

Osborn, H. F. 1929. *From the Greeks to Darwin* (Charles Scribner's Sons, New York), 398 pp.

O'Shea, M. V. 1906. *Dynamic factors in education* (Macmillan, New York), 320 pp.

Ospovat, D. 1974. Embryos, archetypes, and fossils: von Baer's embryology and British paleontology in the mid-nineteenth century (Ph.D. dissertation, Dept. of the History of Science, Harvard Univ.), 352 pp.

Otis, E. M., and R. Brent. 1954. Equivalent ages in mouse and human embryos, *Anat. Rec.* 120: 33–63.

Overhage, P. 1959. Um die ursächliche Erklärung der Hominisation, *Acta Biotheor. Leiden* 8: 1–126.

Owen, R. 1841. *John Hunter's observations on animal development edited, and his illustrations of that process in the bird described* (R. and J. E. Taylor, London), 64 pp.

Parmelee, M. 1912. *The principles of anthropology and sociology in their relations to criminal procedure* (Macmillan, New York), 410 pp.

Pavlov, A. P. 1901. Le Crétace inférieur de la Russie et sa faune, *Nouv. Mém. Soc. Imp. Nat. Moscow* 16: 87.

Pennak, R. W., and D. J. Zinn. 1943. *Mystacocarida, a new order of Crustacea from intertidal beaches in Massachusetts and Connecticut,* Smith. Misc. Coll., vol. 103, no. 9, 11 pp.

Perrier, E., and C. Gravier. 1902. La tachygenèse ou accelération embryogénique, *Ann. Sci. Nat. Zool.* 16: 133–374.

Peter, K. 1955. Palingenese und Cenogenese in der Embryologie, *Gegenbaurs Morph. Jb.* 94: 65–110.

Piaget, J. 1960. *The child's conception of physical causality* (Littlefield, Adams & Co., Paterson, New Jersey), 309 pp.

——— 1969. Genetic epistemology, *Columbia Forum* 12: 4–11.

——— 1971. *Biology and knowledge* (Univ. Chicago Press, Chicago), 384 pp.

Pianka, E. R. 1970. On *r* and *K* selection, *Am. Nat.* 104: 592–597.
——— 1972. *r* and *K* selection or *b* and *d* selection, *Am. Nat.* 106: 581–588.
Pilbeam, D., and S. J. Gould. 1974. Size and scaling in human evolution, *Science* 186: 892–901.
Poppelbaum, H. 1960. *Man and animal, their essential difference* (Anthroposophical Publishing Co., London), 160 pp.
Portmann, A. 1941. Die Tragzeiten der Primaten und die Dauer der Schwangerschaft beim Menschen: ein Problem der vergleichen Biologie, *Rev. Suisse Zool.* 48: 511–518.
——— 1945. Die Ontogenese des Menschen als Problem der Evolutionsforschung, *Verh. Schweiz. Naturf. Ges.* 125: 44–53.
Powell, J. W. 1889. Evolution of music from dance to symphony, *Proc. Am. Assoc. Adv. Sci.*, pp. 1–21.
Prahlad, K. V., and L. E. Delanney. 1965. A study of induced metamorphosis in the axolotl, *J. Exp. Zool.* 160: 137–146.
Preyer, W. 1884. *Die Seele des Kindes* (Th. Grieben's Verlag, Leipzig), 487 pp.
Rabaud, E. 1916. Les phénomènes embryonnaire et la phylogenèse, *Scientia* 19: 270–289.
Rack, G. 1972. Pyemotiden an Gramineen in schwedischen landwirtschaftlichen Betrieben. Ein Beitrag zur Entwicklung von *Siteroptes graminum* (Reuter, 1900) (Acarina, Pyemotidae), *Zool. Anz.* 188: 157–174.
Raikov, B. E. 1968. *Karl Ernst Von Baer 1792–1876: Sein Leben und sein Werk* (J. A. Barth, Leipzig), 516 pp.
——— 1969. *Germanskie Biologi-Evolyutsionisty do Darvina* (Izdat. Nauka, Leningrad), 232 pp.
Raup, D. M. 1966. Geometric analysis of shell coiling: general problems, *J. Paleontology* 40: 1178–1190.
Raymont, T. 1906. *The principles of education* (Longman's Green & Co., London), 381 pp.
Remane, A. 1960. Die Beziehung zwischen Phylogenie und Ontogenie, *Zool. Anz.* 164: 306–337.
——— 1962. Gilt das biogenetische Gesetz noch heute? *Umschau* 18: 571–574.
Rensch, B. 1958. Die Abhängigkeit der Struktur und der Leistungen tierischer Gehirne von ihrer Grösse, *Naturwiss.* 45: 145–154; 175–180.
——— 1959. *Evolution above the species level* (Columbia Univ. Press, New York), 419 pp.
——— 1971. Die phylogenetischen Abwandlungen der Ontogenesen, in G. Heberer, ed., *Die Evolution der Organismen,* vol. 2, *Die Kausalität der Phylogenie* (Gustav Fischer, Stuttgart), pp. 1–28.
Reynolds, V. 1967. *The apes* (E. P. Dutton, New York), 296 pp.
Riddiford, L. M. 1970. Prevention of metamorphosis by exposure of insect eggs to juvenile hormone analogs, *Science* 167: 287–288.
——— 1972. Juvenile hormone in relation to the larval-pupal transformation of the Cecropia silkworm, *Biol. Bull.* 142: 310–325.
——— 1975. Juvenile hormone-induced delay of metamorphosis of the viscera of the cecropia silkworm, *Biol. Bull.* 148: 429–439.

Riedl, R. J. 1969. Gnathostomulida: is there a fossil record, *Science* 164: 856.

Riesenfeld, A. 1969. The adaptive mandible: an experimental study, *Acta Anat.* 72: 246–262.

Rignano, E. 1911. *Upon the inheritance of acquired characters: a hypothesis of heredity, development, and assimilation* (Open Court Publishing Co., Chicago), 413 pp.

Ritterbush, P. C. 1964. *Overtures to biology: the speculations of eighteenth-century naturalists* (Yale Univ. Press, New Haven), 287 pp.

Roark, R. N. 1895. *Psychology in education* (American Book Co., New York), 312 pp.

Roethke, T. 1961. *Words for the wind* (Indiana Univ. Press, Bloomington), 212 pp.

Romanes, G. J. 1896. *The life and letters of George John Romanes,* ed. E. Romanes Longmans, Green, London), 391 pp.

Ross, D. 1972. *G. Stanley Hall: the psychologist as prophet* (Univ. Chicago Press, Chicago), 482 pp.

Roux, W. 1881. *Der Kampf der Theile im Organismus* (W. Engelmann, Leipzig), 244 pp.

——— 1894. 'Einleitung' zum Archiv für Entwicklungsmechanik der Organismen, *Arch. Entwicklungsmech. Org.* 1: 1–42.

——— 1905. *Die Entwicklungsmechanik. Ein neuer Zweig der biologischen Wissenschaft.* Vorträge und Aufsätze Entwicklungsmechanik der Organismen no. 1 (W. Engelmann, Leipzig), 283 pp.

Ruddle, F. H., ed. 1973. *Genetic mechanisms of development* (Academic Press, New York), 383 pp.

Rudwick, M. J. S. 1960. The feeding mechanisms of spire-bearing fossil brachiopods, *Geol. Mag.* 97: 369–383.

Russell, E. S. 1916. *Form and function: a contribution to the history of animal morphology* (John Murray, London), 383 pp.

Ruzhentsev, V. E. 1963. Theory of phylogenetic systematics, part 2, *Int. Geol. Rev.* 5: 915–944.

Sacher, G. A. 1959. Relation of lifespan to brain weight and body weight in mammals, *CIBA Foundation Colloquia on Aging* 5: 115–133.

——— 1966. *Dimensional analysis of factors governing longevity in mammals,* Proc. Int. Cong. Gerontol. (Vienna), 14 pp.

Sacher, G. A., and E. F. Staffeldt. 1974. Relation of gestation time to brain weight for placental mammals: implications for the theory of vertebrate growth, *Am. Nat.* 108: 593–615.

Santillana, G. de. 1961. *The origins of scientific thought* (Weidenfeld & Nicolson, London), 320 pp.

Schindewolf, O. H. 1929. Das Problem der Menschwerdung, ein paläontologischer Lösungsversuch, *Jb. Preuss. Geol. Landesanst.* 49: 716–766.

——— 1936. *Paläontologie, Entwicklungslehre und Genetik: Kritik und Synthese* (Bornträger, Berlin), 108 pp.

——— 1946. Zur Kritik des "Biogenetischen Grundgesetzes," *Naturwiss.* 33: 244–249.

——— 1950. *Grundfragen der Paläontologie* (Schweizerbart, Stuttgart), 506 pp.

Schmalhausen, I. 1927. Beiträge zur quantitativen Analyse der Formbildung: 2, Das Problem des proportionalen Wachstums, *Roux' Archiv Entwicklungsmechanik,* vol. 110.

Schmidt, H. 1909. *Das biogenetische Grundgesetz Ernst Haeckels und seine Gegner* (Neuer Frankfurter Verlag, Frankfurt), 159 pp.

Schneiderman, H. A. 1972. Insect hormones and insect control, in J. J. Menn and M. Beroza, eds., *Insect juvenile hormones* (Academic Press, New York), pp. 3–27.

Schultz, A. H. 1926. Fetal growth of man and other primates, *Quart. Rev. Biol.* 1: 465–521.

——— 1948. The number of young at birth and the number of nipples in primates, *Am. J. Phys. Anthrop.* 6: 1–24.

——— 1949. Ontogenetic specializations of man, *Archiv Julius Klaus-Stiftung* 24: 197–216.

——— 1950. The physical distinctions of man, *Proc. Am. Phil. Soc.* 94: 428–449.

——— 1955. The position of the occipital condyles and of the face relative to the skull base in primates, *Am. J. Phys. Anthrop.* 13: 97–120.

Scott, A. C. 1938. Paedogenesis in the Coleoptera, *Zeits. Morph. Ökol. Tiere* 33: 633–653.

——— 1941. Reversal of sex production in *Micromalthus, Biol. Bull.* 81: 420–431.

Scott, H. M. 1897. *Organic education* (J. V. Sheehan, Ann Arbor, Michigan), 291 pp.

Scott, J. 1958. The cranial base, *Am. J. Phys. Anthrop.* 16: 319–348.

——— 1963. Factors determining skull form in primates, *Symp. Zool. Soc. London* 10: 127–134.

Scow, R. O., and M. E. Simpson. 1945. Thyroidectomy in the newborn rat, *Anat. Rec.* 91: 209–226.

Scrutton, C. T. 1965. Periodicity in Devonian coral growth, *Palaeontology* 7: 552–558.

Sedgwick, A. 1894. On the law of development commonly known as von Baer's law; and on the significance of ancestral rudiments in embryonic development, *Quart. J. Microscop. Sci.* 36: 35–52.

Seeley, L. 1906. *Elementary pedagogy* (Hinds, Noble & Eldredge, New York), 337 pp.

Sehnal, F., and A. S. Meyer. 1968. Larval-pupal transformation: control by juvenile hormone, *Science* 159: 981–983.

Sehnal, F., and V. S. A. Novak. 1969. Morphogenesis of the pupal integument in the waxmoth (*Galleria mellonella*) and its analysis by means of juvenile hormone, *Acta Entomol. Bohemoslov.* 66: 137–145.

Selander, R. K. 1965. On mating systems and sexual selection, *Am. Nat.* 99: 129–141.

Serban, M. 1960. La néotenie et le problème de la taille chez les copépodes, *Crustaceana* 1: 77–83.

Serres, E. R. A. 1824. Explication du système nerveux des animaux invertébrés, *Ann. Sci. Nat.* 3: 377–380.

——— 1827a. Recherches d'anatomie transcendante, sur les lois d'organogénie appliquées à l'anatomie pathologique, *Ann. Sci. Nat.* 11: 47–70.

——— 1827b. Théorie des formations organiques, ou recherches d'anatomie transcendante sur les lois de l'organogénie, appliquée à l'anatomie pathologique, *Ann. Sci Nat.* 12: 82–143.

——— 1830. Anatomie transcendante—Quatrième mémoire: Loi de symétrie et de conjugaison du système sanguin, *Ann. Sci. Nat.* 21: 5–49.

——— 1834. Recherches sur l'anatomie comparée des animaux invertébrés: Que sont par rapport aux vertébrés et à l'homme les animaux invertébrés, *Ann. Sci. Nat.* (ser. 2) 2: 238–248.

——— 1842. *Précis d'anatomie transcendante appliquée à la physiologie,* vol. 1 (Paris).

——— 1860. Principes d'embryogénie, de zoogénie et de teratogénie, *Mém. Acad. Sci.* 25: 1–943.

Severtsov, A. N. 1927. Über die Beziehungen zwischen der Ontogenese und der Phylogenese der Tiere, *Jena. Z. Naturwiss.* 56 (o.s. 63): 51–180.

——— 1935. Modusy Filembriogeneza, *Zool. Zhur.* 14: 1–8.

Shepard, P. 1973. *The tender carnivore and the sacred game* (Charles Scribner's Sons, New York), 302 pp.

Shinn, M. 1900. *The biography of a baby* (Houghton Mifflin, Boston), 247 pp.

——— 1907. Notes on the development of a child: II, the development of the senses in the first three years of childhood, *Univ. Cal. Publ. Education* 4: 1–257.

Shumway, W. 1932. The recapitulation theory, *Quart. Rev. Biol.* 7: 93–99.

Simpson, G. G. 1944. *Tempo and mode in evolution* (Columbia Univ. Press, New York), 237 pp.

——— 1953. *The major features of evolution* (Columbia Univ. Press, New York), 434 pp.

Singh-Pruthi, H. 1924. Studies on insect metamorphosis: 1, prothetely in mealworms (*Tenebrio mollitor*) and other insects: effects of different temperatures, *Biol. Rev.* 1: 139–147.

Slijper, E. J. 1936. Die Cetaceen vergleichend-anatomisch und systematisch, *Capita Zoologica* 7: 1–590.

Smit, P. 1962. Ontogenesis and phylogenesis: their interrelation and their interpretation, *Acta Biotheoretica* 15: 1–104.

Smith, A. J. 1899. Evolution and education again, *Pop. Sci.* 54: 554–555.

Smith, H. M. 1956. Paleogenesis, the modern successor to the biogenetic law, *Turtox News* 34: 178–180; 212–216.

——— 1969. The Mexican axolotl, *Bioscience* 19: 593–597.

Smith, J. P. 1898. Evolution of fossil Cephalopoda, in D. S. Jordan, ed., *Footnotes to evolution* (D. Appleton, New York).

——— 1914. *Acceleration of development in fossil Cephalopoda,* Stanford Univ. Publ. Univ. Series (Stanford, Calif.), 30 pp.

Smyth, J. D. 1969. *The physiology of cestodes* (W. H. Freeman & Co., San Francisco), 279 pp.

Snyder, J., and P. W. Bretsky. 1971. Life habits of diminutive bivalve molluscs

in the Maquoketa Formation (Upper Ordovician), *Am. J. Sci.* 271: 227–251.
Snyder, R. C. 1956. Comparative features of the life histories of *Ambystoma gracile* (Baird) from populations at low and high altitudes, *Copeia,* pp. 41–50.
Soot-Ryen, T. 1960. *Pelecypods from Tristan da Cunha,* Results Norw. Sci. Exp. Tristan da Cunha, no. 49, 47 pp.
Southwood, T. R. E. 1961. A thermal theory of the mechanism of wing polymorphism in Heteroptera, *Proc. Roy. Entomolog. Soc. London A* 36: 63–66.
———, R. M. May, M. P. Hassell, and G. R. Conway. 1974. Ecological strategies and population parameters, *Am. Nat.* 108: 791–804.
Spencer, H. 1861. *Education: intellectual, moral and physical* (G. Manwaring, London), 190 pp.
——— 1881. *First principles of a new system of philosophy* (D. Appleton, New York), 592 pp.
——— 1886. *The principles of biology,* 2 vols. (D. Appleton, New York), 492 pp., 598 pp.
——— 1895. *The principles of sociology,* 3rd ed. (D. Appleton, New York).
——— 1904. *An autobiography,* 2 vols. (D. Appleton, New York), 655 pp., 613 pp.
Spinage, C. A. 1968. Horns and other bony structures of the skull of the giraffe, and their functional significance, *E. Afr. Wildlife J.* 6: 53–61.
Spock, B. 1968. *Baby and child care,* rev. ed. (Pocket Books, New York), 620 pp.
Sprent, P. 1972. The mathematics of size and shape, *Biometrics* 28: 23–37.
Sprules, W. G. 1974a. The adaptive significance of paedogenesis in North American species of *Ambystoma* (Amphibia: Caudata): an hypothesis, *Can. J. Zool.* 52: 393–400.
——— 1974b. Environmental factors and the incidence of neoteny in *Ambystoma gracile* (Baird) (Amphibia: Caudata), *Can. J. Zool.* 52: 1545–1552.
Srivastava, V. S., and L. I. Gilbert. 1969. The influence of juvenile hormone on the metamorphosis of *Sarcophaga bullata, J. Insect Physiol.* 15: 177–189.
Staal, G. B. 1967. Endocrine aspects of larval development in insects, *J. Endocrinol.* 37: xiii–xiv.
Stanley, S. M. 1968. Post-Paleozoic adaptive radiation of infaunal bivalve molluscs: a consequence of mantle fusion and siphon formation, *J. Paleontol.* 42: 214–229.
——— 1972. Functional morphology and evolution of byssally attached bivalve mollusks, *J. Paleontol.* 46: 165–212.
——— 1973. An explanation for Cope's rule, *Evolution* 27: 1–26.
——— 1975. Adaptive themes in the evolution of the Bivalvia (Mollusca), *Ann. Rev. Earth Planetary Sci.* 3: 361–385.
Stanton, W. 1960. *The leopard's spots: scientific attitudes toward race in America, 1815–1859* (Univ. Chicago Press, Chicago), 245 pp.
Starch, D. 1927. *Educational psychology* (Macmillan, New York), 568 pp.
Starck, D. 1954. Morphologische Untersuchungen am Kopf der Säugetiere, besonders der Prosimier, ein Beitrag zum Problem des Formwandels des Säugerschädels, *Z. Wiss. Zool.* 157: 169–219.

——— 1960. Das Cranium eines Schimpansefetus (*Pan troglodytes*) (Blumenbach 1799) von 71 mm SchStlg., nebst Bemerkungen über die Körperform von Schimpansenfeten, *Morph. Jb.* 100: 559–647.

Starck, D., and B. Kummer. 1962. Zur Ontogenese des Schimpansenschädels, *Anthrop. Anz.* 25: 204–215.

Starr, F. 1895. *Some first steps in human progress* (Chautauqua Century Press, Meadville, Pa.), 305 pp.

Stauffer, R. C. 1957. Speculation and experiment in the background of Oersted's discovery of electromagnetism, *Isis* 48: 33–50.

Stearns, S. C. 1976. Life-history tactics: a review of the data, *Quart. Rev. Biol.* 51: 3–47.

Stebbins, G. L. 1973. Evolution of morphogenetic patterns, *Brookhaven Symp. Biol.* 25: 227–243.

——— 1974. *Flowering plants: evolution above the species level* (Harvard Univ. Press, Cambridge), 399 pp.

Steffan, W. A. 1973. Polymorphism in *Plastosciara perniciosa, Science* 182: 1265–1266.

Stortenbecker, C. W. 1967. *Observations on the population dynamics of the red locust,* Nomadacris septemfasciata *(Serville) in its outbreak areas,* Agricultural Res. Rep. 694, 118 pp.

Stover, W. J., and D. J. Greathead. 1969. Numerical changes in a population of the desert locust, with special reference to factors responsible for mortality, *J. Appl. Ecol.* 6: 203–235.

Strickland, C. E. 1963. The child and the race: the doctrines of recapitulation and culture epochs in the rise of the child-centered ideal in American educational thought 1875–1900 (Ph.D. dissertation, Univ. Wisconsin), 350 pp.

Strong, J. 1900. *Expansion under new world-conditions* (Baker and Taylor, New York), 310 pp.

Strong, L. 1970. Epidermis and pheremone production in males of the desert locust, *Nature* 228: 285–286.

Stubblefield, C. J. 1936. Cephalic sutures and their bearing on current classifications of trilobites, *Biol. Rev.* 11: 407–440.

Stunkard, H. W. 1962. The organization, ontogeny and orientation of the Cestoda, *Quart. Rev. Biol.* 37: 23–34.

Sulloway, F. In press. Freud as psychobiologist: a study in scientific revolution and revolutionary ideology.

Sully, J. 1895. Studies of childhood: XIV, the child as artist, *Pop. Sci.* 48: 385–395.

——— 1896. *Studies of childhood* (D. Appleton, New York), 527 pp.

Surlyk, F. 1974. Life habit, feeding mechanism and population structure of the Cretaceous brachiopod genus *Aemula, Palaeogeog. Palaeoclimat. Palaeoecol.* 15: 185–203.

Swedmark, B. 1958. *Psammodriloides fauveli n. gen. n. sp.* et la famille des Psammodrilidae (Polychaeta Sedentaria), *Ark. Zool.* (2) 12: 55–64.

——— 1964. The interstitial fauna of marine sand, *Biol. Rev.* 39: 1–42.

——— 1968. The biology of interstitial Mollusca, *Symp. Zool. Soc. London* 22: 135–149.

Swedmark, B., and G. Teissier. 1966. The Actinulida and their evolutionary significance, *Symp. Zool. Soc. London* 16: 119–133.

Takhtajan, A. 1969. *Flowering plants: origin and dispersal* (Smithsonian Institution Press, Washington), 310 pp.

Tasch, P. 1953. Causes and paleoecological significance of dwarfed fossil marine invertebrates, *J. Paleontol.* 27: 356–444.

Taylor, F. S. 1963. *Science and scientific thought* (W. W. Norton, New York), 368 pp.

Teissier, G. 1955. Allométrie de taille et variabilité chez *Maia squinado, Archs. Zool. Exp. Gén.* 92: 221–264.

Temkin, O. 1950. German concepts of ontogeny and history around 1800, *Bull. Hist. Med.* 24: 227–246.

——— 1959. The idea of descent in post-romantic German biology, in B. Glass, O. Temkin, and W. L. Straus, Jr., eds., *Forerunners of Darwin: 1745–1859* (Johns Hopkins Univ. Press, Baltimore), pp. 323–325.

Thayer, V. T. 1928. *The passing of the recitation* (D. C. Heath, New York), 331 pp.

Thompson, D. W. 1917. *On growth and form* (Cambridge Univ. Press, Cambridge), 793 pp.

——— 1942. *Growth and form,* 2nd. ed. (Macmillan, New York), 1116 pp.

Thorndike, E. L. 1919. *Educational psychology* (Columbia Univ. Teachers College, New York), 442 pp.

Thorson, G. 1965. A neotenous dwarf-form of *Capulus ungaricus* (L.) (Gastropoda, Prosobranchia) commensalistic on *Turritella communis* Risso, *Ophelia* 2: 175–210.

Tihen, S. A. 1955. A new Pliocene species of *Ambystoma* with remarks on other fossil ambystomatids, *Contr. Mus. Paleontol. Univ. Michigan* 12: 229–244.

Tilley, S. G. 1973. Life histories and natural selection in populations of the salamander *Desmognathus ochrophaeus, Ecology* 54: 3–17.

Todd, T. W., and D. W. Lyon, Jr. 1924. Endocranial suture closure, its progress and age relationship: I, adult males of white stock, *Am. J. Phys. Anthrop.* 7: 325–384.

——— 1925a. Cranial suture closure: II, ectocranial closure in adult males of white stock, *Am. J. Phys. Anthrop.* 8: 23–40.

——— 1925b. Cranial suture closure: III, endocranial closure in adult males of Negro stock, *Am. J. Phys. Anthrop.* 8: 47–71.

Treat, A. E. 1975. *Mites of moths and butterflies* (Cornell Univ. Press, Ithaca, N.Y.), 362 pp.

Trueman, A. E. 1922. The use of *Gryphaea* in the correlation of the Lower Lias, *Geol. Mag.* 59: 256–268.

Truman, J. W., and L. M. Riddiford. 1974. Hormonal mechanisms underlying insect behaviour, *Adv. Insect Physiol.* 10: 297–352.

Uhlenhuth, E. 1919. Relation between metamorphosis and other developmental phenomena in amphibians, *J. Gen. Physiol.* 1: 525–544.

Ulrich, E. O., and W. H. Scofield. 1897. *The Lower Silurian Gastropoda of Minnesota,* Minnesota State Geol. Survey Final Rept., vol. 3, no. 2, pp. 813–1081.

Ulrich, H., A. Petalas, and R. Camenzind. 1972. Der Generationswechsel von

Mycophila speyeri Barnes, einer Gallmücke mit paedogenetischer Fortpflanzung, *Rev. Suisse Zool.* 79 (suppl.): 75–83.

Urbanek, A. 1960. An attempt at biological interpretation of evolutionary changes in graptolite colonies, *Acta Palaeont. Polonica* 5: 127–234.

——— 1966. On the morphology and evolution of the Cucullograptinae (Monograptidae, Graptolithina), *Acta Palaeont. Polonica* 11: 291–544.

——— 1973. Organization and evolution of graptolite colonies, in R. S. Boardman, A. H. Cheetham, and W. A. Oliver, eds., *Animal colonies* (Dowden, Hutchinson & Ross, Stroudsburg, Pennsylvania), pp. 441–514.

Uvarov, B. P. 1957. The aridity factor in the ecology of locusts and grasshoppers of the Old World, in *Arid Zone Research: VIII, Human and Animal Ecology Reviews of Research* (UNESCO, Paris), pp. 164–198.

——— 1961. Quantity and quality in insect populations, *Proc. Roy. Entomol. Soc. London* 25: 52–58.

Van Valen, L. 1973. Festschrift, *Science* 180: 488.

——— 1974. A natural model for the origin of some higher taxa, *J. Herpetol.* 8: 109–121.

Vermeij, G. J. 1973. Adaptation, versatility, and evolution, *Syst. Zool.* 22: 466–477.

Vialleton, L. 1916. A propos de la loi biogénétique, *Scientia* 20: 101–114.

Vogel, C. 1964. Über eine Schädelbasisanomalie bei einem in freier Wildbahn geschossenen *Cercopithecus torquatus atys, Z. Morph. Anthrop.* 55: 262–276.

——— 1968. The phylogenetical evaluation of some characters and some morphological trends in the evolution of the skull in catarrhine primates, in B. Chiarelli, ed., *Taxonomy and phylogeny of Old World primates, with references to the origin of man* (Turin), pp. 21–55.

Vogt, C. 1864. *Lectures on man* (Longman, Green, Longman & Roberts, London), 475 pp.

Vorzimmer, P. J. 1970. *Charles Darwin: the years of controversy* (Temple Univ. Press, Philadelphia), 300 pp.

Wake, D. B. 1966. *Comparative osteology and evolution of the lungless salamanders, family Plethodontidae,* Mem. S. Cal. Acad. Sci. 4, 111 pp.

Wald, G. 1963. Phylogeny and ontogeny at the molecular level, in A. I. Oparin, ed., *Evolutionary biochemistry* (Pergamon Press, London), pp. 12–51.

Waloff, Z. 1962. Flight activity of different phases of the desert locust in relation to plague dynamics, *Colloq. Int. Centre Nat. Res. Sci.* 114: 201–216.

——— 1966. *The upsurges and recessions of the desert locust plague: an historical survey, Anti-Locust Mem.* 8, 111 pp.

Walton, A., and J. Hammond. 1938. The maternal effects on growth and conformation in Shire Horse–Shetland Pony crosses, *Proc. Roy. Soc. London B* 125: 311–335.

Walzel, O. 1932. *German romanticism,* tr. A. L. Lussky (G. P. Putnam's Sons, New York), 314 pp.

Ward, J. 1913. *Heredity and memory* (Cambridge Univ. Press, Cambridge), 56 pp.

Wassersug, R. J. 1975. The adaptive significance of the tadpole stage with comments on the maintenance of complex life cycles in anurans, *Am. Zool.,* in press.

Way, M. S. 1973. Population structure in aphid colonies, *Bull. Entomol. Soc. New Zealand* 2: 76–84.

Weidenreich, F. 1904. Die Bildung des Kinnes und seine angebliche Beziehung zur Sprache, *Anat. Anz.* 24: 545–555.

——— 1924. Die Sonderform des Menschenschädels als Anpassung an den aufrechten Gang, *Z. Morph. Anthrop.* 24: 157–189.

——— 1932. Über pithekoide Merkmale bei *Sinanthropus pekinensis* und seine stammesgeschichtliche Beurteilung, *Z. Ges. Anat./Z. Anat. Entwicklungsges.* 99: 212–253.

——— 1941. The brain and its role in the phylogenetic transformation of the human skull, *Trans. Am. Phil. Soc.* 31: 321–442.

Weismann, A. 1875. Ueber die Umwandlung des mexicanischen Axolotl in ein Amblystoma, *Z. wiss. Zool.* 25 (suppl.): 297–334.

——— 1881. The origin of the markings of caterpillars, in A. Weismann, *Studies in the theory of descent,* tr. and ed. R. Meldola (Sampson Low, Marston, Searle & Rivington, London), pp. 161–389.

——— 1904. *The evolution theory,* vol. 2, *The biogenetic law* (Edward Arnold, London), pp. 159–191.

Wells, J. W. 1963. Coral growth and geochronometry, *Nature* 197: 948–950.

Westoll, T. S. 1950. Some aspects of growth studies in fossils, *Proc. Roy. Soc. London B* 137: 490–509.

Weston, R. F. 1972. *Racism in U.S. imperialism: the influence of racial assumptions on American foreign policy, 1893–1946* (Univ. South Carolina Press, Columbia), 291 pp.

White, J. F., and S. J. Gould. 1965. Interpretation of the coefficient in the allometric equation, *Am. Nat.* 99: 5–18.

White, T. H. 1958. *The once and future king* (Berkley, New York), 639 pp.

Whitear, M. 1957. Some remarks on the Ascidian affinities of vertebrates, *Ann. Mag. Nat. Hist.* 10: 338–347.

Whitman, C. O. 1894a. Evolution and epigenesis, *Biol. Lect. Mar. Biol. Lab. Wood's Holl,* pp. 205–224.

——— 1894b. Bonnet's theory of evolution, *Biol. Lect. Mar. Biol. Lab. Wood's Holl,* pp. 225–240.

——— 1894c. The palingenesia and the germ doctrine of Bonnet, *Biol. Lect. Mar. Biol. Lab. Wood's Holl,* pp. 241–272.

Whittaker, R. H. 1969. New concepts of kingdoms of organisms, *Science* 163: 150–160.

Wiedersheim, A. 1879. Zur Anatomie des *Amblystoma Weismanni, Z. Wiss. Zool.* 32: 216–236.

Wigglesworth, V. B. 1936. The function of the corpus allatum in the growth and reproduction of *Rhodnius prolixus* (Hemiptera), *Quart. J. Microscop. Sci.* 77: 191–222.

——— 1940. The determination of characters at metamorphosis in *Rhodnius prolixus* (Hemiptera), *J. Exp. Biol.* 17: 201–222.

——— 1952. Hormone balance and the control of metamorphosis in *Rhodnius prolixus* (Hemiptera), *J. Exp. Biol.* 29: 620–631.

——— 1954. *The physiology of insect metamorphosis* (Cambridge Univ. Press, Cambridge), 152 pp.

——— 1966. Hormonal regulation of differentiation in insects, in W. Beermann, ed., *Cell differentiation and morphogenesis* (North Holland Publ. Co., Amsterdam), pp. 180–209.

Wilbur, H. M. 1971. The ecological relationship of the salamander *Ambystoma laterale* to its all-female, gynogenetic associate, *Evolution* 25: 168–179.

Wilbur, H. M., and J. P. Collins. 1973. Ecological aspects of amphibian metamorphosis, *Science* 182: 1305–1314.

Wiley, R. H. 1974. Evolution of social organization and life-history patterns among grouse, *Quart. Rev. Biol.* 49: 201–227.

Wilford, F. A. 1968. Embryological analogies in Empedocles' cosmogony, *Phronesis* 13: 108–118.

Wilkie, J. S. 1967. Preformation and epigenesis: a new historical treatment, *Hist. Sci.* 6: 138–150.

Williams, A., and A. D. Wright. 1961. The origin of the loop in articulate brachiopods, *Palaeontology* 4: 149–176.

Williams, C. M., and F. C. Kafatos. 1972. Theoretical aspects of the action of juvenile hormone, in J. J. Menn and M. Beroza, eds., *Insect juvenile hormones* (Academic Press, New York), pp. 29–41.

Williams, J. B. 1961. The dimorphism of *Polystoma integerrimum* (Frölich) Rudolphi and its bearing on relationships within the Polystomatidae, *J. Helminthol.* 35: 181–202.

Williams, M. 1976. Sexual selection, adaptation, and ornamental traits: the advantage of seeming fitter (unpublished).

Willis, J. H. 1974. Morphogenetic action of insect hormones, *Ann. Rev. Entomol.* 19: 97–115.

Wilson, A. C., L. R. Maxson, and V. M. Sarich. 1974. Two types of molecular evolution: evidence from studies of interspecific hybridization, *Proc. Nat. Acad. Sci.* 71: 2843–2847.

Wilson, A. C., V. M. Sarich, and L. R. Maxson. 1974. The importance of gene rearrangement in evolution: evidence from studies on rates of chromosomal, protein, and anatomical evolution, *Proc. Nat. Acad. Sci.* 71: 3028–3030.

Wilson, C. B. 1911. North American parasitic copepods, *Proc. U.S. Nat. Mus.* 39: 189.

Wilson, E. B. 1892. The cell lineage of *Nereis*, *J. Morph.* 6: 361–480.

——— 1894. The embryological criterion of homology, *Biol. Lect. Mar. Biol. Lab. Wood's Holl*, pp. 101–124.

——— 1904. Experimental studies on germinal localization: I, the germ regions in the egg of *Dentalium*, *J. Exp. Zool.* 1: 1–72.

Wilson, R. R. 1975. High-energy physics at Fermilab, *Bull. Am. Acad. Arts Sci.* 29: 18–24.

Wirz, K. 1950. Zur quantitativen Bestimmung der Rangordnung bei Säugetieren, *Acta Anat.* 9: 134–196.

Wisniewski, L. W. 1928. *Archigetes crytobothrius* n. sp. nebst Angaben über die Entwicklung im Genus *Archigetes* R. Leuck, *Zool. Anz.* 77: 113–124.

Wolpert, L. 1969. Positional information and the spatial pattern of cellular differentiation, *J. Theoret. Biol.* 25: 1–47.

Wood Jones, F. 1929. *Man's place among the mammals* (Edward Arnold, London), 372 pp.

Wright, S. 1916. *An intensive study of the inheritance of color,* Carnegie Inst. Washington Publ. no. 241.

——— 1967. Comments on the preliminary working papers of Eden and Waddington, in P. S. Moorehead and M. M. Kaplan, eds., *Mathematical challenges to the neo-Darwinian interpretation of evolution,* Wistar Institute Monograph no. 5 (Philadelphia), pp. 117–120.

Wu, T. T., E. C. C. Lin, and S. Tanaka. 1968. Mutants of *Aerobacter aerogenes* capable of utilizing xylitol as a novel carbon, *J. Bacteriol.* 96: 447–456.

Wyatt, I. J. 1963. Pupal paedogenesis in the Cecidomyiidae (Diptera) II, *Proc. Roy. Ent. Soc. London A* 38: 136–144.

——— 1964. Immature stages of Lestremiinae (Diptera: Cecidomyiidae) infesting cultivated mushrooms, *Trans. Roy. Entomol. Soc. London* 116: 15–27.

——— 1967. Pupal paedogenesis in the Cecidomyiidae (Diptera) 3: a reclassification of the Heteropezini, *Trans. Roy. Entomol. Soc. London* 119: 71–98.

Wynne-Edwards, V. C. 1962. *Animal dispersion in relation to social behavior* (Oliver & Boyd, London), 653 pp.

Yazmajian, R. V. 1967. Biological aspects of infantile sexuality and the latency period, *Psychoanal. Quart.* 36: 203–229.

Yezhikov, I. 1933. K teorii rekapityulyatsii, *Zool. Zhur.* 12: 57–76.

——— 1937. Zur Rekapitulationslehre, *Biol. Generalis* 13: 67–101.

Yonge, C. M. 1962. On the primitive significance of the byssus in the Bivalvia and its effects in evolution, *J. Mar. Biol. Assoc. U.K.* 42: 113–125.

Young, J. Z. 1962. *The life of vertebrates* (Clarendon Press, Oxford), 820 pp.

Young, R. M. 1970. *Mind, brain, and adaptation in the nineteenth century* (Clarendon Press, Oxford), 278 pp.

Zimmermann, W. 1967. Methoden der Evolutionswissenschaft (=Phylogenetik), in G. Heberer, ed., *Die Evolution der Organismen,* vol. 1 (Gustav Fischer, Stuttgart), pp. 61–160.

Zimmern, H. 1893. Reformatory prisons and Lombroso's theories, *Pop. Sci.* 43: 598–609.

——— 1898. Criminal anthropology in Italy, *Pop. Sci.* 52: 743–760.

Zuckerkandl, E. In press. Programs of gene action and progressive evolution, in *Molecular Anthropology* (Plenum Press, New York).

Zuckerman, S. 1954. Correlation of change in the evolution of higher primates, in J. S. Huxley, A. C. Hardy, and E. B. Ford, *Evolution as a process* (George Allen & Unwin, London), pp. 347–401.

——— 1955. Age changes in the basicranial axis of the human skull, *Am. J. Phys. Anthrop.* 13: 521–539.

Glossary*

ACCELERATION A speeding up of development in ontogeny (relative to any criterion of standardization), so that a feature appears earlier in the ontogeny of a descendant than it did in an ancestor.

ADULTATION Jagersten's term for the appearance, by acceleration, of adult ancestral characters in the larvae of descendants. A variety of recapitulation.

ADULT VARIATION De Beer's term for the introduction of new features in the adult stage during evolution.

ALLOMETRY Change of shape correlated with increase or decrease in size. The change in size may reflect ontogeny, phylogeny, or merely the static differences among related animals (adults of all mammals, for example).

ALLOMORPHOSIS Allometry based on the comparison of related individuals, rather than upon ontogeny. Systematic change of shape among successive adults of a phyletic sequence, for example, is allomorphosis.

ALTRICIAL A mode of vertebrate ontogeny characterized by large litters, rapid development, short gestations, and the birth of relatively undeveloped, helpless young.

ANABOLY Severtzov's term for evolution by addition of a new feature to the end of the embryonic period of morphogenesis and before the subsequent period in which growth proceeds in geometric similarity.

ANALOGY A similarity between two organisms due to independent evolution of the similar feature by each—for example, the wings of a bat and a butterfly. See homology.

*I include all terms for processes and results of the relation of ontogeny and phylogeny, but not, for the most part, names of organisms or parts of the body.

ARCHALLAXIS Severtzov's term for the addition of a new feature during the early period of embryonic morphogenesis, usually causing a major alteration of subsequent ontogeny.

BIOGENETIC LAW Haeckel's term for his principle that ontogeny recapitulates phylogeny.

BIOGENY Haeckel's term for the history of organic evolution.

BLASTAEA A hypothetical, adult ancestral animal, identical in form to the embryonic blastula and inferred, according to the biogenetic law, solely from the existence of the blastula as a common stage in the ontogeny of higher animals.

BLASTOMERE Any of the embryonic cells produced during the first few cleavages of the fertilized ovum.

BLASTULA The hollow ball of cells produced early in the ontogeny of higher animals, before invagination forms the gastrula.

BRACHYCEPHALY Short-headedness (opposed to dolichocephaly, or long-headedness). Since the head increases in relative length during ontogeny, adult brachycephaly is usually interpreted as a paedomorphic feature.

CARRYING CAPACITY The largest population size that can be maintained by an environment.

CENOGENESIS 1. According to Haeckel, exceptions to the repetition of phylogeny in ontogeny, produced by heterochrony (temporal displacement), heterotopy (spatial displacement), or larval adaptation.
2. According to de Beer, adaptations introduced into juvenile stages that do not affect the subsequent course of ontogeny. (Haeckel's meaning was much broader; de Beer's restricted definition includes just one of Haeckel's categories.)

CLANDESTINE EVOLUTION Evolutionary change introduced and developed in juvenile stages and incorporated into descendant adult stages by paedomorphosis. Since juvenile stages fossilize so rarely, these evolutionary events would go unnoticed (hence clandestine), until "promoted" to a descendant adult.

CONDENSATION A necessary principle if recapitulation is to operate. Rates of development must speed up during phylogeny so that descendants pass through the ancestral part of ontogeny faster than the ancestors did; this condensation may occur by deletion of stages or by accelerated development.

CORM Haeckel's term for a colony of organisms, the highest level of his hierarchy of organization.

CORPUS ALLATUM The endocrine organ that secretes juvenile hormone in insects.

DEVIATION De Beer's name for the morphological results of von Baer's law. Early embryological stages of related forms are very similar. As development proceeds from the general to the special, ontogenetic paths of related animals diverge gradually during embryology.

DIFFERENTIATION The development of organs and body parts in ontogeny from simpler antecedent structures.

DISSOGONY Chun's term for sexual maturation at two separate stages of an organism's life, with an intervening period during which no gametes are produced.

ECDYSONE The molting hormone of insects.

ECTODERM The outer of three germinal layers. See germinal layers.

ENDODERM The inner of three germinal layers. See germinal layers.

EPEMBRYONIC Hyatt's term for stages of ontogeny following birth and the completion of embryonic stages.

EPIGENESIS The idea that morphological complexity develops gradually during embryology from simple beginnings in an essentially formless egg. During the debate of the eighteenth and early nineteenth centuries, epigenesis represented the theory that complexity must be imposed from without by some vital force or directed entelechy working upon an egg that had only the potential (not the inner determinants) for normal development.

EVOLUTION In the eighteenth and early nineteenth-century debate over epigenesis versus preformationism, evolution was used (by Bonnet, for example) as a synonym for preformationism. This use antedates its present meaning. Evolution, as a term for organic change in phylogeny, was introduced by Spencer in the mid-nineteenth century.

EXACT PARALLELISM Cope's term for the precisely equal acceleration or retardation of all organs so that juvenile stages of descendants are exact replicas of adult ancestors (acceleration), or that adult stages of descendants are exact replicas of juvenile stages of ancestors (retardation).

FETALIZATION Bolk's term for paedomorphosis involving the preservation of ancestral fetal stages in adult descendants. Bolk applied his term to human evolution and postulated a general retardation of development, not just the retardation of a few, selected organs.

GASTRAEA Haeckel's hypothetical adult ancestor of all higher animals, inferred from the existence of the gastrula as a common stage in early ontogeny of higher animals.

GASTRULA The early embryonic stage of higher animals produced by invagination of the blastula to form inner and outer germinal layers.

GEMMULE Darwin's term for a hypothetical particle of heredity carried by all cells and capable of moving to sex cells, thus permitting a direct influence of environment upon heredity.

GEOMETRIC SIMILARITY Size increase or decrease with no accompanying change of shape. Two objects of different size and the same shape are geometrically similar. Also called isometry.

GERMINAL LAYERS According to a largely outmoded view of embryology, the initial three layers formed by invagination, which produces the gastrula with ectoderm and endoderm, and by the subsequent development of mesoderm between ectoderm and endoderm; all organs differentiate in a predictable manner from the germinal layers.

GERONTOMORPHOSIS De Beer's term for evolution by the modification of adult stages only. In de Beer's view, gerontomorphosis leads to evolutionary dead ends.

HETEROCHRONY 1. According to Haeckel, displacement in time of ontogenetic appearance and development of one organ with respect to another, causing a disruption of the true repetition of phylogeny in ontogeny. The embryonic heart of vertebrates, for example, now appears far earlier in ontogeny than its time of phylogenetic development would warrant.

2. Cope used the same definition as Haeckel, but viewed heterochrony as support for the biogenetic law. Recapitulation must be defined organ by organ, not in terms of the whole body. The heart may be far more strongly accelerated than other organs, but it is still accelerated, and acceleration is the mechanism of recapitulation.

3. De Beer defines heterochrony as phyletic change in the onset or timing of development, so that the appearance or rate of development of a feature in a descendant ontogeny is either accelerated or retarded relative to the appearance or rate of development of the same feature in an ancestor's ontogeny.

HETEROTOPY Haeckel's term for a phyletic change in the location (in the germinal layers) from which an organ differentiates in ontogeny—thus forming an exception to recapitulation. Reproductive organs must have arisen from ectoderm or endoderm in ancestors, since these were the only original germ layers and all organisms must have reproductive organs. They now originate in mesoderm.

HOLOMETABOLOUS Referring to insects that undergo complete metamorphoses in their ontogeny from larva to pupa to imago.

HOMOLOGY A similarity between two organisms due to inheritance of the same feature from a common ancestor. See analogy. (The recognition of homology is a key to the tracing of evolutionary lineages.)

HYPERMORPHOSIS The phyletic extension of ontogeny beyond its ancestral termination (usually to larger body sizes and increased complexity of differentiating organs)—producing recapitulation as a result because ancestral adult stages are now intermediate stages of a lengthened descendant ontogeny.

IMAGO The adult molt of a holometabolous insect. The adjectival form, imaginal, does not refer to illusory appearance.

INEXACT PARALLELISM Cope's term for the imperfect repetition of entire adult ancestors in descendant ontogeny, caused by unequal rates of acceleration and retardation of various organs.

JUVENILE HORMONE A hormone, secreted by the corpus allatum of insects, that acts with ecdysone to control the developmental results of molting. Transition from larva to pupa to imago occurs as the titer of juvenile hormone drops.

K SELECTION Selection on individuals in populations at or near the carrying capacity of their environments, usually favoring the production of few, slowly developing young, well adjusted in form and function to their (usually stable) environment. *K*-selected individuals are good competitors in conditions of density-dependent mortality.

KYPHOSIS A bending of a part of the body with concave side towards the center of the body. Usually used to refer to the vertebrate backbone; here used to discuss the bending of the cranial axis.

LIFE-HISTORY STRATEGY A selected set of adaptations to local environments, involving such quantitative aspects of life history as fecundity, the timing of maturation, and the frequency of reproduction. Responses to *r* and *K* selection represent two different life-history strategies.

MACROEVOLUTION The study of evolutionary events and processes that require long times for their occurrence or operation—conventionally defined at taxonomic levels involving the origin and deployment of species and higher taxa, not changes of gene frequencies within local populations.

MACROMUTATIONISM A theory that new taxa evolve essentially instantaneously via a major genetic mutation that establishes reproductive isolation and new adaptations all at once.

MECKEL-SERRES LAW Russell's term for the version of recapitulation preached by the Naturphilosophen. The stages of ontogeny run in parallel with the sequence of adults of different species arranged in a chain of being, because a single direction of development pervades all of nature.

MESODERM The middle of three germinal layers. See germinal layers.

METATHETELY The retention of some juvenile characters by an insect imago that has developed through the normal number of molts. Metathetely is a variety of neoteny as defined in this book.

MONERA Haeckel's term for the original hypothetical adult ancestor of all higher forms—a single cell without a nucleus. The term is still used today for the kingdom of procaryotic organisms (bacteria and blue-green algae). Procaryotes lack a nucleus and other cell organelles.

NATURPHILOSOPHIE A romantic movement in late eighteenth and early nineteenth-century German biology—a search for unification of all natural phenomena and processes through such transcendental and developmental beliefs as: the history of the universe is the history of spirit, beginning in primal chaos, striving upward to reach its highest expression in man.

NEO-LAMARCKISM A popular late nineteenth-century alternative to Darwinism, postulating that adaptations arise as characters acquired by active organic responses to environment and are passed on to offspring by heredity.

NEOTENY Paedomorphosis (retention of formerly juvenile characters by adult descendants) produced by retardation of somatic development.

OLD-AGE THEORY Hyatt's whimsical description of his theory of racial life cycles. Evolutionary lineages, like individuals, go through a programmed sequence of youth, maturity, old age, senescence, and death.

ONTOGENY The life history of an individual, both embryonic and postnatal.

ORTHOGENESIS The theory that evolution, once started in certain directions, cannot deviate from its course, even though it leads a lineage to extinction.

ORTHOGNATHY A situation in which the jaw does not project far forward in front of the cranium (called orthognathy because the cranium and jaw lie essentially in a plane). The opposite—a jutting jaw—is called prognathy). Adult orthognathy is paedomorphic in mammals.

ORTHOSELECTION Selection exerted upon a lineage in the same direction for a substantial segment of geological time.

PAEDOGENESIS 1. von Baer's term for parthenogenetic reproduction by insect larvae structurally unable to copulate.
2. De Beer's term for paedomorphosis (retention of formerly juvenile characters by adult descendants) produced by precocious sexual maturation of an organism still in a morphologically juvenile stage. Rejected in this book in favor of its synonym, progenesis.

PAEDOMORPHOSIS The retention of ancestral juvenile characters by later ontogenetic stages of descendants.

PALINGENESIS 1. For Bonnet, the ontogenetic unfolding of individuals already preformed in the egg—referring especially to the "end" stage of "ontogeny," the resurrection of each body at the end of time from a "germ of restitution" preformed in the original homunculus within the egg.
2. For Haeckel, the true repetition of past phylogenetic stages in ontogenetic stages of descendants.

PANGENESIS Darwin's theory of heredity. Somatic cells contain particles that can be influenced by the environment and the activity of organs containing them. These particles can move to the sex cells and influence the course of heredity.

PARABIOSIS The experimental (or natural) joining of two individuals, so that they share a common blood supply.

PARALLELISM Cope's term for the results of acceleration and retardation. The stages of ontogeny run in parallel with the adult stages of phylogeny.

PARTHENOGENESIS Reproduction by the female parent alone, leading to the development of an organism from an unfertilized egg.

PERIGENESIS Haeckel's theory of heredity. Somatic cells are altered by environment and by the activity of organs containing them. These modifications are passed as wave motions to the sex cells, thus permitting the inheritance of acquired characters.

PHEREMONE A chemical, usually a hormone, secreted by organisms and used in communication with other members of the species.

PHYLEMBRYOGENESIS Severtzov's name for his theory of the relationship between ontogeny and phylogeny. See anaboly and archallaxis.

PHYLOGENY The evolutionary history of a lineage, conventionally (though not ideally) depicted as a sequence of successive adult stages.

PHYLOGERONTIC Senescent stages of phylogeny, according to advocates of the theory of racial life cycles.

PLASTIDULE Haeckel's term for the basic units of life. (Defined just this vaguely by him, so don't blame me. Plastidules are, at least, subcellular building blocks.)

POECILOGONY Giard's term for a course of development in related animals, in which adult stages are virtually identical and juvenile stages are highly divergent. (If the adults are homologous, then poecilogony is an exception to recapitulation because juvenile stages cannot represent adult ancestors.)

POLYGENIST The early- and mid-nineteenth-century theory that human races are separate species.

PRECOCIAL A mode of vertebrate ontogeny characterized by small litters, slow development, extended gestation, and the birth of relatively well-developed, capable young.

PREFORMATIONISM The notion that all major structures of the adult are already preformed in the sex cell (egg or sperm, depending on your preference) and that ontogeny is the unfolding ("evolution") of this pre-built complexity.

PROGENESIS Paedomorphosis (retention of formerly juvenile characters by adult descendants) produced by precocious sexual maturation of an organism still in a morphologically juvenile stage.

PROGNATHY A situation in which the jaw projects far forward in front of the cranium. See orthognathy.

PROTEROGENESIS Schindewolf's theory of the relationship between ontogeny and phylogeny. New evolutionary features are introduced suddenly into juvenile stages and slowly work their way forward, by neoteny, towards adult stages.

PROTHETELY The production of an insect imago at an earlier-than-normal molt stage, leaving some features in a juvenile stage. Prothetely is a variety of progenesis as defined in this book.

r SELECTION Selection on individuals in populations well below the carrying capacity of their environments, usually favoring the early and rapid production of large numbers of rapidly developing young. *r* selection generally operates in ecological situations favoring fast increase in population size—either because environments fluctuate so severely and unpredictably that organisms do best by making as many offspring as fast as possible so that a few can weather the storm; or because ephemeral, superabundant resources can best be utilized to build up population size before their inevitable exhaustion.

RACIAL SENESCENCE The theory that lineages, like individuals, pass through programmed stages leading inevitably to phyletic exhaustion and death (extinction).

RECAPITULATION The repetition of ancestral adult stages in embryonic or juvenile stages of descendants.

REGULATORY GENE A gene that controls development by regulating the turning on and off of structural genes that manufacture proteins to build body parts.

REPETITION Morgan's term for von Baer's theory (opposed to recapitulation). Development proceeds from the general to the special. The young embryos of related animals repeat the same stages (hence Morgan's name) until they reach an intermediate stage of embryogenesis, after which they diverge by adding their own particular characters.

RETARDATION A slowing down of development in ontogeny (relative to any criterion of standardization), so that a feature appears later in the ontogeny of a descendant than it did in an ancestor.

STRUCTURAL GENE A gene that controls the synthesis of a protein that builds part of the body.

SUPER-LARVATION Lankester's term for paedomorphosis—a retention of larval stages by adult descendants.

TACHYGENESIS A common nineteenth-century synonym for acceleration.

TECTOLOGY Haeckel's term for the science of the architecture of organic parts.

TERATOLOGY The study of abnormal development.

TERMINAL ADDITION One of the necessary laws of recapitulation (see condensation). New evolutionary features are added to the end of ancestral ontogenies (so that previous adult stages become preadult stages of descendants).

VON BAER'S LAW Development proceeds from the general to the special. The earliest embryonic stages of related organisms are identical; distinguishing features are added later as heterogeneity differentiates from homogeneity. Recapitulation is impossible; young embryos are undifferentiated general forms, not previous adult ancestors.

Index